PICTURING SCIENCE
PRODUCING ART

Faraday's Islands *was created near the end of the twentieth century, and it's made out of machines that span the last hundred years. As we move into the next millennium, machines aren't going to look like this, nor will they behave in quite the same ways. So I tried to use the acoustics of the room to create an impression of time travel. Machines that were nearby would sound sharp and clear, but sounds from across the room would reverberate more and more, almost as though they were echoes from some past event. One time someone came to see a performance of mine in which I was using similar kinds of obsolete projectors and old record players, and afterwards this person compared it to building a car that would run on milk. He meant: you could do it, but why on earth would you bother? Why not just use multi-media software on a computer to do everything? But the problem is that even if you did have digital technologies that were infinitely malleable (which they're not), and even if you could do anything with them (which you can't), you still wouldn't think of doing just anything with them. They'll always lead you in certain specific directions. And as an artist, I think my job is to interrogate these technologies, not just to use them.*

PERRY HOBERMAN

PICTURING SCIENCE PRODUCING ART

CAROLINE A. JONES
PETER GALISON

EDITORS

with Amy Slaton

LONDON AND NEW YORK

First published 1998 by Routledge

Published 2013 by Routledge

2 Park Square, Milton Park, Abingdon, Oxon OX14 4RN

711 Third Avenue, New York, NY 10017, USA

Routledge is an imprint of the Taylor & Francis Group, an informa business

Book Design: Stratford Publishing Services, Inc.

Library of Congress Cataloging-in-Publication Data
Picturing science, producing art / edited by Caroline A. Jones and Peter Galison.
    p. cm.
    Includes bibliographical references and index.
    ISBN 0-415-91911-8 (hardcover). — ISBN 0-415-91912-6 (pbk.)
    ISBN 978-0-415-91912-8 (pbk)

    1. Art and science. I. Jones, Caroline A. II. Galison, Peter Louis.
N72.S3P53   1998
701'.1'05—dc21   97-40521
                                                                 CIP

Cover art: Courtesy of the Bayerische Staatsbibliothek München (Clm. 527, fol. 64v). For further details please see caption for Figure 4, p. 261.

Throughout the book are selected images from Perry Hoberman's installation *Faraday's Islands*, on view from 3–19 November 1995 in a former Cadillac showroom now owned by Boston University. Commentary accompanying the images is drawn from "Culture's Technology, Technology's Culture," a panel discussion held on November 4 at Boston University. Participants were Leo Marx, William Keenan Professor of Cultural History at M.I.T., the editor, Caroline Jones, and the artist, Perry Hoberman.

We dedicate this volume to the memory of
Edward E. Jones and Milton Bluestein.

Contents

Seeing Wonders

Objectivity/Subjectivity

Cultures of Vision

Preface and Acknowledgments

An edited volume such as this one relies, first and foremost, on the generosity of the scholars whose work it contains. Without repeating the roster of participants, we would like to thank each of these busy and illustrious minds for staying with this project from the first workshop in 1995, "Histories of Science, Histories of Art," to the present volume. Comments on the papers were cogently provided by scholars whose insights molded many of our subsequent revisions: architectural historian Keith Morgan (on "Styles"), art historian Patricia Hills (on "The Body"), historian Steven Harris (on "Wonders"), philosopher James Conant (on "Objectivity/Subjectivity"), and art historian Margaret Carroll (on "Vision"). To these analysts we are grateful.

Support for this interdisciplinary foray was generously provided by the Mellon Foundation, the Boston University Humanities Foundation, the Harvard University Provost Fund, and the National Science Foundation. Much of the collaborative planning work was accomplished while both editors were on leave at the Institute for Advanced Study in Princeton; to our colleagues there we are indebted. Harvard University and Boston University provided much in-kind support.

The cluster of events culminating in *Picturing Science, Producing Art* benefited from the assistance of numerous individuals. Pride of place goes to Amy Slaton, then a postdoctoral fellow at Harvard University, whose unflagging wit and good humor smoothed over many a rough spot, and whose intellectual contributions to the shape of both conference and book were substantial. Perry Hoberman, the artist invited to produce a provocative interactive installation in a vast expanse of Boston University real estate (a former Cadillac showroom on one of Boston's busiest streets), complied with a shoestring budget to produce the astonishing, crowd-pleasing *Faraday's Islands* (images of which season these pages). Art history graduate student Rachelle Dermer helped with graphic design and served as Hoberman's pivotal assistant; fellow artist Angela Wyman earned our gratitude with many a tactful assist; art historian Eric Wolin and sculptor Thomas Gudaitis helped navigate *Faraday's* in and out of port.

Many students in Harvard's history of science department and Boston University's art history department helped with the workshop, ancillary events, and the final publication: Deborah Coen, Nani Clow, Michael Gordin, Jaimey Hamilton, Stacey McCarroll, and Stephanie Taylor.

In seeing the current publication into final form, we are warmly appreciative of our editor, William Germano, whose electronic and personal communications achieved just the right blend of warm support, insightful critique, and steely admonishment. Our families put up with all the chaos; particularly during the conference weekend we are grateful to Marion and Jerry Galison and Catherine Rogers. To all those museums, archives, libraries, churches, and private collectors credited on the images that follow, we extend our final thanks. Without the stimulation of their images we would have much less to see in the productive nexus between science and art.

INTRODUCTION

Picturing Science, Producing Art

Caroline A. Jones and Peter Galison

Analytic attempts to distinguish "art" and "science" often founder at the boundaries drawn between them. Do the alligators that hang from the ceiling in the late Renaissance cabinet of wonders at Wurms form part of the history of scientific classification, or part of the history of aesthetics? Are theories of female reproduction in Cinquecento Italy marked more by discourses of medicine, or by contemporaneous casting techniques? Did early photographs of mammals in motion serve primarily to educate the eye, or to provide raw data for physiologists? To bring such questions into a late-twentieth-century frame, is entering an artist's website an artistic or a technological experience? As the chapters in this book demonstrate, the much-vexed inquiry as to whether science and art are incommensurable realms of knowledge is misplaced. What promises more is a view of history that asks: What are the conditions under which objects become visible in culture, and in what manner are such visibilities characterized as "science" or "art?" We are after precisely these boundary conditions.

There are moments in the nineteenth and twentieth centuries when such categorization (as *either* science or art) was itself the point. Coincident with the rise of modernism, and in part constitutive of modernism as a form of knowledge, scientists and artists contrasted their two domains. Each defined the other by a near absolute opposition. Science, the anatomists of the 1860s insisted, began when artistic license was canceled. Art, Baudelaire maintained, began when the deadening industrial-mechanical ethos of science could be forcibly set aside. In the production within laboratories and studios, in the power and ambition of art and science to capture the world, in the variegated and evolving audiences that art and science demanded (or even created), the two realms have been separated, and their resulting relations described variously as markers of the premodern, signposts of the modern, and charged conduits into the postmodern. What much of this focus on "art" and "science" as discrete *products* ignores are the commonalities in the *practices* that produce them. Both are regimes of knowledge, embedded in, but also constitutive of, the broader cultures they inhabit.

Over the last twenty years, scholars have increasingly probed scientific and artistic objects to get at these practices, seeking the historical conditions of possibility that have made them meaningful. Using the resources of history, philosophy, and sociology (as well as art history and the history of science), what are the most current ways and places in which we can think through these two domains? That is the query motivating the essays in this collection.

ART AND SCIENCE AS BINARY ECONOMY

There is a history to the perception of difference between science and art, and a parallel history to the attempt to unify the two. Although "art" is the older term, its emergence as a humanist enterprise in the Renaissance is coeval with the birth of talk about "scientific method." From this point on, each defined and legitimated itself in relation to its shadow term, and the continuing strength of the dyad is reflected in the very structure of a late-twentieth-century undergraduate education featuring "arts and sciences."[1] A peculiar feature of this polar linkage during the twentieth century, as revealed in C. P. Snow's famous inauguration of the "two cultures" debate in 1959, was its unstated assumption of what might be called an economy of the binary. Like all binaries, art and science needed to be yoked together (yet held apart) in order to accrue the strengths of their polar positions: soft versus hard, intuitive versus analytical, inductive versus deductive, visual versus logical, random versus systematic, autonomous versus collaborative, and, like all binaries, at some level, female versus male.[2] The binary production of knowledge (the bifurcation of practices) was equally simple: art invented, science discovered.

Rather than address science and art as if these "opposites" were permanent features of the world, this book aims to explore the intersection of their histories, and to do so in a way that positions methodological and philosophical issues front and center. Though differing in many respects, the essays in this volume do hold certain strategies in common. They are not aimed at identifying universal demarcation criteria that separate science from art, nor are they after a description that might conjoin the two activities under a single broad and unifying rubric. Instead, the effort here is to explore how historians of art, historians of science, philosophers, and cultural historians can learn from one another's methods at the boundaries between their fields, and how the historical inquiry into conditions of artistic and scientific image production can shed light on multiple philosophical and historical issues. The essays are grouped under rubrics formulated as both topics and queries into the productive force of scientific and artistic representations. From a variety of angles, they emphasize the central theme of this book: namely, that art, science, and the hermeneutical concepts that we bring to them are historically and culturally embedded. Neither practice has unique and absolute purchase on "reality," and neither is as alienated from history as its rhetoric might imply.

Although we seek to frustrate the standard binary economy, it is inevitable that as soon as "art" and "science" are mentioned, a host of other projects will come to mind. For clarity it is worth characterizing at least a few of these enterprises, if only to distinguish them from the direction of the present volume.

We begin, not coincidentally, with the late nineteenth century, when the Industrial Revolution was at its peak. It was at this moment, particularly in the most rapidly industrializing nations (e.g., England), that a rich controversy developed over whether art and science had (or should have) distinguishable goals. From John Ruskin and Charles Baudelaire to T. H. Huxley and Thomas Carlyle, the encroaching dominance of industrial technology made it imperative that the energies of an instrumental science be understood (and, possibly, contained).[3] Two things seemed clear: art occupied the domain of the creative, intervening mind, and the scientific ethos seemed to demand precisely the suppression of such impulses. (This was, of course, an intervention specific to its time. As many of the essays in this volume relate, both earlier and later bifurcations were very different.) Particularly in this largely British frame of reference (which C. P. Snow would inherit some decades later), the scientific method became linked inextricably with technology, industrial progress, and class mobility, while institutionalized art and literature came to be associated with the preservation of tradition, social order, and the conservation of rustic values. The special case of the modernist avant-garde defined itself, in one sense, precisely through its opposition to this particular binary. Confronting institutions of art and canonical literature, self-described modernists manifested their opposition to the academy through a pronounced tropism for advanced scientific and technological ideas—from X rays and relativity to radio and airplanes.[4] The perceived difference between the two domains would be mobilized precisely to destabilize the cultural category of "art," through the newly powerful realm called "science."

Along more explicitly psychological lines, various authors of the mid-twentieth century argued for parallels between creativity in art and science. One thinks here of the Gestalt-era psychologists of the 1950s and 1960s, such as Rudolf Arnheim on visual thinking, or Anton Ehrenzweig on the link between abstraction in visual art and science.[5] Along with this Gestalt-psychological tradition, which paid particular attention to the perceptual, there was also the work of practicing physicists such as

Ernst Mach and Harvard physicist-philosopher Percy Bridgman. For these scientists, an interest in sense perception was closely coupled to broader epistemological concerns. It is into this psycho-philosophical tradition that physicist Gerald Holton's influential work on "themata" and scientific creativity in scientific thinking belongs.[6] It is suggestive that many of these theorists (Bridgman is an exception) had emigrated from Europe to the United States during or after the Second World War. They were forced to leave behind their roots in a Central European *wissenschaftlich* approach to learning, where all fields of inquiry had been unified as one systematic investigation into various products of the human mind.[7] They took up influential positions in a pragmatic country in which highly specialized autonomous branches of inquiry were rapidly becoming the norm.

For all their continuities with prior literature, such postwar discussions of art and science had a new vocabulary after 1959 with C. P. Snow's widely discussed and immensely controversial lecture and publication, *The Two Cultures*.[8] Snow's intervention (and the responses to it) had implications that may well be more revealing historically than theoretically. For Snow, the two cultures were not only different, they were unequal: the scientific ethos stood for all that was hopeful, progressive, vigorously heterosexual, and future oriented, while the artistic-literary tradition embodied the profoundly hidebound culture of a decaying elite.[9] Some scholars took aim at the dichotomy, others at the ascendency of the scientific. Whether in appreciation or condemnation, the sudden currency of Snow's phrase revealed how completely and deeply divided the domains of art and science were held to be—at least by some.[10]

Perhaps in response to this sense of a division, a new body of work emerged in the 1960s that sought explicitly to explore the similarities (and admitted differences) between the practices of art and science. These thinkers constructed, in a way, the "anthropology" of the two cultures that Snow had presupposed but never fully explained.[11] When historian and philosopher of science Thomas Kuhn wrote his *Structure of Scientific Revolutions* (1962) and its follow-on essays, he deliberately treated the production of science in a "sociological" way that made both science *and* art the "products of human behavior," demanding a more ethological approach. Indeed, the widespread popularity of Kuhn's book drew in large measure from the seeming universality of its story of normal-crisis-revolutionary developments and paradigm shifts that could be viewed *across* the arts and sciences. But when E. M. Hafner pursued such similarity relations between pictures in art and in science, Kuhn drew the line, arguing that pictures were, on the one hand, essential for artists, whereas, "The scientific illustrations, on the other hand, are at best by-products of scientific activity. . . . In Hafner's striking parallels, an end product of art is juxtaposed with a tool of science."[12] For Kuhn and the scientists with whom he identified, pictures and aesthetic criteria in general were mere means to an end, whereas for artists they were ends in themselves.[13] The binary economy rules Kuhn's argument, with the artist an active agent recording a passive nature, and the scientist a passive recorder of natural flux.

Time and again during the 1960s, this tension between alliance and antagonism emerged. Just as Hafner had grounded his art-historical claims in Kuhn's depiction of science, art historian Ernst Gombrich drew his methodology explicitly from the theory of the scientific method offered by Karl Popper. Popper had aimed to separate the productive domain of true science from the cultural noise of "pseudo-science," and generated his set of "demarcation criteria" toward that end. Was the experimental premise testable, and, through testing, falsifiable? If so (and only so, Popper contended), could scientific explanations participate in the progressivist march of science. Ignoring the obvious—that art could never be "falsifiable" in the strict sense Popper had intended—Gombrich adapted the Popperian program to his theory of schemata, or "making and matching," in which the artist (like the scientist) renders an approximation of the natural world that can be tested, corrected, amended, and improved. Gombrich's *wissenschaftlich* unification of all human activity proved productive, but carried in its wake a problematic exclusion of much of twentieth-century abstraction, from Cubism to Abstract Expressionism and beyond.[14]

These sympathetic endeavors to locate similarities between art and science (while preserving philosophical distinctions between the two) formed as vigorous a tradition as the efforts to map the differences between them. At the present late-twentieth-century moment, anxieties about the divide have diminished. There is little attention paid by the authors in this book to the structural inquiries of previous decades that mapped the parallels and antiparallels perceived between the two types of activity. This is not even to speak of the difficulty seen presently in defending the notion that there are (or ever were) only *two* "activities" in the domains marked "science" and "art." Neither are the issues addressed here reducible to questions of "influence" by one autonomous sphere on another (although clearly the active appropriation and use of various prevailing discourses *can* be found). This distinguishes the present volume from much of the existing scholarship, which presumes the binary economy in order to chart its differential forces.

When presumptions of that binary economy have been at work, the results can be profound—as is best exemplified in the classic essay by Erwin Panofsky establishing Galileo's debt to artistic traditions of *chiaroscuro* for his interpretation of the craters of the moon.[15] Looking in the other direction (from science to art) Linda Dalrymple Henderson has provided sweeping chronicles of artists' reworkings and creative misreadings of non-Euclidean geometry, and Martin Kemp has charted artistic investigations of optics "from Brunelleschi to Seurat."[16] Kemp goes further than Henderson in claiming a deep congruence between "the central intellectual and observational concerns in the visual arts and the sciences in Europe from the Renaissance to the nineteenth century," and does so largely by looking at those moments in which artists seem to him to have "consciously aspired towards goals that we would now regard as scientific."[17]

This anchoring of artistic to scientific practices is mirrored by a large, interesting, and growing body of literature by those who seek to interpret images that remain "unclaimed" by the institutions of art, but are readable as constructions of visual knowledge. James Elkins and Barbara Maria Stafford are among the art historians who have begun to open up this terrain. Elkins's term for this new activity is not art history but "image studies,"[18] a term redolent of other late-twentieth-century academic discourses from "cultural studies" to "visual studies," many of which tap post-structuralist and literary theories of the text. The scientific or other non-art images studied by Elkins and others play a myriad of conceptual roles, from aiding calculation to summarizing data, from the documentation of priority to the conceptualization of models only awkwardly put into analytic or mathematical form. In the study of such problems, art historians join science studies scholars in examining such diverse topics as Feynman graphs and Minkowskian space-time diagrams, images from electron microscopes, X rays, CAT and PET scans, digitalized computer visualizations of data, patent sketches, and the transformation of images from one medium into another.[19] The evident variety and depth of these concerns about the links, interfaces, or gray areas between "art" and "science" (ever more loosely construed) emphasizes the intellectual intensity of current debates over their relations. But rather than searching for brackets to join or wedges to split the vexed dyad, or mining some *terra incognita* between its two (always unequal) halves, we want to set this binary economy aside. The authors here address questions of viewing and knowing in which both artistic and scientific practices are brought into consideration, among many other kinds of cultural practices and productions.

There is nothing monolithic in this assembly; these are variable slices into histories that are themselves characterized by their heterogeneity. Yet there are themes within this diversity, assembled here as a cluster of "sites" for examining the productive work that both scientific and artistic images do, as well as the practices and institutions through which those images are embedded in culture. The representations at issue here are not just the canonical end products of artistic processes (oil on canvas or sculpted stone) or the end products of scientific ones (perfected equations or "golden events"), although these can be found. As authors, we want more broadly to include the iconography of cartoons, scientific images of DNA, particle tracks, anatomical photographs, artists' printed diagrams and poems, instrumental motion studies, fossils, enameled birth trays, concrete factory buildings, illustrated panegyrics, botanical broadsheets, and attempted resolutions of astronomical "monsters." We want, singly and collectively, to ask what work these images do, and what historically specific conditions make it possible for them to count as part of culture.

The "sites" at which we gather to address such images pose five thematic questions, headed by the rubrics **Styles**, **The Body**, **Seeing Wonders**, **Objectivity/Subjectivity**, and **Cultures of Vision**. Each site is opened by an image from electronic-media artist Perry Hoberman's material meditation on the human-machine interface, *Faraday's Islands;* Hoberman's motley aggregates of consumer appliances are themselves representations of the problems posed.

In the first site, **Styles**, we ask: *How are images and practices aggregated, and to whose benefit?* Style is the presumptive tool for such aggregation and disaggregation, and from its common usage in art history, the term has been appropriated to characterize shifts, breaks, or modes of production in the history of science as well. (The view of blenders from Hoberman's installation [page 25] reminds us that "style" is also embedded in the commodity fetish). Yet the authors here would like to "make strange" this common tool, and they question with clarity and precision its unproblematic usage and narrowed definition. In full knowledge of its troubling past, the scholars in **Styles** would propose that we use this framing device only after its outlines have been radically redrawn.

Next, we turn to the implications of specific pictures in **The Body**, asking: *How do images shape body knowledge, and for whom?* What, for example, coexists with the depicted body—how are its divine, mechanical, productive aspects displayed or suppressed? In what sense is the body a "technoscientific" amalgam, as Hoberman's piece might suggest (page 99)? As with the **Styles** section, **The Body** spans a broad historical area to investigate varieties of body knowledge available at different historical moments, from the early codification of Christian dogma to the discourse of cyberspace.

A more narrow historical focus can also be useful. For the sake of such specificity, the **Seeing Wonders** site brings together several essays that focus on the Renaissance and early modern period (with Hoberman's spectacle standing as a later variant of the traditional, highly staged *Wunderkammer* [page 209]). Here, we take aim at a specific epistemological question: *What do we know when we see?* The more "wondrous" the image, the more loaded the question becomes. While their objects ranged from rocks to saints and from bees to peasants, the artists and natural philosophers of the early modern period linked seeing to knowing in revealing ways. The wonders examined here presuppose (and enforce) specific worldviews—located in particular knowledges of the thing seen.

Turning to the later modern period, a parallel site emerges in the historicized binary **Objectivity/Subjectivity** (a binary that Hoberman's installed and projected machinery [page 325] is meant to question). The query here is: *What do images presuppose about (human) nature?* What do discourses of "objectivity" and "subjectivity" produce in the way of images, and how do those images in turn produce knowledge? What types of statements must be marshaled to support their interpretations?

In direct correspondence with the **Objectivity/Subjectivity** site, **Cultures of Vision** calls forth a final epistemological problem, suggested by the forest of projection screens in Hoberman's installation (page 399): *What viewers and processes does the image presuppose?* Not as general as questions of "(human) nature," this site harbors issues of location, mediation, politics, physiology, and attention, all of which enter into the objects and practices that designate science and art.

Looking in some detail at the various essays constituting these sites, we will identify the network of overlapping concerns that animate our contemporary narratives of picturing science and producing art. There can be no doubt that in attempting to locate the conditions of possibility for various historical regimens of seeing and knowing, we reveal something of our own desires to trouble the bifurcation of what are still institutionalized as separate domains.

SITES

Styles: How are images and practices aggregated, and to whose benefit?
In the essay by Carlo Ginzburg that opens this site, there are two notions of the work of art (or product of science). The one is relational (embedded in a specific historical moment), the other absolute (a fixed attribute of form). While the absolute can be understood through the relational (history), the converse is not so. (That is to say, one cannot derive history from form.) Style, *mutatis mutandis*, is both absolute and relative, but only the relative (historicity) of style can explain the other (local production of the absolute). Ginzburg's essay and the others in this section are ultimately about the deployment of "style" as a heuristic device for aggregating production. But each author critiques the felicity of that heuristic, showing that there is always something prior about style—some assumption governing its use—residing, perhaps, in authorial uniqueness, or (more typically) in purely formal relations, theories about ethnic origins, or absolutes of other types.[20] Ginzburg, in interrogating style, shows how the notion can serve both to split and to lump. He concludes with a sharp critique of the very category itself as ideologically laden—"an instrument of exclusion"—and calls for both an acknowledgment of the utter uniqueness of a particular work in its isolation, and for a nonsimultaneous *translation* of the work's singularity into a relationship with history.

Irene Winter is similarly critical of the heuristic of style, but only as it has been reified as distinct from "meaning" through the peculiar divagations of art-historical theory. She makes the useful distinction between "stylistic analysis" as an operation that is clearly located in the viewer, while the more problematic concept of style is positioned as something inherent and identifiable in the work itself. Refusing to relinquish stylistic analysis, her real target is the development within art history of two paths, where style was reserved for form alone, while "iconology" was given to be the bearer of meaning. Through a close reading of objects identified geographically as roughly contemporaneous products of "Syrian" versus "Phoenician" cultures, Winter seeks a composite model of style as containing both elements that are "not-necessarily-conscious" and those that are "consciously deployed"—a historically bounded set of possibilities that are winnowed down in the work itself for reasons that have everything to do with meaning. In Winter's analysis, the ruler's strong arm is inevitably both formal (creating structural and decorative patterns) and meaningful (conveying specific information about power)—yet, crucially, "the potential use and value of style as a concept depends entirely upon the nature of the analytical operation(s) in which it performs."

For Ginzburg, the possibility of the individual object's resistance to aggregation must be held in tension with our (not always progressive) need to make social sense of it. For Winter, who is working with objects that are both divorced from individual makers and unknowable outside social systems, the problem is a different one, a more delicate negotiation in which "reading in" is balanced with a sensitive appreciation for the obdurate peculiarities of historically situated cultural forms. This dialectic between individual (makers, readers, objects) and social (modes of meaning, contextually embedded producers of objects) is intrinsic to the heuristic of style itself. Elsewhere Svetlana Alpers has commented upon the fact that scholars outside art history have been drawn to the discrimination of styles "because it is scientific"—more empirical "than the critical appreciation of and interpretation of individual works."[21] This sense of the "scientific" use of style as erasing or subsuming the individual occurs too in Amy Slaton's essay on technological styles.

Slaton shows how technicians' factory forms, as read by art historians, have been used to suggest stylistic aggregations that work precisely through the *absence* of imagery (in this case, through the absence of "the decorative"). This is a notion of style that no longer depends on links forged between essential qualities of the works themselves, but upon shared modes of providing them with cultural significance.[22] Clearly Slaton's is the same "modal" heuristic that Ginzburg and Winter deploy. She argues that it is important to extend style beyond innovators (Henry Ford, for example, as the "author" of the automobile) and to identify it with the taste of consumers, and the existence of technical practices together with the institutions that enforce them. "Technological style" thus becomes more than a borrowed metaphor, more than an analogue of style in the artistic-architectural sense. Conjoint practices issue in both style-as-technique and style-as-formal-relations. The concrete factory aesthetic emerges from engineering concerns, but its forms inaugurate their own history of signification.

For Slaton, Ginzburg, and Winter, style is a culturally loaded term that brings powerful forces of nationalism, politics, and racism into play in the fields of aesthetics and social interaction. Science is pictured here as actively wielding style (Ginzburg) rather than distantly reflecting or unconsciously manifesting it; style in art and architecture is dismantled to reveal the processes of professional identity formation (Slaton) or the production of national cultural identity by interpretive history itself (Winter). Gudea and Assurnasirpal, Augustine and Feyerabend, cement technicians and architects—all are shown to picture science or produce art in insistently stylized ways. But far from the mysterious attribute of a cultural Zeitgeist, in this book style is viewed as strategically constructed, both in the act of making culture and in the process of interpreting it.

The Body: How do images shape body knowledge, and for whom?

Processes of doing science and making art involve the body, but the book's authors argue that "the body" is most often figured as an *object* of these cultural inquiries, constructed through the parallel and intertwining discursive regimes of (natural) science and (figurative) art. Where interpretation and identity are key issues for the **Styles** section of the book, here the central theme is the *power* of images to instantiate and produce knowledges of, and by, the body. Arnold Davidson begins the section with a minutely historicized account of the iconography of the stigmata, demonstrating that notions of evidence and theological dogma regarding this bodily miracle were fixed in panel paintings and frescoes before they ever appeared in the putatively authorizing Vatican texts. Simultaneously, he shows how the visual iconography of the stigmata itself becomes a parallel tradition that never fully converges with the textual accounts. St. Francis's markings cannot be visualized as the higher "imaginative vision" (identified by post-Augustinian philosophy) that the textual accounts want to emphasize; the fresco paintings of Giotto and his followers inevitably *embody* the miraculous in particularly concrete ways. We suggest that the very materiality of paint (that is, the transformation of the narrative's iron-red blood to iron-red pigment), instantiates a baser "corporeal vision" that was, in Davidson's words, "meant to stabilize the status of the stigmata [as] a singular miracle." "Official" possibilities for the religious body were enlarged in this visual tradition, and subsequent miraculous bodily transformations were experienced—one might say indelibly marked—by the body knowledge such images produced.

Moving from the late medieval period to the eighteenth century, Londa Schiebinger also explores the ways in which verbal and visual discourses construct body knowledge. But by focusing on gender divisions in scientific practices, she also locates the production of what we might call "body ignorance." Like the oral and internal traditions of women's body knowledge that Barbara Duden describes as eliminated by scientific knowledge,[23] Schiebinger posits a variety of "counter-bodies" that Enlightenment science ignores—individual human bodies obscured by racist, sexist, and colonial programs; polymorphously sexual plant bodies gendered and socialized; "native" bodies (and indigenous knowledge) overrun by the expanding discourse of colonial natural philosophy. In Schiebinger's most salient example of such "body ignorance," she examines how the cartographic and classificatory gaze of Enlightenment science erased the experiential knowledge of the Surinamese women whose views had been presented by female naturalist Maria Sibylla Merian. Merian's complex descriptions of the plants of Surinam included political critiques and medicinal lore—specifically, the local knowledge that the seeds of a certain tree worked as an abortifacient (knowledge that worked to confound the plans of Dutch colonial slave traders along the way). The production of ignorance from this matrix of knowledges took place as Merian's descriptions were taken over by British naturalists, who had their own natalist objectives. The indigenous body knowledge of the plant's medicinal use was suppressed. While Davidson shows the way in which visual imagery can serve to "corporealize" knowledge of natural wonders (which we see as parallel to Augustine's lowest category of "corporeal vision"), Schiebinger shows the ways in which the visual and verbal discourses of colonial botany worked precisely to erase such corporeal knowledge (in favor of a type of knowledge we might categorize as analogous to Augustine's third and highest category of "intellectual vision").

With a view of the early decades of the twentieth century, Caroline Jones examines a realm that might be identified with the intermediate Augustinian realm of "imaginative vision." Here, in the practices of a single modernist artist (Francis Picabia), Jones locates the modern body-machine complex, analogous to what Donna Haraway terms the "technoscientific body" (already suggested by the image of Hoberman's installation). Picabia's evocative line drawings navigated certain normative modes of knowing the sexed body and explored the psychological states that were then held to enforce sexual difference. These were modes (conveyed by his own neurologists) that theorized his persistent neurasthenia as a sexual disorder, reparable only through the proper channeling of procreative and electrical energies. The standard model of "influence" that might be used to explain Picabia's work (in which a concept moves from scientific discourse to artistic imagery) is confounded by a closer reading of the images, particularly those dedicated to the very neurologists formulating the psychologized sexual body Picabia inhabited. Picabia's machinic images produce their own renegade forms of knowledge, some appropriate to the neurasthenic subject, and some frankly out of that subject's domain. The newly *visceral* presence of technology in the Picabian body allowed hybrid, hermaphroditic, and synoecious couplings that (like Davidson's visualized stigmata) present an instantaneous visual "tradition" at odds with the dogmas established by textual culture—even the textual culture that might be constructed by Picabia's own accompanying poems.

In her expansive voyage over the terrain mapped by the scientists of "Life Itself," Donna Haraway observes their visual culture—generated by game designers, molecular geneticists, microbiologists, and commercial advertisers—with a mordant yet curiously sympathetic eye. Haraway traces, as does Schiebinger, exclusionary (and largely unconscious) tropes of cartographic delineation and their origin in systems of colonial control. She, too, examines the production of ignorance—in this case accomplished by the reductive "mapping" of the Human Genome Project. Empowered by the rigid yet fragile operation of what she calls "gene fetishism," the technoscientific body of the genome is produced through a variety of discourses. The most visual of these is that "official art of capitalism" (as David Harvey has termed it), the advertisement. The cartoons used to market genetic research technologies constitute Haraway's most powerful object, for here the operation of the fetish becomes an anxious negotiation between the production of ignorance and the body knowledges it would erase. The advertisements' jokes, their very comedic structure, attempts to resolve these negotiations in favor of "Man$_{TM}$," the parthenogenic substitute for diverse lived bodies' narratives, experiences, and subjectivities. As Haraway's analysis shows, the links between these anxious comedic structures and the more official stories of science are profound: the metaphor of the map ensures the systematics of colonial control; the construction of human bodies as husks for "replicators" and "selfish genes" fuels the cultural unconscious that produces the fetish; the lie of the "master molecule" empowers the fetishists in their disavowal of the living in favor of the replicant and the undead.

The implications of Haraway's larger argument connect, as well, with artist Perry Hoberman's contribution to this volume. Hoberman and Haraway would both agree that the gene fetish is related (one is tempted to say "genetically related") to the ideology of cyberspace. Each has argued that cyberspace is falsely theorized (and popularized) as a disembodied realm that leaves the "meat" of the body behind, in exchange for the map-like manipulations of various electronic simulation games (in Haraway's account, primarily the Maxis Corporation's "Sim" games—SimLife, SimCity, SimEarth). Such "deanimations" (as Haraway terms them) are experienced by Hoberman in his role as a sometime producer of virtual-reality technologies (known in the industry as "location-based entertainments," but perhaps more aptly described as "location-erasing entertainments"). The disembodiment of such visual and verbal discourses is always strategic, as Haraway shows (even if it may be the unconscious strategy of the fetish). Hoberman, too, works to materialize the systematics (the marketing ploys and electrical grids) and links to the body (hair dryers, food blenders and mixers, foot massagers) that make technology as cathected as it is. As the work of both Hoberman and Haraway reveals, the cyberbabble cycling around "virtual" reality serves above all to erase other realities, from distant yet specific worlds of colonialist empires, to the more proximate "meat" of migrant workers in the computer-chip industry in Silicon Valley, to the narrowed choices that the rhetoric of "interactive" technology serves to mask. As Faraday's Islands and other works by Hoberman emphasize, technology always operates in an embodied world, where, at the most, we might aspire to inhabit what Haraway calls the "carbon-silicon fused flesh of technoscientific bodies"—hybrids, once again, as the neurasthenic Picabia already imagined us to be.

Seeing Wonders: What do we know when we see?
In the most historically focused and largest of the book's sites, five authors address objects from an age before the production of colonial and cartographic certainty, when boundaries between the natural and the artificial, the seen and the known, the monstrous and the wondrous, were fixed at points far from the contemporary compass. Krzysztof Pomian's essay magisterially tracks a shift from ancient epistemes of equivalence between vision and cognition (seeing *as* knowing, and, in parallel, "to know is to see"), to Enlightenment models of cognition as production, in which seeing (as mediated by "scopes," both tele- and micro-), is productive of a Cartesian "intellectual intuition" only later challenged by Hume. Pomian concludes with a third model, characteristic of the contemporary moment, which he identifies as "indirect cognition," a mode in which "seeing" is knowing-through-technology. Sight in Pomian's ultimate moment has become distant from "mere" ocular vision. Unlike the boundaries that will be traced by the other authors in this section (between nature and art, true and false reproduction, the panegyric and the scientific, the premodern and the modern), all of which involve modes of *visual* representation, Pomian's final regime of knowledge (which could also be called "instrumental cognition") suggests a potentially postmodern frame. Highly mediated, eliding into unbounded, less visual zones in which "nature" is produced purely discursively, such "indirect cognition" produces all the wonders of the universe that we no longer need to "see" to believe.

Pomian's philosophical and historical sweep is focused in subsequent essays on more narrow spans of Renaissance and early modern European natural philosophy (practices conducted by those whom we identify today as "artists" as well as by those now categorized as "scientists"). These other essays illustrate how different the sight-knowledge relation can be from contemporary models (even from those just beginning to evolve). As Lorraine Daston argues, the relation between seeing and knowing often begins with the cognitive side of the equation, and the supposed self-evidence of the seen dissolves with the historically shifting boundaries of belief about the powers of nature set against those of humankind. What is it, Daston asks, that makes a thirteenth-century observer decide that an image-bearing stone (a cameo) is imprinted by nature rather than cleverly carved, while four centuries later another image-bearing stone (a fossil) provokes questions as to whether it is naturally deposited or artificially formed? As Daston insists, these distinctions were not fuzzy at the time—they were fixed firmly and definitively in the thirteenth century, to be redrawn in the seventeenth just as firmly and definitively. The kinds of indirect, postmodern knowledges to which Pomian alludes resonate intriguingly with Daston's analysis: What kind of boundaries are being drawn today between "nature" and "artifice," as postmodern theorists simulate carbon-silicate hybrids and invent ways to store knowledge in a manipulated biomass?

Introducing the theme of "wonder" that threads through this site, Daston explores the ambivalence that greets these unstable objects—the figured fossils, carved cameos, crystal-studded bibelots, and nature-machine amalgams that are seen as marvels in some epochs, kitsch in others. It is precisely ambivalence that fuels Katharine Park's chosen historical moment, as well, but her fifteenth-century Italians experienced an ambivalence tinged with horror—a profound fear of the unchecked power of reproduction in both nature and art. Park pursues the nature/art boundary into the crannies of medical and juridical debate, and chases the "wonder/horror" dichotomy into the anxious terrain already set forth by the earlier section, **The Body**. Relating to both Haraway's examinations of the gene fetish and Jones's look at machinic sex, Park analyzes late-Medieval theories of reproduction in which the visible is, paradoxically, both proof of secure knowledge and product of false knowing. Park's historical subjects harbor anxieties: about the vulnerability of females' reproductive apparatus, and the skills of counterfeiters in altering a newly impressionable Nature. There were strong connections between cuckoldry and counterfeit in the thoughts of Park's Italian clerical elite. Early efforts to dissect the female corpse were linked, she argues, to these anxieties about female and monetary reproduction. The membrane of the female body was held to be permeable and "impressionable," and potent images were uniquely capable of influencing the more fluid female form. The power of the sign in Park's history thus oscillates between passive symbol of prior knowledge, and potent stimulator of new knowledge that may be false, or true. The sign's capacity to shift from miraculous wonder to counterfeit horror has everything to do with the status of representation itself in fifteenth-century Tuscan culture.

The oscillatory relation Park traces between the image as that which registers knowledge and that which produces it also obtains in David Freedberg's analysis. Freedberg's chronicle of the destabilizing power of natural imagery in the sign systems of the later Italian Renaissance is a progressivist narrative (as the Renaissance patrons, panegyrists, and members of the "Lynx-eyed" academy themselves believed). In his specific focus on the iconology of the bee, Freedberg traces a tense, taut line between the knowledge produced by the new technologies of vision (e.g., the microscope) and the symbolic knowledge necessitated by the Medicean reign. The more that "bees" become the subjects of specific natural-historical inquiries, the less they can function as transparent vehicles of Papal flattery. *The more they "know," the less they "represent."* Freedberg celebrates the microscopic accuracy of the engravings prepared for the Barberini Pope (whose escutcheon sported three bees), but argues that such a celebration of optical technology was dangerous at a time when Galileo was being targeted as a heretic. Such micrographical accuracy did not extend, of course, to a correct identification of the head of the hive as the *queen* bee; for the papal panegyrists, the fecund and benign monarch of bee-dom could only be a king, explicitly analogized to the pope himself. Between classical tales of sweet honey and smooth governance, and new microscopic visions of black, hairy, bug-eyed creatures with multiply jointed legs and inhuman sexual practices, an uneasy gap began to open. In a real sense, these bee-studded images promulgated knowledge that their authors became anxious to constrain.

In Joseph Koerner's richly allusive essay, such visually implicated knowledges shift differently. If Freedberg defines an opposition between representing and reporting (or praising and knowing), Koerner shows how such a conscious opposition must itself be seen as a moment in the development of modernism. While imagery may be a maker of knowledge, it is also, for Koerner, a manifestation of a worldview. The image plays a crucial historical role in visually demarcating (for present-day viewers) a premodern ("unknowing") universe of Bosch from an already modern ("knowing") frame of Breughel. The premodern is incapable of referencing itself as a representation, while the modern is powerless to avoid it. Citing Lévi-Strauss's inability to penetrate the savage world he would understand without thereby destroying its very "savagery," Koerner theorizes the historicity of framing itself: Bosch's refusal to "frame," to bracket the wondrous from the horrific or the monstrous from the sacred, stands in contrast to Breughel's consciously framed tableaux. Koerner finds in Bosch and Breughel closely linked yet crucially disparate pictures that "stand at our disposal for apprehending the threshold to an alternative historical reality." His distinctions between "representing" and "knowing" return us again to distinctions among the categories of modernism, its precursor, and its potentially postmodern sequel. These artworks, for Koerner, register the crucial juncture at which the world splits among conflicting worldviews. As in Haraway's discussion of the postmodern "pov" or point-of-view, Koerner traces the move from world as plenum, to "the" world as contingent and discursively framed.

Objectivity/Subjectivity: What do images presuppose about (human) nature?

"Objectivity," in its widespread usage, is one of the most vaunted attributes of science in both popular and scholarly accounts. Some notion of objectivity motivates most analyses of what separates the production of science from the production of art, with "subjectivity" the shadow term that is held to separate art from science. And yet, as the authors of this section demonstrate, neither category is stable or sufficient—not for artists and not for scientists. Peter Galison, building on joint work published elsewhere with Daston, argues that the scientist's pictorial objectivity is, fundamentally, a nineteenth-century concept, exemplified in the discourse of the scientific atlas. Long before the term "objectivity" itself appears, these atlases served as visual compilations and repositories of the basic objects of science—the best and truest depictions of bodies (for example) that could be produced. But in the first of these "true to nature" tomes (which appeared in the eighteenth century), the atlas image was anything *but* a depiction of some specific bit of nature—the very idea was anathema. *True* images at this point were held to be precisely those in which the artist/scientist was able to part the curtains of appearances, and in so doing reveal an inner or hidden reality obscured from sight. Distinct from Pomian's first epoch of "vision *as* cognition," these Enlightenment thinkers found much to mistrust in that which was merely seen. Genius was needed to discern the true from the fleeting. By contrast, Galison contends, the goal of the nineteenth-century natural philosopher became increasingly to *restrain* this individual "genius," and to harness the image-making process to appearances so "mechanically" that it would preclude the possibility—indeed, even the *suspicion*—of any human intervention whatsoever. Not coincidentally (as we have argued), it was also at this moment that the roles of scientist and artist began to congeal into their binary domains. Scientists and their defenders claimed the new automaticity of depiction as objectivity, which itself became a newly valued term. But as Galison reveals, the fate of objectivity did not rest here. In the twentieth century, subjective judgment (which had long been a term of opprobrium for nineteenth-century scientists) became a term of approbation for atlas makers, who chose to celebrate their roles as expert interpreters rather than advertise how closely they confined and policed their artist-collaborators.

Galison's account of the nineteenth-century production of pictorial scientific objectivity as self-effacement and externalization stands in stark contrast to the interiority suggested by Jan Goldstein in her depiction of the simultaneous rise of Cousinian psychology among upper-middle-class Frenchmen. Constructed as a hodgepodge of neo-German Idealist philosophy, Victor Cousin's teachings were taught throughout the Lycée and university systems, coming close to an official philosophy of the (male) bourgeoisie. Front and center stood everything that was subjective, everything associated with a forceful will; Cousinianism was a celebration of the individualistic, morally independent, highly sensible and sensitive *moi*. Because Cousin's hierarchies so privileged the subjective, they might at first appear to be at loggerheads with the nineteenth-century atlas makers Galison describes, whose rallying cry was self-abnegation. But as Goldstein makes clear, the Cousinians saw their task of self-inquiry as one in which, paradoxically, self-sacrifice and asceticism were central moral characteristics. Perhaps one should put it this way: the subjectivism associated with Cousinian individualism, creativity, and force of male character involves the supervaluation of the *moi* (subjectivity), while a different but related type of individual fortitude came to be supervalued in the sciences. The moral profile of the Cousinian ascetic (called "subjectivity") jibes precisely with the willful suppression of the scientist's desire to see a theory confirmed or an expectation realized (termed "objectivity"). The scientist's receptivity to the world is, by the light of the atlas makers, not born of passivity but of triumphant self-restraint.

The notion of objectivity-as-self-restraint produces an intriguing disagreement between historian of photography Joel Snyder, on the one side, and Galison and Daston, on the other. For Galison, the salient feature of objectivity as captured in the nineteenth-century atlas-making tradition is that it is both procedural and moral; it is an attempt by the picture-making scientist to abolish the idealizing, "artistic" interventions of earlier observers. For Snyder, the point of the physiologist-photographer Etienne-Jules Marey's work lies precisely in the fact that it does away with the central role of the "observer" altogether. Snyder argues that (for Marey) it is insignificant whether the process under consideration could be observed accurately by humans, or even at all. Put differently, Marey's instruments construct images entirely unavailable to unmediated human vision (arriving once again at Pomian's category of "indirect cognition"). As with all images, ultimately even the instruments fall away, and only chronophotographic tracings remain. These tracings, not the original photographic subjects (trotting horse, running man) then become the true "subject of investigation." Marey's staccato images do not "freeze" perceptual time, they schematize temporal progression. As Snyder is at pains to emphasize, even before Marey the long-exposures of early photography did *not* show what a human observer saw. Boats passing on the river vanished in virtue of their movement, and streets were voided of their carriages and their *flâneurs*. From considerations such as these, Snyder concludes that whatever else they do, photographs are not aimed uniquely at enhancing sense impressions. At times they create a new domain of the visual, producing at the same time new viewing subjects to make sense of that domain.

The three essays of this section can be structured as follows. For Galison, there is no stake in claiming for the mechanical-objectivists any kind of sense-data impressionism. None of the nineteenth-century atlas makers (nor their eighteenth-century predecessors) grounded their images on what we might see with the unaided eye. In this sense, Snyder's Marey is functioning as a research physiologist, doing precisely what astronomers or anatomists were also doing in *their* laboratories and observatories: correcting the senses with mechanical aids, teaching us just where our senses can lead us astray, and, indeed, constructing entirely new modes of vision through which the world would subsequently be perceived. What is striking in the Marey story, and what connects it back to Goldstein's culture of Cousinianism, is what Marey held to be necessary in replacing the senses: *the imagination*.[24] For most German, British, or American atlas makers of the mid- to late nineteenth century, "imagination" suggested the vagaries of artistic license, a freedom from the constraints of mechanical reproduction; we might recall also that the "imaginative" was only the middle register of Augustine's hierarchy of religious visions, between the corporeal (Marey's senses) and the intellectual. One might speculate, building on Goldstein's work, that the long tradition of Cousinian psychology (with its emphasis on the conciliation of art and science) left a positive valence to the imagination in French physico-physiological research that was absent in the Anglo-Saxon world. The imaginative elided with the intellectual in the French hierarchy of representations. But however one considers the particularities of these instances, the broader lesson is clear: the objectivity/subjectivity axis that has so characterized debates over the domains of art and science was itself a historical entity coeval with those debates. It took its defining form in the nineteenth century, and its history forms the backdrop to our own.

Cultures of Vision: What viewers and processes does the image presuppose?

This final site deals with the logic of "visual culture" and the issue of visuality itself, which together form the subject of inquiry within much of science studies and art history. From her perspective as an art historian, Svetlana Alpers performs a complex reading of representations of the artist's workplace, including genres such as still life and landscape that are not usually read as indexical studio signs. Alpers seeks both to reflect on the relation of artist to reality, and to analogize artists' efforts to those of scientists participating in the mimetic and analytic traditions of experimentation described elsewhere by Galison and Alexi Assmus. As the subsequent essay by sociologist of science Bruno Latour also does, Alpers's contribution underscores the double action that follows from linking art and science. The comparison grants a "seriousness" to artists, rendering them skillful rather than merely moral; at the same time, it brings experimenters out of their isolation in a separate "culture," and in so doing, redefines the epistemic status of what they do. At first pass, one might model the studio on the laboratory, focusing attention on the role of technician-assistant. But Alpers is after the painting's self-promoting status as an indicator of individual experience in general, experience in which the individual's presence in the world is not tangential, but rather central to the activity of making art.[25] And in this respect, the artist in the studio is manifestly *unlike* the scientist in the early modern laboratory. Withdrawing (elsewhere she calls it "retreating") into the studio is a regressive act, one that returns us to a prior experience. As regressive, the view from the studio is colored either as originary (how the child sees) or as precursor to philosophy (how we come to experience through vision). The explorations of the artist are in this sense philosophical and psychological quite as much as aesthetic.

The personal, philosophical, and psychological also enter into Bruno Latour's paper, which thematizes the plurality of "cultures" in this section's title by posing a question that is pressing for science studies, for art history, and for our theories of religious faith. What, he asks, can we learn from the way these vastly different regimes of knowledge use visual techniques to point toward "remote phenomena and absent features?" At this level of abstraction, the painter employs iconology, the scientist symbolic representations, and the theologian one realm of reality to stand in for another. But most importantly, Latour insists, the dynamic of this set of symbols (and symbols of symbols) does not function by directly invoking the final referent, but rather by a complex process of mediation that is itself the bearer of meaning. In the articulation of these systems of mediations, both the historian of art and the historian of science end up showing how complicated it is to put together the elements of a finished piece of work. Varnishes, dealers, assistants, patrons; maps, measuring devices, graphs, charts; angels, saints, monks, worshipers—these chains of mediators constitute the circumstances under which the work of art, science, (or religion) is produced. Here (Latour insists) an asymmetry arises. Constructivism *flatters* the arts because exhibiting mediations works "in the same direction" as the art's own ambition, but the same multiplication of mediators *threatens* a popular construal of science that holds it to be an infinitely direct and immediate reference to the world.

A more sophisticated view, Latour argues, would take science to be that which is held constant through transformations; instead of trying to get at *things* and *mind* directly, he wants to bracket those categories in and of themselves, and get at them through the dynamical transformation of one mediator into another. In the end, Latour wants a language of visual culture rich enough to include many types of mediators, but one in which no type is subsumed by any other. He asks that we bracket out the extremes of *res* and *cogito*, and focus on the "cooking steps" that mediate between.

Simon Schaffer has a similar aim, but his kitchen proffers less heavenly fare. Schaffer wants an understanding of the widely distributed features of popular culture, and the central role they play in defining scientific knowledge. More specifically, he aims to show that the nebular hypothesis in astronomy—the notion that stars and planetary systems formed through the coalescence of clouds of gas in space—was tied root and branch to nineteenth-century battles over evolution, the progress of civilization, and the Irish Question. For both friends and enemies of the nebular hypothesis, progress in the heavens (from chaos to brilliant stars) vouchsafed the idea that there could be progress below (in politics and society).

Schaffer's story, however, is not purely a narrative of abstract ideas. The contest over "progress" in deep space was fought, among other places, in the famed observatory of the Earl of Rosse in Ireland. Rosse (William Parsons) and his second in command, Ulsterman Thomas Romney Robinson, inveighed against papism, materialism, and evolution. Their aims oscillated between process and product. Process encompassed the astronomical display of a factory-like laboratory in which production was explicit, workmanlike, and British (in explicit distinction to the rural Irish surround). Product centered on the content of the observatory's pictures, produced through exquisite draftsmanship and always aiming at the "resolution" of the so-called nebulae into stars. For if such a resolution could be completed, it would (so Rosse and his allies contended) not only refute the nebular hypothesis, but also the broader promise of evolutionary progress (and social responsibility) that it seemed to imply.

The stakes of debates in visual culture are also at issue in art historian Jonathan Crary's essay. Crary, too, is after the dynamics of visual culture and, like Alpers (and Galison, and Pomian), registers a nineteenth-century shift. Crary, however, looks not to changes from mimesis to analysis (Alpers), nor from genial to mechanical to judgment-based objectivity (Galison), nor from "vision-as-cognition" to "vision-as-production" (Pomian). Although these histories can all be linked with his account, he focuses instead on a single thread within the epistemic shifts of modernism. He charts the deep reconceptualizations of *attention* (involving perception, cognition, and aesthetics) that he sees as constitutive of the late-nineteenth-century subject. Put starkly, Crary's account identifies a transformation from classical theories of vision as something mechanical and capable of abstraction from the body (exemplified by the camera obscura), to modernist notions of perception as a process characterized by temporal flux and embedded in a physical body. The newly felt fragility of perception made attention and attentiveness new problems—problems of pressing urgency within both the modernizing workplace and modernist art. No longer was it possible to think of vision as fundamentally passive, a system in which the mind was imprinted by an external world. Crary joins those in science studies who argue against continuity with prior theories of mind: late-nineteenth-century epistemologies foreground the observer and the integrative, *active* observing process; the eye becomes "thick" and the viewing process fundamentally unstable.[26] Conceptually this marked a shift, from representation as a simple trajectory between equals to a relation of inherently unequal forces, from a semiology of perception to a physics of perception. For Crary, the modernist obsession with an aesthetics of "presence" and raptness takes place within this new epistemological field. Our histories of nineteenth-century visual culture must be read against such scientific understandings of perception and attention. They register the fault lines of an emerging modernist episteme, and set the stage for our own late-twentieth-century theories of the spectacular.

In its overarching analysis of the way that representations function in scientific and artistic discourses, *Picturing Science, Producing Art* attempts to present a broader analysis of knowledge production as a whole. By denaturalizing the categories "science" and "art," and by attempting simultaneously to historicize and locate the mechanisms that enable their binary economy to function, we seek to provide more than just a belated corrective to the "two-culture debate" (lingering still in the late twentieth century). The cultural frames and positions available to scientists and artists as producers, and the equally constrained yet movable locations of those who interpret their work, have been our objects of study. By historicizing notions that see science as revealed Truth and art as mere individual statement, we take both realms of knowing more seriously. For the interdisciplinary scholars of this book, science and art are deeply important sources of knowledge, neither transcending the social (as "pure scientific knowledge") nor propelling society from without (as "art of genius"). We have blurred the boundaries in order to demonstrate the ways that both domains *make* culture, revealing how they mark both mind and matter in the process.

Notes

1. For a helpful summary, see Stefan Collini, introduction to C. P. Snow, *The Two Cultures* (Cambridge: Cambridge University Press, 1993), particularly p. xii. Here Collini notes that the term "scientist" was proposed only in the mid-nineteenth century, explicitly by analogy with the word "artist":

    an article of 1834 [reported] on how the lack of a single term to describe "students of the knowledge of the material world" had bothered meetings of the British Association for the Advancement of Science in the early 1830s, at one of which "some ingenious gentleman proposed that, by analogy with *artist,* they might form *scientist,*" though the same report records that "this was not generally palatable."

    Collini's internal quotations are from William Whewell, who may himself have been the "ingenious gentleman" he referenced. See Sydney Ross, "Scientist: the Story of a Word," *Annals of Science* 18 (1962): 65–85.

    A search on the Internet with the limiters "Art and Science" brings up thousands of websites that turn out to be orientation maps for undergraduate college curricula. Interestingly enough, the formerly capacious label "College of Liberal Arts," expanded to include the natural and social sciences, has often experienced a (sem)-meiotic division into its shadow binaries, "College of Arts and Sciences." This conversion was adopted at Boston University in the spring of 1996.

2. As collaborators, we may be accused of exemplifying these binaries, but we also enjoy switching between them and multiplying the terms.

3. Charles Baudelaire, "The Modern Public and Photography," in "The Salon of 1859," *Baudelaire: Selected Writings on Art and Artists,* trans. P. E. Charvet (Cambridge: Cambridge University Press, 1972), pp. 291–98. John Ruskin, *The Eagle's Nest* (New York: J. Wiley and Son, 1873). Also see citations by Collini, introduction, pp. xiv–xv: T. H. Huxley, "Science and Culture" (1880), reprinted in Huxley, *Science and Education: Essays* (London, 1893), pp. 134–59; Matthew Arnold, "Literature and Science" (1882), reprinted in *The Complete Prose Works of Matthew Arnold,* ed. R. H. Super, vol. x (Ann Arbor: University of Michigan Press, 1974), pp. 52–73.

4. The literature on both avant-gardism and technologism is vast. For a good introduction see Renato Poggioli, *Theory of the Avant-Garde,* trans. Gerald Fitzgerald (Cambridge, Mass.: Belknap Press of Harvard University Press, 1968); Peter Bürger, *Theory of the Avant-Garde* (Minneapolis: University of Minnesota Press, 1984); Reyner Banham, *Concrete Atlantis* (Cambridge, Mass.: MIT Press, 1986) and *Theory and Design in the First Machine Age* (Cambridge, Mass.: MIT Press, 1960); Richard Guy Wilson et al., *The Machine Age in America, 1918–1941* (New York: Brooklyn Museum and Harry N. Abrams, 1986); *The Great Utopia: The Russian and Soviet Avant-Garde, 1915–1932* (New York: Solomon R. Guggenheim Museum, 1992). For a view of the conjunction of fascism with avant-gardism and modernism, see also Jeffrey Herf, *Reactionary Modernism* (Cambridge: Cambridge University Press, 1984), and forthcoming studies by Jeffrey Schnapp and Hal Foster. A more optimistic view is reflected in the work of Linda Dalrymple Henderson, for which see below.

5. Rudolf Arnheim, *Art and Visual Perception* (Berkeley and Los Angeles: University of California Press, 1954); Anton Ehrenzweig, *The Hidden Order of Art: A Study in the Psychology of Artistic Imagination* (Berkeley and Los Angeles: University of California Press, 1967): "Psychologically, abstraction in modern scientific thought is not merely reminiscent of abstraction in modern art, but is due to the same phenomenon of dedifferentiation. . . . The need for seeing incompatibles 'together' is more easily discerned in periods of transition when science is still groping for new models to accommodate still existing contradictions and inconsistencies." Ehrenzweig uses here the wave/particle duality as such an "incompatibility" that requires the suspension of secondary Gestalt types of vision in favor of his "dedifferentiated" primary vision (pp. 131, 133).

6. Gerald Holton, *Thematic Origins of Scientific Thought* (Cambridge: Harvard University Press, 1973) and *The Scientific Imagination* (Cambridge: Cambridge University Press, 1978).

7. See Fritz Ringer, *The Decline of the German Mandarins* (Cambridge: Harvard University Press, 1969).

8. For commentary, one might begin with F. R. Leavis's withering critique in *Two Cultures? The Significance of C. P. Snow* (London: Chatto and Windus, 1962). See also John de la Mothe, *C. P. Snow and the Struggle of Modernity* (Austin: University of Texas Press, 1992).

9. For the deep gendering of Snow's account, see, in particular, his comment that the nature of scientific culture is "steadily heterosexual," without literary culture's emphasis on "the feline and oblique." C. P. Snow, "The Two Cultures," *New Statesman* (October 6, 1956): 413. For this and other insights into Snow and his milieu we are indebted to Stefan Collini, our colleague at the Institute for Advanced Study in Princeton during the 1994–95 academic year. See Collini, introduction, p. xxvi.

10. Historian of literature and technology Leo Marx continues to plumb the vitality of the two cultures debate, seeing it as instrumental in isolating discussions regarding the environment to a small group of natural scientists. See his "The Environment and the Two Cultures Divide," in *Science, Technology, and the Environment*, ed. James Fleming and Henry Gemery (Akron, OH: University of Akron Press, 1994).

11. See Leavis's critique in *Two Cultures?*; essay also anthologized in Leavis, *Nor Shall My Sword: Discourses on Pluralism, Compassion and Social Hope* (London: Chatto and Windus, 1972), anthropological issue discussed on p. 50.

12. T. S. Kuhn, *The Essential Tension* (Chicago: University of Chicago Press, 1977), p. 342. Kuhn is responding to E. M. Hafner, "The New Reality in Art and Science," *Comparative Studies in Society and History* 11 (1969): 385–97. Note that Kuhn's title encodes the "tension" between two other binary poles, philosophy and history, but also perhaps implicitly art and science.

13. Kuhn, *Essential Tension*, pp. 342–43.

14. On links between the Vienna Circle and the Dessau Bauhaus, see Peter Galison, "Aufbau/Bauhaus: Logical Positivism and Architectural Modernism," *Critical Inquiry* 16, 4 (1990):709–52. On the implications and impact of Gombrich's account of abstract art, see Caroline A. Jones, "Abstraction and the Leaven of Criticism," in *Eyesight Alone: Clement Greenberg and American Art* (forthcoming).

15. Erwin Panofsky, "Galileo as a Critic of the Arts," *Isis* 47, part 1, no. 147 (March 1956): 3–15; see also Samuel Edgerton, "Galileo, Florentine 'Disegno,' and the 'Strange Spotednesse of the Moon,'" *Art Journal* (Fall 1984): 225–32.

16. Linda Dalrymple Henderson, *The Fourth Dimension and Non-Euclidean Geometry in Modern Art* (Princeton: Princeton University Press, 1983); also see her forthcoming book on Marcel Duchamp. Martin Kemp, *The Science of Art: Optical Themes in Western Art from Brunelleschi to Seurat* (New Haven: Yale University Press, 1990). See also Samuel Y. Edgerton, Jr., *The Heritage of Giotto's Geometry: Art and Science on the Eve of the Scientific Revolution* (Ithaca: Cornell University Press, 1991), and J. V. Field, *The Invention of Infinity, Mathematics and Art in the Renaissance* (Oxford: Oxford University Press, 1997).

17. Kemp, *The Science of Art*, p. 1.

18. James Elkins, "Art History and Images that Are Not Art," *Art Bulletin* 77, 4 (1995): 551–71. See also Elkins, *The Object Stares Back: On the Nature of Seeing* (New York: Simon & Schuster, 1996). Barbara Maria Stafford's books on these subjects include: *Body Criticism: Imaging the Unseen in Enlightenment Art and Medicine* (Cambridge, Mass.: MIT Press, 1991) and *Artful Science* (Cambridge, Mass.: MIT Press, 1994).

19. This literature is so vast, one can only indicate some key starting points. On diagrams, see S. S. Schweber, *QED and the Men Who Made It* (Princeton: Princeton University Press, 1994), especially chapter 8: "Feynman and Space-Time Processes"; also, Schweber, "Feynman and the Visualization of Space-Time Processes," *Reviews of Modern Physics* 58, 2 (1986): 449–509. Galison, "Minkowski's Space-Time: From Visual Thinking to the Absolute World," *HSPS* 10 (1979): 85–121. On indicator diagrams, see Robert Michael Brain, *The Graphic Method: Inscription, Visualization, and Measurement in Nineteenth-Century Science and Culture* (Ph.D. dissertation, University of California at Los Angeles, 1996), and M. Norton Wise, "Fleeming Jenkin Measures Energy and Utility: Indicator Diagrams and Supply-Demand Curves," paper delivered at the Second Annual Harvard-MIT-Princeton Grad-

uate Workshop in the History of the Physical Sciences (Princeton University, May 1997); Robert Brain and M. Norton Wise, "Muscles and Engines: Indicator Diagrams in Helmholtz's Physiology," in *Universalgenie Helmholtz: Ruckblick Nash 100 Jahrnen*, ed. Lorenz Krüger (Akademie Verlag, 1994), 124–45. See also Michael Lynch and Steve Woolgar, ed., *Representation in Scientific Practice* (Cambridge, Mass.: MIT Press, 1990); Bruno Latour's discussion of "immutable mobiles" in *Science in Action* (Cambridge: Harvard University Press, 1987), pp. 227, 236–37 (found also in Latour, "Drawing Things Together," in *Representation in Scientific Practice*, ed. Lynch and Woolgar). Bernard Carlson and Michael E. Gorman, "Interpreting Invention as a Cognitive Process: Thomas Edison, Alexander Graham Bell, and the Telephone," *Science, Technology, and Human Values* 15 (Spring 1990): 131–64. See also Peter Galison, *Image and Logic: A Material Culture of Microphysics* (Chicago: University of Chicago Press, 1997) on instrument-produced images.

20. For further thinking on the subject, see Arnold Davidson, "Styles of Reasoning, Conceptual History, and the Emergence of Psychiatry," in *The Disunity of Science*, ed. Peter Galison and David Stump (Stanford: Stanford University Press, 1996).

21. Svetlana Alpers, "Style is What You Make It: The Visual Arts Once Again," in *The Concept of Style*, ed. Berel Lang (Ithaca: Cornell University Press, rev. ed., 1987): pp. 137–62, quote, p. 138.

22. This resonates with Alpers's call for a turn "from style as historical ordering to the mode of making." (in ibid., p. 162.)

23. Barbara Duden, *The Woman Beneath the Skin: A Doctor's Patient in Eighteenth-Century Germany*, trans. Thomas Dunlap (Cambridge: Harvard University Press, 1991).

24. As Snyder relates, Marey emphasizes the role of the imagination in the final chapter of his last book: "The images . . . appeal rather to the imagination than to the senses." E.-J. Marey, *Movement*, trans. Eric Pritchard (London, 1895), p. 304.

25. For a discussion of the ideology of this construct, see Caroline A. Jones, *Machine in the Studio: Constructing the Postwar American Artist* (Chicago: University of Chicago Press, 1996).

26. See, for example, Tim Lenoir, "The Eye as Mathematician: Clinical Practice, Instrumentation, and Helmholtz's Construction of an Empiricist Theory of Vision," in *Hermann von Helmholtz and the Foundations of Nineteenth-Century Science*, ed. David Cahan, California Studies in the History of Science, 12 (Berkeley and Los Angeles: University of California Press, 1993).

The appliances are supposed to suggest various local species on different islands. I wanted each island to have its own character, so I arranged them in such a way that there would seem to be a dominant appliance on each island. But I didn't want it to be totally pure. For instance, a blender might be a kind of persecuted minority on an island of radios, but over on another island, the blender might be the dominant force—it's everywhere—the blender is king. Formally, certain islands were monocultures, while others were much more heterogeneous. I didn't want to depict some kind of linear progression, but I did try to create the impression that the machines were evolving over time, transforming, adapting, forming a kind of ecology.

PERRY HOBERMAN

Styles

CARLO GINZBURG

Style as Inclusion, Style as Exclusion[1]

In 1605, the imprisonment of two Venetian priests on some petty charges triggered a major diplomatic war between the Republic of Venice and the Holy See. Pope Paul V, relying upon a principle that had been recently argued by some prominent theologians, felt entitled to intervene in the political affairs of the Republic of Venice, and asked for the release of the two priests. A heated debate followed; the juridical and political independence of the Republic of Venice as well as, on a more general level, the relationship between State and Church, were at stake. The Venetian point of view was powerfully argued in a series of writings by the Republic's official theologian, Paolo Sarpi, the Servite friar who later became famous all over Europe as the pseudonymous author of the *History of the Council of Trent*. In 1607, Sarpi was excommunicated; some months later he was assaulted near his convent by five men with daggers. Sarpi, badly wounded, whispered to the physician who was treating his wounds that, as everybody knew, they had been made *"stylo Romanae curiae"*—meaning "by the knife of the Roman Curia" as well as "by the legal procedure [literally, the pen] of the Roman Curia."[2]

Sarpi's splendid, untranslatable pun is an appropriate introduction to a discussion of the political implications of style. As we will see, "style" often has been used as a cutting device, as a weapon, and as a self-defining category. It has also played an important (and insufficiently recognized) role in the acceptance of cultural diversity—as well as in establishing cultural hegemonies. I will explore the unfolding of these ambiguities in the domain of the visual arts. Eventually the relevance of this topic to the history of science will also emerge.

The text I will start from is taken from Cicero's *De oratore* (55 B.C.). Crassus, who represents the author's point of view, introduces his remarks on oratory by recalling Plato's dictum that all intellectual activities are bound together by an internal coherence. But what follows (III, 7, 25; 9, 36) is very unplatonic. In nature, Crassus/Cicero says, there is "in its own kind a multiplicity of things that are different from one another and yet are esteemed as having a similar nature."[3] This apparently obvious principle is then projected by Cicero first into the arts, both visual and verbal, then into oratory, transforming the notion of genre (*genus*) into something close to our notion of individual style. Within a single art, like sculpture, he writes, we have excellent artists like Myro, Polyclitus, and Lysippus, whose extreme diversity is appreciated by everybody. The same can be said about painting (he mentions Zeuxis, Aglaophon, and Apelles) or poetry. Latin poets like Ennius, Pacuvius, and Accius are as different from each other as the Greek poets, Aeschylus, Sophocles, and Euripides: all of them are nearly equally praised "in their various genre of writing" (*in dissimili scribendi genere*). Their excellence is incomparable; perfection, as Cicero shows by giving succinct definitions of the characteristics of various orators, is reached by every artist in his own way. But ultimately, Cicero says, if we could scrutinize all the orators from every place and time, would we not conclude that there are as many genres (*genera dicendi*) as there are orators?[4]

Cicero's emphasis on the importance of specific genres, even to the point of identifying them with single individuals, was inspired by the rhetorical notion of "appropriateness" (in Greek, *to prepon*).[5] Cicero explicitly rejected the notion of an all-embracing genre of oratory that would be appropriate for all causes, audiences, orators, and circumstances. The only advice he gave to his readers was to choose a style—high, low, or middle—that would be appropriate (*accommodatam*) to the legal case they would be dealing with (III, 54, 210–12). This is obviously far removed from Plato's search for a universal idea of Beauty.

Cicero's implications that excellence and diversity were not incompatible were powerfully unfolded by Augustine in a letter addressed to the imperial commissioner, Flavius Marcellinus.[6] Volusianus, the Roman senator, had raised a challenging question: How could God welcome the new Christian sacrifices and reject the old—that is, the Jewish ceremonies? Could He ever change His mind? In his reply Augustine stressed the distinction between "the beautiful" (*pulchrum*) and "the suitable" (*aptum*), which had been the topic of his lost youthful treatise *De pulchro et apto*. "The divine institution of sacrifice was suitable [*aptum*] in the former dispensation," Augustine wrote, "but is not suitable now." This was not a language based on a "jealous" (*Ex.*

34:14; *Deut.* 4:24) approach to truth. In order to articulate the notion that the Old Testament was both true and superseded, Augustine had to rely on a different idiom. He found it in *De oratore*. By a significant shift Augustine reshaped Cicero's aforementioned argument, starting from his introductory remarks on natural diversity, in a temporal perspective. The seasons of the year and the ages of human life show, Augustine wrote, that both nature and human activities "change according to the needs of times by following a certain rhythm, but this does not affect the rhythm of their change." Cicero's basically achronic model, which stressed the variety of roads leading to artistic excellence, was therefore projected into a religious *and* temporal dimension. The rhetorical notion of accommodation allowed Augustine to take simultaneously into account divine immutability and historical change. The long-term impact of this move will not be missed. If the foundations of our notion of historical writing were laid by the Greeks, the foundations of our notion of historical perspective were laid by Augustine, in reflecting on the relationship between Jews and Christians.[7] The diversity of styles, albeit conceived in ahistorical terms, played an important role in the development of historical awareness.

Cicero's argument is echoed in a passage that provides one of the earliest uses of style in the domain of visual arts. It occurs in Baldesar Castiglione's *Il Cortegiano* (The Book of the Courtier), first published in 1528 but written approximately a decade before. The well-known exchange on *sprezzatura* leads to a much debated topic: imitation in literature. Count Ludovico of Canossa, the author's mouthpiece in the dialogue, rejects imitation of ancient models in favor of custom (*consuetudine*), arguing that "excellence can be nearly always achieved through different roads." The implicit allusion to Cicero introduces a reference to contemporary music and then to contemporary painting. In the latter, *"maniera"* and *"stile"* are used as synonyms that give a specific meaning to the generic *"far"* (making):

> Varie cose ancor egualmente piacciono agli occhi nostri tanto che con difficultà giudicar si po quai più lor sono grate. Eccovi che nella pittura sono eccellentissimi Leonardo Vincio, il Mantegna, Raffaello, Michelangelo, Georgio da Castelfranco: nientedimeno, tutti son tra sé nel far dissimili; di modo che ad alcun di loro non par che manchi cosa alcuna in quella maniera, perché si conosce ciascuno nel suo stil essere perfettissimo.[8]

These painters are still part of our canon, as it was built up by Vasari. To establish a hierarchy among them would seem to most of us (as it did to Castiglione) a waste of time. Should we then dismiss Cicero's passage as a mere *topos* or a commonplace? I would regard it instead as a formula that provided an alternative cognitive model: a

Logosformel, we may say, paraphrasing Aby Warburg.[9] This formula was also part, albeit in a rather contradictory way, of Vasari's approach. It was bequeathed to us by— and against—him.

In a still fundamental essay, Erwin Panofsky described Vasari's historical approach as the hybrid result of two antithetical principles: a pragmatic one, which saw each phenomenon as part of a causal process, and a dogmatic one, which saw each phenomenon as a more or less perfect embodiment of a "perfect rule of the art."[10] But for somebody as teleologically oriented as Vasari, an antithesis put in those terms would have been hardly conceivable. He always evaluated each artist and each work for their contribution to the progress of the art. In language echoing (as Panofsky noticed) the scholastic distinction between *simpliciter* and *secundum quid,* Vasari wrote at the end of his work: "I intended to give praise not absolutely [*non semplicemente*] but, as they say, according to [*secondo che*], and with respect for places, times, and other similar circumstances."[11] The evaluation *secundum quid,* far from contradicting the notion of perfection, was in a sense implied by it. He continued:

> In truth, taking the example of Giotto, no matter how highly praised he was in his own day, I do not know what would be said of him and other older artisans if they had existed in Buonarroti's time; moreover, the men of this century, which has reached the peak of perfection, would not have attained the heights they have reached if those who came before had not been as they were.[12]

But Vasari's linear historical construction was in fact undermined by an antithesis, although one very different from that suggested by Panofsky. The first edition of Vasari's *Lives,* published in 1550, did not include a life of Titian, then at the height of his European fame (he had just painted two portraits of Emperor Charles V). At that date, Vasari was already familiar with some of Titian's works; he had even met him in Rome a few years before. The reason for not including Titian's life was given by Vasari at the end of his life of Giorgione, following an elaborate eulogy: "But because he [Titian] is still alive, and his works are under the eyes of everybody, there is no need to speak about him."[13] Michelangelo, whose life concluded Vasari's *Lives,* had to be the only living artist included. Probably Vasari felt that the inclusion of Titian would have spoiled the role of absolute prominence he wanted to give to Michelangelo; he also may have had reason to believe that Michelangelo would not have appreciated the presence of a life of Titian. Whatever the reason, the second edition of Vasari's *Lives,* published in 1568, after the death of Michelangelo, did include a life of Titian, in which great praise was interspersed with criticism. In an often-quoted page Vasari related a conversation he had had with Michelangelo in Rome in 1546, after both of them had seen Titian's *Danae*:

> Buonarroti strongly commended him, declaring that he liked his colouring and style [*maniera*] very much but that it was a pity artisans in Venice did not learn to draw well from the beginning and that Venetian painters did not have a better method of study.[14]

This comment was presumably triggered by the fact that Titian's *Danae* had been inspired by Michelangelo's *Night*.[15] The target of Michelangelo's criticism was not Titian's individual *maniera*, but the intrinsic weaknesses of the stylistic tradition begun by Titian's teacher, Giorgione, the Venitian rival of the Tuscan initiators of "modern style" (*maniera moderna*).[16] Vasari, who obviously shared Michelangelo's attitude, was so open-minded, so unconventional, so undogmatic (*pace* Panofsky) as a critic to provide a memorable description of Titian's mythological paintings: "his last works are executed with such large and bold brush-strokes and in such broad outlines that they cannot be seen from close up but appear perfect from a distance."[17] Here the tension between "style" as an individual phenomenon and "style" in a broader sense, as well as between norm and understanding, is pushed to an extreme.[18]

Vasari's *Lives* provided a model whose impact went far beyond the realm of visual arts—the Whiggish idea of scientific progress being a most notable example. But the intrusive presence of Titian in the second edition (1568) pointed to an unsolved tension. In 1557 the Venitian writer Ludovico Dolce had reacted to the first edition with a *Dialogo della Pittura*, in which the argument put forward by Cicero, and then spread by Castiglione, surfaced again: "one should not think . . . that there is just one kind of perfect painting." But this tactical move ultimately led also to a linear model, albeit opposed to Vasari's. Dolce praised Titian as "divine and peerless," a blending of Michelangelo's "greatness and fierceness [*terribilità*]," Raphael's "attractiveness and grace," and Nature's colors.[19] Two alternative models were emerging, based on, respectively, "drawing" and "color."[20]

This debate went on from the late seventeenth to the early nineteenth century, opposing first Poussin to Rubens, later Ingres to Delacroix. The antithesis was to some extent related to the one between "ancients" and "moderns," the partisans of color being identified with the latter. (When, in the early nineteenth century, it was suggested for the first time that Greek sculptures and buildings had been painted in a variety of colors, many admirers of antiquity were deeply shocked.) In the introduction to his *Parallèle de l'architecture antique avec la moderne* (1650), Roland Fréart sieur de Chambray, a key figure of French classicism, gave a scornful list of some of the arguments raised by the partisans of the "moderns": among them, that art is involved in an endless progress, "adapting itself [s'accommodant] to the mood of centuries and nations, each of them . . . [having] its own criteria of Beauty."[21] In its shift from rhetoric to theology and from theology to history, the notion of accommodation developed an inexhaustible richness, paving the way for the idea of a multiplicity of tastes that could peacefully coexist. One of the earliest and most striking examples of this attitude is the Entwurf einer historischen Architectur in Abbildung unterschiedener berühmten Gebäude des Altertums und fremder Völker by Johann Bernhard Fischer von Erlach (1721), a leading figure of Austrian Baroque architecture, who during his long stay in Rome (1670–1686) was strongly influenced by Francesco Borromini's work.[22] Fischer's lavish illustrations include, among other things, the temple of Salomon (according to the reconstruction given by the Spanish Jesuit Villalpanda); the rocks of Stonehenge; a series of mosques (from Pest to Constantinople); the residence of the king of Siam; the imperial court of Peking; a series of Chinese bridges; and a series of buildings by the author himself. In his introduction, Fischer von Erlach justified this shocking array of different works by connecting them to a larger diversity related to "national tastes [goûts des nations]," which included not only architecture, but dress and food as well.[23] "Taste" was apparently a broader, more flexible notion than style. Fischer accepted even "bizarre" details like Gothic ornaments and Indian-like roofs, insofar as they were part of a domain that—with the exception of a few universal architectural principles, like the rules of symmetry and stability—everything was a matter of taste and therefore subject to dispute. One is reminded of the attitude, explicitly based on the principle of accommodation, held by contemporary Jesuit missionaries toward non-European cultures.[24] In fact, the Entwurf was heavily indebted (albeit without acknowledgment) to the works of the Athanasius Kircher, the Jesuit polymath, whom Fischer had met in Rome.[25]

In a few decades an unprecedented phenomenon emerged: the simultaneous use of different styles, an architectural experiment that was attempted first in gardens—a peripheral space, placed between nature and culture, wilderness and civilization.[26] But the coexistence of Gothic ruins and Chinese pagodas in English gardens could elicit polemical reactions, as a transgression of the rules of taste. "The applause which is so fondly given to Chinese decorations or to the barbarous productions of the Gothic genius," one reads in The World in 1755, "seems once more to threaten the ruin of that simplicity which distinguishes the Greek and Roman arts as eternally superior to those of every other nation."[27]

"Noble simplicity and quiet greatness" is the famous definition of Greek sculpture given by Johann Joachim Winckelmann in his Gedanken über die Nachahmung der griechischen Werke in der Malerey und Bildhauerkunst, also published in 1755. The same qualities, Winckelmann added, were shared by Greek writings of the same period—

those by Socrates's pupils, for instance—as well as by Raphael, as an imitator of antiquity.[28] "There is only one beauty, as there is only one good," Winckelmann once wrote.[29]

All this has a definite Platonic ring. But in his most influential work, the *Geschichte der Kunst des Altertums* (1764), Winckelmann did not insist exclusively on the revelations of eternal Beauty. Rejecting the biographical approach used by Vasari, Winckelmann identified the history of art with the account of "its origins, development, changes and decadence and with the variations of style according to the various peoples, times and artists."[30] He analyzed not only Egyptian, Etruscan, and Greek styles but, within the Greek, four different stages ("severe," "sublime," "beautiful," and "mean"). For the first time, style was identified as the subject of art history and connected to history in general.

In order to analyze stylistic variations, Winckelmann focused on the manifold conditions that shaped them. Besides mentioning the role of climate in rather traditional terms, Winckelmann insisted on the importance of political freedom on the arts, hence on style as a historical index.[31] But a third and much less prominent element, often missed by interpreters, throws an unexpected light on Winckelmann's approach as a whole. In summing up the main features of Etruscan style, Winckelmann remarked that they were shared, to a certain extent, by the Etruscan people as well, and the tendency to delve into excessive details could also be found in their "contrived and artificial" literary style, quite different from the pure clarity of the Romans. The style of the Etruscan masters could still be perceived in the works of their successors, including Michelangelo, the greatest of all; the same features account for the weaknesses of Daniele of Volterra, Pietro of Cortona, and others. Raphael and his school, on the contrary, had been spiritually closer to the Greeks.[32]

The derogatory comparison between Tuscan writings and Tuscan painters did not imply a conscious imitation, but an alleged ethnic continuity between Etruscans and Tuscans. In a rather unexpected direction, this argument developed the reflections of two authors whose works had made a deep (although unacknowledged) impression on Winckelmann: Caylus and Buffon.[33]

In his *Recueil d'antiquités egyptiennes, etrusques, grecques et romaines* (1752 onward), Count Caylus rejected a mere antiquarian approach in favor of a method aiming to consider ancient monuments as "a proof and expression of a taste which dominated either a certain age or a certain country."[34] In a letter to the antiquarian Bianconi, Winckelmann significantly admitted that Caylus, whom he usually tended to put down as a pedant, "deserved the glory of having for the first time started to understand the gist of the style of the art of ancient peoples," although his efforts had been limited by the fact of living in Paris.[35] Caylus had anticipated the need, much more influentially stressed by Winckelmann, to connect the history of art to history in a broad sense.[36] But the alleged continuity between Etruscans and Tuscans went much beyond the idea of national tastes. Here the indirect impact of Buffon is noticeable. On two occasions, 1750 and 1754, Winckelmann made long extracts from Buffon's *Histoire Naturelle*.[37] From Buffon he learned to convey visual observations, based on minute inspection, in a vivid style, which aimed at a sort of classical impersonality.[38] What animal species had been for Buffon's great comparative enterprise, styles were for Winckelmann, who also focused on the species (the style), not on the individual (the single work of art, or the single artist). This analogy may have led Winckelmann to argue that style, as well as being either created or imitated, could also be biologically transmitted—a momentous step, as we will see.

Winckelmann's rediscovery of Greek art had a deep, lasting impact. It ran across an age of political turmoil (the French Revolution, the Napoleonic wars), European expansion overseas (India, the Pacific Islands, Egypt, and so forth) and deep intellectual and social changes. Military conquests, archaeological excavations, and museums unveiled civilizations remote both in space and time; an unprecedented variety of visual documents became accessible to a large European audience.[39] An early, impressive reaction to this latter phenomenon is witnessed by the *Lectures on Sculpture* delivered by John Flaxman at the Royal Academy from 1810 onward, and published after his death (1829).[40] Flaxman, a sculptor himself, was (and is) better known for his illustrations of Homer, Hesiod, Aeschylus, and Dante, notable for a spare, restrained outline drawing seemingly inspired by the artistic principles proclaimed by Winckelmann.[41] But as Goethe promptly noticed, Flaxman's outline drawing echoed not only "Etruscan" (that is, Greek) vase painting but Italian primitives as well.[42] Flaxman's *Lectures on Sculpture* provide a theoretical and historical framework for this dual influence. In his lecture "On Style" he identified a first principle, which he described in his florid prose as

> some well-known quality which originates in the birth of the art itself—increases in its growth—strenghtens in its vigour—attains the full measure of beauty in the perfection of its parent cause—and, in its decay, withers and expires! . . . Such a quality immediately determines to our eyes and understanding, the barbarous attempt of the ignorant savage—the humble labour of the mere workman—the miracle of art conducted by science, ennobled by philosophy, and perfected by the zealous and extensive study of nature.
>
> This distinguishing quality is understood by the term Style, in the arts of design.

Flaxman's approach was obviously hierarchical.[43] But to my knowledge he was the first to include the works of "humble workmen" and even savages under the category of style, which he interpreted as follows:

This term, at first, was applied to poetry, and the style of Homer and Pindar must have been familiar long before Phidias or Zeuxis were known: but, in process of time, as the poet wrote with his style or pen, and the designer sketched with his style or pencil, the name of the instrument was familiarly used to express the genius and productions of the writer and the artist; and this symbolical mode of speaking has continued from the earliest times to classical ages, the revival of arts and letters, down to the present moment, equally intelligible, and is now strengthened by the uninterrupted use and authority of ancients and moderns.

Thus Flaxman projected into a distant past ("the earliest times") what Caylus and Winckelmann in the previous century had written on style in the domain of visual arts. Then he made a step forward:

And here we may remark, that as by the term style we designate the several stages of progression, improvement, or decline of the art, so by the same term, and at the same time, we more indirectly relate to the progress of human mind, and states of society; for such as the habits of the mind are, such will be the works, and such objects as the understanding and the affections dwell most upon, will be most readily executed by the hands.

Style, as a concept connecting mind and hands, could therefore be applied to definite stages of intellectual and social history. From this argument Flaxman drew a remarkable inference:

Thus the savage depends on clubs, spears and axes for safety and defense against his enemies, and on his oars or paddles for the guidance of his canoe through the waters: these, therefore, engage a suitable portion of his attention, and, with incredible labour, he makes them the most convenient possible for his purpose; and, as a certain consequence, because usefulness is a property of beauty, he frequently produces such an elegance of form, as to astonish the more civilized and cultivated of his species. He will even superadd to the elegance of form an additional decoration in relief on the surface of the instrument, a wave line, a zig-zag, or the tie of a band, imitating such simple objects as his wants and occupations render familiar to his observation—such as the first twilight of science in his mind enables him to comprehend. Thus far his endeavours are crowned with a certain portion of success; but if he extend his attempt to the human form, or to the attributes of divinity, his rude conceptions and untaught mind produce only images of lifeless deformity, or of horror and disgust.[44]

Although set within definite boundaries, Flaxman's admiration for the arts of the savages is definitely striking. He praised their "elegance of form" by connecting it to a quality we have already met: "convenient" (*aptum*, πρεπον). But Flaxman interpreted "convenient . . . for [a] purpose" in utilitarian terms ("because usefulness is a property of beauty")—a reminder that he had worked for Wedgwood, making drawings for vases and cameos.[45] The new relationship between art and industry suggested a broader attitude toward the diversity of artifacts produced throughout history, as well as a broader and less parochial vision of history itself. Flaxman's openness to artistic languages that were distant both in space and time is effectively conveyed by the illustrations, based partly on his own sketches, partly on previous books, attached to his *Lectures on Sculpture*. Through his fluid, undulating line Flaxman was able to catch an astonishing range of visual idioms, translating them into his own: reliefs from Wells Cathedral and from Persepolis, statues from archaic Greece and from India; buildings from Mycenae; miniatures from medieval manuscripts—and so forth. By contrast, Flaxman's saccharine version of Michelangelo's *terribilità* seems ludicrous when compared with the works of his great contemporary, Fuseli.

The *Lectures on Sculpture*, a contemporary listener wrote, appealed to their audience for their "John-Bullism."[46] Flaxman did not conceal his admiration for British medieval sculpture, but the political implications of his *Lectures* went much deeper than that. Flaxman's illustrations can be regarded as a remarkable attempt to understand alien cultures, to penetrate them, to translate them, to appropriate them: a visual equivalent of British imperialism.

Approximately in the same years, the greatest living philosopher also addressed his students in Heidelberg and Berlin on the exotic arts of Asian countries. In his posthumously published *Lectures on Aesthetics*, Hegel remarked that the flight from the representation of reality in Chinese and Indian works of art was due to deliberate distortion, not to technical weakness. Those artifacts, he insisted, were both perfect in their specific sphere, and relatively inadequate if compared to the concept of Art and to the Ideal.[47] In this way Hegel developed a major Romantic theme: the emphasis on artistic freedom. But he also avoided its radical implications, graphically expressed by Heinrich Heine in his *Französische Maler*:

> It is always a big mistake when the critic brings up the question "what should the artist do?" Much more correct would be the question "what is the artist trying to do?" or even "what does the artist have to do?" The question "what should the artist do?" comes by the way of those philosophers of art who, though they possessed no sense of poetry themselves, have singled out traits of various works of art and then, on the basis of what there was, determined a norm for what

everything ought to be, and who delimited the genres and invented definitions and rules. . . . [E]very original artist, and certainly every artistic genius,
brings with him his own aesthetic terms and must be judged according to
them.[48]

Heine wrote these words at the beginning of his long Parisian exile. They resonated in a congenial milieu. A distant echo of them can be heard in an article
published many years later (1854) in the *Revue des Deux Mondes* by Delacroix, the
painter who had embodied for decades the rejection of traditional values. In a passionate defense of artistic variety, Delacroix argued that Beauty could be attained in
different ways, by Raphael and Rembrandt, by Shakespeare and Corneille. To take
antiquity as a model is absurd, he insisted, since antiquity itself did not imply a
single, uniform canon.[49] This article may have elicited Baudelaire's poem "Les phares,"
an extraordinary ekphrastic exercise starting with some of the painters praised by
Delacroix in his article (Rubens, Leonardo da Vinci, Rembrandt, Michelangelo)
and ending with Delacroix himself, the only living artist in the series. The *"phare
allumé sur mille citadelles,"* mentioned in the poem's conclusion as a metaphor for
Beauty, becomes a plural in the title—"Les phares"—reinforcing the point conveyed
by the extreme diversity of each strophe. In May 1855, Baudelaire touched on the
same issue in an article published in *Le Pays* on the *Exposition universelle*, in which
Delacroix, who exhibited thirty-five paintings, attained a belated fame. *Le Beau
est toujours bizarre* ("The Beautiful is always strange"), Baudelaire wrote. "Now, how
could this necessary, irreducible and infinitely varied strangeness [*bizarrerie*], depending upon the environment, the climate, the manners, the race, the religion and
the temperament of the artist—how could it ever be controlled, amended and
corrected by Utopian rules conceived in some little scientific temple or other on this
planet, without mortal danger to art itself?"[50] Hence the rejection of any aesthetic
norm: ". . . take one of those modern 'aesthetic pundits,' as Heinrich Heine calls
them—Heine, that delightful creature, who would be a genius if he turned more
often towards the divine. What would *he* say? what, I repeat, would *he* write if faced
with such unfamiliar phenomena?"[51] Heine's paramount target was possibly August
von Schlegel; Baudelaire's target, French democratic rhetoric. "There is yet another,
and very fashionable, error which I am anxious to avoid like the very devil. I refer to
the idea of 'progress'. This dark beacon, invention of present-day philosophizing. . . .
Anyone who wants to see his way clear through history must first and foremost
extinguish this treacherous aid."[52] This passage, absent in the version that appeared in
Le Pays, was added by Baudelaire after the publication of "Les phares." The *"phare
allumé sur mille citadelles"* first evoked, by contrast, the *"fanal obscur"* of progress,
and then, as a sudden coup-de-théâtre, the lines by "a poet" on Delacroix (*"Delacroix,*

lac de sang hanté des mauvais anges. . . .") followed by their interpretation.[53] Heine had written that every great artist has his own aesthetic; Baudelaire pushed Heine's argument to its logical extreme: "Every efflorescence is spontaneous, individual. Was Signorelli really the begetter of Michelangelo? Did Perugino contain Raphael? The artist stems only from himself. His own works are the only promises that he makes to the coming centuries. He stands security only for himself. He dies childless"[54] An emphasis on the multiple elements affecting artistic variety led Baudelaire to reject the very possibility of a historical approach to art. We will come across this tension again.

The historical sequence we have been analyzing thus far apparently shows the victory of stylistic diversity over stylistic uniformity. Nineteenth-century architecture legitimized the coexistence of different styles, a movement later known as Historicism. Gottfried Semper (1803–1879), the most relevant German representative of this approach, wrote an ambitious work dealing with style in a comparative perspective: *Der Stil in den technischen und tektonische Künsten oder praktische Aesthetik* (Style in technical and tectonical arts, or practical aesthetic), two volumes of which appeared in 1860 and 1863 (the third remained unfinished). According to Semper, the very beginning of his project went back to his student years in Paris (1826–1830), when he spent long hours at the Jardin des Plantes looking at the collections of fossil remains assembled by Cuvier. Semper mentioned these youthful memories twice, first in a letter addressed to Eduard Vieweg, the Braunschweig publisher, on September 26, 1843,[55] and then in a lecture delivered in 1853 in London, where he was living as a political exile (he had to leave Germany in 1848 after having taken an active part in the Dresden Revolution).[56] In both cases Semper suggested an analogy between Cuvier's comparative approach and his own. But the transition from the German letter to the English lecture—the first of a series he delivered in a Department of Practical Arts—brought some significant changes. On the one hand, Semper excised all words inspired by Goethe's morphology: *einfachsten Urform* (originary and simplest form), *ursprüngliche Ideen* (originary ideas), *Urformen* (originary forms), *das Ursprüngliche und Einfache* (the originary and the simple). On the other, a neutral reference to *den Werken meiner Kunst* (the works of my art) became a pointed reference to "industrial art," suggested by the International Exhibition of 1851 in which Semper had been involved, as well as by the specific audience he was addressing:

> We see the same skeleton repeating itself continually but with innumerable varieties, modified by gradual developments of the individuals and by the conditions of existence they had to fulfill. . . . If we observe this immense variety and richness of nature notwithstanding its simplicity may we not by Analogy assume,

that it will be nearly the same with the creations of our hands, with the works of industrial art?[57]

Semper had probably a rather vague notion of Cuvier's work, and was certainly unable to take a viable biological model from it.[58] But Cuvier's undeniable impact on him took place on a more metaphorical level. Goethe, who in the debate between Saint-Hilaire and Cuvier at the Académie Française in 1830 took sides with the former, was paradoxically not less important for Semper. The vision of a "method, analogous to that which Baron Cuvier followed applied to art, and especially to architecture [which] would form the base of a doctrine of *Style*" implied a basic continuity with Semper's Romantic roots.

A similar trajectory, from Goethe's morphology to Cuvier's comparative osteology, finally reinterpeted as an allegiance to Darwin's theory of evolution, is provided by another German-educated art historian, slightly younger but not less remote from the mainstream: the famous connoisseur Iwan Lermolieff, alias Giovanni Morelli.[59] Both Semper and Morelli shared a morphological approach to style, but the latter focused on individual artists, Semper on larger cultural units. The difference in scale implied a different method as well. Morelli never abandoned a rigorous internalist perspective; Semper, on the contrary, regarded style as the result of an interaction between internal and external conditions, which were to be analyzed separately. The first part of his doctrine of style was supposed to deal with "the exigencies of the work itself and which are based upon certain laws of nature and of necessity, which are the same at all times and under every circumstance"—a rather obscure expression pointing at the constraints of matter and instruments (the latter being, as Semper himself admitted, subject to historical change). The second part would have dealt with "local and personal influences, such as the climate and physical constitution of a country, the political and religious institutions of a nation, the person or the corporation by whom a work is ordered, the place for which it is destined, and the Occasion on which it was produced. Finally also the individual personality of the Artist."[60]

An item is absent from this list: race. At approximately the same time, George Gilbert Scott, one of the main restorers of Westminster Abbey, spoke of the Gothic revival as "our national architecture, the only genuine exponent of the civilization of the modern as distinguished from the ancient world, of the Northern as distinguished from Southern races." "We do not want," Scott wrote, "to adapt ourselves to mediaeval customs, but to adapt a style of art which accidentally was mediaeval, but is *essentially national,* to the wants and requirements of our own day." Hence his conclusion: "The indigenous style of our race must be our *point de départ.*"[61] This was not an isolated voice. During the course of the nineteenth century the conflation of history, anthropology, and biology had accelerated a parallel conflation of national character, style, and race. In the debates on architectural styles, race gained a prominent place. Semper's silence on this topic is remarkable.[62] He strongly believed in "national geniuses," as well as in his power to comprehend them through humble archaeological remains, which he compared to fossils. Like Cuvier, he could boast that the Nile Pail—the Aegyptian holy vessel—and the Greek Situla—so closely related to the Doric style—gave us access to the architecture in which the two peoples expressed their respective essence (*Wesen*) in monumental form.[63] But in this ambitious enterprise Semper refused to rely upon race as a conceptual shortcut.

Semper spent the last period of his life in Vienna, where his main architectural works—the Hofburg Theater and the Outer Burgplatz—dramatically changed the image of the city.[64] His book on style was widely echoed by archaeologists and art historians. Toward the end of the century a powerful dissenting voice emerged. In his *Stilfragen* (1893), Alois Riegl rejected Semper's deterministic materialism, notwithstanding a repeated (but mostly tactical) distinction between the Semperians and Semper's subtler thought. To Semper's interpretation of artistic development as basically determined by instruments, Riegl opposed an autonomous drive toward decoration and form, which he later named *Kunstwollen* (will to art), emphasizing its historical dimension.[65] In his great book *Spätrömische Kunstindustrie* (late Roman art industry, 1901), Riegl argued that the artistic productions of an age traditionally regarded as decadent—including the bas-reliefs of the Arch of Constantine, which had been scornfully dismissed as clumsy by Vasari—could be interpreted as coherent expressions of a specific, homogeneous *Kunstwollen*, inspired by principles as legitimate as those of classic art, although widely divergent from them.

The links between Riegl's impressive scholarly work and the artistic events of contemporary Vienna have been often emphasized. When he argued that the "geometric style" was not a primitive phenomenon dictated by lack of representational power, as the Semperians had suggested, but the deliberate product of a sophisticated artistic will, one is immediately reminded of Gustav Klimt's nearly contemporary paintings and their geometric decorations.[66] But Riegl's theoretical framework had a different and longer ancestry, as his crucial debt to Hegel suggests.[67] The aforementioned passage on Indian and Chinese art from Hegel's *Lectures on Aesthetics* may have provided the starting point for Riegl's reflections, which focused on European art. Moreover, Riegl shared Hegel's teleological vision, which allowed him to justify late Roman art according to its own criteria and as a necessary transition in the development of world history:[68] in a way, a rephrasing of Vasari's distinction between appreciation *simpliciter* and appreciation *secundum quid*.

As a weapon against materialistic determinism, Riegl's *Kunstwollen* seemingly echoed the Romantic notion of artistic freedom that inspired Heine's question: "*Was will der Künstler?*" But instead of focusing on the individual artist as a subversive genius,

Riegl dealt with collective entities like late Roman and Dutch *Kunstwollen*.[69] To look at styles according to an ethnic perspective (whatever this meant) was also part of the Romantic legacy. As we have seen, race was often mentioned in this context. Baudelaire, for instance, included race in a miscellaneous list of constraints upon art, along with customs, climate, religion, and the artist's individual character. But in the increasingly anti-Semitic atmosphere of fin-de-siècle Vienna, the remarks made by Riegl in his university lectures on the rigidity of the Jewish vision of the world and its resulting "inability to change and to improve" must have struck a deep chord in his audience.[70] Two years before (1897), Riegl had included in his lectures a parallel between early Christianity and modern Socialism, praising the latter because, "at least in its main manifestations, it aims at the improvement of this world."[71] This passage has been convincingly interpreted as a reference to the Christian Socialists, whose anti-Semitic leader, Karl Lüger, had just been elected mayor of Vienna.[72] To what extent Riegl shared the anti-Semitic attitude of the Christian Socialist party, we do not know. But his propensity to take style and race as coextensive entities emerges in a footnote to *Late Roman Art Industry*: the "often overrated" divergence between late Pagan and early Christian art is hardly believable in itself, Riegl wrote, since Pagans and Christians belonged to the same race.[73]

Wilhelm Worringer, the most successful popularizer (albeit at a much lower level) of Riegl's bold ideas, did not hesitate to put them in an explicitly racial framework.[74] In his *Formprobleme der Gotik*, Worringer repeatedly connected different degrees of stylistic purity to an ethnic hierarchy:

> it may be said that France created the most beautiful and most living Gothic buildings but not the purest. The land of pure Gothic culture is the Germanic North. . . . It is true that English architecture is also tinged with Gothic, in a certain sense; it is true that England, which was too self-contained and isolated to be so much disturbed in its own artistic will [*Kunstwollen*] by the Renaissance as was Germany, affects Gothic as its national style right down to the present day. But this English Gothic lacks the direct impulse of the German Gothic.[75]

Hence the conclusion:

> For Gothic was the name we gave to that great phenomenon irreconcilably opposed to the classical, a phenomenon not bound to any single period of style, but revealing itself continuously through all centuries in ever new disguises: a phenomenon not belonging to any age but rather in its deepest foundations an ageless racial phenomenon, deeply rooted in the innermost constitution of

Northern man, and, for this reason, not to be uprooted by the levelling action of the European Renaissance.

In any case we must not understand race in the narrow sense of racial purity: here the word race must include all the peoples, in the racial mixture [*Rassenmischung*] of which the the Germans have played a decisive part. And that applies to the greatest part of Europe. Wherever Germanic elements are strongly present, a racial connection in the widest sense is observable, which, *in spite of* racial differences in the ordinary sense, is unmistakably operative. . . . For the Germans, as we have seen, are the *conditio sine qua non* of Gothic.[76]

The years that have passed since 1911, when these words were written, have given them a sinister patina. Anachronistic readings must of course be avoided. But Worringer's "wide" notion of race, so wide to overcome the narrow meaning of "racial purity," inevitably calls to mind the Nuremberg laws and their punctilious prescriptions concerning the various degrees of *Rassenmischung*. In Worringer's stylistic club all peoples were included—provided they had an appropriate amount of Germanic blood in their veins.

What I have said thus far can provide an appropriate context for the role ascribed to style by a prominent philosopher of science.

In a well-known essay, Paul Feyerabend tried to apply to science Riegl's theory about art, which he opposed to Vasari's attitude.[77] In a footnote to *Against Method* (1970), Feyerabend had suggested that if we assume that science and art share a problem-solving attitude, the only significant difference between them would disappear; therefore we could speak of "styles and preferences for the former, of progress for the latter."[78] With typical mischievousness Feyerabend was using Gombrich (mentioned in the next footnote) against Popper. But this balance between science and art proved to be only a step toward "Science as Art"—the title of Feyerabend's essay on Riegl, and later of the book in which it was included. Riegl was the perfect choice: his work implied (1) a coherent attack on positivism, based (2) on a vision of history composed of a series of discrete, self-contained artistic wills (*Kunstwollen*) that led (3) to a rejection of the notions of decadence and (4) progress. The last point, concerning progress, seems inaccurate, insofar as it misses the Hegelian, teleological component in Riegl's work.[79] The other points, on the contrary, justified Feyerabend's conclusion: "sciences are arts in the light of this [i. e. Riegl's] modern concept of art."[80] Riegl's relativist approach, based on the idea that each age creates an artistic world of its own, ruled by special laws, offered an unexpected support to a relativist approach to science; it allowed one to dispense with referentiality, truth, reality—putting them, so to speak, in quotation marks. Not surprisingly,

Feyerabend remarked that Riegl, with a few others, "had understood the process of acquiring knowledge and the changes within knowledge better than most modern philosophers."[81]

Feyerabend must have discovered Riegl in the early 1980s.[82] But his posthumous autobiography (*Killing Time*, 1995) shows that Feyerabend came across some sub-Rieglian ideas a long time before, in the Viennese milieu in which he grew up.

Killing Time has been presented as an unusually open and candid account. As is often the case with autobiographies, its openness was probably selective.[83] The section on the Second World War, to which the young Feyerabend volunteered as an officer, fighting on the Russian front and becoming a lieutenant, seems highly reticent: "this is what my army records say: my mind however is a blank," he says, commenting on his military career.[84] Perhaps the past erased from the author's memory (or at least from his account) reemerged through the deliberate ambiguity in the book's title. But the section on war includes a remarkable excerpt from some lectures given by the author to his fellow officers in 1944, at Dessau Rosslau, which is directly relevant to the topic I am discussing. Feyerabend's resume is interspersed with quotations (which I put in italics) from the notes he had taken fifty years before:

People have different professions, different points of view. They are like observers looking at the world through the narrow windows of an otherwise closed structure. Occasionally they assemble at the center and discuss what they have seen: "*then one observer will talk about a beautiful landscape with red trees, a red sky, and a red lake in the middle; the next one about an infinite blue plane without articulation; and the third about an impressive, five-floor high building; they will quarrel. The observer on top of their structure (me) can only laugh at their quarrels—but for them the quarrels will be real and he will be an unworldly dreamer.*" Real life, I said, is exactly like that. "*Every person has his own well-defined opinions, which color the section of the world he perceives. And when people come together, when they try to discover the nature of the whole to which they belong, they are bound to talk past each other; they will understand neither themselves nor their companions. I have often experienced, painfully, this impenetrability of human beings—whatever happens, whatever is said, rebounds from the smooth surface that separates them from each other.*"

My main thesis was that historical periods such as the Baroque, the Rococo, the Gothic Age are unified by a concealed essence that only a lonely outsider can understand. Most people see only the obvious. . . . Secondly, I said, it is a mistake to assume that the essence of a historical period that started in one place can be transferred to another. There will be influences, true: for example, the French Enlightenment influenced Germany. But the trends arising from the influence share only the name with their cause. Finally, it is a mistake to evaluate events by comparing them with an ideal. Many writers have deplored the way in which the Catholic Church transformed Good Germans during the Middle Ages and later and forced them into actions and beliefs unnatural to them. . . . But Gothic art produced harmonic units, not aggregates. This shows that the forms of the Church were not alien forms (*artfremd*, a favorite term at the time), and the Germans of that period were natural Christians, not unwilling and cowardly slaves. I concluded by applying the lesson to the relations between German and Jews. Jews, I said, are supposed to be aliens, miles removed from genuine Germans; they are supposed to have distorted the German character and to have changed the German nation into a collection of pessimistic, egotistic, materialistic individuals. But, I continued, the Germans reached that stage all by themselves. They were ready for liberalism and even Marxism. *"Everybody knows how the Jew, who is a fine psychologist, made use of this situation. What I mean is that the soil for his work was well prepared. Our misfortune is our own work, and we must not put the blame on any Jew, or Frenchman or Englishman."*[85]

In his autobiography Feyerabend speaks repeatedly of Jews, of his attitude toward them, of his Jewish friends, of anti-Semitism, of different ways of playing Shylock. His comments often betray an embarrassed, ambivalent tone.[86] He must have been glad to discover that in 1944 he had taught his fellow officers that Jews were not guilty of Germany's corruption. But the text of his lecture suggests a more complex argument. The visual examples Feyerabend chose to illustrate the difficulties of human communication (at that time he intended to become a painter) remind one of paintings by the Blaue Reiter group—Marc, Kandinsky, Feininger—which were on display in the Exhibit of Degenerate Art (Munich, 1937).[87] Each example suggests a different, coherent, self-contained world, comparable to what the Gothic Age, the Baroque, and the Rococo were on a larger scale. The "concealed essence" that unifies each period (each civilization) is of course style. One feels a distant echo of Riegl's aesthetic approach to history, as a succession of self-contained civilizations, based on specific *Kunstwollen* or styles.[88] But in the meantime Riegl's emphasis on style as a coherent phenomenon, based on its criteria, had acquired a new meaning, already visible in Worringer's work. The association between style and race had reached a mass audience through the work of crude ideologists like H. F. K. Günther (*Rasse und Stil*, 1926), who later became an influential expert on racial issues under the Nazi regime.[89] Style had become an effective instrument of exclusion. In a party rally speech on cultural issues delivered on September 1, 1933, at the "Congress of Victory," Adolf Hitler had said:

It is a sign of the horrible spiritual decadence of the past epoch that one spoke of styles without recognizing their racial determinants. . . . Each clearly formed race has its own handwriting in the book of art, as far as it is not, like Jewry, devoid of any creative artistic ability. The fact that a people can imitate an art which is formally alien to it [artfremde Kunst] does not prove that art is an international phenomenon.[90]

In paraphrasing his 1944 notes, Feyerabend wrote: "Gothic art produced harmonic units, not aggregates, [which] shows that the forms of the Church were not alien forms (artfremd, a favorite term at the time)." The lecture delivered by Feyerabend to his fellow officers undoubtedly echoed the ideas about race, culture, and style advocated by the Nazis. If each civilization is a homogeneous phenomenon, both stylistically and racially, Jews and foreigners could not play any intrinsic role in the development of the German nation because by birth they were excluded from it. The implications of these ideas—from Auschwitz to the former Yugoslavia, from racial purity to ethnic cleansing—are well-known.

In Feyerabend's mature work race never was an issue. But the remarks he made in his 1944 lecture were not unrelated to some major themes of his mature work. In his youth he perceived the difficulty of communication between the various worlds as a painful condition, which even the lonely outsider, who has a privileged access to reality, was unable to solve. One may speculate as to whether those early remarks on the "impenetrability of human beings" provided a psychological stimulus to his later theoretical reflections.[91] In any case, in later years he reserved for himself a role somewhat related to that of the "unworldly dreamer": by comparing different (scientific) worlds, each having its own style, he pointed at their incommensurability. This idea was already implicit in Cicero's argument that no hierarchy can be established when (and only when) artistic excellence is involved. This is a far cry from the assumption that anything goes—whatever it can mean.[92]

The Latin word interpretatio means translation. The interpreter who compares different styles of thought in order to stress their intrinsic diversity performs a sort of translation, a word that comes easily in this context, insofar as styles, having being originally related to writing, have been often compared to languages in order to stress their intrinsic diversity.[93] But translation is also the most powerful argument against relativism. Each language is a different and, to a certain extent, incommensurable world: but translations work. Our ability to understand different styles may throw some light on our ability to understand other languages and other styles of thought— and the other way around.

* * *

I will conclude this paper with an exercise in translation, by suggesting the possibility of a dialogue among three individuals who never—not even metaphorically—spoke to each other (at least to my knowledge).

The first is Simone Weil. In 1941, two years before her death, she wrote in her *Notebooks* a comment to Plato's *Timaeus* (28a):

> When a thing is perfectly beautiful, as soon as we fix our attention upon it, it represents unique and single beauty. Two Greek statues; the one we are looking at is beautiful. The same is true of the Catholic faith, Platonic thought, Hindu thought etc. The one we are looking at is beautiful, the others not.

Weil extended the impossibility to compare—often associated, as we have seen, with artistic experience—not only to philosophy, but to religion as well:

> Each religion is alone true, that is to say, that at the moment we are thinking on it we must bring as much attention to bear on it as if there were nothing else; in the same way, each landscape, each picture, each poem etc., is alone beautiful. A "synthesis" of religions implies a lower quality of attention.[94]

The second is Theodor Wiesengrund Adorno, the German philosopher. An aphorism included in his *Minima Moralia* (written in 1944) reads partly as follows:

> *De gustibus est disputandum.* Even someone believing himself convinced of the non-comparability of works of art will find himself repeatedly involved in debates where works of art, and precisely those of highest and therefore incommensurable rank, are compared and evaluated one against the other. The objection that such considerations, which come about in a particularly compulsive way, have their source in mercenary instincts that would measure everything by the ell, usually signifies no more than the solid citizens, for whom art can never be irrational enough, want to keep serious reflection and the claims of truth far from the works. This compulsion to evaluate is located, however, in the works of art themselves. So much is true: they refuse to be compared. They want to annihilate one another. Not without cause did the ancients reserve the pantheon of the compatible to Gods or Ideas, but obliged works of art to enter the *agon*, each the mortal enemy of each. . . . Beauty, as single, true and liberated from appearance and individuation, manifests itself not in the synthesis of all works, in the unity of the arts and of art, but only as a physical reality: in the downfall of art

itself. This downfall is the goal of every work of art, in that it seeks to bring death to all others. That all art aims to end art, is another way of saying the same thing. It is this impulse to self-destruction inherent in works of art, their innermost striving towards an image of beauty free of appearance, that is constantly stirring up the aesthetic disputes that are apparently so futile.[95]

Works of art in a museum had been compared by Paul Valéry to a "tumult of frozen creatures each of which demands in vain the non-existence of all the others."[96] Adorno implicitly referred to this remark by framing it into Hegel's concept of "the death of art": since each work of art aims to truth, and therefore to its own self-destruction, it shares truth's intolerant quality. Through a different, even opposite path Adorno's intellectualism comes to a conclusion that is paradoxically close to Weil's mysticism: each work of art creates an empty space around itself, and therefore must be perceived in isolation.

This was exactly the target of Roberto Longhi, the Italian art historian, in his essay on art criticism (1950):

Here is the argument which will allow us to annihilate the last relics of metaphysics: the idea of the absolute masterpiece in its wonderful isolation. The work of art, from the vase made by the Greek artisan to the ceiling of the Sistine chapel, is always an intrinsically "relative" masterpiece. The work never stands by itself, it is always embedded in a relationship. To begin with, it is at least a relationship with another work of art. A work of art which would be the only one in the world, would not be regarded as a human product; it would be seen either with awe or with terror, as magic, as tabu, as a work made either by God or by a sorcerer, not by man. We have already suffered too much due to the myth of the divine, and *divinissimi* artists—rather than simply human.[97]

In reading those clashing passages one is reminded of Vasari's Aristotelian and Scholastic distinction between *semplicemente* (simply) and *secondo che* (*secundum quid*, according to).[98] Simone Weil and Adorno urged (albeit from different points of view) us to approach works of art as absolute, unrelated entities. Longhi, as Vasari before him, argued that works of art need a historical, relational, *secundum quid* approach. In my view, the two approaches are both necessary and mutually incompatible; they cannot be experienced *simultaneously*. Like the well-known image showing a duck/rabbit, we are not able see the duck and the rabbit at the same time, although they are both there. But the relationship between the two approaches is asymmetrical. We can articulate the "simple," direct, absolute approach through the language of history—not the other way around.

Notes

1. This is a shortened version of the paper I submitted at "Histories of Science, Histories of Art," held at Harvard University and Boston University in November 1995. A fuller version will appear in a collection of my essays to be published soon by Feltrinelli, Milan. Samuel R. Gilbert revised my English; Perry Anderson, Pier Cesare Bori, and Alberto Gajano helped me with their suggestions and critical remarks. I wish to thank them all.

2. Fulgenzio Micanzio, "Vita del padre Paolo (1552–1623)," in P. Sarpi, *Istoria del Concilio Tridentino*, II, ed. C. Vivanti (Turin, 1974), pp. 1348ff. See also *Stilus Romanae Ecclesiae*, s. l. n. d.; H. W. Strätz, "Notizen zur Stil und Recht," in *Stil. Geschichten und Funktionen eines kulturwissenschaftlichen Diskurselements*, ed. H. U. Gumbrecht and. K. L. Pfeiffer (Frankfurt, 1986), pp. 13–67 (the whole collection of essays is relevant). A good, synthetic introduction to the general topic is provided by the entry "style" in L. Grassi-A. Pepe, *Dizionario della critica d'arte* (Turin, 1978), pp. 565–68. See also E. H. Gombrich, "Style," in *The International Encyclopaedia of the Social Sciences*, XV (1968), pp. 352–61; J. A. Schmoll gen. Eisenwerth, "Stilpluralismus statt Einheitzwang—Zur Kritik der Stilepochen Kunstgeschichte," in *Argo. Festschrift für Kurt Badt*, ed. von M. Gosebruch and. L. Dittmann (Cologne, 1970), pp. 77–95; *Beiträge zum Problem des Stilpluralismus*, ed. W. Hager and N. Knopp (Munich, 1977) (Studien zur Kunst des 19. Jahrhunderts, XXXVIII); J. Bialostocki, "Das Modusproblem in den bildenden Künsten" (1961), in *Stil und Ikonographie* (Cologne, 1981), pp. 132–42; H. G. Gadamer, in "Excurs," *Hermeneutik II, Wahrheit und Methode* (Gesammelte Werke, Band 2) (Tübingen, 1986), pp. 375–78; B. Lang, ed., *The Concept of Style* (Philadelphia, 1979); W. Sauerländer, "From Stilus to Style: Reflections on the Fate of a Notion," *Art History* 6 (1983): 253–70.

3. Cicero, *De oratore*, Loeb Series, trans. H. Rackham.

4. "... *quid censetis si omnes, qui ubique sunt aut fuerunt oratores, amplecti voluerimus? nonne for ut, quot oratores, totidem paene reperiantur genera dicendi?*"

5. See M. Pohlenz, "To πρέπον. Ein Beitrag zur Geschichte des griechischen Geistes," in *Nachrichten von der Gesellschaft der Wissenschaften zu Göttingen aus dem Jahre 1933*, Phil.-hist. Kl., Nr. 16, pp. 53–92, especially p. 58ff.

6. S. Aureli Augustini, *Epistulae*, ed. A. Goldbacher (Vindobonae-Lipsiae, 1904), CSEL 44, 3, ep. 138, "ad Marcellinum," 1, 5, p. 130, partially quoted by A. Funkenstein, *Theology and the Scientific Imagination* (Princeton, 1986), pp. 223–24 (the whole chapter on "accommodation," pp. 202–89, is fundamental). For the English translation see *The Works of Aurelius Augustine, Bishop of Hippo*, ed. M. Dods, XII, 2 (Edinburgh, 1875), p. 197ff. On Volusianus see P. Brown, *Augustine of Hippo* (Berkeley and Los Angeles: University of California Press, 1969), pp. 300–03.

7. I will develop this argument in a paper to be included in a forthcoming volume of essays in memory of Amos Funkenstein.

8. B. Castiglione, *Il Cortegiano*, ed. V. Cian (Florence, 1923), I, xxxvii, pp. 92–93 (the allusion to Cicero has been duly identified by the editor). C. Dionisotti, *Appunti su arti e lettere* (Milan, 1995), p. 121, remarks that in Castiglione's passage artists are for the first time mentioned in a literary debate as ultimate authorities. But Cicero's argument (*De oratore*, III, 7, 25ff.) on the diversity of styles started precisely with sculpture and painting. On the inclusion of Mantegna in Castiglione's list see G. Romano, "Verso la maniera moderna," in *Storia dell'arte italiana*, ed. G. Previtali and F. Zeri, II, 2, 1 (Turin, 1981), pp. 73–74. On the absence of Titian see C. Dionisotti, *Appunti*, pp. 120–22.

9. E. H. Gombrich, "The Renaissance Concept of Artistic Progress and its Consequences," in *Norm and Form: Studies in the Art of the Renaissance* (London, 1966), pp. 139, 140, n. 5, speaks of *topos* in identifying an echo of Cicero's passage in a text by Alamanno Rinuccini. A similar approach is shared by M. Kemp, " 'Equal excellences': Lomazzo and the Explanation of Individual Styles in Visual Arts," *Renaissance Studies* I (1987): 1–26, especially pp. 5–6, 14. For a comparison related to a different milieu see M. Warnke, "Praxis der Kunsttheorie: Über die Geburtswehen des Individualstils," in *Idea*.

Jahrbuch der Hamburger Kunsthalle 1 (1982): 54ff. On Warburg's *Pathosformeln* see E. H. Gombrich, *Aby Warburg: An Intellectual Biography* (London, 1970), p. 178ff.

10. See E. Panofsky, "The First Page of Giorgio Vasari's 'Libro': A Study on the Gothic Style in the Judgment of the Italian Renaissance" (1930), in *Meaning in the Visual Arts* (Garden City, N.Y., 1955), pp. 169–235, esp. p. 206ff.

11. G. Vasari, *Le Vite* (ed. 1568), ed. G. Milanesi, VII Florence 1906, repr. 1973, p. 726 VII, p. 75. (J. Conaway Bondanella's and P. Bondanella's translation, slightly modified.)

12. G. Vasari, *Le vite* (ed. 1568), VII, p. 76.

13. G. Vasari, *Le vite* (ed. 1550), ed. L. Bellosi and A. Rossi, Turin 1986, p. 560.

14. G. Vasari, *Le vite*, VII, p. 417–18 (but the whole paragraph is relevant). See also ibid., p. 431, for a similar comment by Sebastiano del Piombo.

15. C. Hope's suggestion that Michelangelo's remark "was perhaps provoked more by prudery than anything else" (*Titian*, [New York 1980], pp. 89–90) can be safely dismissed. The Michelangelo-esque flavor of Titian's *Danae* had been already perceived by Cavalcaselle: see G. B. Cavalcaselle and J. A. Crowe, *Tiziano, la sua vita e i suoi tempi*, II (Florence, 1878, repr. 1974), p. 57. The relationship with Michelangelo's *Night* was pointed at, possibly for the first time, by Johannes Wilde in the 1950s: see his posthumously published lectures, *Venetian Art from Bellini to Titian* (Oxford, 1974), pl. 149 (in the caption, one should read *Night* instead of *Dawn*). F. Saxl argues that Titian's *Danae* "is taken from a sarcophagus representing a similar scene, that of Leda" ("Titian and Aretino," in *A Heritage of Images* [Harmondsworth, 1970], p. 81), a connection that seems even closer in the case of Michelangelo's *Leda and the Swan* or *Night*. Both works are related to Titian's *Danae* by F. Valcanover in *Da Tiziano a El Greco* (Milan, 1981), pp. 108–09.

16. G. Vasari, *Le vite*, (ed. 1568), IV, Florence 1906, repr. 1973, p. 92.

17. G. Vasari, *Le vite*, VII, Florence 1906, repr. 1973, p. 452. On the issues involved in this passage see P. Sohm, *Pittoresco* (Cambridge, 1991).

18. In the light of these passages one cannot accept the view that "Vasari is interested in style rather than in the individual" (S. Leontief Alpers, "Ekphrasis and Aesthetic Attitudes in Vasari's *Lives*," *Journal of the Warburg and Courtauld Institutes* XXIII [1960]: 190–215, especially p. 210). Much more convincing is Alpers's argument that for Vasari "the means of art are gradually perfected while the ends remain constant" (p. 201), although I cannot share her conclusion that this approach, obviously close to Castiglione and Cicero (both mentioned on p. 212) was "frankly untheoretical."

19. L. Dolce, *Dialogo della pittura*, *Trattati d'arte del Cinquecento*, ed. P. Barocchi, I (Bari, 1960), pp. 202 (on Titian's *Assunta* at the Frari), 206.

20. On this (and on Dolce) see D. Mahon, "Eclecticism and the Carracci," *Journal of the Warburg and Courtauld Institutes* XVI (1953): 303–41, especially pp. 311–13.

21. Roland Fréart sieur de Chambray, "Avant-propos," in *Parallèle de l'architecture antique avec la moderne* (Paris, 1650), pp. 1–2. The same keywords can be found in Bellori's critical remarks on Rubens (G. P. Bellori, *Le vite de' pittori, scultori e architetti moderni*, ed. E. Borea, introduction by G. Previtali [Turin, 1976], pp. 267–68). G. Previtali suggests (pp. XXIV-XXV) a convergence between Bellori's *Vite* and Fréart de Chambray's later work *Idée de la perfection de la peinture*.

22. R. Wagner-Rieger, "Borromini und Oesterreich," *Studi sul Borromini*, II (Roma, 1967), p. 221ff.

23. Johann Bernd and Fischer von Erlach, *Entwurf einer historischen Architectur in Abbildung unterschiedener berühmten Gebäude des Altertums und fremden Völker* (Leipzig, 1721), partially quoted by E. Panofsky, *Meaning*, p. 180). See A. Ilg, *Die Fischer von Erlach* (Vienna, 1895), p. 522ff.; J. Schmidt, "Die Architekturbücher der Fischer von Erlach," *Wiener Jahrbuch für Kunstgeschichte*, 1934, pp. 149–56, especially p. 152; G. Kunoth, *Die Historische Architektur Fischers von Erlach* (Düsseldorf, 1956); E. Iversen, "Fischer von Erlach as Historian of Architecture," *Burlington Magazine* C (1958): 323–25; H. Aurenhammer, *Johann-Bernhard Fischer von Erlach* (London, 1973), esp. pp. 153–59.

24. See my paper "Alien Voices: The Dialogic Element in Early Modern Jesuit Historiography," *Ta Historika* XII, 22 (June 1995): 3–22 (in Greek).

25. See G. Kunoth, *Die Historische Architektur*; E. Iversen, "Fischer von Erlach." Borromini, who had built for the Jesuits the Collegio de Propaganda Fide, left in his testament five hundred *scudi* "for the ornament of the altar of St. Ignace" (R. Wittkower, "Francesco Borromini," *Studi sul Borromini* I, Rome: 1967, p. 44).

26. M. Tafuri, *La sfera e il labirinto* (Turin, 1980), p. 54.

27. *The World*, March 27, 1755, quoted by A. O. Lovejoy, *Essays in the History of Ideas* (New York, 1948 [1960]), p. 121 ("The Chinese Origin of a Romanticism"; but see the whole essay, pp. 99–135, as well as "The First Gothic Revival," pp. 136–65). See also K. Clark, *The Gothic Revival* (1928) (Harmondsworth, 1964), pp. 38–40; O. Sirén, *China and Gardens of Europe in the Eighteenth Century* (New York, 1950, 2nd ed., Dumbarton Oaks, 1990).

28. J. J. Winckelmann, "Pensieri sull'imitazione dell'arte greca," in *Il bello nell'arte* (Turin, 1973), p. 32.

29. Quoted by F. Meinecke, *Die Entstehung des Historismus*, ed. C. Hinrichs (Munich, 1959), pp. 291–302.

30. J. J. Winckelmann, *Geschichte der Kunst des Altertums*, ed. W. Senff (Weimar, 1964), p. 7; U. Link-Heer, "Giorgio Vasari und der Uebergang von einer Biographien-Sammlung zur Geschichte einer Epoche," *Epochenschwellen und Epochenstrukturen im Diskurs der Literatur- und Sprachhistorie* ed. H. U. Gumbrecht and. U. Link-Heer (Frankfurt, 1985), pp. 73–88.

31. F. Haskell, *History and Its Images* (New Haven and London, 1993), pp. 217–24.

32. J. J. Winckelmann, *Geschichte der Kunst des Altertums*, pp. 102–03. A belated echo of this passage can be found in O. J. Brendel's comment on the mother from Chianciano (c. 400, now at Museo Archeologico in Florence): she "gazes in a void before her—gloomy ancestress of Michelangelo's sadly prophetic Madonna at Bruges" (*Etruscan Art* [Harmondsworth, 1978], p. 321).

33. See K. Justi, *Winckelmann, Sein Leben, seine Werke und seine Zeitgenossene* (Leipzig, 1872), vol. II, 2, pp. 86–97, which I am developing into a somewhat different direction.

34. A.-C.-P. comte de Caylus, *Recueil d'antiquités egyptiennes, etrusques, grecques et romaines*, I (Paris, 1761, 1st ed., 1752), p. VII.

35. K. Justi, *Winckelmann*, II, 2, p. 87. See also J. J. Winckelmann, *Lettere italiane*, ed. G. Zampa (Milan, 1961), p. 321. On the tense relationship between Caylus and Winckelmann see G. Pucci, *Il passato prossimo* (Rome, 1993), pp. 80–84.

36. Caylus, *Recueil*, I, dedication to the Académie des Inscriptions et Belles-Lettres:
    je ne regardois que du côté de l'art ces restes de l'Antiquité sçavante échappés à la barbarie des temps; vous m'avez appris à y attacher un mérite infiniment supérieur, je veux dire celui de renfermer mille singularitez de l'histoire, du culte, des usages et des moeurs de ces peuples fameux.
    See K. Pomian's enlightening essay "Maffei et Caylus," *Collectionneurs, amateurs et curieux* (Paris, 1987). The importance of Caylus's antiquarian work has been independently and convincingly argued by F. Haskell, *History*, pp. 180–86, and G. Pucci, *Il passato prossimo*, pp. 108–18.

37. K. Justi, *Winckelmann*, II, 2, p. 95.

38. Buffon's famous dictum *"le style c'est l'homme même"* has been often interpreted (better to say, misinterpreted: see L. Spitzer, "Linguistics and Literary History," in *Representative Essays*, ed. A. K. Forcione et al. [Stanford, 1988], pp. 13, 34) as if style would express the writer as an idiosyncratic individual. In fact, Buffon means that scientific discoveries are external to the human kind, *"l'homme"* in a general sense, not the individual writer; they can become a true property of the human kind, and therefore immortal, only through style ("Discours prononcé à l'Académie française le jour de sa réception," in Buffon, *Oeuvres philosophiques*, ed. J. Piveteau, Paris 1954, pp. 500–05).

39. Ph. Junod, "Future in the Past," *Oppositions* 26 (1984): 49.

40. See on them D. Irwin, *John Flaxman 1755–1826* (New York, 1979), pp. 204–15, who concludes that "as Flaxman's *Lectures* contain so many passages that are relevant to his own art, they deserve more

attention than they have tended to receive" (p. 215). They seem to be even more relevant if they are read in a broader framework.

41. M. Praz, *Gusto neoclassico* (Milan, 1990), p. 67. See in general S. Symmons, *Flaxman and Europe: The Outline Illustrations and their Influence* (Ph.D. dissertation, New York and London, 1984).

42. D. and E. Panofsky, *Pandora's Box* (New York, 1962), pp. 92–93; G. Previtali, *La fortuna dei primitivi. Dal Vasari ai neoclassici* (Turin, 1989), pp. 169–70.

43. See, in the same direction, J. Flaxman, *Lectures on Sculpture* (London, 1829), pp. 201–02.

44. J. Flaxman, *Lectures*, pp. 196–99—a passage which did not escape the attention of E. H. Gombrich: see his paper "From Archaeology to Art History: Some Stages in the Rediscovery of the Romanesque," in *Icon to Cartoon: A Tribute to Sixten Ringbom*, ed. M. Terttu Knapas and A. Ringbom (Helsinki, 1995), pp. 91–108, especially p. 96.

45. *Choice Examples of Wedgwood Art* (London, 1879); *John Flaxman, Mythologie und Industrie* (Hamburger Kunsthalle, 1979).

46. D. Irwin, *John Flaxman*, p. 207.

47. G. W. F. Hegel, *Vorlesungen über die Aesthetik* (Frankfurt, 1970), I, pp. 105–06. These lectures, based on notes taken by students between 1817 and 1829, were first published in 1836–1838.

48. H. Heine, "Französische Maler. Gemäldeausstellung in Paris 1831," in *Historisch-kritisch Gesamtausgabe*, eds. J.-R. Derré and C. Giesen, 12, 1 (Hamburg, 1980), p. 24 (*Painting on the Move: Heinrich Heine and the Visual Arts*, ed. S. Zantop [Lincoln and London, 1989], pp. 133–34.) In quoting this passage, R. Bianchi Bandinelli, *Introduzione all'archeologia* (Bari, 1975), p. 100, n. 84, referred it to Friedrich Schlegel. See also W. Rasch, "Die Pariser Kunstkritik Heinrich Heines," in *Beiträge zum Problem des Stilpluralismus*, pp. 230–44; I. Zepf, *Heinrich Heines Gemäldebericht zum Salon 1831: Denkbilder* (Munich, 1980).

49. E. Delacroix, "Questions sur le Beau," in *Oeuvres littéraires*, I (Paris, 1923), pp. 23–36. See also pp. 37–54: "Des variations du beau" (1857).

50. C. Baudelaire, *Curiosités esthétiques, l'Art romantique et autres oeuvres critiques*, ed. H. Lamaitre (Paris, 1962), pp. 215–16 (*Art in Paris 1845–1862*, trans. and ed. J. Mayne [London, 1965], p. 124).

51. C. Baudelaire, *Curiosités ésthétiques*, p. 213 (*Art in Paris 1845–1862*, p. 123). On Baudelaire and Heine see C. Pichois, "La littérature française à la lumière du surnaturalisme," *Le surnaturalisme français* (Neuchâtel, 1979), p. 27. See also the commentary on H. Heine, *Historisch-kritisch Gesamtausgabe*, eds. J.-R. Derré and C. Giesen, 12, 2 (Hamburg, 1984), p. 566.

52. C. Baudelaire, *Curiosités ésthétiques*, p. 217 (*Art in Paris 1845–1862*, pp. 125–26).

53. C. Baudelaire, *Curiosités esthétiques*, p. 238, n. 1, p. 904, note to p. 219 (*recte* 217).

54. C. Baudelaire, *Curiosités esthétiques*, p. 219 (*Art in Paris 1845–1862*, p. 127).

55. It was published as an advertisement attached to G. Semper, *Wissenschaft, Industrie und Kunst* (Science, industry, and art) (Braunschweig, Friedrich Vieweg und Sohn, 1852 [cf. H. F. Mallgrave, *Gottfried Semper, Architect of the Nineteenth Century* (New Haven and London, 1996), pp. 156–157]). The letter to Vieweg announced a *Vergleichende Baulehre* (Comparative doctrine of construction) in two volumes, allegedly under press. It was never published.

56. G. Semper, "London Lecture of November 11, 1853," edited with a commentary by H. F. Mallgrave, preface by J. Rykwert, *Res* 6 (Fall 1983): 5–31.

57. Ibid., p. 8.

58. J. Rykwert, "Semper and the Conception of Style," *Gottfried Semper und die Mitte des 19. Jahrhunderts*, pp. 66–81.

59. J. Anderson, in her postface to G. Morelli, *Della pittura italiana . . . Le gallerie Borghese e Doria-Pamphili in Roma* (Milan, 1991), pp. 494–503, insists on the importance of Cuvier, which I would have missed in my essay "Clues." I agree with her, although she apparently missed both my quotation from Cuvier and the attached comment, *Clues, Myths, and the Historical Method* (Baltimore, 1989), pp. 116–17.

60. G. Semper, "London Lecture of November 11, 1853," pp. 11–12.

61. G. G. Scott, *Remarks on Secular & Domestic Architecture, Present & Future*, 2nd ed. (London, 1858), pp. 11, 16, 263 (italics are in the text).

62. The racial theories of Gustav Klemm (on whom see H. F. Mallgrave, "Gustav Klemm and Gottfried Semper: The Meeting of Ethnological and Architectural Theory," *Res* 9 [Spring 1985]: 68–79) did not have any impact on Semper.

63. G. Semper, *Der Stil in der technischen und tektonischen Künsten*, 2nd ed. (Munich, 1879), pp. 1–5. Semper used the same example in his first London lecture: *Res* 6 (Fall 1983): 9–10.

64. C. Schorske, *Fin-de-Siècle Vienna* (New York, 1980), pp. 101–04.

65. A. Riegl, *Stilfragen* (*Problemi di stile. Fondamenti di una storia dell'arte ornamentale* [Milan, 1963]). On *Kunstwollen* see M. Olin, *Forms of Representation in Alois Riegl's Theory of Art* (University Park, Penn., 1992), p. 72.

66. W. Hofmann, *Gustav Klimt und die Wiener Jahrhundertswende* (Salzburg, 1970) (English translation, *Gustav Klimt* [Salzburg, 1971], pp. 39–41).

67. M. Olin, *Forms of Representation*; M. Iversen, *Alois Riegl: Art History and Theory* (Cambridge, Mass., 1993).

68. A. Riegl, *Industria artistica tardoromana* (Florence, 1981), p. 86.

69. W. Sauerländer, "Alois Riegl und die Entstehung der autonomen Kunstgeschichte an Fin-de-Siècle," *Fin de siècle. Zu Literatur und Kunst der Jahrhundertswende* (Frankfurt, 1977), pp. 125–39 (Italian translation in S. Scarrocchia, *Alois Riegl: teoria e prassi della conservazione dei monumenti*, Antologia di scritti, etc. [Bologna, 1995], pp. 421–32).

70. A. Riegl, *Historische Grammatik der bildende Künste* (*Grammatica storica delle arti figurative*, ed. F. Diano [Bologna, 1983], p. 288).

71. A. Riegl, *Grammatica storica*, p. 261, n. 21. On Lueger see C. Schorske, *Fin-de-Siècle Vienna*, pp. 133–46.

72. See M. Olin, "The Cult of Monuments as a State Religion in Late 19th Century Austria," *Wiener Jahrbuch für Kunstgeschichte*, XXXVIII (1985), pp. 199–218 (translated in S. Scarrocchia, *Alois Riegl*, pp. 473–88). Olin, who mentions without comment a vicious anti-Semitic letter addressed by a pseudonymous reader to the art historian Franz Wickhoff, stresses the presumable gap between Riegl's universalistic and peaceful attitude and the Christian Socialist's racist policy (pp. 196–97).

73. A. Riegl, *Grammatica storica*, p. 77, and *Spätrömische Kunstindustrie* (*Industria artistica tardoromana*, p. 149). On the racial overtones of Riegl's categories see J. von Schlosser's worried remark ("La scuola viennese" [1934], in *La storia dell'arte* [Bari, 1938], p. 125). Gombrich is much more detached: "As a child of his time, Riegl never doubted the influence of racial factors on stylistic developments" (*The Sense of Order* [London, 1979], p. 184).

74. The theoretical inconsistency of Worringer's opposition between *Abstraktion* and *Einfühlung* (see his book with the same title, 1908) has been criticized by E. Panofsky, "Der Begriff des Kunstwollens" (Italian translation, *La prospettiva come "forma simbolica"* [Milan, 1961], p. 175, n. 7). But Worringer's distinction rather simplistically develops an opposition already suggested by Riegl: see for instance *Industria artistica*, p. 113.

75. W. Worringer, *Formprobleme der Gotik* (1911; I consulted the second edition, Munich 1912, p. 97) (H. Read, ed., *Form in Gothic* [New York, 1964], p. 142). See on him N. H. Donahue, *Forms of Disruption: Abstraction in German Modern Prose* [Ann Arbor, 1993], pp. 13–33.

76. W. Worringer, *Formprobleme*, pp. 126–27 (*Form in Gothic*, pp. 179–80; the translation has been slightly modified).

77. P. Feyerabend, "Wissenschaft als Kunst. Eine Diskussion der Rieglschen Kunsttheorie verbunden mit dem Versuch, sie auf die Wissenschaft anzuwenden," in *Sehnsucht nach dem Ursprung. Festschrift für Mircea Eliade*, ed. H. P. Duerr (Frankfurt, 1983). I consulted the Italian translation, *Scienza come arte* (Rome-Bari, 1984), pp. 93–161; a shorter version was published as "Science as Art: An Attempt to Apply Riegl's Theory of Art to the Sciences," *Art + Text* 12–13 (1983): 16–46.

78. P. Feyerabend, *Against Method* (*Contro il metodo* [Milan, 1973], p. 197, n. 219).

79. P. Feyerabend, *Wissenschaft als Kunst*, 1984 ("according to Riegl, there is no progress, only change"; Italian translation, *Scienza come arte*, p. 118).

80. P. Feyerabend, *Wissenschaft als Kunst* (*Scienza come arte*, p. 156).

81. P. Feyerabend, *Wissenschaft als Kunst* (*Scienza come arte*, p. 51, n. 29). The list includes, besides Riegl, E. Panofsky, B. Snell, H. Schäfer, and V. Ronchi. Riegl and Panofsky are added as a postscript; the last three names are mentioned in connection with *Against Method* (1970) in P. Feyerabend, *Killing Time* (Chicago, 1995), p. 140.

82. In the essay "Quantitativer and qualitativer Fortschritt in Kunst, Philosophie und Wissenschaft" (*Kunst und Wissenschaft*, ed. P. Feyerabend and Ch. Thomas [Zürich, 1984], pp. 217–30, and P. Feyerabend, *Wissenschaft als Kunst*), Vasari is interpreted, so to speak, as a forerunner of Riegl, who is not mentioned.

83. The way in which he tells his reaction to the news of Germany's surrender is revealing: "I was relieved, but I also had a sense of loss. I had not accepted the aims of the Nazis—I hardly knew what they were—and I was too much contrary to be loyal to anyone" (P. Feyerabend, *Killing Time*, p. 55). None of these statements is believable in the light of what the author said before: that he read aloud *Mein Kampf* to his assembled family, had ambivalent attitudes toward the Nazis, and dreamed to join the SS.

84. P. Feyerabend, *Killing Time*, p. 45.

85. Ibid., pp. 47–50. He remarked: "I still have the complete text of the lectures—forty pages of a six-by-eight inch notebook. This is truly miraculous, for I am not in the habit of assembling memorabilia."

86. See also J. Agassi, "Wie es Euch gefällt," in *Versuchungen. Aufsatze zur Philosophie Paul Feyerabends*, ed. H. P. Duerr, I (Frankfurt, 1980), pp. 147–57, and Feyerabend's reply (*Scienza come arte*, pp. 83–85). I found Feyerabend's remarks on Auschwitz very disturbing.

87. See the catalogue of the 1993 exhibit at the Los Angeles County Museum. On Feyerabend's desire to become a painter see *Killing Time*, p. 43.

88. W. Sauerländer, "Alois Riegl," p. 432.

89. H. F. K. Günther, *Rasse und Stil. Gedanken über die Beziehungen im Leben und in der Geistesgeschichte der europäischen Völker, insbesondere des deutschen Volkes* (Munich, 1926) (Worringer's *Formprobleme der Gotik* is mentioned on p. 56). On Günther see the long eulogy by L. Stenel-von Rutowski, "Hans F. K. Günther, der Programmatiker des Nordischen Gedankens," *Nationalsozialisische Monatshefte* 68 (November 1935): 962–98; 69 (December 1935): 1099–114.

90. A. Hitler, "Die deutsche Kunst als stolzeste Verteidigung des deutschen Volkes," *Nationalsozialistische Monatshefte* 4, 34 (October 1933): 437, partially quoted by Saul Friedländer, *Nazi Germany and the Jews*, I (New York, 1997), p. 71. A great part of the October 1933 issue of the *Nationalsozialistische Monatshefte* focused on art in the Third Reich.

91. P. Feyerabend, *Against Method* (Italian translation, *Contro il metodo* [Milan, 1973]).

92. See the somewhat embarrassed remarks in P. Feyerabend, *Scienza come arte*, pp. 31–33.

93. Styles in architecture, T. L. Donaldson said around 1840, can be compared "to languages in literature. There is no style, as there is no language, which has not its peculiar beauties, its individual fitness and power—there is not one which can be safely rejected" (quoted by H. F. Mallgrave, *The Idea of Style: Gottfried Semper in London* [dissertation, 1983], p. 199). The parallel has a long story, since the aforementioned passage by Castiglione (*Il Cortegiano*, pp. 92–93).

94. See S. Weil, *Cahiers*, II (Paris, 1972), p. 138 (*The Notebooks of Simone Weil*, I, trans. A. Wills [London, 1976], pp. 244–45). The importance of these passages has been stressed by P. C. Bori, *From Hermeneutics to Ethical Consensus among Cultures*, trans. P. Leech, Ch. Hindley, and D. Ward (Atlanta, Ga., 1994), p. 18.

95. T. W. Adorno, *Minima moralia* (London, 1974), trans. E. F. N. Jepheoth, n. 47, "De gustibus est disputandum." The whole aphorism is pertinent.

96. P. Holdengräber, " 'A Visible History of Art': The Forms and Preoccupations of the Early Museum," *Studies in Eighteenth-Century Culture* 17 (1987): 115.

97. R. Longhi, "Proposte per una critica d'arte" (1950), *Critica d'arte e buongoverno* (Opere complete, XIII) (Florence, 1985), pp. 17–18.

98. The distinction, based on Aristotle, *On Interpretation*, 18a, 9–18, was spread by Boethius's translation. I analyzed both Aristotle's passage and Boethius's comment in "Mito," *Noi e i Greci*, ed. S. Settis (Turin, 1996), pp. 201–05. In Longhi's aforementioned passage—"rather than simply human"—the adverb "semplicemente" is qualified by the adjective "umani."

I RENE J. W INTER

The Affective Properties of Styles: An Inquiry into Analytical Process and the Inscription of Meaning in Art History*

I n the wider arena of the way(s) in which the arts *and* the sciences generate appropriate terms and concepts to be used as instruments in analytical operations, the term/concept "style" occupies a rather special place: applicable *both* to the ways in which the operations are undertaken[1] and to describable characteristics of the objects of analysis. In the present chapter, I wish to pursue, on the one hand, the *lack* of discreet boundaries between "style" as it is manifest in a work and subject matter— hence, content and meaning—and, on the other hand, the hermeneutic problems raised by attempts to correlate style and meaning through "stylistic analysis" as operationalized in art history.

I have chosen my terms carefully to mirror the language used for certain mathematical operations, as I believe the analogy holds well, and in the hope it will raise questions of methodology common to both the sciences and the humanities. To the extent

* The general issues dealt with in this paper were presented in a College Art Association panel in 1987. Although the case studies used have not changed since then, I am most grateful for this opportunity to reformulate the problem. I would also like to thank a number of graduate students, now colleagues, who over the years have put their good minds to nuanced definitions of style; many will see echoes of themselves in what is presented here. In particular, I would cite Jülide Aker, Jak Cheng, Harry Cooper, Marian Feldman, Elizabeth Herrmann, David Joselit, Brandon Joseph, Leslie Brown Kessler, Michelle Marcus, Steven Nelson, Scott Redford, John Russell, Ann Shafer, and Yuejin Wang. In addition, my thanks to Garth Isaak for help with respect to the mathematical metaphor I was seeking, and to Robert Hunt for acute and critical comments on an early draft.

that style is initially an artifact of the hand of a maker, it is applicable, no less than a plus sign, to the entire *domain* of (all possible) figures; but for a specific task, from that broad domain, individual *items* are selected (i.e., certain figures); then, with the addition of "style" to those figures, that is, the hand of the maker, there results from the *range* of possibilities the specific *image*.

Much of what one can do *with* style depends upon how one *defines* style; and I would assert from the beginning that there is no absolute definition of style, but rather, a range of operative definitions varying with user and analytical task to be performed. Art historians generally revert to Meyer Schapiro's basic definition of 1953, as "the constant form—and sometimes the constant elements, qualities, and expression—in the art of an individual or group."[2] This definition implies that a given style is characterized by a particular attribute or attributes observable in a work or group of works, which in turn permits the construction of "sets" and boundaries on the basis of the presence or absence of defining variables. What Schapiro did not fully account for, however, was the element of agency in the manipulation and organization of form; nor did he engage the issue of the necessity of style in the materialization of content.[3] What is more, neither Schapiro's basic definition nor his extended discussion takes on the question of whether style in fact *inheres* in a work, or rather is made to *adhere* to the work as a product of description, comparison, and classification undertaken by an external analyst.

For my purposes, I would take the position that once there is anything in the work we can call form, then there is also style, but that it is also important to keep what is intrinsic to the work distinct from what is extrinsic to it, by consistently referring to post-hoc determinations as the products of "stylistic analysis." In that way, style is a function of a period, place, workshop, or hand; it is inherent in the work, and it is thus what is apparent to the perceiver. Stylistic analysis then introduces the conscious observation, selection, and articulation of manifest properties to the act of perception. That distinction having been made, what I wish to bring to discussion is the fiction that the so-called style of a given work is merely a passive by-product, an artifact of "making" as divorced from "meaning."

To pursue this issue, it is crucial to see art history within the larger picture of European intellectual history. In that larger picture, it becomes possible to see *why* art historians at a particular moment in the history of the discipline needed to separate the act(s) and sign(s) of making from the range of cultural meanings attached to the finished work (and equally, why this is no longer either necessary or desirable).[4] And it is also possible to pursue the degree to which the *how* of representation enters into the domain of choice—whether consciously as a tool deployed by individuals and/or cultures, or subconsciously as generated from/by a body of ideas and attitudes. This is surely no less true of verbal or musical art forms than of the visual. And, as Joseph

Koerner has demonstrated for the writing of German philosopher Hans Blumenberg,[5] it should be subject to analysis in all forms where construction is integrally intertwined with content.

Art-historical analysis to date has tended to privilege subject matter as the vehicle by which meaning is conveyed to an audience through the work of art, and hence "iconography" as the analytical procedure by which to arrive at an understanding of meaning. In two cases from the ancient Near East, however—one of Phoenician and North Syrian ivory carving of the early first millennium B.C., the other of Mesopotamian royal sculpture of the third and first millennia—it is possible to explore two distinct ways in which style may be said to enter into the arena of meaning. In the first case, I shall argue that questions of style intrude as less conscious expressions of underlying cultural attitudes and patterns; whereas in the second, style actually functions as a consciously deployed strategic instrument with specific rhetorical ends. In both cases, one may match meaning from subject matter in a particular cultural context with expressive content in style. The resultant correlations imply that style in fact plays an important role in complementing or even *activating* the more overt message(s) provided by content, and as such, (1) cannot be divorced from meaning in any study of the affective properties of the work, and (2) should be considered in any historical analysis *of* meaning.

CASE I: STYLE AND CULTURAL MEANING

The first issue, that of style as an expression of underlying attitudes and patterns, arises from a comparison of two groups of ivory carvings of roughly the eighth century B.C., found at the Assyrian capital of Nimrud and a number of other sites in the ancient Near East.[6] Neither group is native to Assyria, and the original objects of which the remaining plaques were clearly components must have been part of the impressive booty and tribute in ivory attested in Assyrian text and depicted on Assyrian reliefs.[7]

One group of ivories has been identified as the product of Phoenician work, congruent with the modern Levantine coast, while the other group has been located within a region centering around northern Syria and southeastern Anatolia—both areas in which the Assyrian army mounted massive military campaigns.[8] When one juxtaposes a "Phoenician"-style furniture plaque depicting a winged female sphinx with a "Syrian" example of the same motif, one is immediately struck by the lack of "Egyptian" features in the Syrian work (Figure 1). None of the decorative details—crown, headcloth, and pectoral or uraeus bib—known from Egyptian representations and quite accurately reproduced in the Phoenician work, is included on the Syrian plaque. Very different also is the extremely round face, puffy cheeks, and broad nose of the Syrian

Figure 1.   Top: Sphinx, North Syrian style. Ivory plaque, Fort Shalmaneser, Nimrud. Bottom: Sphinx, Phoenician style. Ivory plaque, Fort Shalmaneser, Nimrud. Both collection of Iraq Museum, Baghdad.

sphinx as opposed to the more oval face and delicate features of the Phoenician; or the elongated slender and long-legged body of the Phoenician sphinx as compared with the heavy proportions and short legs of the Syrian. In addition, on the Phoenician plaque, the sphinx's wings curve up in a delicate arc over the back, creating a counter-rhythm to the horizontal body and vertical head, with the embarrassing juncture of wing and shoulder hidden by the pectoral; on the Syrian plaque, the wings jut up at an awkward angle, and are folded back parallel with the body, so that one's eye does not move when looking at the piece, but rather is fixed on the massive block of the head and body. The Syrian carver, then, seems to have chosen to emphasize the sense of massive power in the image of the sphinx, at the expense of elegance, grace, and detail of design.

This overall impression is reinforced further by the surrounding space. On the Phoenician plaque, an interplay of filled and empty space has been achieved; the animal is well planned into and comfortably contained within its borders, and the plant elements are spaced to fill the voids between legs and between the head and wing. The Syrian sphinx, by contrast, presses up against the limits of its plaque, as if it were simply too large to be contained within. Very little extra space exists within the rectangle, and even the curving tendrils of a tree stump at the right press up against the animal's body. In short, when one looks at the Phoenician piece, one is struck by the balance, elegance, and careful design of the plaque as a whole. We may infer from allusions in textual sources and from the context of usage in representation that this composite mythological creature had symbolic significance beyond its decorative function;[9] but for the Phoenician sphinx, by virtue of the attention given to its design, and the more remote profile view, one is a step further removed from the impact of the motif than with the Syrian representation, where everything conspires to confront the viewer with the power vested in the sphinx as semidivine being.

Many of these same distinctions can be made in comparing yet another motif: that of a male figure slaying a griffin—a theme linked with myths of the youthful hero-god, Ba'al, promoting fertility and life through his victory over the destructive powers of the sun. Once again, Phoenician and Syrian examples separate themselves quite readily (Figure 2). The Phoenician griffin-slayer is shown with an accurately rendered Egyptian-style wig. Both hero and griffin are winged, and the spacing of the two sets of wings is complementary; their elegant upward curve creates a rhythm that would have contrasted with the downward diagonal of the spear, were the plaque complete. All four of the griffin's feet are planted on the ground, and only his head twists back to receive the spearpoint. The horizontal form of his body balances the diagonals created by the forward-leaning stride of the hero. No individual detail is allowed to predominate, and the delicate rhythm of opposing elements, as with the Phoenician sphinx, creates a sense of harmony and elegance. By contrast, the Syrian griffin-slayer wears headgear borrowed from—but sadly misunderstanding—an Egyptian royal crown.

Figure 2. Left: *Griffin-slayer, North Syrian style. Ivory plaque, Fort Shalmaneser, Nimrud. Museum of Fine Arts, Boston.* Right: *Griffin-slayer, Phoenician style. Ivory plaque, Fort Shalmaneser, Nimrud. Iraq Museum, Baghdad.*

Only the griffin is winged here; yet at first glance, it is difficult to determine to whom or where the wings are attached. The hind legs of the griffin are thrown up against the sides of the plaque, and this diagonal body, plus the bent knee of the hero, expresses the force necessary to subdue such an adversary. This force is further emphasized by the strong diagonal thrust of the sword, which seems to dominate the whole piece. Man and animal are packed into the rectangular space of the plaque, and seem to fight against the very borders, as if ready to burst out of the frame.

These distinctions, whether applied to ivories or to other media, presumably allowed the discerning ancient to identify place of origin, just as today they allow the archaeologist and art historian to divide the finds from Nimrud and elsewhere into two coherent stylistic groups, Phoenician and Syrian, even in cases where motifs are common to the two groups. Their attribution to place of origin is then fixed by comparison with large-scale fixed stone monuments in the two adjacent geographical regions. The consistency of these elements brings to mind Heinrich Wölfflin's well-known pairs of attributes that permitted a distinction between Renaissance and Baroque—his a distinction in time, mine a distinction in space.[10] In fact, Wölfflin's terms, "calm, complacent, graceful, still, in a state of being," for the Renaissance, versus "restless, overwhelming, pathological, in a state of becoming," for the Baroque, sound significantly like a description of the differences between Phoenician and North Syrian.

In my initial study of these ivories, the foregoing distinctions made it possible to define distribution patterns of the two groups, and so to speak of distinct regions of production and economic spheres of interaction for what was in its time the most important luxury-good-cum-artwork in the ancient Near East. In that regard, I was using stylistic analysis, according to Meyer Schapiro's 1953 definition, as an archaeologist: one for whom style represents a diagnostic feature or a series of symptomatic traits permitting one to locate the work spatially or temporally.[11] The operation was based upon a series of premises: (1) a material work possesses (visual) properties; (2) those properties are observable and describable; (3) they are then applicable to other works, according to which, groups or clusters can be established; and (4) they therefore permit of generalization. Furthermore, since the observable properties are inherent in the object, one can come to them without specific insider knowledge. In short, one does not have to control local rules in order to create meaningful clusters.

This approach has not been limited to archaeologists; many of Wölfflin's contemporaries and successors have used style in this way within the discipline of art history. As perspective on such usage, two glosses on Wölfflin's contribution are important here. First, however reductively individual pair-bonds of his descriptive terms have been employed by subsequent practitioners of art history—e.g., separating out single elements, such as "painterly" to be opposed to "linear," "open" as distinct from "closed" forms—Wölfflin himself was explicit that these were not independent variables, but rather existed as sets consisting of co-varying elements—the consideration of all of which was necessary in order properly to assign membership to one group or another. Indeed, the use of *sets of attributes* comes very close to scientific methodology in multivariate analysis, frequently employed in the study of archaeological materials; and there, too, it has been shown that techniques that take account of more than one variable, and particularly the associations between variables, give far stronger (i.e., more informative) results.[12]

Second, the era that generated Wölfflin's distinctions was entirely consistent with their being used "archaeologically." At the time, the world of scholarly inquiry was sufficiently cognizant of the important contributions of Linnaeus and other natural scientists that *classification* was a primary intellectual endeavor, both scientific and historical. The period also followed closely upon late-eighteenth-century analyses of commodity and capital, a significant consequence of which had been the reification, if not fetishization, of *property*—which was quickly translated into the art market, where value, especially in painting, was tied to recognition of a particular master's hand. Style—or, more properly, "stylistic analysis"—was thus used largely as a diagnostic tool, à la Giovanni Morelli,[13] in order to establish period, place, workshop, and/or artist; in short, it was concerned less with meaning than with attribution.

In the present climate of scholarly inquiry, however, other questions of, hence other considerations of and possibilities for, the concept of style and stylistic analysis intervene. Important for our particular case is that Syria and Phoenicia were two contemporary cultural entities, with closely related languages, a common pantheon and a shared mythological tradition. How then to get beyond classification, to account for the differences in what Schapiro called "the meaningful expression" carried by differing styles used to render like motifs?

This is the point, of course, at which "outsider" observations must be augmented by independent, localized evidence that will keep in check projections grounded in mere non-disconfirmation, and instead sustain a hermeneutic supported by data. In the specific instance of the ivories, evidence abounds to demonstrate that the cities of Phoenicia were linked historically to Egypt in ways that the cities of Syria were not; they also looked out geographically on the Mediterranean seaways and socially onto a more multicultural world in ways that the inland cities of Syria did not. When these data are put together with the descriptive properties of the ivories, it becomes possible to contextualize the affective properties of the stylistic characterizations made above. If the Phoenician works are at once more elegant and more removed, with greater balance between *plein et vide*, while the Syrian works are more intense, with greater dynamic impact, then it might be hypothesized that some bearers of Phoenician culture—at least, those producers and users of elite objects—embodied more sophisticated and at the same time less *engagé* attitudes and emotional states than their counterparts in Syria.

Such claims, precisely because they are relatively consistent with our own cultural distinctions, were easier to make in the days before the past was identified as a foreign country and the act of "essentializing" whole cultures or stereotyping subsets had itself been essentialized as a tool of hegemonic discourse. Indeed, ever ahead of his times, Schapiro had already noted the problems attendant upon attempting to read "racial" and cultural worldviews from styles.[14] Not surprisingly, therefore, the scientifically based processual archaeologists of the 1960s and 1970s insisted that we were ill equipped to be "paleo-psychologists,"[15] and argued that such interpretive exercises should not be part of the ancient historian's purview. And yet, there *are* patterns of culture, just as there are consistencies, if not correlations, in cultural/historical styles, as Schapiro himself acknowledged, and as Winckelmann had observed as early as the mid-eighteenth century.[16] The tricky part is the attachment of cultural meaning to those patterns by adducing non-retrojective, non-anecdotal sources of evidence in support of such assignments, and the location of those meanings in the appropriate societal band (whole culture, elite, identifiable subculture, etc.).[17]

For our case, interesting parallels have been suggested by William Rathje in his study of ports of trade and contemporary inland sites of the pre-Columbian New World. According to Rathje's analysis, seacoast towns in general are characterized by more cosmopolitan culture, are less single-minded or committed to any particular religious or social pattern, and, as a reflection of their domination by trading interests, are open to many eclectic influences; inland areas, by contrast, tend to be both more religious and more intensely committed to definite cultural patterns.[18] However one hesitates to characterize for fear of moving from description to caricature, it was certainly possible to find these same traits in the respective cultural identities of modern, pre-1967 Lebanon and Syria, home to the ancient carving centers of Phoenician and Syria; and I believe there is sufficient evidence from the literary and historical record to suggest that similar patterns of social interaction and cultural adaptation pertained in the area during the early first millennium B.C. as well.

The implications of the Phoenician/North Syrian case for the history of art are not trivial, for the case suggests first, that attributes may be identified that convey through style the "meaningful expression" of a work or group of works, and second, that a "meaningful correlation" may be established between a given style and the broader cultural outlook of a region or social group. It then forces upon us the dual methodological problems of how to identify significant attributes—i.e., what unites the hands of Michelangelo and Raphael as manifestations of the Italian Renaissance, rather than what distinguishes them as individual artists[19]—and then, how to read style effectively as a barometer of underlying cultural, regional, group, or personal attitudes appropriate to particular moments of time or states of mind, without engaging in massive retro-jections of value and meaning existing only in the observer?

With this latter question, one evokes not only Schapiro's definition of style as "inner content," but also Riegl's *Kunstwollen*, or "artistic volition," that underlying drive toward a *particular* style in a particular time and place.[20] Panofsky, in his analysis of the concept of *Kunstwollen*, articulated it as, in part, the "psychology" of a given period, in which the collective will becomes manifest in the artistic creation[21]—and indeed, it is possible to see early art history, especially in Germany, as an extension of nineteenth-century studies of social psychology, influenced strongly by Dilthey and others.[22] Although neither Riegl nor Panofsky makes it explicit, what Riegl was reaching for was a concept of "culture" with a small "c," operative in the social sciences as a *system* with certain definitions and certain boundaries, within which both the artist and his work were to be situated, however strong the mark of the individual; and not "Culture" with a capital "C," as the *products* of (certain elements in) a society. It is this that I believe also underlies Wölfflin's observation (Wölfflin, whom subsequent art historians have reified for the articulation of variables that have led to the decontextu-alization of formal analysis!) that "not everything is possible at all times in the visual arts." By this I understand him to be saying that there is a degree of *historical*, if not *cultural* determinism in any given period; that, in his words, we can determine the "feel-ings" (the *Lebensgefühl*) of a period from its style; and that "a new *Zeitgeist* [or, period spirit] demands a new form" (i.e., style).[23] This I would amend for the case of the contemporary Phoenicians and Syrians to: "a different *Kulturgeist* [or, cultural spirit] demands a different form."

This having been said, it is nonetheless true that attaching historically accurate "feelings" to a given visual manifestation is a major problem for the historian, as is the recovery of historically meaningful interpretations of those feelings within a cultural sphere. And obviously, these tasks become increasingly difficult the further one is in time and place from the making, coding, and culture of the original. An early caution came from Dante Gabriel Rosetti, in his "The Burden of Nineveh," where, occasioned by his confrontation with ancient Assyrian art, he warned against unwarranted inter-pretation of the character of a civilization. Legend has it that Rosetti was emerging from the British Museum in 1851, just as a great winged gateway lion from Nimrud was being hoisted up the front steps (see *Illustrated London News*, February 28, 1852), whereupon he was prompted to meditate upon the constraints placed on understand-ing a distant culture known only through isolated remnants (although when we see an individual standing alongside one of these great colossi, 4 to 5 meters in height, it is hard to escape a response to its scale, whatever one's culture, or not to ascribe size as one of its meaningful attributes in antiquity).

Nevertheless, there is risk involved in any attempt to correlate visual attribute with

sociocultural interpretation. Not very fruitful attempts were made by several anthropologists in the 1950s, who used then recently developed "achievement-motivation"
tests for individuals emerging from our own culture as measures of whole cultural traditions (for example, the ancient Greeks and the Maya). The goal was to determine
when the phase of greatest drive toward achievement had occurred—largely on the
basis of the frequency of certain extremely simplistic diagnostic traits, such as diagonal
lines, in the artifactual assemblages of the particular tradition.[24] These studies led to a
dead end because their categories were far too gross, and the underlying assumption
that the meanings attached to particular diagnostic categories were universal and
therefore could be universally applied was never tested, hence never confirmed. The
same criticism of assumed universality could be applied to the cross-cultural search for
binary opposition in the structuralism of the 1960s;[25] but what structuralism contributed was the formulation that underlying cultural patterns were manifest and could be
discerned in the material products (namely, "art") of a given cultural universe.

This brings us back to what Henri Zerner has referred to as Riegl's "radical historicism"—his total rejection of normative aesthetics toward the culturally and historically specific.[26] It is certainly true that, following upon post-structuralist and
postmodernist critiques, we now understand cultures (and historical moments) to be
less discretely bounded and homogeneous than initially perceived, and selected
voices—as in cultural and political elites—differentially recorded.[27] But unless we as
art historians wish to concede the impossibility of any historically grounded knowledge, the result of these critiques must be to raise the standards of argument and evidence, rather than to relinquish any hope of explanation. A major problem lies in
confirmation. It is not possible to know for certain whether the emotional and/or
experiential values attributed to the coherent variables that constitute the Phoenician
as opposed to the North Syrian style are valid for the first millennium B.C. in the
ancient Near East. One can certainly seek, and even think to find, corroborating con-
textual information and/or historical analogy that can be brought to buttress interpretations of the work. But even to do so, one must be operating under an initial premise
that there be a meaningful correlation between the manifest elements of style and the
experiential nexus from which they derive—a premise that has not to date been subject to hypothesis formation and testing.

CASE II: STYLE, AGENCY, AND AGENDA

The Phoenician/North Syrian case suggests that certain visual attributes derive
from the special geographical and/or historical situation of the producing culture, and
that they represent not-necessarily-conscious reflections of worldview and experience
held by at least some members of that culture. My second case, that of certain elements of style employed in Mesopotamian royal sculpture, requires briefer discussion,

but raises the important issue of the degree to which a style may be *consciously deployed* as an active agent in constructing a worldview, for particular rhetorical-cum-ideological ends.

Panofsky left this possibility open in his 1920 study of Riegl's *Kunstwollen*, when he referred to the collective will of a period, manifest in artistic creation and apprehended either *consciously or unconsciously*.[28] There is certainly little disagreement in the field that the decorative programs devised for public spaces can embody very conscious constructs. The palaces of Neo-Assyrian rulers from the ninth through the seventh centuries B.C. prove no exception. In the Northwest Palace of Assurnasirpal II (883–858 B.C.), for example, the four epithets of ideal royal attributes found in the king's Standard Inscription, carved on every slab of his palace reliefs, find their exact counterparts in the four ways in which the king is represented: "attentive prince" shows him seated in an ancient posture related to rendering good judgment; "keeper of the gods" shows him attendant upon the sacred tree, under the aegis of the god Assur; "fierce predator" shows him in battle with wild bulls and lions; and "hero in battle" shows him victorious over enemy citadels.[29] But the power invoked in his epithets is also manifest in the size, proportion, and musculature of the human body, as seen not only in the king's own figure, but also in those of his protective *genii* that flank the doorways and mark the vulnerable corner spaces of the palace (Figure 3).[30]

In many respects, the way of rendering the ruler in early Neo-Assyrian art shows continuity with the preceding Middle Assyrian period of the second millennium B.C. For example, on a carved altar of Tukulti-Ninurta I (1244–1208 B.C.), the earlier king carries a similar mace and wears a similar wrapped garment, while in other Middle Assyrian representations, the royal tiara is similar.[31] What has shifted by the early first millennium is precisely the heavier proportions of the royal figure, and, most clearly manifest on the kilted *genii*, the emphasis on massive musculature that is characteristic of early Neo-Assyrian "style."[32]

I would argue that this is not a random shift. It corresponds to an equal intensification in contemporary texts of references to might and power, coequal with the extraordinary military expansion of the Assyrians in the period. In short, the power invoked in subject matter is the power manifest in style is the power at issue in the state.

The well-known statues of Gudea of Lagash (ca. 2110 B.C.) constitute a parallel situation some fifteen hundred years earlier (see Figure 4). They are recognizable by the ruler's characteristic headgear, cylindrical body, clasped hands, and enlarged, staring eyes. While these are all diagnostic features of Gudea statues, and thus serve to date unexcavated works and associate all with the Neo-Sumerian period, it is only when the statues are seen in the context of the inscriptions that accompany virtually every one of the nearly twenty extant works, that we understand we are in the presence of a true confluence of style and meaning. For one of Gudea's chief epithets in Sumerian, indeed, written directly upon one of his statues in a lengthy dedicatory text,

Figure 3. Protective Genius. Limestone relief from the Northwest Palace of Assurnasirpal II, Nimrud. British Museum, London.

*Figure 4. Gudea of Lagash, Diorite statue,
found at Tello. Musée du Louvre, Paris.*

is a₂ sum-ma ᵈNin-dar-a-ke₄ ("arm strengthened [lit., given] by the god Nindara"), and I would argue that this specific quality appropriate to rulership, rather than the random development of a more "realistic" or plastic mode of rendering, is what best accounts for the massive musculature of Gudea's uncovered right arm. We are further told that he possesses wisdom, which in Sumerian translates literally as one "of wide ear" (geštu-dagal), and is regarded with a "legitimizing gaze" by his god, upon whom he is to concentrate in return, thus accounting for the enlarged ears and eyes.[33] In short, as with the Assyrian reliefs, form (style) has been used to convey intended meaning: the power, authority, and appropriate attributes necessary to rule, possessed by virtue of representation by one claiming the right to rule![34]

The covariance apparent in the Gudea and Assurnasirpal statuary between intended message/content and style of rendering strengthens the generalizable relationship between form and meaning suggested above, by adding conscious choice to the construct. This relationship has been cognized more readily in literary studies, and is increasingly evident in the field of literary as well as art history in recent years, from Gary Saul Morson's study of social realism in the Soviet novel to David Summers's study of contrapposto, among many others.[35] The capacity of style to carry value, and to be purposely deployed in order to represent specific values, has even been recognized by art historians whose approach has been largely "formalist"—as, for example, Sidney Freedberg in his study of the stylistic revolution occurring in Italian painting around 1600, in which he noted that the "manner of employing basic elements of a style may be altered . . . to accord with the artist's sense of the nature of his subject."[36] Such a situation of conscious choice served as the core of the important dissertation of Leslie Brown Kessler with respect to the work of Domenichino and Lanfranco, where she showed that differing aims on the part of contemporary artists could call forth not only different subjects, but also distinct styles considered appropriate to those subjects.[37] Although the textual (and cultural) record of the ancient Near East neither identifies individual artists nor includes conscious exegeses on art-making, we can nonetheless observe (and highlight) those related instances—largely court art executed within a domain of political ideology—in which style may be seen to carry value and therefore convey meaning, as well as instances when, as in the later Neo-Assyrian period, styles have been altered in order to accord better with rhetorical ends.[38]

While these observations will come as a surprise to no one, I do believe it is important to put them into the context of current issues in the practice of art history. The initial isolation of "style" from "iconography" as two discrete tools of analysis—the one related to form, the other to content and meaning—served the field well, up to a point. The analysis of style came to serve as the means by which authentication or attribution could be attached to a given work, and was privileged by some practitioners of the history of art, thereby leaving iconography to another set of practitioners, with each subgroup subject to intellectual fashion.[39] Yet the unit, and the unity, is ultimately the *work* as a whole, of which "style" and "iconography" are analytical subsets. In any individual undertaking, therefore, isolated stylistic or iconographical analyses can only be partial. Too often, one or the other has been taken for the whole: the whole of the work, or the whole of the art historical endeavor.

At a moment more than one hundred years since these analytical tools were developed for use in art history, it is important to keep in mind that they have been constructed *by* us, to serve for particular procedures. Often, as is the case with advances in technology, such tools are discovered to possess properties that permit other analytical operations not thought of when the tools themselves were invented. If the division of style and iconography as discrete analytical tools in art history initially became equated with a comparable division between form and meaning, suggesting that meaning was to be revealed through the iconographic enterprise and not through an analysis of style, it is now time to reconsider that division. The degree to which it is no longer sufficient is the degree to which we insist more on the many ways in which an artwork can "mean," along with a better understanding of the various contextualizations of the work today—plural in the face of a postmodern awareness of positionality and polyvalence, but still allowing for more than the fact of the work's production in assessing the cultural and historical climate of its production and subsequent reception.

In this respect, we have come to assume Riegl's negative attitude toward any theory that severed art from history.[40] As concerns the historical divide between form and meaning, it is also apparent that some creative and analytical art historians had pointed out the theoretical limitations of this division quite early. In particular, I would cite Robert Klein, whose work has been too little considered since his untimely death in 1967.[41] Indeed, as noted above, it has been repeatedly demonstrated for individual cases that *both* style and iconography in fact carry meaning; that often the meaning they carry is either identical or complementary; and that when it is not, *we* must account for the discrepancy—purposeful subversion, incongruence—by further analysis of meaning. Therefore, I emphatically underscore once again the importance of the challenge to the exclusion of style from investigations into the domain of meaning—not just in particular cases in the art-historical literature, but as a general principle.

This has been perhaps best understood in studies of clothing styles, from A. L. Kroeber and Roland Barthes to Dick Hebdige and Kennedy Fraser,[42] and is apparent today in both clothing store windows and advertising layouts, where a whole universe of value is subsumed within the category of "taste." Display in advertising and in shop windows—lighting, color, accessories, posture, and grouping of models—serves to set up emotional linkages to merchandise that itself manifests particular properties of style and is embedded in a vast nexus of signification.[43] The acculturated individual who then chooses to dress in a certain style has elected to signal the attendant meanings and values conveyed by the signs upon her/his body.[44] Potential consumers who react to fashion store windows and/or viewers who react to an individual's dress style also represent insiders who know and operate under understood sets of coded references. They should in principle be directly analogous to contemporary audiences for Phoenician or North Syrian ivories or the reliefs of an Assyrian throne room, whose responses would equally be determined by their familiarity with and sensitivity to the full range of associations afforded by the visual stimuli.

The key to "style-as-meaning" lies, I would argue, in cultural context and in the emotional response invoked/provoked by the work. Here I would build upon an essay by James Ackerman, in which the impact of a work is seen as the result of a combination of intellectual knowledge *plus* sensory perception.[45] It is style, I would argue, that sets up the parameters for and the emotional linkages of affective *experience*, via the culturally conditioned sensory motors of visual perception. And in that respect, issues of style engage *both* properties of the work *and* functions of response. In short, style both inheres in a work and lives in the eye of the beholder.

With this, we may return to the aims of the present volume. For at the level of sensory perception, the observation and experience of style as a manifest cluster of attributes links the humanist to the scientist, the historian of art to the historian of science.[46] And furthermore, as an analytical tool, stylistic analysis functions like any scientific attribute analysis, requiring description, classification, and systemic contextualization—goals of the scientist no less than of the humanist.[47]

I have argued for a further component in understanding style, however—one that requires moving from description and classification to experience. It is therefore implied that the analyst of style in any given historical manifestation not only replicates certain scientific procedures in the course of analysis, but also functions as a social cientist in the attempt to capture historicized experience, just as the contemporary experiencer of stylistic properties can only do so as a social being.

I have further suggested that it is only in the unity of "form-*plus*-content" that a given work of visual art realizes its ontological identity—whether for its own original time and place, or for the viewer/analyst at a distance. Since subject matter must be given physical form in order to convey itself visually, the very *act* of making produces a *way* of making; and if one accepts that that way of making is manifest as style, then it is style that not only gives form but also "affective agency" in the psychological sense to the meaning of the subject matter. Or, put another way, style itself then becomes a sign existing between the maker and the world, to be processed no less than subject matter. If it is easier to describe the physical properties of a style than it is to assess their affective value, that is not a license to ignore the latter, or to avoid developing methodologies that will permit access to them.

Schapiro closed his 1953 article on "style" with the statement that, "a theory of style adequate to psychological and historical problems has still to be created." That statement remains true more than forty years later, although we can be said to have made some progress. On the basis of issues raised here, I would propose that at the very least a theory of style must (1) consider the proposition that there is a nonrandom relationship at the macrolevel between a style and the culture/period within which it is produced, before one ever gets to the relationship between a given style and the psyche of a specific individual (i.e., artist) making "art"; (2) acknowledge that style is closely allied with the psychological stimulus known as "affect," and as such is an integral component in the communication of meaning, hence in the response that the work elicits; (3) take account of the fact that the potential use and value of style as a concept depends entirely upon the nature of the analytical operation(s) in which it is employed; and (4) move toward methodologically sound ways to test the hypotheses generated to explain style and/or to explain the relationship between style and other aspects of culture.

Throughout all of the above, it is essential to keep in mind that the concept of style gave rise to its use as an analytical tool, and therefore to place both the concept and its subsequent deployment squarely within the broader history of ideas. To the extent that all analytical concepts can—indeed, must—be scrutinized both as products of a particular moment or moments in history and within the context of a particular set of tasks to be accomplished, the concept of style for the art historian then takes its place with comparable analytical concepts in the history of science.

Notes

1. See, for example, A. C. Crombie, *Styles of Scientific Thinking in the European Tradition*, 3 vols. (London, 1994).
2. M. Schapiro, "Style," in *Anthropology Today*, ed. A. L. Kroeber (Chicago, 1953), pp. 287–312. The subsequent literature is enormous; only a limited selection will be referred to below; however, many of the discussions themselves contain extensive bibliography.
3. In gross terms, this leads to the question of whether there can be a work without style, as it is sometimes said of individuals, who "have no style." To certain architectural historians, as well—Viollet-le-Duc, for example—a building *not* executed within the terms of a certain "order" can indeed be said to be without style. I would argue that this is exactly a case in point of an "operative" (and limiting) definition of the term.
4. On this, see W. Sauerlander, "From Stilus to Style: Reflections on the Fate of a Notion," *Art History* 6 (1983): 254, where style is referred to as a "highly conditioned . . . hermeneutical 'construct,' worked out at a distinct moment in social and intellectual history."
5. J. L. Koerner, "Ideas about the thing, not the Thing Itself: Hans Blumenberg's style," *History of the Human Sciences* 6, 4 (1993): 1–10.
6. See, among others, the primary publications of R. D. Barnett, *A Catalogue of the Nimrud Ivories* (London, 1957); M. Mallowan and G. Herrmann, *Furniture from SW 7 Fort Shalmaneser* [Ivories from Nimrud, fasc. III] (Aberdeen, 1974); and G. Herrmann, *Ivories from Room SW 37 Fort Shalmaneser* [Ivories from Nimrud, fasc. IV] (London, 1986), where extensive bibliography is provided through the date of publication.
7. Narrative reliefs include both the removal of furniture as booty from captured citadels and the subsequent use of such identical furniture, as in the well-known scene of Assurbanipal and his queen in a

garden (A. Paterson, *Assyrian Sculptures: Palace of Sinacherib* [The Hague, 1915], passim; R. D. Barnett, *Sculptures from the North Palace of Ashurbanipal at Nineveh* [London, 1976], Pl. LXIV).

8. A full analysis is presented in I. J. Winter, "Phoenician and North Syrian Ivory Carving in Historical Context: Questions of Style and Distribution," in *Iraq XXXVIII* (1976), pp. 1–26, including a history of these distinctions, along with the social context in which the schools of carving functioned.

9. See, for example, the reference in the Hebrew Bible to a Phoenician master craftsman being employed in the construction of the temple of Solomon plus the decoration of the cella with sphinxes (Hebr. *cherubim*), (I Kings 6:23–30) or the presence of pairs of sphinxes along with griffins and bulls as heraldic pairs in a ritual scene painted on the courtyard wall of a second millennium palace at Mari on the middle Euphrates (A. Parrot, *Mission Archéologique de Mari, II (2). Le palais: peintures murales* [Paris, 1958], pl. XI).

10. H. Wölfflin, *Principles of Art History: The Problem of the Development of Style in Later Art*, trans. M. D. Hottinger (New York, 1956, orig. in German, 1915); *Renaissance und Barock* (Munich, 1888).

11. Schapiro, "Style," p. 287. "Archaeology" has changed appreciably since Schapiro wrote, however, and archaeologists have come to use style as a measure and expression of a variety of sociocultural factors, no longer merely for purposes of plotting distribution—see, for example, D. D. Davis, "Investigating the Diffusion of Stylistic Innovations," in *Advances in Archaeological Method and Theory*, vol. 6, ed. M. B. Schiffer (New York, 1983), pp. 53–89, and the essays included in M. W. Conkey, *Style in Archaeology* (New York, 1989).

12. S. J. Shennan, *Quantifying Archaeology* (Edinburgh, 1988), cited in M. Roaf, "Pottery and p-values: 'Seafaring Merchants of Ur?' Re-examined," *Antiquity* 68 (1994): 776.

13. G. Morelli, "Principles and Methods," in *Italian Painters: Critical Studies of Their Works*, vol. I, trans. from the German by C. Ffoulkes (London, 1900), pp. 1–63.

14. Schapiro, "Style," pp. 306–10.

15. L. R. Binford, "Archaeological Systematics and the Study of Cultural Process," *American Antiquity* 31 (1965): 203–10.

16. J. Winckelmann, *Histoire de l'art chez les anciens*, vol. I (Amsterdam, 1766), pp. 12, 63, who first described the principal characteristics of Egyptian art, and then insisted upon the degree to which those characteristics were to be found in *all* Egyptian works, regardless of date.

17. Michael Baxandall's "period eye," as developed in his study, *The Limewood Sculptors of Renaissance Germany* (New Haven and London, 1980), esp. pp. 143–63, was an attempt at this, although since criticized as too "normative" and monolithic, but at least historically grounded. Similar interpretive forays are encouraged by the recent school of post-processual archaeologists, for example, Ian Hodder, *Reading the Past: Current Approaches to Interpretation in Archaeology* (Cambridge, 1986). Unfortunately, however, as applied to date, in many cases the interpretations put forward are undersupported by evidence. But see now the exploratory piece by Margaret W. Conkey, "Making Things Meaningful: Approaches to the Interpretation of the Ice Age Imagery of Europe," in *Meaning in the Visual Arts: Views from the Outside*, ed. I. Lavin (Princeton, 1995), pp. 49–64.

18. W. Rathje, "The Last Tango in Mayapan: A Tentative Trajectory of Production Distribution Systems," in *Ancient Civilization and Trade*, ed. J. A. Sabloff and C. C. Lamberg-Karlovsky (Albuquerque, 1975), pp. 409–45.

19. Another case of the relationship between style and meaning—one outside the concerns of the present paper—would be when the style in fact reveals the hand of a particular artist (as per Sauerländer's distinction of *stilus*, *Art History* 6, cited above), and the "meaning" of the work therefore takes on special significance *as* the work of that artist and no other (on which, see J. L. Koerner, *The Moment of Self-Portraiture in German Renaissance Art* [Chicago, 1993]). In such cases, the "sign" of the artist—her/his "hand"—assumes an indexical value, independent of subject matter.

20. A. Riegl, *Stilfragen* (Munich, 1977, orig. publ. 1893, 2nd ed., 1923). And see on this, H. Zerner, "Alois Riegl: Art, Value and Historicism," *Daedalus* 105 (1976): 177–88.

21. E. Panofsky, "Das Problem des Stils in der bildenden Kunst," republished in *Aufsatz zu Grundfragen der Kunstwissenschaft* (Berlin, 1964, orig. publ., 1915); also, "Der Begriff des Kunstwollens," *Zeitschrift*

für Aesthetik und allgemeine Kunstwissenschaft XIV (1920), recently translated by K. J. Northcott and J. M. Snyder, "The Concept of Artistic Volition," *Critical Inquiry* VIII (1981): 17–33.

22. Discussion in L. Dittmann, *Stil, Symbol, Struktur: Studien zur Kategorien der Kunstgeschichte* (Munich, 1967), esp. chapter 1. In part, Wölfflin's work was in answer to more deterministic notions of style deriving from material and/or technique—although Heather Lechtman has revived the issue of interdependence between imagery and production techniques ("Style in Technology: Some Early Thoughts," in *Material Culture: Styles, Organization and Dynamics of Technology*, eds. H. Lechtman and R. Merrill [St. Paul, Minn., 1977], pp. 3–20).

23. See, for example, Wölfflin, *Principles*, p. 11. Michael Podro (*The Critical Historians of Art* [New Haven, 1982], pp. xxiv–xxv) provides the background for this in his reference to Wölfflin's first work, *Renaissance und Barock*, as an attempt to "explain changes in style as changes in *attitude*," and Wölfflin's later development, based upon the work of Adolf von Hildebrand, of the "reciprocal adaptation of subject matter and the material of visual representation" (i.e., style).

24. For example, A. F. C. Wallace, "A Possible Technique for Recognizing Psychological Characteristics of the Ancient Maya from an Analysis of their Art," *American Imago* 7 (1950): 239–58; E. Aronson, "The Need for Achievement as Measured by Graphic Expression," in *Motives in Fantasy, Action and Society*, ed. J. W. Atkinson (Princeton, 1958), pp. 249–65. Nor were anthropologists the only proponents of such readings. J. N. Hough, as early as 1948 ("Art and Society in Rome," in *Transactions of the American Philological Association* 79 [Philadelphia], p. 341), posited correlations of artistic techniques with Roman social structure: i.e., between rigidity of design and the rigidity of society, between relief sculpture on a single plane and autocracy, and between spatial illusionism and social participation in government. The work of E. Davies ("This is the way Crete went—not with a bang but a simper," *Psychology Today* 3 [1969]: 43–47), is a by-product of that era, in which the psychological variables needed for achievement were concretized into the number of diagonal lines, s-shapes, and unattached forms manifest in early Greek pottery.

25. C. Lévi-Strauss, *Anthropologie structurale* (Paris, 1958); *La pensée sauvage* (Paris, 1962), etc., and the many analytical studies of structuralism, e.g., Dan Sperber, *Le structuralisme en anthropologie* (Paris, 1968).

26. Zerner, *Daedalus* 105: 185.

27. E.g., James Clifford, *The Predicament of Culture* (Cambridge, 1987).

28. Panofsky, in *Critical Inquiry* VIII: 20, emphasis mine.

29. I. J. Winter, "Royal Rhetoric and the Development of Historical Narrative in Neo-Assyrian Reliefs," *Studies in Visual Communication* 7 (1981): 2–38.

30. J. Curtis and J. E. Reade, eds., *Art and Empire: Treasures from Assyria in the British Museum* (New York, 1995), p. 43; J. B. Stearns, *Reliefs from the Palace of Ashurnasirpal II* (Graz, 1961), pls. 44–51.

31. Reproduced in A. Moortgat, *The Art of Ancient Mesopotamia* (London and New York, 1969), fig. 246.

32. Compare, for example, the rendering of such a genius figure on Middle Assyrian cylinder seals, e.g., ibid., pl. K: 6 and 7, with the Neo-Assyrian palace figures.

33. See extended discussion of this in I. J. Winter, "The Body of the Able Ruler: Toward an Understanding of the Statues of Gudea," in *Dumu-É-dub-ba-a: Studies in Honor of A. W. Sjöberg*, ed. H. Behrens et al. (Philadelphia, 1989), pp. 573–83.

34. One could continue making this case. For example, as one moves to the end of the Neo-Assyrian sequence, to the seventh-century reliefs of later kings such as Assurbanipal, the apparent stylistic shifts covary with developments in narrative complexity: the proliferation of information and the expansion of the visual field reduce emphasis on the king himself, whose figure then becomes both smaller and less massively proportioned—see discussion in Winter, *Studies in Visual Communication* 7, pp. 27, 30.

35. G. S. Morson, "The Socialist Realist Novel and Literary Theory" (unpubl. ms.), p. 15; D. Summers, "Contrapposto: Style and Meaning in Renaissance Art," *Art Bulletin* 59 (1977): 336–61. For the general issue of the impossibility of a neat division between form and content in poetics, see A. C. Bradley's "Oxford Lectures on Poetry," in *The Problem of Aesthetics: A Book of Readings*, ed. E. Vivas

and M. Krieger (New York, 1960), p. 569. See also Jeffrey Ruda, "Flemish Painting and the Early Renaissance in Florence: Questions of Influence," *Zeitschrift für Kunstgeschichte* 2 (1984): 210–36, where the argument is made explicitly that it is essential to look at the *relationship* between style and iconography (p. 236).

36. S. Freedberg, *Ca. 1600: A Revolution of Style in Italian Painting* (Cambridge Mass., 1983), p. 8.

37. L. B. Kessler, "Lanfranco and Domenichino: The Concept of Style in the Early Development of Baroque Painting in Rome" (Ph.D. dissertation, University of Pennsylvania, 1992).

38. Additional cases within the art of the ancient Near East could be cited to buttress those presented here. For example, the so-called "Court Style" in Achaemenid Persian seal carvings of the sixth to fourth centuries B.C. has been suggested to represent a conscious response to demands made upon carvers to construct a "visual language of control and empire" (M. B. Garrison, "Seals and the Elite at Persepolis: Some Observations on Early Achaemenid Persian Art," *Ars Orientalis* 21 [1992]: esp. p. 17).

39. The intellectual climate of the ascendancy of iconographical analysis, following upon the work of Panofsky, was the stimulus, for example, for George Kubler's *The Shape of Time: Remarks on the History of Things* (New Haven, 1962), as he attempted to resurrect the validity of a notion of style in understanding art and cultures. Henri Focillon's *The Life of Forms in Art* (New York, 1943, orig. *La Vie des Formes*), which articulated the premise that certain forms give rise to other forms *as such* without the necessary mediation of culture, helped to keep alive a sharp divide in the field between adherents of one approach or another.

40. Zerner, *Daedalus* 105, p. 179; and see also, Riegl, "Das holländische Gruppenportrait," *Jahrbuch der Kunsthistorischen Sammlung des Allerhochsten Kaiserhauses* 23 (1902), pp. 71–278, esp. p. 73, where he states that "the task of art history is to . . . decipher . . . the essential character" of the period.

41. R. Klein, *Form and Meaning: Essays on the Renaissance and Modern Art*, trans. M. Jay and L. Wieseltier, with a foreward by H. Zerner (New York, 1979, orig. *La Forme et l'intelligible* [Paris, 1970]). See esp. "Thoughts on Iconography," p. 149, where Klein refers to conscious quotations of antique style in nineteenth-century painting (e.g., Gros's *Napoleon at Eylau*) that provide "nonexplicit, wordless meaning"—i.e., one *not* carried overtly through subject matter.

42. A. L. Kroeber, *Style and Civilizations* (Ithaca, 1957); R. Barthes, *The Fashion System*, trans. J. Ward and R. Howard (New York, 1983); D. Hebdige, *Subculture: The Meaning of Style* (London, 1979); Kennedy Fraser, "On and Off the Avenue: Feminine Fashions," a series of articles for the *New Yorker* from November 1970 into the early 1980s. (See, in particular, the beautiful piece on Bianca Jagger in the issue of February 24, 1973, in which "style" itself becomes an attribute, going beyond description to the ascription of positive "affect.")

43. B. Means, "Clothing Store Windows: Communication through Style," *Studies in Visual Communication* 7 (1981): 64–71.

44. The emotional component, and the individual identity constructed in association therewith, distinguishes this view of (choice in and associations with) clothing style from many anthropological studies that emphasize instead an emotionally de-cathected set of attributes permitting the conveyance of information regarding social group membership and maintenance and/or serving as signs in the construction of social boundaries. See, for example, M. Wobst, "Stylistic Behavior and Information Exchange," in *For the Director: Research Essays in Honor of James B. Griffin* (Anthropology Papers, Museum of Anthropology, University of Michigan, no. 61), ed. C. E. Cleland (Ann Arbor, 1977), pp. 317–42, with specific reference to clothing styles. In addition, see M. W. Conkey, "Style and Information in Cultural Evolution: Toward a Predictive Model for the Paleolithic," in *Social Archaeology: Beyond Subsistence and Dating*, ed. C. L. Redman et al. (New York, 1978), pp. 61–85 and R. N. Zeitlin, "A Sociocultural Perspective on the Spatial Analysis of Commodity and Stylistic Distributions," in *The Human Uses of Flint and Chert*, ed. H. de G. Sieveking and M. H. Newcomer (Cambridge, 1986), pp. 173–81, where the notion of style as marker of social and political affiliation is extended to other types of goods/works.

45. J. S. Ackerman, "Interpretation, Response: Suggestions for a Theory of Art Criticism," in *Theories of Criticism: Essays in Literature and Art* [Occasional Papers of the Council of Scholars of the Library of Congress, no. 2] (Washington D.C., 1984), pp. 33–53. This is the sort of problem pursued more by perceptual psychologists than art historians: see, for example, Salek Minc, "Significant Form and Physiological Stimuli in Art Perception," *Australian UNESCO Seminar: Criticism in the Arts, University of Sydney, May 1968* (Canberra, 1970), pp. 68–80, who based his distinction on the work of Clive Bell [*Art* (New York, 1957)], which by that time was largely out of date. However, what he (Minc) was after, the relationship of visual stimulus to the human autonomic response system, not merely the cultural system, is the problem that Ackerman was also raising—one that has yet to be systematically explored.

46. Herein lies the answer to the problem raised by David Topper ("The Parallel Fallacy: On Comparing Art and Science," *British Journal of Aesthetics* 30 [1990]: 311–18): the illogical propositions that create a false analogy between "art" and "science" become logical once one compares "art *history*" and "science" as parallel analytical traditions.

47. The related issues of the *assumption* of discrete boundaries in descriptive sets and the *authority* assumed by the analyst who, in the very act of determining what variables will be privileged in the making of those boundaries, inscribes exclusion into the analytical process, have not been pursued here, but should be acknowledged as problematic. I would argue that it is important to take them into consideration in specific analytical undertakings.

AMY SLATON

Style/Type/Standard: The Production of Technological Resemblance

INTRODUCTION

This essay reconsiders a famous stylistic absence: the departure of ornamentation, traditional design motifs, and idiosyncratic profile from a broad swath of American architecture after 1900. That these features of earlier architectural styling are missing from many commercial, civic, and large-scale residential buildings erected over the course of the twentieth century is evident to the casual observer. It is the notion of absence itself that I want to examine. It has served as the primary analytic instrument for historians who examine the roots of modernist architecture in America. The initial embrace of austerity and uniformity by many American builders between 1900 and 1930, when it is considered at all, is treated by historians as a renunciation of stylistic self-consciousness. Perhaps because utilitarian buildings of this period were frequently designed by engineers rather than architects, this early "functionalism" (a term I will examine) has come to represent a sort of default mode for architecture—engaged when building designers choose to serve commerce rather than the more traditional master of high culture. Historians grant later manifestations of functionalism (the mid-century buildings designed by architects) greater aesthetic sophistication but attribute this development to American receptivity to International Style design precepts imported from Europe rather than to any indigenous appreciation of simplified form.[1] The origins of the twentieth-century American commitment to the standardized undecorated building remain wholly negative phenomena—rooted in the conservation of effort and money, the rejection of expressive possibility, the paring away of intention.

This paper recasts these absences as presence: of designing engineers' intentionality and authority, and thus of cultural meaning and social consequence for utilitarian buildings. By looking at the first expressions of this building mode in the United States after 1900—the thousands of undecorated, virtually identical concrete-frame factories that swelled industrial neighborhoods between 1900 and 1930—I will identify a complex of positive forces behind the American embrace of utilitarian building design. Far from being the products of technical personnel answering the demands of industry with some pre-ordained set of design solutions (pre-ordained by what or whom, we would have to ask), the factories were created with tremendous awareness of cultural and market forces. Their appearance not only prefigures that of much later American architecture, but reflects the potential of a cultural enterprise—here, architecture—to be mutually determinative with technology and commerce, and thereby very directly a cause of social change. It is this relationship that makes the idea of an absence so unsatisfactory as an explanation for American modernism: it elides what can only be called the political genesis of these artifacts.

To retrieve the historical meaning of functionalist industrial architecture, we must first see these buildings as ambitious examples of industrial production. In many respects the factory buildings were like the goods made within: undifferentiated in form and produced with modern, streamlined procedures. Catalogs put out by factory builders between 1900 and 1930 show a remarkably homogenous collection of offerings, the buildings varying in size but in few other ways (see Figures 1 and 2). The typical reinforced-concrete factory building erected between 1900 and 1930 was rectangular, usually from 50 to 75 feet wide and from 100 to 900 feet long. Most were from four to eight stories high, without brick cladding or ornamentation to disguise their reinforced-concrete skeleton frames. Where ornamentation was used it was usually in the form of a simple cornice, or very occasionally, a tower that housed stairways and bathrooms. So great is their uniformity that factory buildings of virtually identical appearance held industries ranging from shoe manufacturing to hose weaving, from the production of rubber gloves to the processing of breakfast cereals. The factories display an ingenious application of contemporary tenets of industrial standardization.

But while the economic imperatives of mass production may have contributed to the popularity of this building style for its builders and buyers, they are not sufficient to explain the proliferation of these structures. As Reyner Banham points out, builders of hotels and hospitals of this period also sought economies but those buildings look very different from the factories, and, we might add, from one another.[2] Functionalism, which I take here to mean an expressive emphasis on the characteristics of mechanized production (simplicity and repetition of form) is similarly unhelpful as an explanatory term. A more foundational question must be asked: How did uniformity

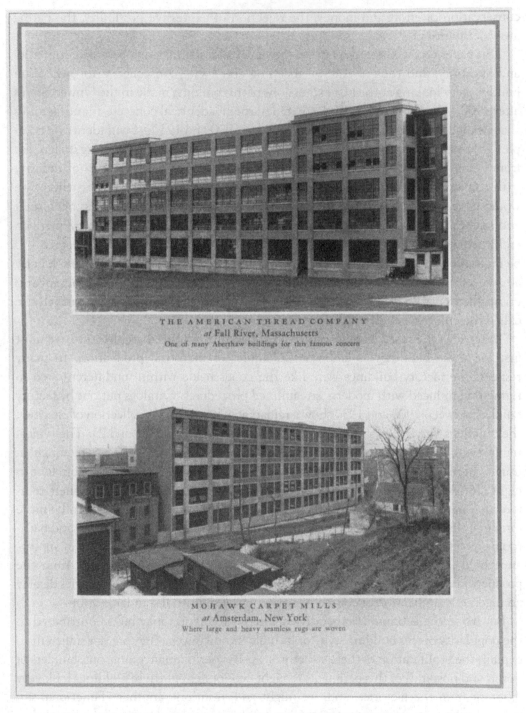

THE AMERICAN THREAD COMPANY
at Fall River, Massachusetts
One of many Aberthaw buildings for this famous concern

MOHAWK CARPET MILLS
at Amsterdam, New York
Where large and heavy seamless rugs are woven

Figure 1.   Illustration of factories from "Built by Aberthaw," Catalog of the Aberthaw Construction Company, Boston, Massachusetts, 1926.

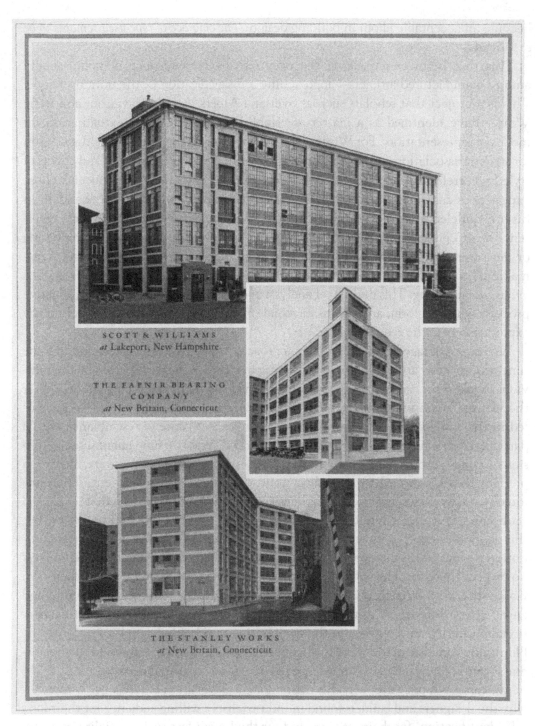

SCOTT & WILLIAMS
at Lakeport, New Hampshire

THE FAFNIR BEARING
COMPANY
at New Britain, Connecticut

THE STANLEY WORKS
at New Britain, Connecticut

Figure 2. More factories, from "Built by Aberthaw," Catalog of the
Aberthaw Construction Company, Boston, Massachusetts, 1926.

become an acceptable idiom in building design? *The embrace of type itself remains to be explicated.*

This task involves unpacking the occurrence of resemblance or commonality among manufactured artifacts—approaching a disaggregation of the notion of style. This is a project that scholars such as Svetlana Alpers, Arnold Davidson, and Irene Winter have identified as a matter of establishing historical contextualization for modes of representation.[3] For Winter especially the goal is to presume no single reason for stylistic associations among objects but rather to pinpoint the sources and effects of style by correlating patterns of expression with the experiences of producers. Those experiences can include technical aspects of design activity. Michael Baxandall, in his unparalleled linking of expressive style and experience, explicitly connects representation to perceptions of technical skill and knowledge (as deployed in rendering or measurement, for example) shared by artists and patrons. He thus connects art-making, by way of technical knowledge, also to patterns of influence and prestige in a given social setting. This multilevel analysis of expressive convention—as productive process, symbolic form, and means of social organization—serves as a model for my exploration of architectural change.[4]

We are encouraged through such an approach, as Svetlana Alpers recognized some time ago, to study artifacts without "choosing in advance the parts played by the individual maker, his community, certain established modes of perceiving the world, or the viewer."[5] Considered in such dynamic terms, the celebration of type—as embodied in the standardized factory buildings—emerges as a genre of resemblance with a particular social history. It is a history that reflects at least two fundamental social tensions of mass production. First, the factory designers and builders were offering a product that conformed to prevailing norms of industrial management. The reinforced-concrete factories were simplified and standardized objects that exploited economies of scale, savings deriving from the repetition of forms and processes. But this economization put its suppliers at risk of obsolescence: Would not true standardization do away with the need for experts, for the designers and planners themselves? To protect their standing with clientele, factory designers and builders cast their own work—the work of crafting and implementing standards, of typology—as a rare and elevated competence. A status then accrued to the standardizers, and their occupational authority was bolstered in the competitive world of industrial operations. The blunt functionalism of the reinforced-concrete factory buildings expressed the heightened status of new technical knowledge to a receptive industrial market.

We must ask, of course, why this epistemological elevation of typology "worked"—why the designers and builders found in industrialists a willing audience for their lofty self-identification, for their services, and for the higher fees such specialized services might command. Here we find the second, and perhaps larger, political significance of functionalist design. The elevated status of "standardizers" dovetailed with a vast redistribution of skills, credit, and opportunity in the industrial workplace after 1900, by which many rank-and-file workers found themselves laboring without intellectual reward or occupational mobility; others, like the factory designers, ascended to secure planning or managerial roles. Factory owners shared the stratified vision of productive labor embodied in the builders' self-concept. The same patterns of social change undergirded the production (i.e., mass production) of a modern, utilitarian architecture, *and* industrialists' enthusiasm for a functionalist building style.

Design, always a blend of social and cultural operations, is here specifically a product of and a signifier of technical expertise, each role supporting the other in an involution of technical practice and reputation. The new factory buildings may appear to have been, and indeed were, simpler in form than their predecessors, but standardization was as richly determined and promising a stylistic choice for its promoters as more individualized aesthetic gestures were for conventional architects. This essay considers the origins and consequences of that commitment to uniformity.[6]

PRODUCTION OF THE REINFORCED-CONCRETE FACTORY BUILDING

To arrive at this historic contextualization of architectural uniformity we need first to map the ways in which labor—conceptual and physical—was organized in the creation of these buildings. A handful of prominent architects created notable innovations in reinforced-concrete factory building technology and design after 1890. Ernest Ransome[7] and Albert Kahn,[8] in particular, have garnered the attention of historians. Other architectural firms that achieved celebrity for their industrial commissions in these years include Purcell and Elmslie, Pond and Pond, and Schmidt, Garden and Martin. However, reinforced-concrete factories were often built without the involvement of well-known architects or any architects at all. The vast majority of these buildings were designed and erected within a world of commercial transactions rather than cutting-edge engineering or name architects. Lesser-known firms learned of new technologies and designs through trade publications and professional organizations and through patents taken out by leading designers, and then disseminated the structures to locales around the country.

In the first decades of the twentieth century, services of the factory designers reached the market in three ways. The industrialist commissioning a plant could employ his own forces for all construction work. He would in this case enlist an engineer or architect to draw up plans, hire subcontractors for specialized work, and assume all responsibilities for erecting a plant.[9] A second option involved the factory owner soliciting plans and specifications for a factory building from an engineering firm and then submitting them to prospective building concerns or general contractors for bids. The engineering firm would coordinate the work of the winning contractors. This approach was substantially easier on the owner than taking on supervisory tasks himself, but still entailed a fairly close involvement.

A third option removed the building owner most thoroughly from the construction process and showed the greatest growth in popularity among industrialists who bought factories at this time. This was the hiring of building firms that incorporated an engineering division equipped to design factory buildings and a construction division able to erect the buildings from start to finish. Such firms usually maintained separate departments for promotion, drafting, estimating, accounting, purchasing, expediting, and construction. With these facilities a building firm could select the best site for a client after having its own staff study local geographic, supply, and labor conditions, and then coordinate every aspect of construction from excavation to final painting.

A number of the engineering firms that operated along these lines were very successful. Perhaps best known today are the international concerns Lockwood, Greene Company and Stone & Webster (see Figure 3). Other firms of slightly smaller size performed similar services on a regional basis, particularly in the Midwest and Northeast where the growth of manufacturing industries was substantial between 1900 and 1930. Their functional departmentalization made the engineering/building firms kin to other streamlined mass-production industries of the day and no doubt helped create their appeal for industrial clients. Not only were the complexities of dealing with bids and subcontractors eliminated for factory owners who turned to the modern building firms, but the costs added as each contractor and subcontractor sought profit were also removed. In his 1931 report on American construction trades, William Haber summarized the advantages that the integrated engineering/building firm held for factory owners. Purchasing, planning, and expediting were each conducted by a specialized department with the latest methods and machinery. Such unification and centralization allowed the multi-function construction company to exploit economies of scale and the emerging art of coordinating production tasks.[10]

Significantly, Haber concludes his discussion of the integrated engineering/building firms with a further indictment of smaller-scale methods of project management:

> No study has been made of the amount of time lost by workmen through failure in material deliveries, but from the meager evidence available it seems to be tremendous. With the same modern scientific organization in charge of construction, the contrast between its operations and those of the "broker" contractor becomes more striking.[11]

It is not simply the large size of the integrated firms that brings them success, but their "scientific" nature; in Haber's use of the word "broker" there is an intimation of *undeserved*

Paramount Knitting Co., Kankakee, Ill.
Combined warehouse and manufacturing space. Built 1914

Aluminum Goods Mfg. Co., Newark, N. J.
Distributing warehouse and branch factory. Built 1916

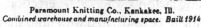

McAvity, T., & Sons, Ltd.		St. John, N. B.
McNeal Machinery Co.		Joplin, Mo.
Mechanical Rubber Co.		Chicago, Ill.
Merchants Terminal Warehouse Co.		New Bedford, Mass.
Montgomery Ward & Co.		St. Paul, Minn.
Morgan & Wright		Detroit, Mich.
Nashua Mfg. Co.		Nashua, N. H.
Nassau Smelting & Refining Co.		Tottenville, N. Y.
Naumkeag Steam Cotton Co.		Salem, Mass.
New Bedford Cotton Mills Corp.		New Bedford, Mass.
New Departure Mfg. Co.		Bristol, Conn.
New England Waste Co.		Revere, Mass.
N. Y. Belting & Packing Co.		Passaic, N. J.
Otis Co., The	1898	Ware, Mass.
Pacific Mills (Hampton Department)		Columbia, S. C.
Pacific Mills (Print Works Department)		Lawrence, Mass.
Pacolet Manufacturing Co	1900	Pacolet, S. C.
Palmolive Co., The		Milwaukee, Wis.
Paramount Knitting Co.		Kankakee, Ill.
Passaic Print Works	1890	Passaic, N. J.
Pepperell Manufacturing Co.		Biddeford, Me.
Pickett Cotton Mills		High Point, N. C.
Planters & Mfrs. Warehouse Corp.		Baltimore, Md.
Poe, F. W., Manufacturing Co	1895	Greenville, S. C.
Quinebaug Co.		Danielson, Conn.
Rice, Barton & Fales Machine & Iron Co.		Worcester, Mass.
Saco-Lowell Shops		Lowell, Mass.
State Board of Control of Wisconsin		Waupun, Wis.
Staten Island Shipbuilding Co.		Port Richmond, N. Y.
Studebaker Corporation, The		South Bend, Ind.
U. S. Cartridge Co.		Maurer, N. J.
Vassie & Company, Ltd.		St. John, N. B.
Walker, Wm H., & Co.		Buffalo, N. Y.

Massachusetts Cotton Mills, Lowell, Mass
Cotton storehouse of 40,000 bales' capacity. Built 1909

Naumkeag Steam Cotton Company, Salem, Mass.
Concrete storehouse built 1914 following the Salem fire

Figure 3.   Illustrations of factories and partial list of buildings erected by
the Lockwood, Greene Company, from their 1925 Annual Report.

profit. He later associates the work of small firms and independent contractors with "excessive competition" that "puts a premium on astuteness and disloyalty rather than engineering skill."[12]

This disparagement of independent contractors and small building concerns was not unique to Haber. The makers of one brand of steel reinforcing for concrete advertised in 1920 that they "would not license contractors or materials dealers." They wished to place their products in the hands of acknowledged experts only. The practice of obtaining free plans for factory buildings from steel suppliers and even from insurance companies was also losing favor among factory owners. The erection of the efficient, economical factory building was coming to be associated with firms that were at once specialists in this type of product and comprehensive, integrated enterprises able to handle every aspect of factory construction.[13]

KNOWLEDGE IN THE MARKETPLACE: THE CONSTRUCTION SITE

To understand the success of the factory-building firms, we need to see these characteristics as parts of a consciously pursued program. The identity of the reinforced-concrete building firm was a doubled one. It contained a promise of rationalized, streamlined operations that would bring clients efficiency and savings. Such savings were predicated on a minute and hierarchical division of labor, as might be found in the most carefully organized manufacturing enterprises of the day. But the reputation of concrete construction firms also held intimations of a rarified and subjective knowledge about factory building—a body of knowledge not subject to division. This doubled character reflected the builders' devotion to a particular social organization of labor that not only retained a place for their own expertise, but also represented a social vision shared by the industrialists who sought their services. We will first consider the industrialized features of concrete factory construction, and then the claims that seemed to bring factory builders an immunity to the very deskilling and routinization they advocated for other forms of technical labor.

In many ways the conditions of concrete construction resembled those of contemporary factory operation far more closely than they did conventional building methods. In the early part of the twentieth century, the most common procedures for reinforced-concrete construction involved bringing the raw materials of concrete to the construction site, mixing them there to create the pourable medium, and then filling reusable wooden or metal molds, or forms, that had metal reinforcing rods pre-positioned within them. When the concrete in the forms had hardened, the forms were removed and relocated, and a successive floor or section of the building was erected in a virtually uninterrupted sequence. This procedure echoed emerging methods of factory production that sought to replace batch, or unit-based, production with continuous "flow" processes.

Further, as had rapidly become common practice in manufacturing contexts, a division of physical and conceptual labor emerged in reinforced-concrete construction. On the majority of concrete building projects the actual handling of materials—the erection of forms, the preparation and placement of reinforcing rods, and the delivery, distribution, mixing, and pouring of concrete—was accomplished by a large body of relatively untrained workers, distinct (with the exception of some carpenters who built wooden forms) from the established body of experienced (and often unionized) workers commonly employed in masonry and wood construction at this time. Supervising these workers were managers employed by building firms or contractors. These managers might themselves be hierarchically divided into field supervisors who were in turn directed by office-based designers and administrators, but however organized internally this conceptual realm remained a purview distinct from the physical labor of concrete construction.

Carrying forward the rationalization process, builders gradually removed many aspects of concrete construction from the building site after 1900. The construction of forms and assembly of reinforcing rods increasingly were taken over by outside suppliers.[14] These auxiliary businesses, located off the construction site, mass-produced materials that previously had been individually fabricated in the course of building. Some intricate types of forms and reinforcement continued to be fabricated by workmen on the building site, but enough were standardized and mass-produced to effect substantial economies. These products were often called "systems" by their promoters, suggesting that their use also would save builders the conceptual tasks of understanding and planning the use of reinforcement.[15]

A second set of managerial initiatives in reinforced-concrete construction after 1900 achieved the tightened control of work that remained on the site, and reiterates the close relationship of productive process and social vision on the part of construction firm managers. The daily supervision of construction forces was brought about through the use of elaborate administrative procedures—ranging from mnemonically coded work orders to studies of workers' hygiene habits.[16] Building firm operators also lavished attention on problems of quality control on the construction site. Because concrete construction was operated as a flow process, costs incurred by faulty or wasted materials could accelerate rapidly. From its inception as a favored material for large commercial projects, concrete was subject to testing in the field. The scope and means of testing derived from university-based materials science programs. Academics, working as members of professional associations and as paid consultants to the building and materials industries, developed an elaborate body of field inspection and testing procedures.

In keeping with the building firms' modern managerial approach, tasks of quality control remained always outside the purview of the concrete laborer. Only consulting and mid-level salaried engineers performed the work of cement and concrete testing, visiting the site as needed. Building firm managers deemed quality control to be largely a conceptual, rather than physical, task.[17]

So was the majority of work defined and organized on the concrete building site of the early twentieth century. Technical knowledge was treated by building firm operators as a commodity. No technical task escaped division and delegation in the firms' search for fast, efficient, predictable factory construction—except, that is, those tasks that firm owners and operators claimed as their own. *Their* work, their technical knowledge, somehow occupied a universe impervious to the economizing, reductive trends of industrial routinization. Industrialists seeking new plants accepted this description of factory-building expertise and willingly paid the costs of employing its claimants. Standardization stopped here. We can now ask why and how this protective encapsulation of factory-building expertise came to be.

KNOWLEDGE IN THE MARKETPLACE: THE BUSINESS OF BUILDING

The essence of the high status obtained by the factory-building experts was a further commodification of knowledge and technique, but one that strategically defined their work as necessarily comprehensive, or, indivisible. While they defined and stratified the labor needed to erect factory buildings, the firms successfully combatted the idea that standardization or mechanization of construction could effectively be applied without their oversight. This involved celebrating, in a number of ways, the subjectivity of standardization as an intellectual project. As we will see, in achieving their monopoly of reinforced-concrete factory construction through such means, the experts also brought about the high cultural valuation of their product: the functionalist industrial building.

The concrete builders' campaign for secure occupational status took the form of a vast promotional initiative. One major tactic used by the integrated factory-building firms was to distinguish the expertise of the specialized factory designer and builder from that of the building's owner. One engineer reminded manufacturers in 1911 that when they selected established engineers to design and erect their plants, it would be clear that "the creative work of the industrial engineer has to do with such matters as are not usually included in the routine experience and work of owner or operator."[18] The founder of a large factory-engineering/building firm, promoting his company in 1919, cast the relationship between industrialist and industrial engineer as similar to that of client and attorney. The analogy suggests that the knowledge of the engineer was necessary to ensure effective business operation, of the highest professional caliber, and most interestingly in this sphere of otherwise routinized production, *not* a matter of repetition (see Figure 4).[19]

Pertinent Facts

WE originated this service.

We have fifteen years' experience in Industrial Structures.

We eliminate the general contractor and save you his profit.

We take only one commission, and that is paid to us by our client.

We do not sell plans. We have no standard plans.

We have never designed two of our plants alike, and never expect to.

Our service is sold as a whole and not in part.

Our references are the firms listed on the opposite page and many others for whom we have built plants, or prepared reports, made investigations or appraisals.

Our whole story may be summarized in the statement that we give you a maximum building in a minimum time and at a minimum cost.

FRANK D. CHASE, Inc.

Industrial Engineers

645 N. Michigan Ave., Chicago - Whitehall Bldg., New York

Figure 4.   Frank D. Chase, "A Better Way to Build Your New Plant"
(promotional publication), 1919.

This meant that effective factory construction required, at the very least, a certain level of expertise. Clayton Mayers, an engineer for the successful Aberthaw Construction Company, described in detail the possible errors that could occur in beam design. He warned of excessive and inappropriate reinforcing practices, specifying that "these errors are not errors in computations, but are errors of careless design and the result is dire waste of materials."[20]

In this declaration, Mayers blamed uneconomical results on selected practitioners, rather than on the existence of specialized practitioners. In so doing he made a careful distinction between the theory and the practice of reinforced-concrete construction. This distinction became a cornerstone of factory specialists' defense of their expert standing. As did other engineers of the day, Mayers pointed to the need for the assistance of knowledgeable professionals in following the growing body of codes and standards for concrete. Standards, and the whole body of standardized systems of reinforcing and concrete construction, posed dangers to the factory designer and owner because, Mayers believed, each building presented "new problems." Only by careful study could the designer achieve effective and economical application of standardized products and procedures. With such rhetoric, concrete experts embedded a practical complexity in a theoretically simplified technology.

In forwarding this type of argument, engineers were beginning to appropriate for their own knowledge and experience the commercial cache that might have attached instead to technical objects themselves. Engineers and other factory design specialists commonly declared that "materials alone do not constitute a system." A. J. Widmer, a consulting engineer who specialized in reinforced concrete, wrote in 1915:

> A staff of experienced engineers is a most essential feature of a true system. The furnishing of reinforcing steel of correct types cannot constitute a system unless the design of the structure is complete in the hands of engineers experienced in the application of those particular types.[21]

According to the engineers, savings were to be had from eliminating the need for skilled labor on the worksite, not from eliminating consulting engineers. Advertising text combined claims for the efficiency and speed of building systems with invocations of "proved experience." As another prominent engineer put it in his 1911 prescriptions for construction using standardized elements:

> the assembling of these materials into final structures and the installation of the equipment would be under the direct control of those who know the exact reason for the provision of every single feature; and their knowledge of future operating conditions enables them to exercise an intelligent discretion that should

result in a more harmonious whole than could result solely through a literal adherence to the most elaborate specifications.[22]

A call for discretion, and a deep knowledge of the "reasons" for technical specifications, welded a systematized and standardized construction method to an almost inchoate set of intellectual abilities purportedly possessed by the engineers issuing that call.[23]

All such rhetoric conveys the tension technical experts perceived between the value of rationalized production methods to their industrial clientele and the potential loss of occupational authority that might follow from the implementation of those methods. We can see the irony of marketplace demands with which the experts contended. Surely the combination of standardized materials and methods with customized applications could have struck the industrialist buying a factory building as paradoxical. Why did building systems and standards exist if not to do away with the necessity for (costly) specialized expertise and for the very presence of inchoate abilities on the construction site?

In the trade literature of the early 1900s, the specialized experts preempted this argument with a sophisticated and exquisitely self-serving conception of how unrationalized (that is, undivided) technical knowledge could work for commerce. Consulting engineer Willard Case articulated the relationship of standardized technologies and engineering expertise. He noted

> a logical and healthy tendency from several causes toward *type classification*, and this has embraced not only the form of design and character of construction, but the exterior architectural treatment as well.[24]

Invoking the notion of type classification was a powerful gesture. It cast standardization and the work of the factory design and construction as tasks of taxonomic distinction. According to this formulation, while construction could involve the same organizational methods that manufacturers used, *specific* bodies of knowledge were not necessarily transferable between different production situations. To refer to engineering and construction work in this way elevated them to the status of scientific pursuits and made standardization seem not a reductive simplification of labor but a complex analytical undertaking. This definition supported the claims of factory specialists that industrial plants "are now based on a logical scientific method of analysis" and that "the business of the engineer is the science of building."[25]

As Case's words indicate, the uniformity of the "typed" factories (again, their resemblance was undisguised by any distinguishing decoration) celebrated this set of skills. The outward form of the buildings asserted the qualifications of their builders, their vital competencies in a commercial context prone to the devaluation of technical skills. The reinforced-concrete factory buildings were both the product of the modern organization of conceptual and physical labor and public symbols of that organization. Not only do variety, idiosyncracy, and historicizing reference lose their status in such a calculus of architectural expression, but uniformity becomes a hallmark of intellectual achievement, occupational success, and social influence. The buildings' resemblance was actually constitutive of *occupational difference*. Spreading this ideology was not so much an ironic as a necessary, and possibly brilliant, gesture by the ambitious professionals who designed, built, and marketed industrial architecture in the new century.

THE AESTHETICS OF TYPE

This discussion has tried so far to demonstrate the centrality of engineering practice to the form assumed by the American industrial landscape in this century. By rooting the emergence of modernist factory design in the organization of construction and design work it has proposed a highly specific association between two spheres of cultural activity. Such specificity is recommended by Peter Galison in his discussion of architectural and philosophical participation in the European *Aufbau* movements of the first half of this century.[26] Examining the attitudes of Bauhaus architects and of proponents of logical positivism, he suggests that links between the two "arenas of culture" arise from a set of cultural meanings shared by the two factions—a set of powerful, if not fixed, images and aspirations that reflect a common vision of contemporary technologies. In a slight variation on Galison's formulation, I attribute to one group—the factory-designing engineers—a pair of cultural meanings for technology. One meaning was grounded in the practical sphere, one in the expressive sphere, and in the discourse of cultural accomplishment each meaning could be used to bolster the other. Therefore, to complete our picture of the early-twentieth-century factory designers at work, we should note that as they pursued the efficacious use of concrete for affordable factory buildings and bolstered their own standing in the marketplace, these men also claimed an explicitly aesthetic significance for their products.

We can first note that the building firms and consultants who designed and erected the reinforced-concrete factories were neither ignorant of contemporary architectural fashion nor dismissive of its demands. Promoting the minimal use of traditional materials and ornamentation in these buildings, builders of reinforced-concrete factories crafted aesthetic arguments for functionalist design against a backdrop of vigorous critical debate in the architectural press. The specific terms of ideological exchanges among critics and architects of 1900 ranged from the formalist to the moral. Advocates of modern, utilitarian design and promoters of eclectic, historicizing architecture accused one another of aesthetic ineptitude, antisocial behavior, and even antidemocratic intent. The sweeping nature of their concerns grew from their conceptions of

how industrialization would transform American life and culture. Critics, public figures, and professionals of all kinds assessed American prospects in the new century.
They rooted the progress or the imminent demise of American culture in the growth
of mass production and mass consumption and encroaching subordination of all other
endeavors to these goals. Depending on the interpreter, American arts and letters—
including architecture—and the pursuit of an orderly modern society could be
expected to flounder or flourish in tandem amid these changes.[27]

Builders of reinforced-concrete factories entered the critical fray to praise the aus-
tere structures to critics and the larger audience of potential factory buyers. Although
promotional literature produced by factory-building firms never failed to mention the
efficiency and economy of concrete construction it also offered explanations, praise,
and justification for the appearance of the factories in answer to prevailing critical
debates. The factory builders joined those analysts who claimed a favorable prognosis
for American culture in the new era of mass production. Their buildings would be part
of modern culture and challenge the rear-guard assumption that only conventional
academic practice could yield buildings of architectural significance. The factory
builders' arguments addressed all the generalities in which contemporary architectural
experts trafficked, listing advantages to the modern factory that included the "intrinsic value" of a well-designed building and the benefits of health and contentment for
factory workers.[28]

At this point we might begin to see a link between the two sets of cultural meanings
given to the new factories by their creators: a conservatism unites the builders' practical and aesthetic agendas. Each advantage to utilitarian factory design mentioned
above indicates a portion of the factory builders' ideology of modernity, yet in no way
did their program challenge foundational precepts of aesthetic accomplishment in the
United States. First, creating the case for the visual "pleasure" that a well-designed
factory might bring to "the discerning,"[29] factory builders offered self-justifying discussions of factory design in publications of the cement trade and factory management.
The content of this literature might be described as an association of the reinforced-
concrete factory's constitutive elements—the exposed concrete column, the standardized steel-sash window, and all the other simplified, repetitive forms typically used in
this type of construction—with traditional architectural values of visual beauty and
harmony.

Similarly, factory builders and architectural critics were formulating new ideas of
what constituted good design and, more broadly, what constituted contributions to
American "taste" or culture, but as they did so they extended an old aesthetic
premise—that certain kinds of architectural forms were appropriate for buildings of
certain functions—to a contemporary situation. For architects and critics, this aspect
of "realism" was largely a matter of taste. A commentator writing in American Architect
in 1909 explained bluntly that "a free use of intricate detail or expensive materials in a

soap factory would be mere affectation."[30] Factory builders, on the other hand, saw a second, and distinctly conservative, reason to express through a building's form "the purposes for which it is intended." Both groups believed that the material nature of a building can have as full an expressive meaning as any other architectural convention, but factory builders also believed in the "advertising value of a handsome plant in the path of national travel." That value stemmed from the factory's identification with industrial processes it contained. If the appearance of the factory conveyed economical and repetitious production methods, unencumbered by superfluous detail or disguise, anyone encountering the structure might see in it the modern attitudes of the building's operators, and thus deduce the nature of the work conducted within. Such buildings would have a "definite effect for good . . . upon customers and as an advertisement to those who pass it."[31]

CONCLUSION

In this happy blending of culture and commerce the outward form of the reinforced-concrete factory building reiterates the organization of labor under which it was created and with which it operates. We see a political compatibility to the technical, commercial, and aesthetic aspirations of the factory engineers.

We also see that "functionalism" can comprise not only a frank architectural expression of the material nature of a building and an expression of a building's function, but also the builders' (hoped-for) function within a market or community.

To summarize: as they solved the practical problems of an expanding production sector, the technical occupations forwarded a hierarchical vision of American industry that reduced the autonomy and opportunities of the great majority of industrial employees while establishing a secure niche for their own services. The introduction of simplified and standardized production processes and goods displaced established productive trades, created a broad stratum of low-paid positions that offered little hope of training or advancement, and at the same time brought employment to university-trained engineers.[32] The enthusiasm of these experts for an overt expression of new technologies and materials in the outward forms of buildings and manufactured goods was not an accidental by-product of economic expediency and technical problem solving but an expression of this new social order—certainly modern in a narrow high-cultural sense, but not necessarily progressive in any broader cultural sense. This originary aspect of the modernist aesthetic reflects what David Harvey refers to as modernism's "real nether side," which lay, he writes,

> in its subterranean celebration of corporate bureaucratic power and rationality, under the guise of a return to surface worship of the efficient machine as a sufficient myth to embody all human aspiration.[33]

To give less weight to the reinforced-concrete factory buildings, and other seemingly mundane products of modern industrial enterprise, would be to shortchange drastically the ambitions of their creators, and deflect any possibility of understanding the social alterations wrought by that enterprise.

Notes

1. See Reyner Banham, *A Concrete Atlantis* (Cambridge, Mass.: MIT Press, 1986); and Terry Smith, *Making the Modern: Industry, Art, and Design in America* (Chicago: University of Chicago Press, 1993), pp. 57–92. Banham sees an aesthetic appreciation of utilitarian structures arising after Le Corbusier's 1927 enthusiastic appraisal of American grain elevators and factory buildings. Banham writes of the reinforced-concrete factory buildings that "hard-nosed patrons and the architects who served them . . . had no ideological axes to grind, no revolutionary postures to maintain" (p. 53). Revolutionary perhaps not, but ideological without question.

2. Banham, *A Concrete Atlantis*, p. 6.

3. Arnold I. Davidson, "Styles of Reasoning, Conceptual History, and the Emergence of Psychiatry," *The Disunity of Science*, ed. Peter Galison and David Strump (1996, Stanford University Press), pp. 75–100; Irene J. Winter, "The Affective Properties of Styles: An Inquiry into Analytic Process and the Inscription of Meaning in Art History," in this volume.

4. Michael Baxandall, *Painting and Experience in Fifteenth Century Italy* (Oxford: Oxford University Press, 1972), pp. 1–34.

5. Svetlana Alpers, "Style is What You Make it: The Visual Arts Once Again," in *The Concept of Style*, ed. Berel Lang (Philadelphia: University of Pennsylvania Press, [1987] 1979), p. 114.

6. Guidance for this project can also be found in recent work in the history of science and the history of technology that considers the implementation of standardized practices and the creation of standardized forms (for instruments, tools, or finished goods). Both disciplines have found complexities within processes of industrial rationalization; see Michael Nuwer, "From Batch to Flow: Production Technology and Work-Force Skills in the Steel Industry, 1880–1920," *Technology and Culture* 29 (1988): 808–38; Geoffrey Bowker and Susan Leigh Star, "Knowledge and Infrastructure in International Information Management," *Information Acumen*, ed. Lisa Bud-Frierman (London: Routledge, 1993), pp. 187–213; Robert Kohler, *Lords of the Fly: Drosophila and the Experimental Life* (Chicago: University of Chicago Press, 1994); Karen Rader, "Making Mice: C. C. Little, the Jackson Laboratory, and the Standardization of *Mus Musculus* for Research" (Ph.D. dissertation, University of Indiana, 1995). See also: Gail Cooper, "Custom Design, Engineering Guarantees, and Unpatentable Data: The Air Conditioning Industry, 1902–1935," *Technology and Culture* 35 (July 1994): 506–36; Amy Slaton, "Origins of a Modern Form: The Reinforced-Concrete Factory in America, 1900–1930" (Ph.D. dissertation, University of Pennsylvania, 1995).

7. Ransome (1844–1917) began his career in California in the 1870s addressing the difficulty of building to resist earthquakes. In 1885, he constructed a flour mill with floor slabs cast integrally with supporting beams, each beam reinforced in its tension zone with a single rod. By 1889, he had developed a homogenous system of floor construction in which girders, beams, and slabs of a given bay were cast as a unit on concrete columns (Ernest L. Ransome, "Reminiscence," *Reinforced Concrete Buildings* [New York: McGraw-Hill, 1912]; reprinted in Howard Newlon, Jr., ed., *A Selection of Papers on Concrete* [Detroit: American Concrete Institute, 1976], pp. 291–93; Henry J. Cowan, *The Master Builders*, vol. 1, *Science and Building: Structural and Environmental Design in the 19th and 20th Centuries* [New York: John Wiley & Sons, 1977], pp. 36, 79).

8. Albert Kahn (1869–1942) established his practice in Detroit. In addition to industrial commissions, he worked on residential, institutional, and civic buildings. His brother, Julius Kahn (1874–1942), founded the Kahn Trussed Concrete Steel Company in Detroit, and established affiliates of this

business around the world while working closely with Albert. For information on the Kahns, see Grant Hildebrand, *Designing for Industry: The Architecture of Albert Kahn* (Cambridge, Mass.: MIT Press, 1974); Smith, *Making the Modern*; Peter Conn, *The Divided Mind: Ideology and Imagination in America, 1898–1917* (Cambridge: Cambridge University Press, 1983), pp. 214–17; and Federico Bucci, *Albert Kahn: Architect of Ford* (New York: Princeton Architectural Press, 1993).

9. Charles Day, *Industrial Plants: Their Arrangement and Construction* (New York: Engineering Magazine, 1911), pp. 79–80; Harold V. Coes, "Better Industrial Plants for Less Money," *Factory: The Magazine of Management* 17 (January 1917): 22.

10. William Haber, *Industrial Relations in the Building Industry* (Cambridge: Harvard University Press, 1930) (reprint edition: New York: Arno and the *New York Times*, 1971), p. 58.

11. Haber, *Industrial Relations*, p. 59.

12. Haber also claims that smaller firms were prone to getting involved in "unsound" competitive practices and that for this reason building trade unions preferred to work with the large integrated building firms (ibid., pp. 61, 72).

13. Advertisement for Barton-Spider Web System, *Sweet's Catalog* 1920, p. 220.

14. As one engineer summarized in 1906, the essence of economy in concrete was to be found in the duplication of forms and the elimination of architectural details that complicate form construction (Ross F. Tucker, "The Progress and Logical Design of Reinforced Concrete," *Concrete Age* 3 [September 1906], p. 333).

15. In 1906, *Cement Age* published a review of ten commercial systems of reinforcement; by 1914, dozens of firms advertised systems of preassembled reinforcement in *Sweet's Catalog*. Mass-produced steel reinforcement became so affordable by the early 1910s that even large construction firms ended production of their own reinforcing rods. See "The Story of Aberthaw" (unpublished manuscript in the archives of Aberthaw Construction Company, North Billerica, Mass.). The costs to building firms of machine-fabricated forms or reinforcement were offset by savings in labor on the construction site and by the prevention of excessive wood or steel consumption. By 1903 builders could make use of precast concrete beams, columns, and floor slabs, as well, avoiding the exigencies of erecting forms and of pouring concrete above ground level. Elements were cast on the ground with reinforcement in place and once set, rapidly assembled by relatively low-paid, little-trained workers.

16. See Slaton, "Origins of a Modern Form," Chapter 4: "The Business of Building: Technological Choices and Organization of the Concrete Construction Firm," pp. 192–261.

17. Trade literature of the concrete industry encouraged this division of labor. Engineers wrote of the ineptitude of field workers and the trustworthiness of their own profession, and thereby placed blame for structural failures on laborers and credit for technical success on their own proficiency. See Slaton, "Origins of Modern Form," pp. 117–44.

18. Day, *Industrial Plants*, p. 18.

19. Frank D. Chase, *A Better Way to Build Your New Plant* (Chicago: Frank D. Chase, Inc., 1919), p. 4.

20. Clayton Mayers, *Economy in the Design of Reinforced Concrete Buildings* (Boston: Aberthaw Construction Company, 1918), p. 5.

21. A. J. Widmer, "Reinforced Concrete Construction" (Illinois Society of Engineers and Surveyors, 1915), p. 148. Widmer advocated the Kahn System by name. This raises the question of whether engineers had a certain fidelity to a given system and whether there was any mutual commitment on the part of the fabricating company.

22. Day, *Industrial Plants*, p. 96.

23. See, for example, advertisement for the Roebling Construction Company, *Sweet's Catalog* 1907–1908, p. 148. Ironically, the concrete engineers' message was reinforced by other building trades that felt threatened by the introduction of systemized building methods. The president of the Bricklayers,' Masons' and Plasterers' International Union said in 1905:

> Concrete construction is a dangerous undertaking and requires the most skilled and intelligent direction. Our organization has contended that the adoption of the various concrete systems

used in construction are experimental and uncertain at best, and the work requires the most skillful mechanics.

(President Bowen, *Bulletin*, Building Trades Association, New York, September 1905, p. 228, cited in Haber, *Industrial Relations*, p. 39.)

24. Willard L. Case, *The Factory Buildings* (New York: Industrial Extension Institute, 1919), p. 254, emphasis added.

25. Day, *Industrial Plants*, p. 4; Case, *Factory Buildings*, p. 255. On the role of generalized knowledge in the creation of modern professions, see Theodore Porter, "Objectivity as Standardization: The Rhetoric of Impersonality in Measurement, Statistics, and Cost-Benefit Analysis," *Annals of Scholarship* 9 (1992): 30.

26. Peter Galison, "The Cultural Meaning of *Aufbau*," *Scientific Philosophy: Origins and Developments*, ed. F. Stadler (Dordrecht, Boston: Kluwer Academic Publisher, 1993), pp. 90–91.

27. Richard Wightman Fox and T. J. Jackson Lears, eds., *The Culture of Consumption: Critical Essays in American History 1880–1980* (New York: Pantheon, 1983); and Conn, *Divided Mind*.

28. These phrases appear in "Utilitarian Structures and Their Architectural Treatment," *American Architect* 96 (November 10, 1909): 183; the sentiments are echoed throughout contemporary discussions of factory design in architectural and technical journals and in books on factory operation.

29. "Utilitarian Structures," p. 185.

30. "Utilitarian Structures," p. 183. In 1921 another critic wrote more calmly that "A building should indicate by its exterior treatment and design something of the purposes for which it is intended. The indiscriminate use of decoration and color should be avoided in the design of an industrial buildings" (Arthur J. McEntee, "Recent Development in the Architectural Treatment of Concrete Industrial Buildings," *Architecture* 43 [January 1921]: 18).

31. "Utilitarian Structures," p. 186; *The Library of Factory Management*, vol. 1, *Buildings and Upkeep* (Chicago: A. W. Shaw Co., 1915), p. 87.

32. For detailed discussions on the stratification of labor in the United States after 1900, see David Noble, *America By Design: Science, Technology, and the Rise of Corporate Capitalism* (Oxford: Oxford University Press, 1977); David Gordon, Richard Edwards, and Michael Reich, *Segmented Work, Divided Workers: The Historical Transformation of Labor in the United States* (Cambridge: Cambridge University Press, 1982); and Harry Braverman, *Labor and Monopoloy Capital: The Degradation of Work in the Twentieth Century* (New York: Monthly Review Press, 1975).

33. David Harvey, *The Condition of Postmodernity* (Cambridge, Mass.: Blackwell Publishers, 1989), p. 36.

Many of these machines are gendered, if not sexed.
With product names like "Handy Hannah foot mas-
sager" and "Stenorette," they are feminized machines
that project the desire to become female for the male
who controls them. Others are neuter, but made for
the domestic sphere controlled by women. There's a
whole sub-history of gender in these machines.

CAROLINE JONES

When I first showed Faraday's, I had the idea that people would be overwhelmed by
all the noise and chaos—I thought they'd want to get out of there as fast as possible.
But most people stayed for quite a while, trying each appliance out, learning to
choreograph the switch matting—they'd stay no matter how noisy it got. I didn't
choose that; in fact, I actually wanted to make something that would seem, well,
a little more frightening. And I do think there are still certain eerie moments, when
you're there by yourself—moments of uncertainty. You're not always sure that
you're triggering the machines; sometimes they seem to have lives of their own.
So maybe we're not always the ones controlling our own technology; we're just
part of a system, a circuit. And your body completes the circuit.

PERRY HOBERMAN

The Body

ARNOLD DAVIDSON

Miracles of Bodily Transformation, or, How St. Francis Received the Stigmata[1]

In this paper I hope to show how the texts and images of St. Francis of Assisi's stigmatization built on one another to provide a persuasive representation of this miracle, a representation, that is, that would actually persuade thirteenth- and fourteenth-century readers and viewers of its reality. A detailed examination of the techniques and modalities of persuasion employed by these writers and artists can help us gain access to a set of profound and wide-ranging stakes that were at issue in these representations and were located at every level of culture. Thus, studying the strategic intervention of discourse and painting in this historical context allows us to understand why the battles fought around St. Francis's stigmatization were so intense and long-lasting, and why so many different resources of rhetorical and pictorial persuasion were deployed around this miracle.[2] No less historically significant, since Francis's stigmatization crucially contributes to making theologically and culturally possible a whole new range of bodily miracles, understanding its representations is a cornerstone in helping us articulate a changing medieval sensibility.

The stigmatization of St. Francis of Assisi allegedly took place on September 14, 1224. As a result of the fact that, and the way in which, this event has become so firmly lodged in the history of Western culture, it is all too easy to forget how extraordinary, exceptional, and even unique an event it was initially considered to be. First of all, it should be remembered that the vast majority of miracles found in the lives of saints are healing miracles.[3] Considered overall, the miracles of saints are generally represented as falling into characteristic types, the prototypes of which are found in the Bible, which increases the authority of the miracle.[4] However, there is no biblical prototype for St. Francis's stigmatization. The word "stigmata" appears only once in the New Testament, in Galatians 6:17, where Paul proclaims, "I bear on my body the marks of Christ" (*"ego enim stigmata Iesu in corpore meo porto"*). Whether or not one interprets this remark as referring to actual physical marks of ill treatment, there is no evidence that Paul is referring literally to the five wounds of Christ. The context of Paul's declaration makes it clear that the marks of Jesus he bears are not to be taken simply as outward impressions, like circumcision, but rather show symbolically that the world has been crucified to him and he to the world.[5] What is central is "the new creation," the fact that Paul belongs to Christ, and these are what his "stigmata" mark; they are not themselves Christ's wounds nor are they in any way miraculous.

St. Francis's stigmatization was represented, both textually and iconographically, as a unique miracle, indeed a miracle greater than any other miracle. It marked, one could say, a new stage in the history of the miraculous. Its purported novelty, its supposed status *sui generis*, provoked deep hostility and incredulity by many different groups of people. Other early-thirteenth-century cases of purported stigmatization were unequivocally rejected by Church authorities, attributed to self-infliction, surrounded by an air of scandal and even heresy.[6] To counter the doubts and denials concerning Francis's stigmatization no fewer than nine papal bulls were issued between 1237 and 1291, three of them in 1237 by Gregory IX, the great patron of the Franciscans, who canonized Francis in 1228.[7] In his bull of April 11, 1237, *Usque ad terminos*, Gregory IX condemned a Cistercian bishop in Bohemia who had expressly denied the stigmatization of St. Francis and prohibited its iconographical representation. The bishop had claimed that "only the son of the eternal Father was crucified for the salvation of humanity and the Christian religion should accord but to his wounds alone a suppliant devotion."[8] In censuring this bishop, Gregory IX referred to Christ's adornment of Francis as "the great and singular miracle" (*"grande ac singulare miraculum"*), words repeated by Alexander IV in 1255.[9] Nor did papal defenses of the stigmata, in response to widespread hostility, end in the thirteenth century. When the Dominicans, unable to counter the official approval of Francis's stigmata, put forth some of their own members as having received this divine gift, they threatened the uniqueness of the miracle worked on Francis's body. In the bull *Spectat ad Romani* of September 6, 1472, Sixtus IV thus was led to prohibit the representation of St. Catherine *"cum stigmatibus Christi . . . ad instar beati Francisci."* ("with the stigmata . . . in the likeness of blessed Francis.")[10] As late as 1522, the author of the *Dialogo del Sacro Monte della Verna*, Mariano da Firenze, was still defending the reality, the uniqueness, and the singularity of Francis's stigmatization. To a doubting Thomas's citation of Galatians 6:17, invoking Paul as a prior case of stigmatization, the author responded that Paul was not speaking literally. This could be established from the fact that Paul is *never* painted with the stigmata: "Paul was painted without them: but as for Francis you see him with the stigmata."[11]

These few examples already indicate how central visual representations were to debates about the stigmatization. Artistic representations played an important role in the diffusion of the theme of Francis's stigmata, and opposition to the stigmata often took the form of opposing such representations or mutilating those that already existed.[12] Chiara Frugoni's remark in an article on the relation between iconography and female mystical visions can be applied as well to the specific case of Francis's stigmatization: "Precisely because the multitude of people are nourished on images and not books—they go to church, look at paintings, hear the exegesis of them in the sermons, but don't directly read the Bible—it is a world of images that is the nourishment of their spiritual life."[13] In order to understand Francis's stigmata and their role in the history of miracles of bodily transformation, we must make use of both images and texts.

I will argue here, although the argument could be extended at even greater length, that as Franciscan hagiography of St. Francis developed, representations of the stigmatization focused on its unparalleled and wondrous character and had the effect of heightening its miraculous status. In response to recurrent doubts and denials, as well as to more general hagiographical and political pressures, these representations were meant to stabilize the status of the stigmata, dispelling any hesitations about its being a singular miracle, special even within the category of the miraculous. The production of these textual and visual depictions culminated in a virtual divinization of Francis, portraying him as a figure whose stigmatization marked him out as distinct even among saints, viewing him as a new Christ, an *alter Christus*.[14] In turn, the presentation of Francis as a new Christ could not but provoke further hostility and incredulity.

The first description of the stigmata themselves, although not of the stigmatization, occurs in the *Epistola Encyclica* of Brother Elias of October 3, 1226, announcing the death of Francis:

> And now I announce to you a great joy, a new miracle. The world has never heard of such a miracle, except in the Son of God, who is Christ our Lord. A little while before his death, our brother and father appeared crucified, bearing in his body the five wounds, which are truly the stigmata of Christ. His hands and feet were as if punctured by nails, pierced on both sides, and had scars that were the black color of nails. His side appeared pierced by a lance, and often gave forth droplets of blood.[15]

Starting with the claim that this is "a new miracle," Brother Elias unambiguously identifies Francis's wounds with the true stigmata of Christ, thus at once demarcating Francis's uniqueness in terms of his bodily conformity to Christ. Bodily similitude is here inextricably linked to proof of Francis's status. Although the passage from Galatians is alluded to, we see that from the very beginning Francis's stigmata are interpreted to have no precedent "except in the Son of God, who is Christ our Lord." Elias's description clearly implies that only Francis's side wound bled, while the apparent nail wounds in his hands and feet have in themselves little of the miraculous about them, appearing as blackened scars that might look like nails. Although the description is certainly framed in terms of the greatness of the miracle, it does not itself invoke the miraculous structure of the wounds that will be so prominent a part of later descriptions.

Tommaso da Celano's *Vita Prima S. Francisci*, the first biography of Francis, written between 1228 and the beginning of 1229, contains an extensive description of both the stigmatization and the stigmata.[16] Here are the most important relevant passages:

> When he was staying in a hermitage, called Alverna from the place where it stood, two years before he gave his soul back to heaven, he had a vision from God. There appeared to him a man, like a Seraph with six wings, standing above him, with his hands extended and feet joined, fixed to a cross. Two wings were raised above his head, two were extended for flight and two covered his whole body.
>
> When the blessed servant of the most High saw these things, he was filled with the greatest wonder but he did not understand what this was supposed to mean to him. Still he rejoiced very much, and was exceedingly happy because of the kind and gracious look with which the Seraph looked at him, whose beauty was beyond estimation, but at the same time he was frightened in seeing him fixed to the cross in the bitter pain of suffering. Francis arose, if I may say so, sad and happy, such that joy and grief alternated in him. He anxiously meditated on what the vision could mean, and for this reason his spirit was greatly troubled.
>
> While he was unable to come to any understanding of it and his heart was entirely preoccupied with it, this is what happened: the marks of the nails began to appear in his hands and feet just as he had seen them before in the crucified man above him.
>
> His hands and his feet appeared to be pierced in the center by nails, whose heads were visible on the inner side of his hands and on the upper part of his feet, while the pointed ends protruded from the opposite sides. The marks on his hands were round on the inner side and elongated on the outer, and small pieces of flesh looked like the ends of the nails, bent and beaten back and rising above the rest of the flesh. In the same way the marks of the nails were impressed on his feet, and raised above the rest of the flesh. His right side was also pierced as if with a lance, and covered over with a scar, and it often bled, and his tunic and his undergarments were often sprinkled with his sacred blood.[17]

Thomas's description of the stigmata also states that only the side wound bled, but, unlike Elias, his representation of the nail wounds takes on a truly extraordinary character. The wounds themselves assume the appearance of nails, the nail heads and points seeming to come out of the flesh. But not wanting his readers to think that actual nails were driven through and left in Francis's hands and feet, he later makes it clear that "it was wonderful to see in the middle of his hands and feet, not the holes of nails, but the nails themselves formed from his flesh and having the color of iron."[18] One is led to believe that a glance at Francis's hands and feet would produce the impression that real nails protruded from him, but on closer examination one would see that his flesh was miraculously configured into the shape of nails.

Let me turn immediately to the representation of the stigmatization itself, making only a few points that are most central to my arguments. First, I want to emphasize, as other commentators have, that, according to Thomas, Francis's stigmata begin to appear in his hands and feet after the disappearance of his vision "that he had seen a little before in the crucified man."[19] Second, Francis was standing when he received the stigmata—he "arose." Third, Francis did not understand the meaning of his vision; its significance was made known to him by the appearance of the marks of the nails themselves. Fourth, Thomas gives us no causal account whatsoever, natural or supernatural, of the appearance of the stigmata. He describes the vision, Francis's state of mind, and the appearance of the marks of the nails. Nothing he says allows us to make an attribution as to the proximate cause of the stigmata, and, specifically, he does not designate the seraph as the cause. Finally, let me very briefly take up Thomas's representation of Francis's vision.

The vision is of a man who appears as a seraph, his hands extended and feet joined together, in a standard iconography of crucifixion, and he is affixed to a cross. The six wings of the man-seraph are arranged so that two of them are extended above his head, two are extended for flight, and two are wrapped around his whole body. The most obvious source for the vision of a seraph is Isaiah 6, where Isaiah's vision of the Lord on his throne includes seraphs who "stand in attendance of Him." Without here tracing the narrative and iconographical convergences and divergences between Isaiah's and Francis's visions, I want simply to recall that although the New Testament never mentions a seraph, Pseudo-Dionysius' Celestial Hierarchy places the seraph at the head of the first rank of heavenly beings, consisting first of seraphs, then of cherubs and thrones. The seraph has "the highest place because he is placed immediately next to God, and thanks to this proximity he receives divine revelations and initiations."[20] Pseudo-Dionysius tells us that the seraph that appeared to Isaiah "was able to elevate him to the sacred contemplation that allowed him to see, to speak in symbols the highest essences placed under, next to and around God"[21] and that, specifically, "the angel that had imparted the vision to him transmitted, as far as possible, his own knowledge of the sacred mystery."[22] Furthermore, the seraph is "the principle that comes immediately after God of all sacred knowledge and of all imitation of Him" and thus seraphim are the highest transmitters of divine illumination.[23]

In his *Ecclesiastical Hierarchy*, Pseudo-Dionysius describes the seraphim as standing in assembly around *Jesus*, looking upon him and receiving his spiritual gifts.[24] The appearance of a seraph to Francis would have been a sign of a truly exalted divine vision, a vision conveying the highest divine illumination. Moreover, the derivation of "seraph" from "burning," which indicates "their fiery nature," and which can be found in both Pseudo-Dionysius and Gregory the Great, will play an important role in the mystical interpretation and symbolism of the stigmatization of Francis.[25] The appearance of the seraph in Francis's vision is thus theologically overdetermined, and we shall see that the iconography of the stigmatization raises even further issues of interpretation.

The first pictorial representations of Francis receiving the stigmata occur on two enamel reliquaries from Limoges in 1230.[26] In every respect, except for the absence of the seraph's cross, these earliest depictions faithfully reproduce the verbal account of the stigmatization found in Tommaso da Celano's *Vita Prima*, the only account written before 1230. In these enamels we notice first that unlike the vast majority of depictions, the physical milieu of the stigmatization is not that of a mountainside; this detail is explained by the fact that Tommaso speaks directly only of the "hermitage" called Alverna, nowhere referring to the mountainside that appears in later accounts. Moreover, St. Francis is standing, as in Thomas's account, his posture and gestures those of the *orans* position of prayer. Francis faces the viewer, his head tilted upward and toward the left, and he is obviously not looking at the seraph, who is placed directly overhead. The seraph is in the sky, the celestial space being indicated by the clouds and heavenly bodies that surround him. The seraph is depicted with six wings arranged as Thomas describes them; he has four wounds, on his hands and feet but not on his side, and he is not affixed to a cross. Francis bears all five wounds of Christ, represented by red dots, and, in contradistinction to the visual depictions that were immediately to follow, the side wound is clearly visible. Most importantly, the artist of these earliest images has tried to indicate that the vision of the seraph and Francis receiving the stigmata are not contemporaneous. Not only the placement of the seraph overhead, but, even more significantly, the fact that the scene of the seraph is separated from that of St. Francis by a red line etched in the metal, serves to represent the temporal separation of the vision and the imprinting. This separation of the two scenes, and the arrangement of the two figures, follows precisely Thomas's account. We see no causal interaction between the man-seraph and St. Francis, and so no depiction of the precise cause of the stigmata. These images articulate knowledge, but they have their gaps.

The early and mid-thirteenth century produced a significant number of panel paintings of the life of St. Francis. Here I will only briefly comment on the earliest panel painting of the life of St. Francis, signed by Bonaventura Berlinghieri and dated 1235 (a detail showing the stigmatization scene is in Figure 1).[27] This painting, done for the church of San Francesco in Pescia, contains six scenes from the life of St. Francis, including the first known paintings of Francis preaching to the birds and receiving the stigmata. The background of the stigmata scene contains the hermitage mentioned in Tommaso da Celano, but the physical surroundings are those of a mountainside. It is possible that Berlinghieri knew that Alverna was a mountain, but, more likely, the depiction of Francis on a mountainside was used to convey deep symbolic significance. Three crucial events in Christ's life took place on mountains: the Transfiguration on Mt. Tabor, the Agony in the Garden on the Mount of Olives, and the Crucifixion on Mt. Calvary. References to all three of these events were implicitly, and sometimes explicitly, incorporated into the paintings of St. Francis receiving the stigmata. In this case, I believe that I can show that the most obvious reference is to the Mount of Olives. The seraph is depicted as described in Thomas, his wings red and brown, but he is not fixed to a cross. He is looking straight ahead, not down at Francis, and there is no real interaction or even emotional connection between the seraph and Francis. However, a viewer who did not know the details of the story would have to have concluded that the appearance of the seraph and the receiving of the stigmata were contemporaneous, since Berlinghieri has telescoped the two separate scenes without giving any indication that they were temporally distinct. This simultaneous depiction of the seraph and Francis, the mountainside, and even Francis's praying posture makes this scene an unmistakable iconographical reference to Christ's Agony in the Garden. A thirteenth-century viewer of this painting would have easily made this reference, recognizing the adaptation of this scene to the Agony in the Garden as specifically narrated by Luke. In the Lucan account, when Jesus goes to the Mount of Olives to pray to his Father, he is described as "having knelt down and prayed" and "Then there appeared an angel from heaven to strengthen him"; finally, "gripped by anguish he prayed more intensely; and his sweat became like drops of blood that fell to the ground" (Luke 22:41–44).[28] Thus Francis kneeling and praying on a mountainside when an angel appears to him, followed by an extraordinary physical transformation, directly evokes this scene in Jesus' life that occurs immediately before his betrayal, arrest, and crucifixion.[29] Moreover, Francis is not standing in this scene. His prayer gesture, kneeling with hands (almost) joined, is a posture that was not common until the thirteenth century.[30] The primary meaning of the joined hands, of recollection and of offering oneself in concentrated surrender to God, especially in conjunction with kneeling, was used to express intense devotion to the presence of Christ in the Eucharist.[31]

Thus Francis's posture would indicate a great intensity of prayer. Francis's head is raised and tilted toward the right, his eyes rolled back as if in devout meditation. He is not looking at the seraph, but seems to be recollecting himself and giving himself up to

Figure 1. Bonaventura Berlinghieri, stigmatization scene from Francis of Assisi
(detail from upper left), 1235, wood panel painting, church of San Francesco, Pescia.

God, exactly as Christ does at the Agony in the Garden. Furthermore, only during the thirteenth century did the prayer gesture of kneeling with the hands joined become common in the iconography of the Agony in the Garden, as opposed to the earlier representations of Christ's prayer showing him with hands outstretched rather than joined. Greater focus on this episode in Christ's life in the thirteenth and fourteenth centuries is, no doubt, related to the increased theological reflection on, and devotion to, Christ's humanity. While the Agony in the Garden, with the angel who comforted Christ, served to humanize him, the stigmatization served to divinize Francis. Berlinghieri contributes to this divinization by having Francis's prayer on Alverna parallel the iconography of Christ's prayer on the Mount of Olives, thus brilliantly adapting the appearance of the Isiahean seraph to a New Testament theme. The miracle of the stigmata is the culmination of Francis's life as he reenacts the life of Christ.

The representation of the stigmata themselves, by four black dots on Francis's limbs, is relatively understated compared with later paintings, although the marks are unmistakably visible. This painting, while visually representing the fact of the stigmata, frames it by an interpretation of the whole event of the stigmatization, maintaining this physical fact within the spiritual significance of the event. Equally visible is the absence in Francis's right side of any wound whatsoever. Indeed, none of these early panel paintings depicts a wound in Francis's side, despite Thomas's description. This absence, I believe, itself carries deep symbolic significance, having to do with the symbolic import of Christ's own side wound. Following Tommaso da Celano but adding their own innovations, artists' early representations of the stigmatization exhibit the attitude of unparalleled importance that surrounded this miracle, an attitude that would eventually make Christ the only possible parallel for Francis.

In light of what I have said about these early texts and images of the stigmatization, how could one further increase its status as a miracle? How could one depict it even more miraculously than these early representations did? An answer to this question can be found in the writings of Bonaventure and in the paintings of Giotto.

Bonaventure was commissioned to write a biography of St. Francis in 1260 at the General Chapter of Narbonne. This biography was completed by 1263 and in 1266, at the General Chapter of Paris, Bonaventure's biography was officially approved. Moreover, a decree was passed ordering the destruction of all earlier biographies. Bonaventure's *Legenda Maior* was decisively to influence almost all future representations, both textual and visual, of St. Francis. Nowhere is this effect more evident than in Bonaventure's discussion of the stigmatization. Here is the passage from Bonaventure that parallels the one I have already cited from Tommaso da Celano. After describing the "seraphic ardor of the desires" of Francis, Bonaventure writes:

On a certain morning about the feast of the Exaltation of the Holy Cross, while Francis was praying on the mountainside, he saw a Seraph with six fiery and shining wings descend from the height of heaven. And when in swift flight the Seraph had reached a spot in the air near the man of God, there appeared between the wings the figure of a man crucified, with his hands and feet extended and fastened to a cross. Two of the wings were lifted above his head, two were extended for flight and two covered his whole body. When Francis saw this, he was overwhelmed and his heart was flooded with a mixture of joy and sorrow. He rejoiced because of the gracious way Christ looked upon him under the appearance of the Seraph, but the fact that he was fastened to a cross pierced his soul with a sword of compassionate sorrow.

He wondered exceedingly at the sight of so unfathomable a vision, realizing that the weakness of Christ's passion was in no way compatible with the immortality of the Seraph's spiritual nature. Eventually he understood by a revelation from the Lord that divine providence had shown him this vision so that, as Christ's lover, he might learn in advance that he was to be totally transformed into the likeness of Christ crucified, not by the martyrdom of his flesh, but by the fire of his soul.

As the vision disappeared, it left in his heart a marvelous ardor and imprinted on his body marks that were no less marvelous. Immediately the marks of nails began to appear on his hands and feet just as he had seen them a little before in the figure of the man crucified. His hands and feet seemed to be pierced through the center by nails, with the heads of the nails appearing on the inner side of the hands and the upper side of the feet and their points on the opposite sides. The heads of the nails in his hands and his feet were round and black; their points were oblong and bent as if driven back with a hammer, and they emerged from the flesh and stuck out beyond it. Also his right side, as if pierced with a lance, was marked with a red scar from which his sacred blood often flowed, moistening his tunic and his undergarments.[32]

Unlike Thomas, Bonaventure describes Francis as *praying* on a *mountainside*, and does not describe him as standing when he received the stigmata. Like Thomas, Bonaventure writes that the vision disappeared before the stigmata began to appear on Francis's body ("As the vision disappeared . . ." ". . . just as he had seen a little before"). As for the stigmata themselves, Bonaventure follows Thomas in describing them as "formed from the flesh itself," and even further increases, in ways I shall not discuss here, their miraculous configuration.[33]

Turning now to the most important differences between the *Vita Prima* and the *Legenda Maior*, in the latter the subjective cause of the stigmata is the fire of Francis's love consuming his soul ("the fire of his soul").[34] Bonaventure, for the first time, also attributes a causal role to the vision, which acts as the, so to speak, objective cause of the stigmata: "As the *vision* disappeared, it left in his heart a marvelous ardor and *imprinted on his body marks* that were no less marvelous" (*"et in carne non minus mirabilem signorum impressit effigiem"*). Thus Bonaventure's causal attribution has two components: the subjective state of Francis's soul and the objective nature of the vision itself that, in some unspecified way, impresses the stigmata on Francis's body. As regards the vision, Bonaventure does not speak merely of a seraph and a crucified man, but, absolutely decisively for the later representations of the stigmatization, identifies this crucified man with Christ himself. Thomas of Celano's "He rejoiced very much and was exceedingly happy because of the kind and gracious look with which the Seraph looked at him" is transformed into "He rejoiced because of the gracious way Christ looked upon him under the appearance of the Seraph." The language of Bonaventure's description is extremely important; the Latin uses the words *"Christo sub specie Seraph."* This phrase is highly significant because it echoes the language of the real presence of Christ in the Eucharist, which became dogma in 1215. In the *Tree of Life*, Bonaventure refers to Christ "*sub specie panis*" and Aquinas explains that although Christ is really present in the Eucharist, he is seen not under his own proper species (*sub propria specie*) but rather "*sub specie panis et vini.*"[35] Thus representing Francis as having seen Christ "*sub specie Seraph*" reinforces the idea that Francis had a vision of the real presence of Christ, even if "under the appearance of the Seraph."

As Carlo Ginzburg has argued, in speaking of the Eucharist after 1215, one should not merely speak of a contact with the divine, but of a *presence* of the divine in the strongest possible sense of the word, a "*sur-presence.*" Next to this presence, other manifestations of the sacred paled in comparison.[36] In light of Francis's own devotion to the Eucharist, as expressed for example in the first *Admonition* ("And as he showed himself in the true flesh to the holy apostles, so also he now shows himself to us in the consecrated bread"), and of Bonaventure's insistence on the intensity of this devotion ("His very marrow burned with fervor for the sacrament of the Lord's body . . . tasting, as if intoxicated in the spirit, the sweetness of the spotless Lamb, he was often rapt in ecstacy"), the description of Francis's vision as of Christ *sub specie Seraph* serves to emphasize the reality of the vision, exactly as if Christ were present "in the true flesh."[37]

The new description of Francis's vision and the claim that the vision itself was an agent of Francis's stigmatization are reflected in the iconographical transformations that came in the wake of the *Legenda Maior*. Giotto (or Giotto and his assistants—I leave problems of attribution aside) produced three paintings of St. Francis receiving the stigmata: a fresco in the fresco cycle in the upper church of Assisi, an altarpiece with predella for the Church of San Francesco in Pisa, now in the Louvre, and a fresco in the fresco cycle for the Bardi Chapel in Santa Croce in Florence. All three paintings merit detailed discussion, especially as regards their differences, but for my purposes here I shall focus on the Assisi fresco, which is Giotto's first such painting, is based directly on Bonaventure, and served as a prototype for many later depictions of this scene (Figure 2). The Assisi fresco shows, I think it is fair to say, a perfect representation of Christ *sub specie Seraph*. (The other paintings decrease this impression.) The six wings of the seraph are arranged in the standard manner, although more of the upper body is exposed, making it clear that there is a human form beneath the wings. Although the face of the man is now faded, it is clearly Christ, his beard and hair as traditionally depicted and his halo fully visible. As if to dispel any doubt whatsoever about the nature of the vision, the caption to the fresco tells us that Francis "*vidit Christum in specie Seraphim crucifixi.*" We are told in this caption that Francis was praying on the side of Mt. Verna when he received the stigmata, although his posture here is not that of any traditional prayer gesture. His hands appear to be in an *orans*-type position, although he is kneeling on one knee. All commentators interpret this posture, and especially the position of the hands, as that required by the way in which

Francis's reception of the stigmata is depicted. But they fail to remark that his hands exhibit the gesture of wonder toward a miracle, and are in this respect an exaggerated form of the gesture found in Giotto's painting of Francis and the cross of San Damiano, where the moment depicted is that of Francis hearing the miraculous voice that descends from the image of the crucifixion. Not only are the vision and the receiving of the stigmata contemporaneous (as in the earlier iconographical tradition), but the vision of Christ (under the appearance of the seraph) is also shown to be what I have called the objective cause of the stigmata. The caption again tells us that it was Christ under the appearance of the crucified seraph who *"impressit in manibus et pedibus et etiam in latere dextro stigmata crucis"* and it goes on to identify these stigmata as those of Jesus Christ. This is the first painting to depict the physical process of stigmatization, five rays of light descending from Christ's stigmata to produce Francis's stigmata. There is no *textual* precedent at all for the depiction of these rays of light. They are, I believe, a complete innovation of the artist.[38] Although the luminosity of saints is often used to represent a divinization of the soul and although some depictions of the Transfiguration show rays of light descending to the disciples, it is unprecedented to see these divine rays of light being used to, in effect, divinize Francis by wounding him with the stigmata.[39] As extraordinary as these rays are, it is difficult to know how else one could visually represent the impression of the stigmata on Francis's body by the Christ/seraph. They are a modality of transmission that accurately captures a sense of *impressit*, while at the same time emphasizing pictorially that these impressions are supernatural. From this time forth, this objective cause of the stigmatization will be continually depicted, while the subjective cause, Francis's burning love for Christ crucified, will recede into the background, at least as far as visual representations are concerned. And since one is here trying to depict the miraculous, moreover a new and singular and disputed miracle, then the visible, indeed tangible, manifestation of the supernatural is necessary. To depict the stigmatization after the vision had disappeared, as the texts describe it, would decrease the effect of the painting as an unambiguous representation of the miraculous. And to fail to imagine the modality of the transmission would allow doubts or questions about precisely how Francis received the stigmata, doubts that are thoroughly dissipated by this painting. By depicting Christ supernaturally and materially transmitting his stigmata to Francis, the miraculous character of the stigmatization is made the focus of the painting. A viewer of this painting could not have failed to have been filled with the wonder of this miracle. The visual innovations of this fresco successfully and magnificently served this purpose. Before I consider one further aspect of this painting, let me note that all five stigmata are visible on Francis's body, including the wound in the right side seen through the opening in his tunic.

Another major innovation in this painting, which also will have profound conse-

*Figure 2.   Giotto, Basilica of St. Francis of Assisi,
ca. 1300, fresco, church of San Francesco, Assisi.*

quences for later representations, consists of the figure in the lower right-hand corner, Brother Leo, a witness present at the very event of Francis's stigmatization. Again, there is no textual precedent at all for the presence of anyone but Francis at the stigmatization. All of the texts have Francis praying alone on the mountainside. Of course, in his insistence on the reality of the stigmata, Bonaventure was greatly preoccupied with the question of witnesses. In *Legenda Maior* he had written:

> Now, through these very certain signs not only corroborated sufficiently by two or three witnesses, but superabundantly by a multitude of persons, God's testimony about you and through you has been made overwhelmingly credible, removing from unbelievers any veil of excuse, strengthening believers in faith, lifting them with trustworthy hope, inflaming them with the fire of charity.[40]

Bonaventure, alluding to Deuteronomy 19:15 and Matthew 18:16–17, both of which require the evidence of two or three witnesses to sustain a charge, transposes the concern with witnesses from criminal law to the authentification of miracles. In the case of Francis's stigmata, we have confirmation not merely by two or three witnesses, but rather confirmation "superabundantly by a multitude of persons."

In his sermon on St. Francis, preached in Paris on October 4, 1255, Bonaventure refers to the *plurality*, the *authority*, and the *holiness* of the witnesses, and he goes on to give a detailed explanation of why these stigmata could only have been miraculous. In speaking of the plurality of witnesses, he tells us that "more than one hundred clerics corroborated with their testimony" these marks on Francis's body.[41] Bonaventure is not overly preoccupied with distinguishing between those witnesses who saw the stigmata on Francis while he was alive, those witnesses who saw the stigmata on his body after Francis's death, and any witnesses who might have seen the process of stigmatization itself. He does, however, give us examples of the first and second categories of witnesses, but nowhere mentions anyone who would have been an example of the third type.[42] Since Bonaventure considered the very form of the stigmata to be miraculous, seeing them should have been sufficient to convince one that a miracle had transpired, for one would have seen nails formed from Francis's own flesh. But even given this miraculous form, how much more compelling would have been a witness to the very event, testifying to the appearance of the Christ/seraph and to the transmission of the stigmata, serving vicariously, as it were, to allow us to witness the event.

In fact, strictly speaking, Brother Leo is not the first depicted witness of the stigmatization. In a painting from around 1280, done by a follower of Guido da Siena, Francis is shown kneeling on both knees, receiving the stigmata from a seraph (not depicted as Christ) who is nailed to a cross. To his right are two small bears. One of them seems undisturbed by the event, but the second bear is unequivocally depicted as a witness to the stigmatization. Although his back is toward Francis, he has turned his head as far as possible toward the left and is looking over his shoulder at the apparition of the seraph. There is no way to interpret the unnatural posture of this bear except to say that he is turning toward the event, straining his head to look at something that has roused him.

Giotto's Assisi fresco does, however, give us the first depicted human presence (besides Francis) at the stigmatization. Brother Leo is in a position to be a confirming witness of what happened during Francis's stigmatization; he fulfills the role of the most proximate possible witness to the event, present while it takes place. It is as if in addition to Bonaventure's claims about the plurality, authority, and sanctity of the witnesses, Giotto has added a claim of proximity on behalf of Francis's closest companion and confessor. But even while the fresco incorporates the most possible proximate witness, the function of this witness remains ambiguous. Were Leo to look in front of him, he would see Francis receiving the stigmata; were he to look directly above, he would see the upper part of the Christ/seraph. But he is not watching the event; he is reading. He has thus become a potential or virtual witness, present at the stigmatization and so *capable* of seeing it as it happens, yet absorbed in reading, at least at this precise moment apparently oblivious to the event.

Although I believe I could show that Leo's reading carries profound symbolic significance, I will not here traverse the detailed hermeneutical path necessary to uncover all of the layers of significance. Most generally, the contrast between Francis praying and Leo reading invokes the contrast between prayer and the study of sacred theology made by Francis in his letter to Anthony of Padua.[43] (The most plausible hypothesis is that Brother Leo is reading the Gospel.) Furthermore, Bonaventure has Francis contrast reading and studying with prayer "after the example of Christ of whom we read that he prayed more than he read."[44] As in Christ's life, prayer takes precedence over reading, so Francis prays on the mountainside while Leo reads, and Francis's praying culminates in his stigmatization, while Leo's reading distracts him from a vision of the supernatural.

At a more abstract level, the iconology of this scene contrasts prayer and lack of watchfulness, which can be represented either by reading or by sleeping. In some later paintings Leo quite literally sleeps, while in others the postures of sleeping and reading are combined. So in the predella to Bellini's Pesaro altarpiece, the witness to the stigmatization has his book propped up, but his head, heavy with sleep, rests on his hand and his eyelids are closed. Lack of watchfulness, represented by sleeping, clearly associates Leo, Francis's disciple, with the disciples of Christ, who slept during the episode of the Agony in the Garden, and who, in the Lucan account of the Transfiguration are also said to be "weighed down by sleep." So on the one hand, while the sleeping or reading of the witness compromises his status as a witness, on the other hand, these very postures identify him with the disciples of Christ. Therefore, the praying Francis is even further identified, by contrast or in opposition to the disciple, with Christ Himself, of whom he becomes a living effigy.

As one might expect, it did not take long for Leo's virtual witnessing to be transformed into actual witnessing. In Sassetta's often copied painting, for example, we see Leo still with a book in his hand; but he is now watching the event of the stigmatization: no longer distracted by reading, his right hand raised in wonder, one of the traditional signs of witnessing a miracle. One could produce a multitude of examples of depictions of the actual witnessing of the stigmatization: seated witnesses, standing witnesses, witnesses hiding from Francis yet still viewing the event, witnesses spatially contiguous to Francis, and witnesses depicted at some distance from him. What these depictions have in common is the representation of an individual who sees what Francis sees, what we see depicted in the painting, and who reacts with the surprise and awe that one would expect, precisely the emotions that the paintings are intended to arouse in their viewers. Furthermore, if there could be one witness to the stigmatization, nothing should prevent there from being more than one. And so, for instance, in Domenico Ghirlandaio's fresco of the stigmatization in Santa Trinità, Florence, proximity of witnessing and plurality of witnesses have been, as it were, joined, so that we see a number of witnesses viewing the stigmatization from different positions and different distances. All of these variations on the theme of witnessing, even with all of their significant differences, have as their overarching aim to attest to the reality of the miracle, to witness it and to allow us to witness it, and to convey symbolically Francis's uniqueness as the image of Christ.

Another important conclusion that we can draw from this iconography concerns the nature of the vision itself. According to a typology that goes back to Augustine, visions are divided into corporeal, imaginative, and intellectual. Corporeal visions involve an external sensible form; imaginative visions are sensible visions completely circumscribed within the imagination; intellectual visions involve a supernatural consciousness that is produced without the aid of internal or external impressions or forms.[45] None of the texts on Francis's stigmatization make direct reference to this typology of visions. There is no doubt that the vision does not conform to the model of an intellectual vision, but there has been much dispute about whether it should be classified as an imaginative or corporeal vision. Many commentators have agreed with Octavian Schmucki that the vision "did not affect the external but only the internal senses, and therefore it neither had nor could have had true eyewitnesses."[46] Although I believe that Bonaventure's text describes the vision as a corporeal one, since only a vision of that kind could *impress* the marks of the stigmata on Francis's body, it would take a great deal of detailed exegesis to establish that conclusion.[47] The iconography of the stigmatization much more directly depicts the vision as a corporeal one. An imaginative vision, being produced in the beholder's imagination, could not be seen by other people. If more than one person sees the vision, then it must be a corporeal vision, whereby the object seen exists outside the people beholding it.[48] Thus the witnessing of the stigmatization by persons other than Francis testifies to the corporeal nature of the vision. Here we have another reason to attend to the significance of the description of the vision as "Christ under the appearance of the Seraph." It was widely argued that after Christ's ascension to heaven, he no longer appeared bodily, since that would have required him to leave heaven.[49] He either appeared imaginatively (or intellectually, of course) or under a species other than that of his own body, as when he appears in the Eucharist under the form of bread and wine. Thus there would be no theological problem in having Christ appear to Francis corporally "under the appearance of the Seraph" since although bodily present, Christ is not so *sub propria specie*. As the iconography of the stigmatization develops and we find representations of the vision that depict Christ, with little and sometimes no indication whatsoever of the figure of the seraph, and that also incorporate actual witnesses, we are confronted with a theological paradox. For either it is an imaginative vision, for which there could be no witnesses, or it is a corporeal vision, and so cannot be a vision of Christ Himself under the figure of his own body. To represent other people witnessing Francis's vision, which would require that the vision be corporeal, while at the same time making this a vision of Christ *sub propria specie*, is theologically incoherent. But then there is no reason why we should assume that the iconography must be subject to all of the rigid conceptual constraints of the theology. This situation makes Giotto's Assisi fresco all the more brilliant in its combination of a (virtual) witness and of Christ unequivocally "under the appearance of the Seraph."

We can arrive at a similar conclusion about the corporeal nature of the vision by examining Giotto's Bardi Chapel fresco of the stigmatization. In this painting there is no witness to the vision (except perhaps the falcon in the upper left-hand corner) and the Christ/seraph has become much more Christ-like and much less seraph-like. (He is dressed as Christ was after the crucifixion, and his human bodily features seem to take precedence over the angelic form represented by the wings.) Here the figure of Francis itself attests to the corporeal nature of the vision. Francis is turning toward the vision; the position of his legs and body indicates that he was praying with his back toward the direction of the vision; at the moment depicted he is in the process of turning his entire body counterclockwise to face the vision. As the rotation and placement of his right leg show, it is exactly as if, being disturbed by something behind and above him, he has been caught in the awkward position of still turning to confront the vision.[50] An imaginative vision would not provoke such an odd bodily posture; occurring within the imagination, it would not have required Francis to turn in this abrupt way. But if something were bodily present, and Francis were turning to see what it was, the position of his own body is easily understandable. Here again, even without represented witnesses, the iconography of the stigmatization helps us answer a crucial question about the event: What kind of vision was it taken to be?

St. Theresa, in recounting her transverberation in chapter 29 of her autobiography, clearly takes the representations of St. Francis's stigmatization as the background literary model. The angel that appears to her is described as a seraph ("one of those angels very elevated in the hierarchy, who seems to burn completely with divine ardor"), although she calls it a cherub, admitting that the angel did not tell her its name and that there are many differences between angels that she does not know how to express.[51] But she is very insistent and unhesitant in emphasizing that this was a corporeal vision:

> I saw next to me, on the left, an angel in corporeal form, something that I could not see except in rare circumstances. Even though in fact angels often appeared to me, I did not see them corporeally, but as in the vision of which I spoke before. In this vision it pleased the Lord that I see the angel in such a way [i.e., corporeally].[52]

This account is also good evidence that certain kinds of physical transformations (transverberation and stigmatization are often linked, so much so that Teresa is frequently placed on the list of those who have received the stigmata) were typically represented as produced by corporeal visions, even though from a theological point of view, corporeal visions are not considered as elevated as imaginative or intellectual visions.[53]

One reason why the representation of the vision as corporeal turns out to be so significant is that various attempts to deny the miraculous status of the stigmata depended on describing the vision as imaginative and then giving, in effect, a purely psychological interpretation of the vision and its effects. Thus Petrarch in a letter to Tommaso da Garbo from November 9, 1366, writes:

> Concerning the stigmata of Francis, this is certainly the origin: so assiduous and profound was his meditation on the death of Christ that his soul was filled up with it, and appearing to himself to be also crucified with his Lord, the force of that thought was able to pass from the soul into the body and leave visibly impressed in it the traces.[54]

Strictly speaking, Petrarch leaves the vision entirely out of account, and attributes the stigmata to the power of Francis's thought. But his description allows no possibility for any type of vision other than an imaginative one, and given widespread views about the powers of the imagination, a psychologically interpreted imaginative vision would have only contributed to the passage of the thought into the body. Even fifty years before Petrarch, Petrus Thomae had to refute the arguments of those who saw in the stigmata only the effects of Francis's *vehemens imaginatio*.[55] This kind of proto-psychological explanation, typically invoking the power of the imagination, has its culmination in Pomponazzo's *De incantationibus*, where he claimed that even if one admits that Francis had the stigmata, they would not have been the result of a miracle, but of the natural forces of an unbridled imagination.[56]

Such interpretations were made so much more inevitable by the fact that Bonaventure's description of the stigmatization makes explicit reference to Francis's "fire of his spirit" and his "marvelous ardor." The *Fioretti*, following Bonaventure, invokes Francis's "fervor," "mental fire," and "extreme ardor and flame of divine love."[57] Although these descriptions were an essential part of what I have called the mystical interpretation of the stigmatization, it was all too easy to reinterpret psychologically these mystical states and to consider them as nothing more than excesses of the imagination. Mystical claims about the transformative power of divine love could thus be detached from their theological context and refashioned with the aim of undermining the miraculous nature of the stigmatization. Of course, correctly theologically interpreted, such claims were a crucial part of the account of the stigmatization; Francis's mystical state constituted what I referred to as the subjective cause of the stigmata. Without entering into the details of this mystical interpretation, one can understand how a tension might develop between the mystical and the miraculous interpretations of the stigmata, the result of too exclusive a focus on either the subjective or objective causes.[58]

Consider Giovanni Bellini's spectacular painting of St. Francis, now in the Frick Collection in New York (Figure 3). Bellini's painting is, I believe, an *exact* representation of the moment when the stigmata begin to appear as described in the hagiographical texts. It accurately represents the "extreme ardor and flame of divine love" left in Francis's heart as the disappearing vision left the stigmata imprinted on his body. It perfectly portrays the mystical state that was the subjective cause of the stigmata. It contains no seraph and, *a fortiori*, no representation of the causal process of stigmatization, and there is no depiction of Brother Leo as a witness. All of these features are in complete agreement with the description of the moment of stigmatization found in the texts. But precisely because of the absence of the seraph, many historians have felt it necessary to deny that this is a painting of Francis receiving the stigmata.[59] (I believe that a reference to the just-disappeared Christ/seraph can be found through an examination of the shadows in the painting, but I shall not discuss that here.) The painting has received three different titles: *San Francesco nel deserto*, *San Francesco in estasi*, and *San Francesco riceve le stimmate*. It is as if the power of the iconographical tradition has made it almost impossible to see the textual accuracy of this painting. Thus most art historians have focused on the depiction of the landscape or on Francis's facial expression without seeing how the painting could be related to the receiving of the stigmata. The only truth behind this reaction is to be located in the fact that however accurate the painting is to the texts, it does not have the specific effect of underlining the miraculous status of the stigmatization, its supernatural causation. A person could view Giotto's Assisi fresco without knowing the textual details about Francis's stigmatization, and be certain that he was witnessing the representation of a miracle. Viewing Bellini's painting in ignorance of the texts, the spectator is certainly moved and perhaps even recognizes, through Francis's countenance alone, that something of divine significance is transpiring, but he does not see the direct divine intervention that authorized and guaranteed the special status of Francis's stigmata.

Figure 3.   Giovanni Bellini, Saint Francis, *ca. 1480, tempera and oil on poplar panel,*
49" x 55⅞"; 124.4 cm. x 141.9 cm. The Frick Collection, New York.

Bellini's painting allows doubts and uncertainties that Giotto's does not, and the two paintings visually exemplify the tensions that could result between the mystical and miraculous interpretations of the stigmata.

A satisfactory historical and philosophical interpretation of the stigmata would require taking into account both interpretations, since the mystical and miraculous dimensions of the stigmata both are central to understanding its full significance. But even putting aside compositional problems about the simultaneous visual representation of these two dimensions, during the Middle Ages and Renaissance when there were so many persistent doubts about the reality of the stigmata, a painting such as Giotto's—with Francis, a witness, the Christ/seraph, and the causal interaction between Francis and the vision producing the stigmata—was most effective in addressing these doubts directly. For the unlettered, the doubts could be countered by the forceful visual details of the painting itself. As Giacomo da Vitry wrote, "to lay people it is necessary to show everything concretely, as if they had it before their eyes."[60] Moreover, by incorporating a recognizable iconography of the life of Christ into the representation of the stigmatization and its consequences, as Giotto also does in other frescoes in the Assisi series, it was possible to emphasize Francis's uniqueness and his special proximity to Christ, as exemplified, above all, by the fact that they, and they alone, bore the stigmata of the passion on their flesh. Furthermore, if we are to take the textual descriptions *literally*, Francis's stigmata were unlike any other future stigmata. They were unique in character, never to be encountered again, miraculous even among stigmata.[61] Later descriptions of other stigmata, as well as later iconography, do not rival Francis's from the point of view of the miraculous. Not all stigmata have been created equal, and Francis, both historically and theologically, remains the model to which all other examples must be compared.

Notes

1. The longer, original version of this paper was written in Italian and is forthcoming in a volume on the history of mysticism and bodily transformations, edited by Paolo Santonastaso, to be published by Marsilio Editori. Illustrations for all of the images discussed in this paper can be found in that version. All of the references to Franciscan texts in this version are from *Fonti Francescane, Editio Minor* (Assisi: Editrici Francescane, 1986), and the standard abbreviations used in the notes that follow are those listed at the beginning of this edition. In providing English translations here I have consulted some already existing English versions as well as the Latin originals. I am indebted to many people for reactions to and comments on versions of this paper. I would especially like to thank Mario Biagioli, Diane Brentari, Caroline Bynum, James Conant, Lorraine Daston, Peter Galison, Carlo Ginzburg, Caroline Jones, Claudio Leonardi, and Hilary Putnam.
2. My focus on the strategic dimensions of discourse and painting is indebted to some much-underappreciated ideas of Michel Foucault. For discussion of these ideas, see my introduction, "Structures and Strategies of Discourse: Remarks Towards a History of Foucault's Philosophy of Language," to the volume *Foucault and His Interlocutors* (Chicago: University of Chicago Press, 1997), especially pp. 2–5.
3. For the predominance of healing miracles, see P. A. Sigal, *L'homme et le miracle dans la France mediéval* (Paris: Cerf, 1985).
4. Badouin de Gaiffier, "Miracles bibliques et vies des saints," in *Études critiques d'hagiographie et d'iconologie* (Brussels: Societé des Bollandistes, 1967).
5. See, among others, Antonio Royo Marin, *Teologia della perfezione cristiana* (Milan: Edizioni Paoline, 1987), p. 1094, and Pierre Adnès, "Stigmates," *Dictionnaire de spiritualité*, t. XIV, col. 1211–1212. See also the terminology used in 2 Corinthians 4:10–11.
6. André Vauchez, "Les stigmates de Saint François et leurs détracteurs dans les derniers siècles du moyen âge," *Mélanges d' Archéologie et d'histoire*, vol. LXXX (1968). See pp. 598–99 for a discussion of earlier rejected cases of stigmatization.
7. The best discussion of this topic remains Vauchez, "Les stigmates." See also Rona Goffen, *Spirituality in Conflict, Saint Francis and Giotto's Bardi Chapel* (University Park: Pennsylvania State University Press, 1988), especially chapter 2.
8. Cited in Vauchez, "Les stigmates," p. 601.
9. The words of Gregory IX are cited in Goffen, "Spirituality in Conflict," p. 97, footnote 24. For Alexander IV's bull, see *Le stimmate di santo Francesco dagli scritti del XIII e XIV secolo*, A cura di Marino Bernardo Barfucci (Arezzo: Edizioni La Verna, 1975), pp. 36–37.

10. Cited by Vauchez, "Les stigmates," p. 611, footnote 4. Millard Meiss has argued convincingly that although St. Francis's stigmatization was the model for St. Catherine's stigmatization, her vision more resembled the paintings of the subject than the texts. See *Painting in Florence and Sienna after the Black Death*, (Princeton: Princeton University Press, 1951), p. 117.

11. Cited by Goffen, *Spirituality in Conflict*, pp. 20–21.

12. See Vauchez, "Les stigmates," p. 624.

13. See Chiara Frugoni, "Le Mistiche, le visioni, e l'iconografia: rapporti ed influssi," in *Temi e problemi nella mistica femminile trecentesca* (Todi: Centro di Studi Sulla Spiritualitá Medievale, 1983), p. 152. I completed my research on Francis's stigmatization before the publication of Frugoni's *Francesco e l'invenzione delle stimmate*, and so I haven't been able to take this important work fully into account.

14. See Vauchez, "Les stigmates," pp. 621–23. Also see Stanislao da Campagnola, *L'Angello del sesto sigillo e l'* "*Alter Christus*" (Rome: Ed. Laurentianum, 1971).

15. LFE 5 (*Epistola Encyclica* of Brother Elias) For a discussion of Elias's description of the stigmata, see H. Berger, "La Forme des stigmates de Saint François d'Assise," in *Revue d'histoire ecclésiastique* 35 (1939). For an exhaustive bibliography and discussion, see Octavian Schmucki, *The Stigmata of St. Francis of Assisi: A Critical Investigation in the Light of Thirteenth Century Sources* (New York: The Franciscan Institute, St. Bonventure University, 1991). See especially chapter 5.

16. I shall not discuss Frate Leone's remarks about the stigmata, written by him on the parchment that contained Francis's "Lodi di Dio altissimo" and "Benedizione a Frate Leone," since this was not intended to be a public document. Despite its historical importance, for my limited purposes here it can be left aside. For the document see *Fonti Francescane*, p. 134, footnote 1.

17. 1 Cel. (Tommaso da Celano, *Vita Prima* 1), pp. 94–95.

18. 1 Cel., p. 113.

19. See Etienne Gilson, "L'interprétation traditionelle des stigmates," *Revue d'histoire franciscaine*, vol. II, 1925. As Gilson argues, the Latin makes clear the temporal separation between the vision and the appearance of the stigmata.

20. Ps. Diogini l'Areopagita, *Gerarchia celeste* (Celestial hierarchy), in *Gerarchia celeste, Teologia mistica, Lettere* (Rome: Città Nuova, 1986), chapter VII, 1, p. 47. Pseudo-Dionysius is speaking here of the entire first order of angels, at whose head stand the seraphim.

21. Ibid., chapter XIII, 4, p. 73.

22. Ibid., chapter XIII, 4, p. 74.

23. Ibid., chapter XIII, 3, p. 72.

24. *Gerarchia ecclesiastica* (Ecclesiastical hierarchy), 480B–480C.

25. For Pseudo-Dionysius, see *Gerarchia celeste*, chapter VII, 1, p. 47, and chapter XII, 4, p. 74, which contains the quoted phrase. For Gregory the Great, see *Hom in Evang*, I, hom. 34, nn. 10, 12, in *Patrologia Latina* 76, 1251, 1254.

26. The most complete discussion of these reliquaries remains H. Matrod, *Les Stigmates de Saint François. Leur plus ancienne représentation connue* (Paris: Association Franciscaine, 1906).

27. On Berlinghieri, see E. B. Harrison, "A New History of Bonaventura di Berlinghiero's St. Francis Dossal in Pescia," in *Studies in the History of Medieval Italian Painting* 1, 2 (Autumn 1953). Harrison is very helpful on the history of the painting, but has little to say about its iconography.

28. The accounts in Matthew 26 and Mark 14 both omit the angel and give a different description of the posture of Jesus praying.

29. There is no purely textual way to determine whether Luke considered this bloody sweat to be supernatural. For discussion see Antonio Royo Marin, *Teologia*, pp. 1101–03.

30. Gerhardt Ladner, "The Gestures of Prayer in Papal Iconography of the Thirteenth and Early Fourteenth Centuries," in *Images and Ideas in the Middle Ages*, vol. I (Rome: Edizioni di storia e letteratura, 1983).

31. Ibid., pp. 214, 234.

32. LegM. XIII (Bonaventure, *Legenda Maior*), p. 3.

33. LegM. XV, p. 23.

34. This aspect of Bonaventure's account develops 2 Cel. 135 (Tomasso da Celano's *Vita Secunda*), a topic to which I shall return at the end of this chapter.

35. Bonaventure, *Tree of Life*, p. 16. Thomas of Aquinas, *Summa Theologiae*, 3, q. 75, q. 76.

36. Carlo Ginzburg, "Représentation: Le Mot, L'Idée, La Chose," *Annales ESC* (November–December 1991), p. 1230.

37. For Francis, see Am. I (*Admonition*); for Bonaventure, see LegM. (*Legenda Maior*) IX, 3. See also the last sentence of IX, 2.

38. I see no reason to invent lost textual sources to account for the innovations in this painting, as does Umberto Milizia in his *Il ciclo di Giotto ad Assisi* (De Rubeis Editore, 1994) pp. 115–16. There is every reason to believe that Giotto was as great an innovator as Bonaventure.

39. For the significance of luminosity, see Antonio Royo Marin, *Teologia*, pp. 1125–28.

40. LegM. xiii, p. 9.

41. Parts of this sermon are reproduced in Marino Bernardo Barfucci, *Le stimmate di santo Francesco*, pp. 157–62. The quotation is from p. 159.

42. See, for example, LegM. xiii, p. 8 and xv, p. 4.

43. LAn. (Letters to Anthony of Padua) reproduced in *Fonti Francescane*, p. 125.

44. LegM. XI, 1.

45. See Antonio Royo Marin, *Teologia*, pp. 1064–70.

46. Octavian Schmucki, *The Stigmata of St. Francis*, p. 196. But see also p. 219.

47. The sermon of October 4, 1255, reproduced in Barfucci, appears to describe the vision as a corporeal one. See Barfucci, *Le stimmate*, p. 161. See also Jacques Bougerol, *Francesco e Bonaventura. La Legenda Major* (Vicenza: Edizioni L. I. E. F., 1984), p. 45. In addition, see Legm. (*Legenda Minor*) VI, lecture II, which describes the vision "*apparebat exterius*" and whose analogy with the impression of a seal seems to require an external impression. The description of the vision in Tommaso da Celano is more ambiguous, but I think that even in this description there is some evidence that the vision was considered corporeal.

48. See Aquinas, *Summa Theologiae* I, Q51, A2.

49. Ibid., 3a, Q76, A8. See also Antonio Royo Marin, *Teologia*, p. 1068; and Adolphe Tanquerey, *The Spiritual Life: A Treatise on Ascetical and Mystical Theology* (Belgium: Society of Saint John the Evangelist, n.d.), pp. 701–02.

50. It is all too easy to misdescribe Francis's posture. In his superb book *La Raison des gestes dans l'occident medievale* (Paris: Gallimard, 1990) Jean-Claude Schmitt has said, "*le saint pivote sur lui-même comme s'il voulait éviter les effets de l'apparition et s'enfuir*" (p. 318).

51. Teresa d'Avila, *Libro della mia vita* (Rome: Edizioni Paoline, 1975), pp. 258–59.

52. Ibid.

53. On transverberation, see Pierre Adnès, "Transverbération," *Dictionnaire de spiritualité*, t. XV, col. 1174–84.

54. Petrarca (Petrarch), *Lettere senili* (Florence: Le Monnier, 1868), VIII, letter 3, p. 465. For discussion see Vauchez, "*Les stigmates*," pp. 624–25, and André Vauchez, "La stigmatizzazione di San Francesco d'Assisi. Significato e portata storica," in *Ordini mendicanti e società italiana. XIII-XV secolo* (Milan: Arnoldo Mondadori, 1990), p. 55.

55. See G. E. Mohan, "Petrus Thomae on the Stigmata of St. Francis," *Franciscan Studies* 8 (1948).

56. The relevant text is quoted in Vauchez, "Les stigmates," p. 625.

57. Fior (*Fioretti*), "Della terza considerazione delle sacre sante istimate," author unknown.

58. Many of my preceding and following remarks can be read, in part, as a response to Gilson's brilliant article, "*L'interprétation traditionelle*," which I believe focuses too exclusively on the mystical interpretation of the stigmata. The account of François de Sales, *Traité de l'amour de Dieu*, livre VI, chapter 15, has more to recommend it than Gilson allows.

59. A thorough discussion of the painting can be found in J. V. Fleming, *From Bonaventure to Bellini: An Essay in Franciscan Exegesis* (Princeton: Princeton University Press, 1982). I do not agree with Fleming's overall interpretation of the painting. I think the most compelling account of it can still be found in Millard Meiss, *Giovanni Bellini's St. Francis in the Frick Collection* (New York: The Frick Collection and Princeton University Press, 1964).

60. Cited by Frugoni, *"Le Mistiche,"* p. 152.

61. See, for example, the remarks of R. Biot, *L'énigme des stigmatisés* (Paris, 1955), cited by E. Longpré, *François d'Assise* (Paris: Beauchesne, 1966), p. 158; and H. Thurston, "Some Physical Phenomena of Mysticism: Stigmatization," *The Month* 134 (1919), p. 152.

LONDA SCHIEBINGER

Lost Knowledge, Bodies of Ignorance, and the Poverty of Taxonomy as Illustrated by the Curious Fate of *Flos Pavonis*, an Abortifacient

In a moving passage in her magnificent 1705 *Metamorphosis insectorum Surinamensium*, the German-born naturalist Maria Sibylla Merian records how the African slave and Indian populations in Surinam, then a Dutch colony, used the seeds of a plant she identified as the *flos pavonis*, literally "peacock flower," as an abortifacient:

> The Indians, who are not treated well by their Dutch masters, use the seeds [of this plant] to abort their children, so that their children will not become slaves like they are. The black slaves from Guinea and Angola have demanded to be well treated, threatening to refuse to have children. In fact, they sometimes take their own lives because they are treated so badly, and because they believe they will be born again, free and living in their own land. They told me this themselves.[1]

This passage is remarkable for several reasons. First, it was written by a rarity—a European woman traveling on her own to record the bounty of nature. Women naturalists were rare in the rush to know exotic lands; we know of only a few examples: Jeanne Baret sailed with Louis-Antoine de Bougainville around the world disguised as the male valet of Philibert Commerson, the ship's botanist and her fiancé.[2] "A little virgin" saved the English slave trader Richard Ligon's ship and crew by spinning thread from a cargo of cotton to mend the sail.[3] Other women, like Lady Charlotte Canning, collected as a sidelight to their main occupations as colonial wives, traveling where their husbands happened to take them, but these, again, were rarities.[4]

Merian's passage is also remarkable for what it reveals about the global politics of plants in the early modern period—specifically the culturally induced loss of certain craft-botanic knowledge traditions. In the explosion of knowledge generally associated with the scientific revolution and global expansion, European awareness of herbal antifertility agents, such as Merian's *flos pavonis*, declined dramatically. Contrary to other trends, where naturalists assiduously collected local knowledge of plants for medicines and potential profit, there was no systematic attempt to introduce into Europe new and exotic contraceptives and abortifacients gathered from cultures around the globe. Mercantilist policies guiding global expansion did not define trade in such plants as a lucrative or desirable business, nor did the great East and West trading companies often place women in the field.

The history of Merian's *flos pavonis* is interesting for what it reveals about contemporary European systems of botanical nomenclature. Historians of botany for many years focused almost exclusively on the rise of systematics (scientific nomenclature and classification) and underplayed the importance of economic, medical, and other types of applied botany. More recent history, by contrast, looks at enhancing our appreciation of the connections between natural history and national economies, exploring also botanists' attitudes toward non-European cultures.[5] In the eighteenth century, while economic and medical botanists tended to value and collect vast stores of local knowledges along with specimens from diverse cultures around the globe, system builders tended to discard local names of the plants, preferring to devise European names and conceptual schema also for exotic plants. This development is epitomized, as we shall see, in the linguistic history of *flos pavonis*. In the course of the eighteenth century, the variety of names for Merian's peacock flower—many of them East Indian and emphasizing the plant's beauty—was reduced to a single term still used internationally, *Poinciana pulcherrima*, a name commemorating a seventeenth-century governor of the French Antilles. As European taxonomists focused their attention increasingly and exclusively on the abstract morphology and anatomy of plants, cultural and geographic connections were often abandoned.

FLOS PAVONIS: COLONIAL CONNECTIONS

Maria Merian was indeed bold to travel to Surinam in search of exotic insects. Moral and bodily imperatives kept the vast majority of Europe's women close to home; the German anthropologist Johann Blumenbach was typical in warning that white women taken to very warm climates succumbed to "copious menstruation, which almost always ends, in a short space of time, in fatal hemorrhages of the uterus."[6] There was also the often expressed fear that women giving birth in the tropics would deliver children resembling the native peoples of those areas. The intense African sun, it was thought, produced black babies regardless of the mother's complexion.

Figure 1. *Maria Sibylla Merian with exotic specimens brought from Surinam and displayed at the Stadthaus in Amsterdam, engraving based on portrait done in 1715. By permission of the Öffentliche Kunstsammlung, Kupferstichkabinett Basel.*

Despite warnings from the mayor of Amsterdam, who had lost four daughters in Surinam, Maria Merian (Figure 1) deposited her will and set sail in 1699 at the age of 52, only a decade after political upheavals in that colony left the governor dead, shot by his own soldiers. Maria was accompanied by her daughter, Dorothea, trained from an early age to work as her mother's assistant. Maria Merian was not schooled, as many of Linnaeus's students would be, to be sent into the field, nor had she been commissioned to make the journey by a trading company or scientific society, as were many of the botanists in this period. Her interest was self-generated and largely self-supported, part of her lifelong quest to find another variety of caterpillar as economically significant as the silkworm. For two years she collected, studied, and drew the insects and plants of the region.[7]

Despite her rarity as a woman naturalist, Merian's practices in the field were by and large similar to those of her male colleagues. Like Hans Sloane, her contemporary and a future president of London's Royal Society, she was keen to collect "the best information" concerning the exotic plants and insects she encountered from "books and the local inhabitants, either European, Indian or Black."[8] Like the astronomer Peter Kolb, who wrote an early ethnology of the Africans at the Cape of Good Hope, Merian developed deep friendships with several Amerindians and displaced Africans in Dutch Guiana who served as her guides to desirable specimens and provided access to dangerous, often impassible regions.[9] Merian also followed the practice common up to that time of retaining native names and recording much else that native peoples told her about the plants and animals she studied. In the introduction to her *Metamorphosis*, which she advertised as the "first and strangest work done in America," she wrote: "the names of the plants I have kept as they were given by the natives and Indians in America."[10]

Reliance on local peoples and their knowledge made sense as European trading companies sent naturalists into Africa, India, China, Japan, and the Americas (Figure 2). It is therefore curious that in this atmosphere, where voyagers often directly transcribed plants' native names, Merian chose to continue to use the Latin name for this plant, *flos pavonis* (Figure 3).[11] Given that Merian recorded the personal experience of

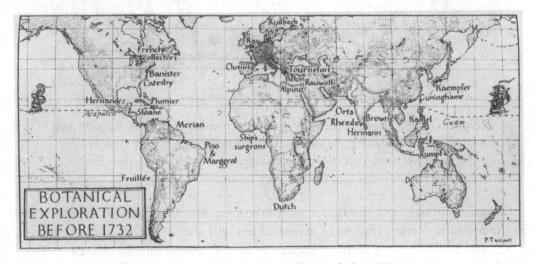

Figure 2.   Major botanical collectors before 1732.
From William Stearn, "Botanical Exploration to the Time of Linnaeus,"
Proceedings of the Linnean Society of London 169 (1958): 177.

Figure 3.   Maria Sibylla Merian's flos pavonis, which she describes as a nine-foot-tall plant with brilliant yellow and red blossoms. Merian, Metamorphosis, plate 45.

her informants in vivid detail, why did she not report a local Arawak or transplanted Angolan or Guinean name for the plant? We do not know whether her informants had a name for it, or whether the enslaved populations learned the name *flos pavonis* from Dutch settlers or from Portuguese or Spanish traders. We do not know whether slave women brought the plant (or its seeds) with them from their homelands, or simply found it again in the Caribbean.

The gap between Merian's professed purposes and her naming practice raises interesting questions about how plants and knowledge of plants circulated during the golden age of European mercantile expansion. The original biogeographic distribution of the *flos pavonis* is not known, though there are many complex and convoluted possibilities. The seventeenth-century traveler Richard Ligon reported having brought seeds of the plant from St. Jago, in the Cape Verde archipelago off the west coast of Africa, to Barbados in the West Indies; the nineteenth-century Swiss botanist Augustin-Pyrame de Candolle claimed that it had its origins in India and was subsequently transported to the Caribbean. A 1991 book, *Flora of Ceylon*, suggests that it was brought to southwest Asia from the Americas.[12] Resolving this question of origins is encumbered by the fact that we are not always sure to what plant a given name applies.

How the seeds of this plant actually traveled, whether drifting by sea or on board a merchant ship, we do not know; nor do we know how the knowledge of its uses spread. Seeds and plants of various sorts were shipped for purposes of commerce, curiosity, medicine, and food in this period. Dutch botanists in Ceylon, for example, shipped chestloads of specimens (often in separate vessels to ensure safe arrival) to Dutch botanical gardens from late in the seventeenth century until late into the eighteenth century.[13] Europeans carried seeds of dietary staples everywhere they settled; even their revictualing stations (the Cape of Good Hope, St. Helena's, Mauritius) were stocked with imported European plants and livestock. Slaves were also sometimes allowed to bring with them plant stocks used as foods or medicines. Renegade seeds also traveled in the fodder of livestock or the soils of plants taken for cultivation.[14]

Maria Merian may have chosen the name *flos pavonis* because she had seen this tropical tree in Amsterdam's ostentatious (by standards of the time) botanical garden, the so-called Hortus Medicus. Specimens had been cultivated there from seeds shipped from the West Indies as early as 1684.[15] The plant was known (though apparently not as an abortifacient) in Europe since the 1660s and perhaps earlier. Most of the European names for this brilliantly flowering plant associated it with the peacock. Jakob Breyne, a Danzig merchant and sometime botanist, reported that in Ambon, an island of Indonesia, the luxuriant tree was called *crista pavonis*, "crest of the peacock," for its "distinguished stamen . . . that bursts forth to form the proud crest of the peacock."[16] This flaming red, yellow, and orange flower was also called the *flore pavonino* (peacock flower) and *flos Indicus pavoninus*.[17] The Dutch living in the East Indies called the plant "peacock tails" (*paauwen staarten*) and the Portuguese labeled it the *"foula de pavan."* Less poetically, the plant was sometimes known by the Latin *frutex pavoninus*, or "peacock bush."[18]

The peacock flower enjoyed other, even more exotic, names. Merian, whose knowledge of Latin was weak, employed Casper Commelin, a friend and director of the botanical garden in Amsterdam, to add bibliographical references to the text of her *Metamorphosis* to place the Surinamese plants and insects she so elaborately recorded and illustrated into the world of European classical learning. What Commelin added to her paragraphs discussing the *flos pavonis* was the term *"tsjétti-mandáru,"* a Latinization of the Malayalam name for the flower that also associated it with the peacock.[19] Commelin drew his information from the *Hortus Indicus Malabaricus*, a magisterial twelve-volume work compiled by Hendrik Van Reede tot Drakenstein describing 740 plants of Malabar (the region of southwest India where Vasco da Gama landed in 1498) published in Amsterdam between 1678 and 1693.[20] In addition to the Malayalam term *tsjétti mandáru* cited by Commelin, Van Reede and his team presented names in "Brahmanese" or Konkani (transcribed as *tsiettia*), Arabic, Portuguese, and Dutch (Figure 4). Paul Hermann, a German medical officer who served in Ceylon for the Dutch East India Company and later taught botany at Leiden, also reported its colorful "Zeylonese" (Sinhalese) name: *monarakudimbiia.*[21]

Van Reede's volumes are intriguing because, like Merian, Reede was keen to record, compare, and contrast information about plants from diverse cultures and traditions. Van Reede strived accurately to transcribe Malayalam and Arabic names because he was eager to profit from older patterns of trade centered in the Indian Ocean and not yet dominated by Europeans.[22] Production of Reede's massive work was driven not by "a love of plants over riches," as Linnaeus would express his ideal of botanical researches a half century later, but by economic and political needs specific to Van Reede's situation.[23] Van Reede was not a botanist (something for which he felt compelled to apologize in the third volume of his *magnum opus*); he was a military man and colonial administrator, commissioned by the Dutch East India Company to seize Malabar from the Portuguese (Figure 5). As governor of the region from 1670 to 1677, he secured local contracts for trade in pepper, pearls, coconuts, rice, the areca palm (the nuts and leaves were used for betel chewing), cardamom, ginger, bananas, teak, and sandalwood, leaving him little time to pursue his botanical interests.[24] His authority as governor, however, was crucial to the success of this "big science" project; only an administrator of Van Reede's stature could command the necessary resources, contacts, and personnel to mount a venture of this magnitude.

Van Reede's text presents a wealth of information about plants, ranging from how they smell (the *flos pavonis* smells like honey) to how they grow, to the history of their names and—for him of crucial importance—their value to commerce and medicine. For Van Reede, local medicines were of vital importance to the Dutch occupation of

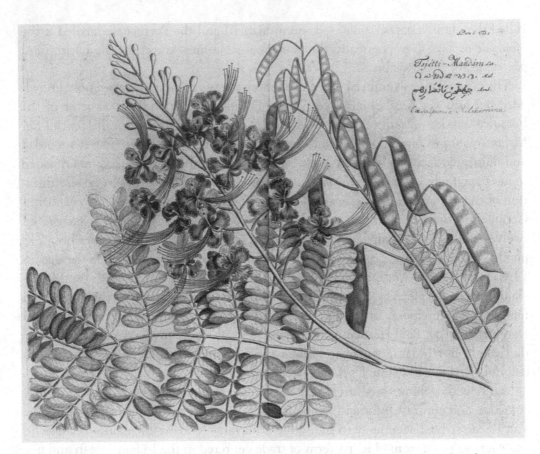

Figure 4. Hendrick Van Reede's tsjétti-mandáru, *the Malayalam name for what Merian called the* flos pavonis. *The name is given also in Aryazuth and Arabic script. (The handwritten Latin name,* Caesalpinia pulcherimma, *was probably added in the twentieth century.) The text includes names in "Braminese" or Konkani, Portuguese, and Dutch. Van Reede,* Hortus Malabaricus, *vol. 6, plate 1.*

India. Management of the colonies required medical expertise and reliable pharmaceuticals, and medicines shipped from Amsterdam were expensive and often spoiled in transit. European *materia medica* often proved ineffectual against the beriberi, dysentery, cholera, jaundice, and malaria plaguing white colonists in tropical areas. Van Reede's major goal was to document local remedies that might prove useful against Europeans' ailments. His *Hortus Malabaricus* also served him as a weapon against his Dutch East India Company rival, Rijklof van Goens, Sr., the governor in Ceylon. Van Goens aspired to make Ceylon a Dutch colonial power second only to Java. Van Reede's *Hortus Malabaricus* was calculated to convince Company officials that Malabar, not Ceylon, was the more profitable part of South Asia.[25]

Figure 5. The imperial face of botany. Hendrik Van Reede tot Drakenstein is portrayed as Commissioner-General of the Dutch East Indies Company in Asia. He is shown in full armor and wears a curled wig. Van Reede, Hortus Malabaricus, *vol. 9.*

To compile his complex text, Van Reede employed at least twenty-five men from many distinct cultures and two different continents. His pursuit of economic and medicinal botany led him to three "venerable" Brahmans, "Gymnosophists by birth and religion," who had collected "through their slaves" the names, medicinal powers, and virtues of the plants described in their book *Manhaningattnam*. The botanist K. S. Manilal, working in Calicut (the Kerala seaport from which calico takes its name) has been unable to locate this or any other medical text predating the *Hortus Malabaricus*.[26] Reede also contracted with a Vaidyar physician by the name of Itty Achuden (belonging to the lowly Chogans, a caste known as "tree climbers") to provide information regarding the medicinal powers of the plants from local ayurvedic practices; Achuden selected the plants that were to be drawn for the book, and reported their names and uses.[27] For conversing with these diverse local experts, Van Reede retained a Portuguese employee of the Dutch East India Company as the official interpreter for the project. The Dutch used Portuguese to converse with the Malayali; Malayali of mixed Portuguese descent and Malayali Christians rarely spoke Dutch.[28] Van Reede also engaged a number of Europeans (mostly Dutch) in both Malabar and Amsterdam to illustrate, order, and edit the manuscript, to render it into Latin, and to provide references to classical European and Arabic botanical sources. As the historian Richard Grove has recently argued, the *Hortus Malabaricus* was "a profoundly indigenous text," a compilation of South Asian botany without equal.[29]

Van Reede's *Hortus* was ranked by Linnaeus as one of the two greatest works contributing to his own work in systematics (the other was the Oxford botanist Dillenius's *Hortus Elthamensis*).[30] Despite this accolade, the wealth of culturally local knowledges embodied in Van Reede's project—and typical also of Merian's contemporaneous text—was not to become the central focus of European high botany. In the process of creating "universal" systems of botany, botanists often dislodged plants from deep cultural matrixes.

Maria Merian's and Hendrik Van Reede's purpose, we have to keep in mind, was to collect for the sake of medical and economic utility, not to classify for the sake of establishing a universal "system." Merian expressly refused to "classify" her plants. Discussing her *Metamorphosis*, she wrote, "I could have given a fuller account, but because the views of the learned are so at odds with one another and the world so sensitive, I have recorded only my observations."[31] In 1694, Merian's *flos pavonis* was included within Joseph Pitton de Tournefort's abstract typology—the classification widely regarded today as one of the forerunners of modern systematics. Tournefort, director of Jardin du Roi in Paris, placed the plant in his Class 21, Section 5, encompassing "trees and shrubs with red flowers and seed pods." As was typical of the new schema, Tournefort's classification focused on the physical characteristics of the plant, in this case the corolla and the fruit. The plant's Asian connections and its medical uses—both of which had played a significant role in earlier European accounts—were not discussed.

A long-standing narrative in the history of botany has emphasized a kind of liberation from the practical, usually medical, focus of premodern botany. William Stearn, for example, describes the rise of modern botany as the notion that "knowledge about plants as plants has a value of its own apart from economic or medical considerations."[32] Several botanical traditions coexisted in the eighteenth century and later became distinguished more sharply into applied botany, including economic and medicinal botany but also horticulture and agriculture, and what we today call theoretical botany, especially nomenclature and classification. In fact, however, these traditions often merged in a single botanist. Tournefort and Linnaeus, celebrated as "fathers of modern botany," also collected abroad. Tournefort gathered some 1,356 plants, including wild madder, marigolds, violets, valerian, dwarf cherries, exotic irises, and dragonhead, while traveling through Levant on a pilgrimage to study the reputed marvels of Mount Ararat (where it was believed Noah's Ark came to rest).[33] Linnaeus's enthusiasm for the fauna and flora of Lapland is well known. He also expended considerable energy trying to grow economically profitable plants, such as Chinese tea, in Sweden to enrich the coffers of his "fatherland."[34] Nomenclature and classification were not, in other words, the cardinal interest of early modern botanists.

Ordering principles were necessary, of course, to make sense of the many new materials flooding Europe. The number of plants known to Europeans quadrupled between 1550 and 1700; Linnaeus alone catalogued some six thousand species in his *Species plantarum*. The question was what form that "sense" would take. European classification developed along a trajectory that relied primarily on morphology (in his 1737 *Hortus Cliffortianus*, Linnaeus distinguished five varieties of *Poinciana* according to leaf shape) and sexual distinctions (in later texts Linnaeus included the *Poinciana* within the class *Decandia*, having "ten husbands" or stamen, and the order *Monogynia*, "one wife" or pistil). And Latin became the international language of abstract systematic botany.

William Stearn has suggested that Latin was chosen for international communication between scholars precisely because few women read it.[35] The claim may confuse cause and effect, but it is hard to deny that the Latin developed by botanists could have been different. Classical Latin was made and remade in this period—new terms introduced, others stabilized—to suit botanists' purposes. Botanical Latin might have incorporated customary names from other cultures as plants from those cultures entered Europe. But this did not happen. In the process of anchoring Merian's *flos pavonis* (Van Reede's *tsjétti-mandáru* and Hermann's *monarakudimbiia*) in the European world, Tournefort devised a wholly new name, *Poinciana pulcherrima*—the name that celebrates his countryman and governor of the French Antilles, Louis de Louvilliers Poinci.[36] Linnaeus approved of this name, and it is still in use today.[37]

Tournefort's name thus celebrated French colonial rule in the Caribbean rather than the plant's own virtues, its East Indian heritage, the peoples who used it, or those who "discovered" it or supplied Europeans with information about it—all of which were featured in other names given at one time or another for the plant. Following Tournefort's lead, Linnaeus mentions only that the plant grows in the Indies (apparently both East and West) and under the sign of Saturn, for its woody character.[38]

In his effort to stabilize botanical nomenclature, Linnaeus in 1737 ruled that "generic names not derived from Greek or Latin roots are to be rejected."[39] Expressly targeting Van Reede's *Hortus Malabaricus*, Linnaeus declared all foreign names and terms "barbarous" (though for some reason he preferred these barbarous names to what he considered the "absence of names" in Merian's account of the plants of Surinam, the other text he mentioned).[40] Linnaeus's extensive rules for botanical nomenclature banished many things: European languages except for Greek or Latin; religious names (he did allow names derived from European mythology); foreign names; names invoking the uses of plants; names ending in *-oides*; names compounded of two entire Latin words; and so forth. Linnaeus retained "barbarous names" only when he could devise a Latin or Greek derivation, even one having nothing to do with the plant or its origin. *Datura* (a genus in the potato family) he allowed, for example, for its association with *dare* from the Latin "to give, because it is 'given' to those whose sexual powers are weak or enfeebled."[41]

To fill the void created by his many expulsions, Linnaeus promoted "as a religious duty" generic names designed to preserve the memory of botanists who have served well the cause of science. Men immortalized in the Linnaean system included: Tournefort (*Tournefortia*), Van Reede (*Rheedia*), the Commelins (*Commelina*), and his own modest self (the *Linnaea* is a small flowering plant indigenous to Lapland). Discussing this practice, Linnaeus asserted that such men were martyrs to science, having suffered wearisome and painful hardships in the service of botany. First of the beleaguered "officers in flora's army" was himself: "In my youth I entered the deserts of Lapland. . . . I lived on only water and meat, without bread and salt. . . . I risked my life on Mount Skula, in Finmark, on icy mountains and in shipwreck."[42] Linnaeus also promoted generic names celebrating European kings and patrons who had contributed to the cost of oceanic voyages, botanical gardens, and textual illustrations. There were, of course, exceptions. Linnaeus named the genus Quassia after the African slave in Surinam who successfully developed it as a medication against fevers. And Linnaeus derived the family name, Monsonia, to honor Lady Anne Monson for her contributions to botany.[43]

For the most part, however, in his reform of botanical nomenclature Linnaeus broke the ties with other cultures that naturalists such as Van Reede and Merian had established. Linnaeus's nomenclature highlighted instead the deeds of great men of European botany. The French botanist Michel Adanson, working some years after Linnaeus, pointed to the absurdity of Linnaeus's naming a colonial plant *Dillenia* after Oxford's Johann Dillenius rather than retaining one of its traditional names.[44]

IGNORANCES EMBODIED

The naming and renaming of Merian's *flos pavonis* involved a complex politics of which she herself was largely oblivious. She was, however, very much aware of another aspect of the politics of this plant: its role as an abortifacient. Merian penned her report of the abortive qualities of the *flos pavonis* at a time when knowledge about abortifacients and contraceptives within Europe was under attack. This body of knowledge—long a trust that passed among midwives, wise women, mothers, daughters, and neighbors—was not destined to become a part of academic botany or medicine as these disciplines developed in the eighteenth and nineteenth centuries. Knowledge of antifertility agents became more secretive, discussed in euphemisms and in code, and almost always behind closed doors.[45]

In her passage about abortion, Merian tells us that she learned about the abortive virtues of the *flos pavonis* directly from the enslaved females of Surinam. Interestingly, Hans Sloane, working in Jamaica a decade before Merian's voyage to Surinam, also reported the abortive qualities of a (different) plant he called the "flour fence of Barbados, wild sena, or Spanish carnations." He mistakenly took this plant to be the same as that which Merian described, and cited her work in an appendix to his book (the flat, broad seedpods of the two plants are quite distinct).[46] Sloane should perhaps not be taken too severely to task for his error; the history of the *flos pavonis* is fraught with ambiguities: a 1981 botanical atlas lists two Latin and up to forty-two common names used within Central America for this particular plant.[47]

Sloane compared his "flour fence" to savin (*Juniperus sabina*), a shrub widely regarded at that time as the most powerful herbal abortifacient in Europe. How did he procure information about its uses? Apparently not from a text: Sloane does not cite previously published sources, such as Van Reede's 1678–1693 *Hortus Malabaricus*.[48] Curiously, Van Reede's work, specifically designed to document the medicinal virtues of plants, did not mention the peacock flower's role as an abortive. The *tsjétti mandáru* (now more commonly transcribed as *settimandaram*) is known today as an abortifacient in Malabar, where it is the bark and not the seeds (as Merian reported) that are prepared for this purpose.[49] The twenty-five men working on Van Reede's project may not have had access to this information, though the slaves involved (sex not specified) may well have; much of the collecting and cataloguing for Garcia de Orta's well-known 1563 *Coloquios dos simples e drogas . . . da India*, for example, was done by a Konkani slave girl, known only as Antonia.[50] Most likely, Sloane received an independent report of the abortive qualities of his "flour fence" from the inhabitants of Jamaica or one of the other islands he visited. Certainly, the *flos pavonis* is still today known in Central and South America as an emmenagogue (medication that induces the menses) and abortifacient—here the flowers are considered the effective part.[51]

Sloane may well have encountered slave women who had aborted their embryos, a practice sufficiently common in the Caribbean to alarm plantation owners. West Indian slave populations generally did not reproduce themselves in this era; plantation owners were continually forced to purchase new slaves from Africa.[52] The low rate of natural increase among slaves seems to have been due, among other things, to amenorrhea and sterility among female slaves caused by hard labor and poor living conditions. The disruption and separation of families must also have given slaves little desire to bear children. Abortion and contraception were also recognized as a form of resistance among slave women. As early as the sixteenth century, Spanish friars recounted how enslaved Indian women killed the infants in their wombs by means of "well-known plant poisons."[53] John Stedman, the inveterate observer of colonial Surinam, recorded that slaves used green pineapple to induce miscarriage and spite their masters.[54] Abortifacients used in the West Indies included yam, papaya, mango, Barbados pride (yet another name for Merian's *flos pavonis*), wild passion flower, and wild tansy. The cotton root was sometimes used by slaves in the southern United States for such purposes, though fertility rates there were not remarkably low.[55] Abortion and infanticide among slaves was considered so damaging to plantation property and profits that *all* slave medicines were outlawed in French possessions in the 1760s. Birth control was only one of several issues here—slaves had also been known to poison the water supplies of their masters.[56]

While both Merian and Sloane mentioned abortifacients, only Merian emphasized the importance of this plant for the physical and spiritual survival of the slave women of Surinam. Slaves in Surinam endured extreme brutality: Stedman (in the 1770s) reported a "revolted negroe" hung alive upon a gibbet with an iron hook stuck through his ribs, two others chained to stakes and burned to death by slow fire, six women broken alive upon the rack, and two slave girls decapitated.[57] While Sloane was well aware that slaves "cut their own throats" to escape such treatment, he did not see his "flour fence" in this context. The future president of the Royal Society of London wrote rather drily, "it provokes the Menstrua extremely, causes Abortion, etc. and does whatever Savin and powerful Emmenagogues will do."[58]

Sloane's discussion of abortion reveals the growing conflict between doctors and women seeking assistance in this matter. Concerning his service as physician to the governor in Jamaica, he wrote:

> In case women, whom I suspected to be with Child, presented themselves ill, coming in the name of others, sometimes bringing their own water, dissembling pains in their heads, sides, obstructions, etc. therby cunningly, as they think, designing to make the physician cause abortion by the medicines he may order for their cure. In such a case I used either to put them off with no medicines at all, or tell them Nature in time might relieve them without remedies, or I put

them off with medicines that will signifie nothing either one way or other, till I be furthered satisfied about their malady.[59]

He finished with a strict warning: "if women know how dangerous a thing it is to cause abortion, they would never attempt it. . . . One may as easily expect to shake off unripe Fruit from a tree, without injury or violence to the Tree, as endeavor to procure Abortion without injury or violence to the Mother." The few learned men who did discuss antifertility herbs in the seventeenth and eighteenth centuries usually did so in order to warn about their dangerous consequences.[60] Sloane himself noted that when an abortion was absolutely necessary to save the life of the mother, "the hand" was generally preferable to herbal preparations.

It is unclear who might have sought out Sloane's services in this regard. Caribbean plantations generally had a hospital for slaves run by a female of this class (who employed medical traditions carried with her from Africa), several younger aides (mostly female), and a midwife (either slave or free). These hospitals were commonly supervised by a local white surgeon who visited only twice a week.[61] It was commonly known that the "herbs and powders" slave women used for abortion were obtained from healers known as "obeah men and women." Concerning slave abortions in Jamaica in 1826, Reverend Henry Beame wrote, "white medical men know little, except from surmise."[62]

I do not want to make too much of the contrast between Sloane and Merian. Merian, to my knowledge, discussed only one abortifacient. Her chief interest was insects, and she described plants primarily in their relationships to them (in the passage cited at the beginning of this paper, she devoted an entire paragraph to the caterpillars living off the plant's leaves). Whether women "do science differently" is currently a topic of heated debate; distinctions, however, should not be drawn too sharply between individual men and women scientists. Many European women—plantation owners or governors' wives, for example—had little interest in their newly adopted countries, and most came and went without collecting any information from the indigenous populations or cultivating any special sympathies toward the women of the region.

Larger historical forces, however, can make gender an important factor. Although they differed in their attitudes toward abortion, Merian and Sloane were unusual in providing knowledge about abortifacients from abroad.[63] Colonial administrators such as Van Reede were most often interested in medicines that could protect traders, planters, and Trading Company troops—among whom few women were found. In the colonies, abortion among slave populations was seen by colonial administrators as a clear threat to plantation property. Even in Europe, mercantilist expansion mandated pro-natalist policies celebrating children as "the wealth of nations, the glory of kingdoms, and the nerve and good fortune of empires."[64] In such climates, agents of botanical exploration—trading companies, scientific academies, and governments—had little interest in expanding Europe's store of antifertility pharmacopoeia. Moreover, customary divisions in physical and intellectual labor within Europe had long left fertility control in women's hands.[65] Though physicians such as Sloane occasionally reported on abortifacients, few had intimate knowledge of such practices. Effective use of the plants required knowledge of the parts of the plant appropriate for use (its root, sap, bark, flowers, seeds, or fruits), the proper time for harvesting, when to administer the drug within the woman's cycle and in what relation to coitus, in what amounts and with what frequency, and so forth. Male physicians also may not have had easy access to women abroad, who were usually the keepers of this knowledge.

As medical men gradually displaced midwives across Europe, the use of herbal abortives and contraceptives declined among the general population.[66] Pregnant women lost their traditional prerogative to judge for themselves when "ensoulment" took place—that is, when they truly were with child.[67] States began to overturn the tradition Aristotelian notion that early abortion was acceptable, even encouraged, when the mother's health was in danger.

Though threatened, the use of herbal antifertility agents did not disappear entirely. Despite priestly admonitions and legal warnings, these practices continued—though more and more hidden from public view. Court records in early modern Italy speak of aborted embryos pushed into cracks in church walls or thrown into cemeteries.[68] An unusual set of records gathered in seventeenth-century Lancashire, England, reveals an abortion rate varying between ten to thirty per one thousand live births; the rate of unrecorded abortions would most certainly be higher.[69] Common abortifacients (rue, savin, squirting cucumber, and pennyroyal) were increasingly discussed in code as "menstrual regulators," as herbs to "promote the menses," "bring down the flowers," "purge the courses," or "restore menses obstructed."[70] While knowledge about antifertility agents was dying in Europe, it was still available to women, at least behind closed doors.

Merian's *flos pavonis* participated in both a revolution in the history of botany and a transformation in the history of the body. At a time of rapid expansion of science more generally, European knowledge of antifertility agents waned. Gender politics lent recognizable contours not to a distinctive body of knowledge[71] but, in this instance, to a distinctive body of ignorance. Ignorance is often not merely the absence of knowledge but, as Robert Proctor has suggested, the project of protracted cultural struggles.[72] Bodies of ignorance, in turn, can mold the very flesh and blood of real bodies. European women's loss of easy access to contraceptives and abortifacients curbed their reproductive and often professional freedoms. An image of upper- and middle-class women developed that celebrated them as both angels in the home and fecund beings hopelessly subservient to the beck and call of nature. The curious history of the *flos pavonis* shows how voyagers selectively culled from the bounty of nature knowledge responding to national and global policies, patterns of patronage and trade, developing disciplinary hierarchies, personal interests, and professional imperatives. In the process, much useful knowledge was lost; many bodies remained ignorant, and still other bodies, ignored.

Notes

1. Maria Sibylla Merian, *Metamorphosis Insectorum Surinamensium* (1705), ed. Helmut Deckert (Leipzig: Insel Verlag, 1975), commentary to plate no. 45.
2. Renée-Paule Guillot, "La vraie 'Bougainvillée': La première femme qui fit le tour du monde," *Historama* 1 (1984): 36–40.
3. Richard Ligon, *A True and Exact History of the Island of Barbados* (London, 1657), pp. 120–21.
4. Ann Shteir, *Cultivating Women, Cultivating Science: Flora's Daughters and Botany in England 1760–1860* (Baltimore: Johns Hopkins University Press, 1996).
5. Lisbet Koerner, "Purposes of Linnaean Travel: A Preliminary Research Report," in *Visions of Empire: Voyages, Botany, and Representations of Nature*, ed. David Miller and Peter Reill (Cambridge: Cambridge University Press, 1996), pp. 117–52.
6. Johann Blumenbach, *The Natural Varieties of Mankind* (1795) trans. Thomas Bendyshe (1865; New York: Bergman, 1969), p. 212n2. Blumenbach codified notions long current in the culture.
7. Londa Schiebinger, *The Mind Has No Sex? Women in the Origins of Modern Science* (Cambridge: Harvard University Press), chapter 3; Natalie Davis, *Women on the Margins: Three Seventeenth-Century Lives* (Cambridge: Harvard University Press, 1995), pp. 140–202.
8. Hans Sloane, *A Voyage to the islands Madera, Barbadoes, Nieves, St Christophers, and Jamaica; with the Natural History*, 2 vols. (London, 1707–1725), vol. 1, p. xlvi.
9. Peter Kolb, *The Present State of the Cape of Good Hope*, trans. Guido Medley (London, 1731).
10. Merian, *Metamorphosis*, introduction, p. 38.
11. This Latin term was used in both the Dutch first edition and the Latin translation.
12. Ligon, *History of the Island of Barbados*, p. 15; Augustin-Pyrame de Candolle, *Prodromus Systematis Naturalis Regni Vegetabilis* (Paris, 1825), p. 484; and M. D. Dassanayake and F. R. Fosberg, eds., *Flora of Ceylon* (New Delhi: Amerind Publishing Co., 1991), vol. 7, pp. 46–48.
13. J. Heniger, *Hendrik Adriaan van Reede tot Drakenstein and Hortus Malabaricus* (Rotterdam: A. A. Balekema, 1986), pp. 76–77.
14. William Stearn, "Botanical Exploration to the Time of Linnaeus," *Proceedings of the Linnean Society of London* 169 (1958): 173–96, esp. 193.
15. D. O. Wijnands, *The Botany of the Commelins* (Rotterdam: Balkema, 1983), p. 59; Heniger, *Hendrik Adriaan van Reede*, p. 162. The plant was also known in Paris by 1666 and, despite the relative proximity of Paris and Amsterdam, has a somewhat distinct linguistic history.
16. Jakob Breyne, *Exoticarum aliarumque minus cognitarium plantarum centuria prima* (Danzig, 1678), pp. 61–64.
17. Hans Sloane, *Catalogus Plantarum quae in Insula Jamaica* (London, 1695), p. 149.
18. Breyne, *Exoticarum*, p. 61.
19. Merian, *Metamorphosis*, pl. 45.
20. Hendrik Adriaan van Reede, *Hortus Indicus Malabaricus*, 12 vols. (Amsterdam, 1678–1693), vol. 6, pp. 1–2. This book was edited by Casper Commelin's uncle Jan from 1678 until his death in 1692; Casper himself prepared an analytical index, which he published separately as *Flora Malabrica sive Horti Malabarici Catalogus* (Leiden, 1696).
21. Paul Hermann, *Horti Academici Lugduno-Batavi Catalogus* (Leiden, 1687), p. 429. On the scripts used in the *Hortus Malabaricus*, see Heniger, *Hendrik Adriaan van Reede*, pp. 148–49.

22. Richard Grove, *Green Imperialism: Colonial Expansion, Tropical Island Edens and the Origins of Environmentalism, 1600–1860* (Cambridge: Cambridge University Press, 1995), p. 83.

23. Carl Linnaeus, *Critica botanica* (Leiden, 1737), no. 238.

24. Indeed Van Reede complained: "All those who have gone there [to Malabar] under the auspices of the Illustrious East India Company are compelled so much to perform their office accurately that they have no leisure to undertake this [a study of plants], even if they wished to." Van Reede, *Hortus Indicus Malabaricus*, vol. 3, p. vii. See also Marian Fournier, "Hortus Malabaricus of Hendrik Adriaan van Reede tot Drakestein," in *Botany and History of Hortus Malabaricus*, ed. K. S. Manilal (Rotterdam: Balkema, 1980), pp. 6–21 and Heniger, *Hendrik Adriaan Van Reede*.

25. J. Heniger, "Van Reede's Preface to Volume III of Hortus Malabaricus and its Historical and Political Significance," in *Botany and History of Hortus Malabaricus*, ed. K. S. Manilal, pp. 35–69.

26. K. S. Manilal, "The Implications of Hortus Malabaricus with the Botany and History of Peninsular India," in *Botany and History of Hortus Malabaricus*, ed. K. S. Manilal, p. 3. Van Reede provided an extensive description of how the text was compiled in *Hortus Malabaricus*, vol. 3, pp. iii–xviii. See also Van Reede, *Hortus Malabaricus*, vol. 1, preface.

27. Heniger, *Hendrik Adriaan van Reede*, p. 43; Grove, *Green Imperialism*, p. 89.

28. Fournier, "Hortus Malabaricus," pp. 13–14.

29. Grove goes so far as to claim that "the existence of European printing, botanical gardens, global networks of information and transfer of materia medica seem to have facilitated the diffusion and dominance of a local epistemological hegemony alongside the erosion of older European and Arabic systems." He suggests that the *Hortus Malabaricus* remains the "only faithful textual record of the accumulate Ezhava botanical knowledge of the seventeenth century." *Green Imperialism*, pp. 78, 89–90.

30. Carl Linnaeus, *Genera plantarum*, 5th ed. (Stockholm, 1754), p. xii.

31. Merian, *Metamorphosis*, p. 38.

32. Stearn, "Botanical Exploration," p. 165. Much of the history of botany has been written as the rise of systematics.

33. Marguerite Duval, *The King's Garden*, trans. Annette Tomarken and Claudine Cowen (Charlottesville: University Press of Virginia, 1982), pp. 42–53.

34. On Linnaeus and his botanical interests, see Lisbet Koerner, "Carl Linnaeus in His Time and Place," in *Cultures of Natural History*, ed. Nicholas Jardine, James Secord, and Emma Spary (Cambridge: Cambridge University Press, 1996), pp. 145–62; and Koerner, "Purposes of Linnaean Travel," pp. 135–36.

35. William Stearn, *Botanical Latin* (1966; Newton Abbot, Devon: David and Charles, 1992), p. 7.

36. Joseph Pitton de Tournefort, *Elemens de botanique* (Paris, 1694), vol. 1, pp. 491–92; vol. 3, pl. 391. See also Jean Baptiste Du Tertre, *Histoire generale des Ant-isles* (Paris, 1667–1671), vol. 1, pp. 125–26.

37. Carl Linnaeus, *Hortus Cliffortianus* (Amsterdam, 1737), p. 158.

38. Carl Linnaeus, *Species plantarum* (Stockholm, 1753), vol. 1, p. 380.

39. Linnaeus, *Critica botanica*, no. 229. This stands in contradistinction to more recent international codes of botanical nomenclature that allow "the genus name . . . [to] be taken from any source whatever." *International Code of Botanical Nomenclature*, ed. W. Greuter (Konigstein: Koeltz Scientific Books, 1988).

40. Linnaeus, *Critical botanica*, no. 218.

41. Linnaeus, *Critica botanica*, no. 229. Linnaeus did accept some well-established generic names derived from non-European languages, including *Coffea*, *Datura*, *Tulipa*, *Zombia*, *Camassia*, *Vanada*, and *Yucca*.

42. Linnaeus, *Critica botanica*, no. 238. See Heinz Goerke, *Linnaeus*, trans. Denver Lindley (New York: Charles Scribner's Sons, 1973), p. 108.

43. William Grimé, *Etho-Botany of the Black Americans* (Algonac: Reference Publications, 1976), p. 167; Shteir, *Cultivating Women*, pp. 48–50.

44. Michel Adanson, *Familles des Plantes* (Paris, 1763), vol. 1, p. clxxiii. On this point, see Joseph Needham, Lu Gwei-Djen, and Huang Hsing-Tsung, *Science and Civilization in China* (Cambridge: Cambridge University Press, 1986), vol. 6, pt. 1, pp. 19, 168.

45. For antifertility agents within Europe, see John Riddle, *Contraception and Abortion from the Ancient World to the Renaissance* (Cambridge: Harvard University Press, 1992).

46. Sloane cites Merian's work in an addendum to his text (*Voyage,* vol. 2, p. 384).

47. Julia Morton, *Atlas of Medicinal Plants of Middle America* (Springfield, Ill.: Charles Thomas, 1981), pp. 284–285.

48. The only medical use for *tsjétti mandáru* given in the *Hortus* is a cure for digestive disorders: "the tincture of the leaves [mixed] with the leaves of scedanga, as much as fills a small gourd, drunk, takes away the colick, especially if the sick lifts up his hands to heaven, standing straight up." Van Reede, *Hortus Malabaricus*, vol. 6, pp. 1–2; Sloane's translation, *Voyage,* vol. 2, p. 50.

49. R. N. Chopra, S. L. Nayar, and I. C. Chopra, *Glossary of Indian Medicinal Plants* (New Delhi: Council of Scientific and Industrial Research, 1956), p. 198; Sudhanshu Jain and Robert DeFilipps, *Medicinal Plants of India* (Algonac, Mich.: Reference Publications, 1991), p. 210.

50. Garcia de Orta, *Coloquios dos simples e drogas . . . da India* (Goa, 1563), preface.

51. Morton, *Atlas of Medicinal Plants*, pp. 284–85.

52. On fertility rates among slaves, see Barbara Bush, *Slave Women in Caribbean Society: 1650–1838* (Bloomington: Indiana University Press, 1990), chapter 7. Bush refutes those who blamed low fertility rates of slaves in this area on adverse sex ratios among slaves, unstable mating patterns, promiscuity, and venereal disease.

53. Bush, *Slave Women*, pp. 137–42. Prolonged maternal breast-feeding was also blamed for low birth rates. Proper recovery periods between births, however, are generally beneficial to both mother and child.

54. Cited in Bush, *Slave Women*, p. 142.

55. John Crellin and Jane Philpott, *Herbal Medicine Past and Present* (Durham: Duke University Press, 1990), vol. 2, p. 176; see also John Riddle, *Eve's Herbs: A History of Contraceptive and Abortion in the West* (Cambridge: Harvard University Press, 1997), chapter 6.

56. John Stedman, *Narrative of a Five Years Expedition against the Revolted Negroes of Surinam* (1796; Baltimore: Johns Hopkins University Press, 1988), p. 266. See also Paul Brodwin, *Medicine and Morality in Haiti: The Contest for Healing Power* (Cambridge: Cambridge University Press, 1996), pp. 41–42.

57. Stedman, *Narrative of a Five Years Expedition*, pp. 26, 271–72.

58. Sloane, *Voyage*, vol. 2, p. 50.

59. Sloane, *Voyage*, vol. 1, p. cxliii. Other great "dissemblers" of illness, according to Sloane, were servants, "both Whites and Blacks."

60. Riddle, *Contraception and Abortion from the Ancient World to the Renaissance*, p. 160.

61. Brodwin, *Medicine and Morality*, pp. 28–32.

62. Cited in Bush, *Slave Women*, p. 139.

63. My larger project investigates to what extent fertility and antifertility agents were collected by early modern European naturalists from abroad.

64. Joseph Raulin, *De la Conservation des enfans* (Paris, 1768), vol. 1, "épitre au roi."

65. See Agnus McLaren, *Reproductive Rituals: The Perception of Fertility in England from the Sixteenth to the Nineteenth Century* (London: Methuen, 1984).

66. Jean Donnison, *Midwives and Medical Men: A History of Inter-Professional Rivals and Women's Rights* (London: Heinemann, 1977); Riddle, *Contraception and Abortion*; and Hilary Marland, ed., *The Art of Midwifery: Early Modern Midwives in Europe* (London: Routledge, 1993).

67. Barbara Duden, *Disembodying Women: Perspectives on Pregnancy and the Unborn*, trans. Lee Hoinacki (Cambridge: Harvard University Press, 1993).

68. Nadia Filippini, "The Church, the State and Childbirth: The Midwife in Italy during the Eighteenth Century," in *The Art of Midwifery*, ed. Marland, p. 157.

69. A priest reported that in 1659, six hundred women in Paris had confessed to having suffocated the fruit in their womb; another observer suggested that the number would have been much higher if it had included those who took early precautions, before "ensoulment," or quickening. Angus McLaren, *A History of Contraception from Antiquity to the Present Day* (Oxford: Basil Blackwell, 1990), pp. 159–60.

70. McLaren, *Reproductive Rituals*, pp. 102–06; see also Riddle, *Contraception and Abortion* and *Eve's Herbs*.

71. Sander Gilman, *Difference and Pathology: Stereotypes of Sexuality, Race and Madness* (Ithaca: Cornell University Press, 1985); Ludmilla Jordanova, *Sexual Visions: Images of Gender in Science and Medicine between the Eighteenth and Twentieth Centuries* (Madison: University of Wisconsin Press, 1989); Londa Schiebinger, *The Mind Has No Sex?*; Thomas Laqueur, *Making Sex: Body and Gender from the Greeks to Freud* (Cambridge: Harvard University Press, 1990); Claudia Honegger, *Die Ordnung der Geschlechter: Die Wissenschaften vom Menschen und das Weib* (Frankfurt: Campus Verlag, 1992); Felicity Nussbaum, *Torrid Zones: Maternity, Sexuality, and Empire in Eighteenth-Century English Narratives* (Baltimore: Johns Hopkins University Press, 1995); and Anne Fausto-Sterling, "Gender, Race, and Nation," in *Deviant Bodies: Critical Perspectives on Difference in Science and Popular Culture*, ed. Jennifer Terry and Jacqueline Urla (Bloomington: Indiana University Press, 1995), pp. 19–48.

72. Robert Proctor, *Cancer Wars: How Politics Shapes What We Know and Don't Know About Cancer* (New York: Basic Books, 1995), p. 8 and chapter 5.

Caroline A. Jones

The Sex of the Machine: Mechanomorphic Art, New Women, and Francis Picabia's Neurasthenic Cure

INTRODUCTION

> Almost immediately upon coming to America it flashed on me that the genius of the modern world is in machinery and that through machinery art ought to find a most vivid expression. . . . The machine has become more than a mere adjunct of life. It is really a part of human life . . . perhaps the very soul. . . . I have enlisted the machinery of the modern world, and introduced it into my studio.
>
> —Francis Picabia, 1915[1]

Picabia's vision of machines as "the very soul" of human life characterizes both the eighteenth-century search for a perfect automaton as well as the late-twentieth-century tropism toward the utopian cyborg (and, per Donna Haraway's essay in this volume, the digitized wellspring of Life Itself). In this essay, however, I want to look at something considerably baser than the soul: I want to question the presumptive *sex* of the machine, the construction of "knowing" machines that are imagined to function down to the level of corporeal reproduction.

The central problem I want to examine is not the experience of living bodies as they intersect with, generate, or labor through the machine; at issue here is instead the sexing of machines in the twentieth-century cultural imaginary. It is a premise of this paper that relations of power, labor, and capital are played out in the realms of machines, men, and women on an internal and "capillary" level.[2] As in all such capillary dynamics, the capillary level of the machinic imaginary is powerfully inflected by the differential relations of sex and gender, and it is something of a truism that technology has largely been constructed in Western society as male, which is to say, technology "expresses and consolidates relations among men."[3] At the same time, specific machines are experienced or fantasized as women, and the seductively female Android has increasingly replaced the lumbering Golems and Frankenstinian male monsters of yore. Such basic coordinates map the terrain in which I want to operate. My hope is to open the cultural imaginary of the machine to close analysis by examining a special case: early-twentieth-century artistic constructions of the female or ambiguously gendered machine.

Within this early-twentieth-century art world, I will be focusing primarily on a reading of some early works by modernist Francis Picabia (1879–1953) that reveal an instability in the role of technology in culture. Rather than a fixed relation, these show a shifting, heterogeneous, hybrid system of interconnections and productive metaphors. Much of this art emerged during Picabia's treatments for acute neurasthenia, presenting a key axis of my inquiry. Were the instabilities in the sex of Picabia's machines a *symptom* of his neurasthenia? Or were they representative of the new imaginary necessitated by the neurasthenic cure, *products*, as it were, of his temporarily medicalized identity? That Picabia's sexed machines might be hermaphroditic, homoerotic, or functionally female—sometimes at different moments, sometimes all at once—problematizes even the strategic essentialisms that would position the machine as the property of the powerful, and "nature" as the only ground on which the Other might stand. They offer possibilities lost in the later codifications of modernism, possibilities that may prove useful if explored anew today.

BINARIES: A BEGINNING

Since the turn of the century brought us Heinrich Wölfflin and modern art history, those attempting to see history in art, or art in history, reflexively use two slides; in written texts, two adjacent images serve the same purpose (Figure 1). The convenient visual binary is intended to summarize an extended historical argument, to convert the complex matrix of humans' visual culture into a linear progression that can be seen "as plain as the nose on your face." I invoke the nose advisedly, given much of the imagery we will see here—but for now let's talk about plainness.

We could play connoisseur with these photographs. One is folded, torn, heavily shadowed; the other's tonality is less developed, its identifying title and "signature" seemingly not the artist's own. But clearly there is only one image shared between these pictures; they are obviously multiples of a sort (despite collector/dealers' descriptions of them as "unique"). These are not merely faithful photographic replicas of an original masterpiece (the ideology undergirding Wölfflin's pioneering pedagogy), nor even the de-auratized "art in the age of mechanical reproduction," as Walter Benjamin's formulation has been translated.[4] Specifically, what we have here are images without an *Ur-Objekt* whose aura they can implicate: two vintage "art" photographs, each the product of the same single negative, exposed and printed by Man Ray (born Emmanuel Radnitsky), the American modernist who teamed up with Parisians Marcel Duchamp and Francis Picabia to ignite the brief and incendiary moment that was New York Dada.

Like Wölfflin, I want to suggest an historical argument here. The first of these prints was made around 1917–18, the second in 1920. Although produced from the same negative, they are presented as two very different works of art. That difference resides explicitly in their social and textual construction, through *différance* and the verbal mechanism of their titles. The work produced on the heels of the Great War (Figure 1, left) was titled *L'Homme*, in Man Ray's beginning French. Its manifest content is a depiction of an eggbeater, but given the title it reads metaphorically as a mechanized, pendulous phallus that throws its hard-edged metallic shadow on the wall. Seen by subsequent interpreters as speaking to Man Ray's own penchant for beating his ovular wife (whom he had left as he began an intense relationship with the Parisian artist Marcel Duchamp), it has also been viewed as an emblem of onanism.[5] But what is Man Ray's second picture? This second print (Figure 1, right) was dated 1920, probably sent to fellow Dadaist Tristan Tzara for publication in Europe. This time, the title is *La Femme*, a different "work of art."[6] This doubling, this mapping of different genders and/or sexes onto seemingly identical machines, is what frames my problematic. The historical question regards the possibility of an instability in the sex of the early-twentieth-century machine—an instability later eradicated by fascism (among other masculinist technocracies), and one we might profitably reimagine now.

Why sex, and not gender? As I'll argue here, what seems to be operating in these mechanized bodies are not only the social roles of gender, but the biological roles of sex (even if we now question the fixity of both categories). At issue for the artists in question was, in the final analysis, *reproduction*—how the male machine might reproduce commodities, or how the female machine might reproduce the male (or the male's labor). But although I speak of sex, gender obviously enters into these constructions of technology, and the messages conveyed about technology's sex are meant in turn to reify new configurations of gender in the social frame.[7]

Let us return, then, to Man's *Femme*. Its new sex allows the eggbeater to reassert an association with the female machines of domestic life, but that association is clouded both by the echoes of *L'Homme*, which still cling to it, and by the psychosexual scenarios opened up by its new female identity. When the eggbeater was *L'Homme*, it fit fairly well into a standard trope of technology as active and masculine. As *La Femme*, however, this image of a readymade threatens to cut the other way: as the blades turn in our imagination, this female machine casts a darker shadow, open steel strips

Figure 1. Left: Man Ray, L'Homme, 1918. *Photographic print,*
50.7 cm. x 38.5 cm. "Exemplaire unique." Vera and Arturo Schwartz
Collection, Milan. Right: Man Ray, La Femme, 1920. Photographic print.
Musée national d'art moderne, Centre Georges Pompidou, Paris.

149

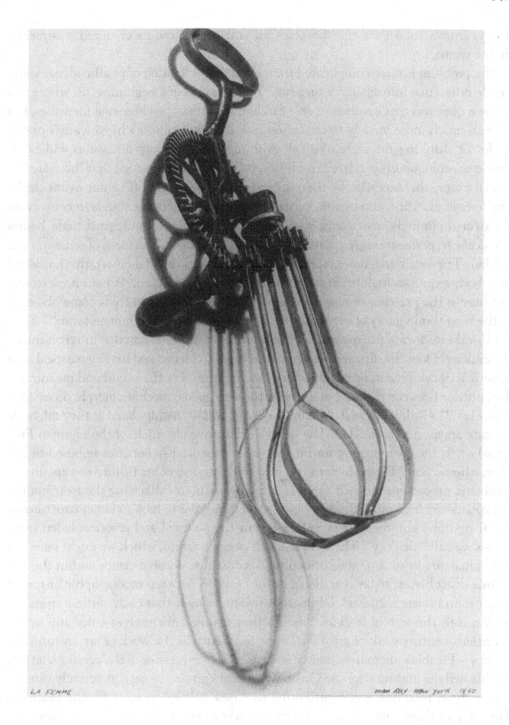

LA FEMME   man Ray · new york · 1920

closing into a solid form that becomes an emblem for the mechanized, castrating, phallic woman.

The problem is more complicated than even this collapsing of phallic identity and female difference into *différance* suggests.[8] This essay is but a beginning, an attempt to define a question that involves issues of individual artists' psychosexual identities, but extends much more broadly to characterize a dominant culture's historically situated modes of thinking the technological. Although much of my discussion will be of female or ambiguously gendered machines, it should be emphasized that these are the special cases, the exceptions—the purview of a deliberately off-center avant-garde. The female machine who manifests herself in Man Ray's *Femme* stands in contrast to the overwhelmingly masculinist discourse of hardened, technologized male bodies that come to permeate early-twentieth-century modernism. The focus of scholars such as Klaus Theweleit and, more recently, Jeffrey Schnapp and Hal Foster, this hardened male body expresses itself in literature, art, film, theater, and war.[9] What I hope to suggest here is the presence of *other* formations that have subsequently become obscured by the near total victory of a masculinized "metallization" of the human form.[10]

How do we locate the specificity of a different or resistant practice in representing technology? Does the slippage between Man Ray's *L'Homme* and his *Femme* speak to a historical development, or merely a random variation? For the origin and meaning of the contrast between the male and female (or ambiguous) mechanomorph, do we look to the level of individual psychoanalytic configurations, manipulated as they might be by state apparatus (as in Klaus Theweleit's exhaustive chronicle of the German *Freikorps*) or, as I want to suggest for Picabia, by the medical systematics imposed by the neurasthenic cure? Do we look to the internal discourse of art history to explain the fetishistic precedents for such works? Or finally, without exhausting the near infinity of explanatory frameworks for any object, will it profit us to look to larger structures of social signification, themselves imbricated in the political and emotion-laden complexes we call "ideology"? These three levels of explanation, which we might label the individual, art-historical, and sociocultural, constitute divisive camps within the discipline of art history today (paralleled by the conflicts between monographic history of science, and science studies). Ultimately, I want to argue that each of these strands is woven into the web of fears and desires that manifest themselves culturally in the twentieth-century work of art. I also want to assert that the work of art, in turn, has agency—Picabia's alternative may be seen as merely expressive of the gender relations established during and after the Great War, but it can also be seen as actively interrogating those relations, and contributing to a new cultural imagery for the machine. The play of these multiple readings only confirms my preliminary observation that there can be no fixity to the sex of the machine, only momentary—but potentially strategic—configurations in a system predicated on motility and flux.

NEW YORK DADA AND THE *FEMME NOUVELLE*

The international avant-garde movement later called "Dada" took its most mechanistic form in New York. Generations of migrants found that New York demanded a
new, technologically mediated art—from Man-Ray-of-Philadelphia to the Parisians
Duchamp and Picabia (driven Westward by the war). In the course of completing his
"New York Interpreted" series of futuristic tableaux, for example, the Italian-born
immigrant Joseph Stella exclaimed "New York is my wife!" The quintessential modern
city became a demanding mechanical spouse whose brash sexuality was seen to be
expressed in the lights of Broadway, the straining spires of skyscrapers, and the soaring
suspension cables of the Brooklyn Bridge.[11]

The Great War was of course a determining contributor to the emergence of Dada
and to its appearance in New York. Duchamp managed to get declared unfit for duty
because of a heart condition, but Picabia drifted into military service, avoiding combat
only through unauthorized mobility (he went "AWOL" in New York while on a military supply mission) and then through a crippling mental disease then diagnosed as
"neurasthenia." As his wife later commented: "he profited by a temporary discharge
which, from medical board to medical board, carried him to the end of the war." [12]
Because the Great War was a conflict of unprecedented industrial scope where the
only victor seemed to be mechanized warfare itself, traditional affiliations between
men and machines were troubled, to say nothing of relations between fully mobilized
men and suddenly professional women. The power of the machine (and, arguably, of
women) had become unassailable by the early 1920s, but artists predisposed to question authority were ambivalent about that power. That ambivalence expressed itself in
a problematization of the sex of the machine, most insistently in the New York Dada
productions we are examining here.

As historians have shown, views of the modernist "new woman" mutated after the
Great War, congealing in a range of negative reactions against supposedly mannish,
efficient females in dark and unconfining clothes, wearing heavy makeup, perhaps, but
possessing brazen desires to vote, to smoke, and to control their own sexuality and
reproductive lives.[13] The dominant tendency to belittle the political and legal struggles of suffragism by linking its adherents to a sensationally liberated sexuality is amply
evident in Man Ray's portrait of his most important patron, Katherine Sophie Dreier.
The assemblage sculpture was titled *Catherine Barometer*, completed in 1920. Dreier's
appurtenances are brought together by Man Ray as follows: from a base of steel wool, a
washboard rises, its front labeled with the work's title—and the subject's name—
together with a placard advising the user to "shake well before using." Out of this
vibratory base comes a thin rod encircled by wire; the measurement of its presumably
ascending energies is calibrated by a color chart mounted on wood. In addition to
returning Dreier to a lower-class-female's domestic sphere (coded by steel wool and

washboards), the portrait of the suffragist and art organizer as a "barometer" suggests that her passions changed with the weather. Beyond that, the wool links itself to Dreier's own wiry hair, cushioning the vibratory mechanism of the washboard in a thinly veiled equivalence between the woman and her most private sexual parts. The apparatus brings to mind Terry Castle's 1987 speculations on "The Female Thermometer," as well as Picabia's 1924 drawings of the *Thermomètre Rimbaud*.[14] In one of these images (published in the artist's own Dadaist journal, *391*), a thermometer protrudes from between the legs of an androgynous nude embraced by a fishtailed male lover; the other shows a naked man sucking or blowing a thermometer-as-flute for the pleasure of an androgynous muse. As Castle argues, the origins of such medical devices were linked to the search for a mechanical model of human nature. The thermometer or "weather-glass" (human barometer) was initially offered as a novelty for measuring female passions, and only later became generalized through psychology to "a universalist model of emotional flux."[15] As if echoing anecdotes about the inventor of the device, who supposedly set the standard for 100 degrees by taking the temperature of his aroused female lover, Picabia's vision of the poet Rimbaud's thermometer fixates on its oral and anal modes. As in Man Ray's barometer, the machine devised to measure the female passions becomes conflated with the passionate female. The iconographic program becomes dedicated to reducing the woman to a female sex part (or, in Picabia's more intriguing version, dissolving her in nonproductive *jouissance*).

But there are phallic elements to the *Catherine Barometer* of Man Ray, as well as in Picabia's *Thermomètre Rimbaud*. The wand of Catherine's ostensible "barometer" extends its slender erection all the way up the color scale. And if the "female thermometer" conflates the object meant to penetrate the female orifice with the female herself, then the woman *becomes* the phallus. These objects thus function as visual oxymorons, like the oxymoron we have already met in Man Ray's contemporaneous *Femme*: the phallic woman. Clearly these works participate in individual psychological frameworks: Man Ray's conflicted relationship to one of his major patrons, and Picabia's evidently elegiac relationship to phallic manhood. And, like all artworks worth their salt, they also participate in art-historical discourses (Duchamp's readymades, in the case of Man Ray, and Aubrey Beardsley's erotic drawings, in the case of Picabia). But, as my argument suggests, these objects can also be viewed within a larger context—the male hysteria circulating around the "*femme nouvelle*," and, in the case of Picabia, the gender negotiations epitomized by neurasthenia.

As Mary Louise Roberts, Debora Silverman, and other scholars have shown, the emergence of the "new woman" was accompanied almost immediately by derisory shadow categories that dogged her liberatory march of progress. *Femmes nouvelles* in the 1890s were stigmatized from the outset as "*hommesses*," linked to technology and described by contemporary males as having an "active, public, mobile, and agitated character . . . associated with the tension and new electrical energy of the city streets and the 'brand new sparks' of the century of technological inventions and 'eternal motions.' "[16] Many deplored the growing association of women and the new technologies, moaning over the dangerous "inversion" fostered by the bicycle (which the *femme nouvelle* seemed to be invariably mounting). One critic put it simply in 1895, in a proscriptive conclusion that would not sit badly with Man Ray some thirty years later: **"A woman exists only through her ovaries."**[17] The fin-de-siècle turn against the *"hommesse"* was a subset of the larger obsession with the *femme fatale*, but the more general model of the evil seductress underwent subtle changes in her conversion to the New Woman. Largely through her conjunction with technology, the fatal *femme* became hardened and masculinized, the manipulative temptress in the shadows converted to a public, phallic woman.

In the post-World War I context more proximate to Man Ray, Duchamp, and Picabia, the new woman was rejected again by male critics, this time not as *"hommesse"* but as *"la garçonne"*—infantilization now added to the masculinization already inflicted on her by those wary of her kind. Gender anxieties may have functioned to mask other conflicts, as Joan Scott has convincingly theorized, but such anxieties proved to have their own trajectory as far as the fate of actual and fictive women was concerned.[18] As Roberts shows, in the novels of veterans writing after the Great War, the *femme nouvelle* bore the brunt of post-conflict rage.[19] Writing in 1927, Pierre Drieu La Rochelle articulated the veteran's feelings of universal loss: "This civilization no longer has clothes, no longer has churches, no longer has palaces, no longer has theaters, no longer has paintings, no longer has books, no longer has sexes."[20] Others would tie such losses explicitly to the invasion, and inversion, of *la garçonne*.

That these ideas had some resonance for noncombatants such as Duchamp, Man Ray, and Picabia is suggested by elements of their work during and immediately after the war. Nancy Ring has noted Duchamp's cryptic reference to his avoidance of armed service in his notes for his major assemblage *The Large Glass*, where he identifies the "bachelor apparatus" as "the cemetery of uniforms and liveries," celibate manhood conflated with the death of military forms. These notions of postmilitary bachelor machines are tied directly (in the manner of an oscillatory mode of being) to the gender reversal performed by Duchamp's seductive alter ego, Rrose Selavy.[21] Picabia, sidelined by desertion and acute neurasthenia, revealed his own ambivalence about femininity during wartime. In his poem titled "Soldats," written in 1917, he concluded his analysis of credulous soldiery with the stanza *"folie / avide / Des attitudes désespérées / le mur / malade / du sexe Féminin."*[22]

Pinning the war on a "sick wall of feminine sex" may have helped solidify the general anger directed at *la garçonne*, but although the discourse was French, the *garçonne* herself was seen to be entirely the product of American influence. "The innocent young thing of yesterday," wrote one French journalist in 1925, has given way to the *garçonne* of today. . . . Add to this sports, movies, dancing, cars, the unhealthy need to be always on the move—this entire Americanization of old Europe, and you will have the secret to the complete upheaval of people and things.[23]

World War I, with its automated regiments dedicated to a single military function, was the first Taylorized war, just as the "Tiller Girls" were the first Taylorized dance troupe.[24] The women on whose bodies the new postwar society was being mapped were seen as similarly "Americanized," *garçonnes* produced in an aggressive, uncontrollable social realm rather than a fantasized patriarchal domestic order from before. Images Picabia produced in 1915 and 1917 (Figures 2 and 3) portray the *garçonne* explicitly as a mechanized *Américaine*, their pert mechanical verticality coding for the emerging Jazz Age "flapper." A commercial illustration of an industrially produced lightbulb, the 1917 *Américaine* (Figure 3) is a transparent vessel, a container whose shape evokes the womb, the breast, the rounded body. The vessel is constricted at its base, however, sealed off and rendered phallic by the metallic cap and threaded base necessary for the bulb to become male (to screw its socket). And the bulb's transparency reveals the duplicity of the Americanized *femme nouvelle*. Within—or is it on the surface?—the bulb's reflective glass are visible the words "Flirt/Divorce," and the same upended as if in a funhouse mirror on the other side. The American lightbulb of Edison and Broadway, labeled a flirt and a hardened woman with too much experience (seduction and then divorce being the presumed temporal trajectory), displays precisely that conjunction of engineering and activism that had so troubled French critics writing three decades earlier.

The extent of this French discourse on mechanical American flirts is made clear by such powerful precedents as Auguste Villiers de l'Isle-Adam's popular 1885 novel *L'Eve futur*, first serialized in the French periodical *La Vie moderne*.[25] More than simply an eerie parallel, Villiers's novel indicates the extensive appeal of these visions of Americanized, androgynous, mechanomorphic women. Villiers tells of the American inventor Edison (maker of the lightbulb in Picabia's *Américaine*), who produces an Android named Hadaly (Persian for "Ideal") to replace the empty flirt who has claimed a young lord's heart. Hadaly/Ideal is an instantiation of two compelling Western philosophies: the Aristotelian binary in which woman is impressionable matter, man impressive force (for the power of this configuration, see Katharine Park's essay in this volume), and the Cartesian mechanical model of the universe that saw its apogee in Julien Offray de la Mettrie's 1748 treatise *L'Homme Machine*. Hadaly will be "imbued with . . . two wills, united in her; she is a *single* duality" when animated by living humans. A "human machine," she is a new "electro-human creature," as Edison describes her, "who with the aid of ARTIFICIAL GENERATION (already very much in vogue during recent years) seems destined within a century to fulfill the secret purpose

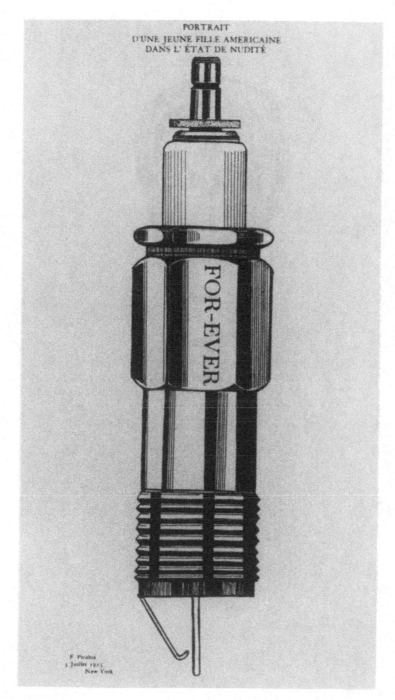

Figure 2. *Francis Picabia,* Portrait d'une jeune fille américaine dans l'etat de nudité, *(Portrait of a young American girl in a state of nudity), July 5, 1915, New York. Line Drawing reproduced in the artist-run journal 291.*

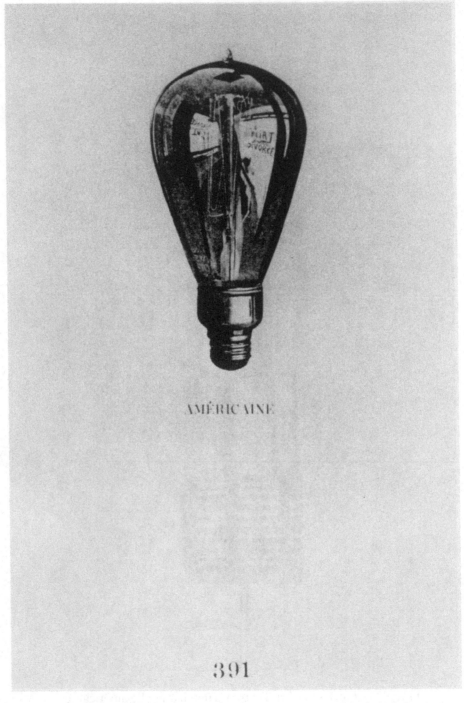

Figure 3.   Francis Picabia, Américaine, 1917, as reproduced on the cover of Picabia's journal 391, 14½" x 10¼". Original was a photograph of Edison's lightbulb retouched in ink.

of our species."[26] Tied immediately to *sex* (in the sense of the biologico-mechanical processes of "generation"), this machinic imaginary is a female, but also an androgynous and *dual* creature. Above all, for Villiers and Picabia alike, these ideal electro-human *femmes* could only be born in America—if not an actual America then the phantasmagorical one in which Picabia found himself after first crossing the Atlantic in 1913.

THE NEURASTHENE AND THE *FILLE NÉE SANS MÈRE*

Our young artist had been born thirty-four years earlier in Paris (1879), the son of a Spanish father and a French mother, and named "François" Marie Martinez Picabia. His father was the descendent of a Cuban planter who had become a Spanish railroad builder; his mother was the daughter of a wealthy businessman who was also a photographer-ally of Daguerre. When Picabia was seven, his mother died, and thereafter he was raised by servants of the household, with the authoritative presence (or intermittent absence) of his father, bachelor uncle, and photographer-grandfather. By 1911 he was making competent post-Impressionist paintings that clearly exhibit the fin-de-siècle fascination with the *femme fatale*, a figure that would elide smoothly into the *hommesse/garçonne*, as we have seen. The dramatic shift into a more advanced nonobjective style began for Picabia with his exposure to Cubism and to Marcel Duchamp, who gave Picabia the first of his eroticized machine-paintings, *The Bride*, shortly after completing it in 1912. (Like Villiers's Edison, Duchamp secured the bonds of male friendship through the exchange of an ambiguously feminine "electro-human" Ideal/ Bride). Although deeply affected by Duchamp's gesture (and by the formal vocabulary of alchemical retorts and mysterious plumbing that Duchamp's *Bride* displayed), Picabia's move toward a fully mechanomorphic abstraction appeared only after his first trip to New York a year later.[27] Self-styled ambassador for European modernism at the 1913 Armory Show, he had come intending to stay for two weeks, but lingered for six months, producing publications, works on paper, an exhibition, and a score of press interviews from his suite at the Hotel Brevoort.

During his American sojourn, Picabia developed a form vocabulary initially linked to Duchamp's. In symbolic abstract portraits of specific African-American musicians and one Russian-born "exotic dancer," he produced some evocative visual phrases: phallic nozzles emit slender probes, which slip between cushiony forms to move toward shapes that are bulbous and uterine (lightbulb-like in shape), orifices and ova proliferating in a delirious display of reproductive excess. This chemico-mechanico-biological *mélange* appears again in the drawing Picabia titled *Fille née sans mère* [FNSM] (Figure 4), translated as *Daughter Born without a Mother*. It is by all accounts the first of Picabia's many incarnations of this provocative theme, and forms the template for all of his subsequent sexed machines.

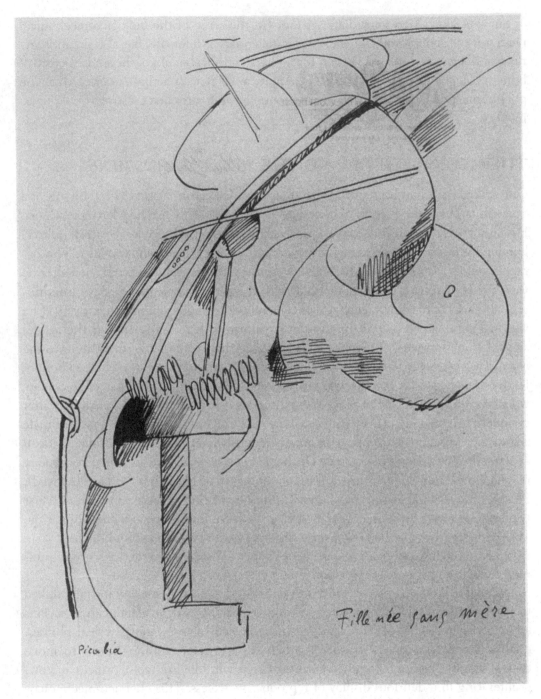

Figure 4. Francis Picabia, Fille née sans mère, (*Daughter born without a mother*),
1913 or 1915; pen-and-ink drawing on verso of hotel letterhead, 10⅜″ x 8½″,
reproduced in the artist-run journal 291 in 1915. Original in the
Metropolitan Museum of Art, Alfred Steiglitz Collection.

Drawn on the back of a sheet of letterhead from his suite at the Brevoort, the small sheet may date from Picabia's first exhilarating trip to the United States in 1913—or it may have been completed during his second trip two years later, when it was published in the avant-garde journal *291*.[28] In this enigmatic sketch, various rods and piston forms seem to move up into a realm of slightly pendulous orbs that read as breasts, buttocks, or eyes. What is significant about this, the earliest of the *FNSM* series, is that it pursues the previous mechanomorphic form vocabulary, but does so with a seemingly transparent linearity, limning an apparent *interior* to the body or bodies that it explores. Read against the gleaming metallic skins and hardened carapaces depicted by other modernist artists, Picabia's perspective is instead that of the doctor/inventor (or the lover)—one who would parse the body's hidden secrets with an instrumental, Roentgen-like gaze.[29]

On the personal and psychoanalytic level, the *"fille"* here could of course be shadowed by its masculine inversion, *"fils,"* describing Picabia's own motherless state. In this interpretation, the soft forms of the upper part of the drawing appear less penetrated by machinery than propped up by it, the kind of relationship made classic in Theweleit's analysis of the technologically hardened scaffolding (endo- or exo-skeleton) that serves to protect the shapeless ego of the not-yet-fully-born. Alternatively, the bulbous fleshy forms that seem to be escaping from the drawing's upper right may be the mother herself, the *fils's* own body a stunted device of frozen gears and flimsy pistons that attempts to capture, reenter, or penetrate the maternal form.

For the art-historical context that constitutes my second level of proposed analysis for the "knowing" and sexed machine, we should look in the first instance to the remarkable avant-garde journal *291*, where *FNSM* was published in June of 1915. But in the machine portraits Picabia prepared for publication with the *Fille*, a very different aesthetic presented itself, as we have seen already in a brief glance at one of his *garçonnes*, the *jeune fille Américaine* (Figure 2). In this and other "portraits," Picabia replicated the cool draftsmanship of the engineer (found also in the technological *garçonne* of 1917 in Figure 3). In this elegantly simple image of a *young American girl in a state of nudity* (copied from the pages of *The Motor*, a popular science magazine), the female machine has been reduced to her essentials, a fresh and irrepressible spark plug whose naive promise, "For-Ever," is belied by her status as an expendable, interchangeable, and replaceable part.[30] Like the Android of *L'Eve Futur*, the *jeune fille* is entirely reproducible, yet herself reproduces only labor (and not Life Itself)—and does that "with the aid of ARTIFICIAL GENERATION," controlled, presumably, by the master of her technology. Despite such limitations, the "electro-human" spark plug is no less ideal. As Edison explained to his incredulous friend, beneficiary of just such a young *fille*, "You see, *she is an angel!* . . . if indeed it's true, as the theologians teach us, *that angels are simply fire and light!* Wasn't it Baron Swedenborg who went so far as to add that they are 'hermaphrodite and sterile?' "[31]

Here we get to the heart of the matter, or rather, the sex of the machine. Surely the spark plug girl is a phallic woman (which is to say a metaphoric hermaphrodite). Yet she is rendered quite explicitly unthreatening by her very "nudity" and controllability—by our recognition that she stands naked of the larger apparatus that controls her sparking, and by our knowledge that she is identical to the tens of thousands like her in combustion engines throughout the United States. Although spark plugs could be found in any combustion engine, perhaps their strongest association was (and is) with the automobile, itself personified increasingly among the French as "*L'Américaine*," in an age when "Fordism" was perceived as one of the United States's most powerful exports to Europe (and the inspiration behind the Taylorized dancing of the interchangeable "Tiller Girls"). The sparky American *fille* is "like an angel" in her strippeddown functionalism and clean lines; Picabia's vision of the plug's erotic potential is suggested by his statement that he chose the spark plug for his girl because she was a "kindler of flame."[32] Like the flirtatious lightbulb that would appear two years later, this *jeune fille* presents the amalgamation of technology, America, and the new woman that saturated Picabia's imaginary at the time. The connection of all these discourses to the reign of neurasthenia is the nexus to which I now turn.

The spark plug and lightbulb flirts both present ambiguous, but putatively "three-dimensional" forms, not the interior probings that characterized the first of the *FNSM* images (Figure 4). There are several other mechanical portraits that bear the *FNSM* title; most of these present the smooth, patinated surfaces of the standard modernist "metallicized" body.[33] But in Picabia's final infatuation with his *Fille*, a book of fifty-one poems and eighteen drawings published in Lausanne, he pursued the original *FNSM*'s interiority. This book's transparent, mysterious, sexed machines float on pages adjacent to Picabia's pithy, disjunctive, Dadaist poems. Found in a few art libraries and largely forgotten by most scholars of early modernism, Picabia's book is beginning to claim a new audience since it was reprinted in Paris in 1992. Both poems and drawings are studded with barely connected textual bits, entries in a bizarre atlas of nouns, puns, and body parts. The impact of the slim volume is sustained and cumulative. Its rhythms are the meditative ones of boredom and dreams, produced in the first three months of a neurasthenic cure.

Having arrived in New York the very day that the United States entered the war, Picabia was forced to leave the city once again after a recurrence of his debilitating mental illness. Prohibited by his doctors from painting, he went first to Spain and then to Switzerland, pursuing the travel regimen that was posited as one of neurasthenia's primary therapies—but doing so in neutral countries that would not further exacerbate his nervous collapse. Apparently drawing and writing poetry could be accomplished within the narrow compass of the therapeutic regime, which required rest, isolation from prior activities and companions, and healthful diversions. Picabia's book was published in April of 1918 with the title *Poèmes et Dessins de la Fille Née Sans*

Mère. It was dedicated, appropriately enough, to the author's three neurological doctors—a Dr. Collins in New York, a Dr. Dupre in Paris, and a Dr. Brunnschweiller in Lausanne.[34] Significantly for a project conceived within a therapeutic frame, *Poèmes et Dessins* contains the most extended and hermetic of Picabia's analyses of the sexed machine.

Neurasthenia, the "disease of civilization," had been popularized by an American neurologist, Dr. George Beard, in a series of clinical and popular texts that culminated in his 1881 credo, *American Nervousness*. Emerging as if from nowhere to afflict tens of thousands of urban workers, its etiology lay (as Beard described it) in the perilous increase of "steam power, the periodical press, the telegraph, the sciences, and the mental activity of women"—a curious list of stresses affecting both men and women, all seen to be exacerbated by the booming American metropolises in which neurasthenia exclusively occurred.[35] The neurasthene was plagued by the kind of exhaustion, obsessive behavior, and sleep disruption that might today be called "depression," or "neurosis"; when typed into an electronic library server such as Harvard's Hollis program, it is rendered equivalent to the contemporary ailment "chronic fatigue syndrome." As historians of medicine always remind us, however, correlating disease categories across the ages is a faulty and unproductive enterprise, and certainly the turn-of-the-century neurasthenic patient's suffering took a form that was highly appropriate to its time, attributed to an overwhelming "nervous bankruptcy"—the depletion of overtaxed storehouses containing the body's naturally generated nerve force.[36] Like a battery or "Edison's electric light," Beard proposed:

> The force in this nervous system can . . . be increased or diminished by good or evil influences, . . . and when new functions are interposed in the circuit, as modern civilization is constantly requiring us to do . . . the amount of force is insufficient to keep all the lamps actively burning—this is the philosophy of modern nervousness. [37]

Beard's formulation dominated the neurological literature until Freud's ascendancy, and his mechanistic model of nervous exhaustion drew on a number of previous thinkers. As historians of medicine Francis Gosling and Charles Rosenberg suggest, "Herbert Spencer [and] Thomas Edison" were generalized patron saints. In addition, "Du Bois-Reymond supplied proof of the electrical nature of the nervous impulse, Helmholtz and Mayer their work in thermodynamics, Marshall Hall and others the concept of the reflex."[38]

These mechanical models for Picabia's disease are suggestive; indeed, "suggestion" was seen as the most powerful aspect of the neurasthenic cure. Picabia's doctor in New York (where, as Beard would have predicted, neurasthenia first struck the young Parisian) was doubtless Joseph Collins, an experienced clinician at City Hospital, and professor of nervous and mental diseases in the New York Postgraduate Medical School. His published analysis of several hundred of his clinical cases fit Picabia very well: 55 percent male, average age mid-thirties, 79 percent of an indoor occupation. Of etiology, Collins wrote, "The effect of overwork and masturbation (under which are included for convenience' sake other irregular forms of sexual indulgence) is generally recognized as being very important. Our statistics corroborate this view."[39] Together with other international specialists on neurasthenia (who ranged from physicians such as Adrien Proust and Gilbert Ballet in France, to the sociologist Emile Durkheim), Collins clearly believed that although neurasthenia might originate with a disposition, it only appeared as a disease in the presence of acute social stress exacerbated by immoral pursuits.[40] Isolation, a change of scene, and "psychical or moral therapy" were held by Collins to be most effective:

> the physician may do much by emphasizing how necessary it is to inculcate habits of obedience and self-repression, eradication of egotism and selfishness, restraint of temper and capriciousness, and the development of moral courage and of physical and mental self-confidence.[41]

Needless to say, the patient was to avoid alcohol or drugs, especially if they were part of the "irregular forms . . . of indulgence" held to be responsible for the onset of the nervous disease. (Picabia certainly would have been admonished to stop abusing drugs and alcohol, which were frequent companions.)[42] While Collins held it to be somewhat less important than in Beard's day, electrotherapy was still occasionally useful—largely through that "power of suggestion" already implicated in the neurasthenic cure: "[Electricity's] unknown nature, its wondrous manifestations, its attributed health-restoring capacities, all tend to impress the patient with its potency for benefit. . . . The form that appeals most powerfully to the patient's emotion and the form that is given from the most complicated and elaborate apparatus . . . is the one that will act most beneficially."[43] To current-day readers the apparatus of coils, conducting plates, and electric brushes (and the places to which they were applied) convey a scene of torture rather than "the best means to restore the nerve-tonus," but they were doubtless effective in implanting the electrical metaphor as a constitutive aspect of the neurasthenic subject.[44]

Did Picabia receive electrotherapy before departing to engage the mechanomorphic *fille* one last time? Given Collins's own judgment of its waning efficacy, it is unlikely he received it in New York. But the involvement of the clinic patient in a system of belief relying on electro-mechanical models of the human interior, I would argue, is more than sufficient to be implicated in the renderings of the *FNSM*. Within the metaphor of "nervous bankruptcy" was twined the long association of neurasthenia with a kind of moral and electrical profligacy—for what depleted the male's "storehouses" and "reserves" more dramatically than unproductive *jouissance*? Although Collins had begun to disdain the efficacy of electrotherapy, in Paris as late as 1910 doctors held that the best cure for "*asthenie genitale*" was still "*électrisation*."[45] The homology between sexual and electrical impulses was compelling for these men of medicine; pursuing such metaphors to their logical conclusion, they reasoned that the conduit for biological generation should "naturally" parallel those mechanical conduits for "Artificial Generation" fantasized by Villiers, and, further, that the neurasthenic cure should involve the curbing of "Copulative Excesses" together with the electrical "replenishment" of the body's reserves. As one New York physician wrote in a 1912 handbook on *Neurasthenia Sexualis*:

> The mechanism of sex-activity may thus be compared with the charge of a Leyden-jar. The generative organs must first be charged, like the jar, with a certain material turgescence and with nervous energy in order to evoke the impulse of de-tumescence. Just as the charge of the Leyden-jar with electricity is of a longer duration, compared with the instantaneous discharge at its contact with the earth, so is the charge of the organism with nervous sex-tension usually of longer duration in comparison with the short duration of the discharge. . . . Sexual activity, therefore, consists in the charging and discharging of the vital fluids and nervous tension.[46]

Needless to say, this was a male model of sudden discharge: "repeated orgasm . . . must lead to nervous disorders."[47] The doctor's concern over "Copulative Excesses" seemed tailored (Taylored?) to the male physique:

> Excesses in copulation are not so harmful as excesses in masturbation. . . . Masturbation is . . . more injurious because it is generally effected through the influence of an exalted imagination. Thus excesses in masturbation harm the generative organs not directly only, but also indirectly by first harming the individual's entire nervous system. . . . [N]o other erotic stimuli cause *such a consumption of nerve power* as this gratification of the impulse of contraction by tactile manoeuvres.[48]

Whether or not the philandering Picabia (who moved both wife and mistress to Zurich before beginning a new affair in Barcelona) was lectured by his doctors on the subject of "excessive venery," we can be sure they assumed that something of the sort had been going on. As one doctor wrote: "The patients who seek medical advice for their neurasthenic troubles are those who have . . . freely and immoderately indulged in the unnatural modes of sensualism, whence their troubles originate. The real continent individuals who avoid any kind of erotic practices remain sound and healthy and do not require medical help."[49] Women, of course, received a different diagnosis based on the overarching etiology of the "disease of civilization;" their excesses lay in what Beard had identified as "the mental activity of women." (Clearly, the mental activity of women can also be seen as a problem *for men*, as historians of suffragism have chronicled.) Neurological specialists in particular spoke out against the New Woman, who inappropriately diverted so much nervous force to her brain that the "central telegraphic office" of her genital organs was starved, generating that nervous bankruptcy about which we have learned so much.[50] The therapies for "American nervousness" were thus intensely gender-specific, with females urged to stop thinking, and men, to stop doing. Women were to become more womanly, men, more feminine (in their enforced modesty and withdrawal from the world). Clearly, the motherless *fille*, that creature of Picabia's neurasthenic convalescence, was his partner in neurasthenia. As a figure for the New Woman, her phallic worldliness matched his "excessive venery." Both modes of behavior were keyed to the modern world, yet disrupted the old order of things, plunging both oversensitive male artists and overambitious New Women into neurasthenic collapse. *Fille* and *fils* alike were in the thick of it, as Picabia's obsessive project reveals.

This partner, the *"fille née sans mère"*: What was her role in the book that bears her name? The art historian William Rubin briefly mentions the book's title in his massive volume on Dada and Surrealism, where he translates it as *Poems and Drawings* by *the Daughter Born without a Mother*. The more usual translation would be *Poems and Drawings* of *the Daughter Born without a Mother*, but Rubin's choice reinforces my earlier observation about the possibility of Picabia's identification with the *Fille* through her inversion/analogy with the *Fils*. Since there are no drawings or poems within the book that are given any part of the title *Fille Née Sans Mère*, none seem to depict (or be "of ") the *fille*; it seems clear that Picabia wished, in this volume, to elide his identity as an author with hers, presenting *her* as his authorial voice in delineating these neurasthenic visions of an eroticized electro-machinic phylum.[51]

This assumption of female identity in authorship has ample precedent, of course (proximately in the compelling example of Duchamp's Rrose Selavy). It would be thematized later, as well, within Surrealism—presumably partly in response to Picabia's example (see, for example, Max Ernst's *Rêve d'une petite fille qui voulut entrer au Carmel* from 1930, where the "dream"—visible only to the dreamer—is "remembered" by the artist, who thereby assumes the *petite fille*'s point of view). Apparently Picabia moved closer to identification with the *fille* over the course of producing the book, for it was originally to be titled *Décapuchonné*, with the *FNSM* functioning as a subtitle. Without the feminine ending, the French word *décapuchonné* describes something that has happened to a male; with the originally intended subtitle, the "decapuchonned" male must be seen to make poems and drawings *to* the FNSM.

The book's original title *décapuchonné*, taken at face value, means "unhooded," or, more colloquially, "defrocked"—the Capuchin monk's cowl removed as a sign of his beginning a secular life (to begin, one supposes, a more raffish existence with the motherless daughter of his dreams). Such a juxtaposition would not be without precedent in Picabia's work, and was common in the violently anticlerical mood of early modernism (witness Ernst's eroticized Carmelite novitiate).[52] The most extensive dictionaries give "defrocked" as a rare definition, however; more common are the range of associations that cluster around "taking off a hood," from the falconist's preparation of his raptor for flight to the writer's removal of a pen's protective top. One dictionary reference uses *"décapuchonner"* in a specifically mechanistic way, comparing the action of *décapuchonnent* to the circular mechanical movement needed to fuel rockets; others evoke a more personal gesture open to manipulations of desire.[53] For Picabia, the range of such associations for the book's original title were all appropriate. The sense of the *FNSM* volume as both "uncapped" (as in liberated) and "defrocked" (as in booted out of religion) presented his ultimate answer to the conservative Catholic natalist movement then on the rise in France. It also, of course, opened on to a world of potentially mechanistic actions, seemingly possessed by a male but played out by the eponymous daughter, the *fille née sans mère*.

In this necessarily brief essay, only a few of the images from the book can concern us, and a few of the poems. The bulk of the line engravings return to the open, linear, elliptical style of Picabia's first drawing for the *FNSM*: a spare iconography reminiscent of hand-drawn genealogical charts, sketchy anatomical diagrams, or even Freud's contemporaneous schematic illustrations of the human psyche.[54] In addition to illustrating parts of machines (and, indeed, Picabia drew extensively from reproductions he found in the popular engineering journal *La Science et la vie*),[55] the drawings are machine-like in another sense: they exhibit the dry line and sober tone of what are called, in English, "mechanical drawings," that is, commercial line drawings made with compass, rule, and mechanical drawing pen. At the same time, Picabia's line is both less and more than mechanical. Less, because unlike the disciplined pen of true mechanical drawing (see Figures 2 and 3 for examples), Picabia's line here refuses to complete itself: it stutters and repeats across the page, it fails to reveal crucial details of the "mechanisms" involved, and it requires elaborate textual inscriptions to explain itself. More, because the same line wanders into the interiors of these machines— limning not merely the cross-section of their motors, but the unexpected soul, id, and furry reproductive parts that Picabia/the *fille* discovers deep within.

A drawing titled *Mammifère*, for example, charts the parts of a mammiferous body, juxtaposing precise medical terminology (*"l'utérus"*) with small furry animals that might be associated with hair-covered erogenous zones of the human female (*"chauve-souris,"* bat, and *"ouistiti,"* marmoset). Another drawing titled *Haricot*, by contrast, is a male device, depicting a sketchy apparatus definitively identified with the label *"Du Mâle,"* of the male, or more colloquially, of his cock. Obligingly enough, a ladder climbs up from the base of this phallic structure, avoiding both "Madagascar" and a poisonous cloud (*"nuage poison"*) on its ascent up the page. Most intriguingly, this *haricot*/beanstalk, while rooted in *"Du Mâle,"* is also marked by the phrase *"destinées,"* thus seemingly fated to return to a plural feminine destination that might be trans-lated as "the intendeds" (as in multiple marriage partners), or "the destined women." The poem on the facing page evokes a therapeutic scene: "she washes herself and binds the hand / smiling always. / She rules the science of chaining / the degrees of water. / . . . I am the monarch warbler variety / [with the] modesty of spermatozoid pas-sivity. / Inaesthetic sailor wan / near the lake without sun."[56]

Haricot's incorporation of female elements into male mechanisms appears again, reaching thematic proportions in other drawings and poems of the book. The drawing *Égoïste* relates to a "convalescing narcissist" (*Narcis convalescent*—doubtless the neurasthene himself), attended by a thin probe labeled "doctor" (*médecin*) on one side, and phrases evoking female landscapes (*femmes paysages*) and *Américaines* on the other.[57] Recall the enthusiasm of Picabia's doctor, Joseph Collins, for inculcating "habits of obedience and self-repression, eradication of egotism and selfishness" in his patients—an American prescription that may have chafed the self-reflective habits of a male Parisian raised in the Cousinian culture of the *moi* (for which see Jan Goldstein in this volume). In his mournful poem "Vivre" (two pages before the drawing *Egoïste*), Picabia/*fille* reflects on the boredom of the neurasthene's regime:

> *Conquering egoism amuses a fool*
> *A lover waits for good times*
> *Affairs of appearances*
> *Me I've never seen*
> *Those who bring them off*
> *The unknown have no theories*
> *Of dissipation*
> *Along the shipwrecked river*[58]

Similarly, in the poem that confronts the drawing in question, the author laments: "The truth of the soul / Is the great cowardice of academic pride / My eyes in your eyes / I am content / In my forgotten solitude"[59] In the same poem, Picabia and the *fille* admit *"J'aime que l'on plie les yeux / Des ennuis"* (I like what bends the eyes / From boredoms), and the multiply sexed and gendered interior views of *Égoïste* and other drawings sug-gest just how the eyes might have been bent from the task at hand.

Still other drawings feature other stand-ins for the convalescing narcissist. The "young Sable" (*"jeune zibeline"*) whose capacious, multifaceted body is mapped in the drawing *Polygamie* incorporates both patriarchal, penetrating "Mormons" (iconograph-ically linked to the "médecin" of *Égoïste*) and a "spring vagina" (*"vagin printanier"*). Both penetrants and opening are connected to the ambiguous body of the sable (the hair of which, it should be noted, is a chief constituent of the paintbrushes forbidden to the neurasthene). The *vagin printanier* is verbally associated with another work from the same year, *Brilliant Muscles/Vagin Brillant*, where the *vagin* is labeled *"mécanique de la region sacrée."*[60] The machine on which the drawing *Polygamie* is based, appropriately enough, is a fan for a gas meter named "Duplex."[61] The *vagin's jouissance*, it seems, is not the quintessentially nonproductive (and hence subversive) labor of a Sadean, but the productively rhythmic, repetitive, metrically measurable and mechanical energies of the androgyne participating in this "Polygamy."

Just as he had discovered machines at "the very soul" of human life, then, Picabia also discovered machinic sex: *vagins, regions sacrées*, and penetrating patriarchs at the heart of his mechanical drawings. The drawings *de la fille* function complexly and intertextually, meanings and associations building up through accretion, enigmatic forms echoing in other drawings with clearer clues, words reappearing in different con-figurations suggesting multiple interpretations. The title for the drawing *Male*, for example, is spelled without the circumflex that appears "correctly" inside the body of *Haricot*.[62] Without the diacritic, it shifts from being a simple cognate for "male," and may instead drift toward a fictively feminized adverb for "bad" (*mal*). The wiry coil that threads through the drawing curls over and over on itself, forming a chain of "elleelleelle"s in a cursive French hand.[63] This Malelle (can s/he be other than Picabia/the *fille*?) appears in the shape of a wobbly hourglass, the word "*hermaphrodism*" emerging from the center of the enigmatic device directly opposite the label "*le chat*." The two words, cat and hermaphrodism, converge at the most constrictive passage of the apparatus, accompanied by the spiraling coil (of electrical wire?) that runs from nowhere to nothing. In these drawings, ladders, constrictions, coils, and conduits become figures for a mechanical cathexis. The pulsing of blood and sperm are linked inextricably to the rush of electrons through a mechanical coil, itself a figure for the neurasthene's nervous energy flowing back into the battery of the ego's emotional reserves.

These themes of hermaphrodism, electricity, and pulsing love machines come into focus in one of the book's most elaborate images, *Hermaphrodism* (Figure 5), which I would argue is also its most revealing and important page. Here some of the cryptic forms of drawings printed earlier in the book become more clear: the wiry coil is explicitly electrical, appearing twice, with one end tipped by a plug and the other labeled "*sperme*." Both feed into (or emanate from) a sexual apparatus ("*appareil sex-uel*"), which bears the shape of many of Picabia's female machines—the disk or hole—again penetrated or activated by a slender rod. In this image, the rod protruding from the sexual apparatus is positioned as actively phallic (not merely metaphorically so). It reaches down to probe an "*oviducte*" studded with egg-like rivets, seeming to deposit its vital electrical/spermatazoid forces in a collecting zone of *mâle haché*. What is being

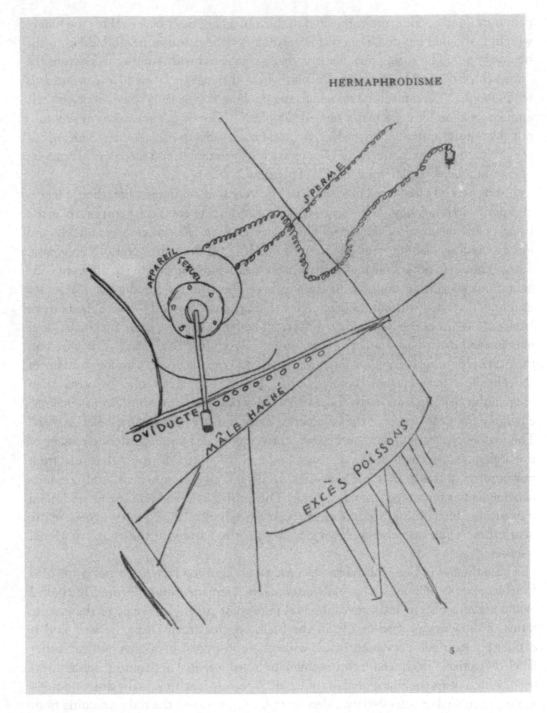

Figure 5. *Francis Picabia*, Hermaphrodisme, *line engraving from Picabia*, Poèmes et dessins de la fille née sans mère, *published fall 1918 in Lausanne, Switzerland (p. 63).*

collected here—the essence of "mixed-up male" or "chopped male"? This minced or mixed substance drains away down lines in the drawing, just so much "excess fish"— animated, fluid creatures whose French name might play on the word "poison," but whose piscine character resonates with the whip-tailed animacules of energizing sperm.

On the page facing *Hermaphrodism*, Picabia/the *fille*'s poem concludes: "It's the nervous system / for the personal imagination / of the pleasures of testing / impossibility." [64] And indeed, the neurasthene's personal imagination *was* a nervous system (in all senses of the phrase), suffused with slipping glimpses of impossible pleasures and new regimes of channeled electrical and sexual impulses. The fully supplied equipment of the FNSM may have been painful to experience for the masculine half of the dyad (the "*mâle haché*"). But the humbled "convalescing narcissist" seems to have taken something from his *fille*; his oscillating and ambivalent identification with the phallic New Woman seems to have produced a new vision of the possibilities and pleasures of "testing impossibility."

In these highly sexed mechanisms, of which *Hermaphrodism* is exemplary, we find the culminating conundrum of Picabia's art at the point of his constitution as a neurasthenic subject. Neither essentially male nor "naturally" female, the hermaphroditic machine presents the personal psychological equivalent of the merger between the motherless *fille* and the orphaned *fils*, the too-active male and the too-mental female. Although it might be supposed that the neurasthenic cure might be intended to eradicate such conflation by enforcing more appropriate sexual and gender behaviors, Picabia's project shows otherwise. Negotiating the role of neurasthenic subject for his doctors, Picabia is interpellated as an author of his own "psychical and moral" therapy. The hermaphroditic machine is what he takes to be his appropriate neurasthenic self. That it exists primarily as a two-dimensional reproduction of a line drawing does not make it any less potent in structuring the psyche.

As befits such a merged and complex identity, let us propose four hands for a concluding analysis of Picabia/*fille*'s hermaphroditic machine: on the one hand, as we have seen, s/he is a phallic little thing, slender rods and hardened disks the very instantiation of the bodies or body parts that are made rigid with discipline or desire. On the other, s/he is a transparent, permeable, trembling membrane, ruler of the "science of chaining . . . water." On the proliferating third hand, s/he offers the inexhaustible mechanism of *jouissance* in a framework of self-love: equipped with sperm conduit and oviduct, s/he is a self-lubricating being, both "wan sailor" and tender nurse, "content in my forgotten solitude." On the fourth hand (and why not go on like Vishnu?) s/he is completely social, "only a machine," controllable by man, produced by him and for him, man's own "daughter born without a mother." This last hand reaches to Pygmalion's Galatea, and then, of course, to Villiers's *L'Eve futur*: the ideal "daughter" born, quite literally, "without a mother," provided as if by God for the pleasure of man and the (re)production of his labor. [65] But the hermaphrodism of Picabia's devices provides a slightly different take on the subject—or at least offers the possibility for more readings than Villiers's frankly misogynist fiction provides.

The crux of Picabia's electrified, organic, hermaphroditic machine is its offer, paradoxically, of a way out of sex—a way to internalize and incorporate difference, together with an open acknowledgment of the slippage and deferral of meaning we call, since Derrida, *différance*. But like Surrealist Andre Breton's comment (itself a modification of one of Picabia's *aperçus*), "I wish I could change my sex as I change my shirt," Picabia's offer needs to be examined closely.[66] "Changing one's sex" implies a transformation more thorough than changing one's shirt—the former suggests a willing exchange of sexual identity, the latter merely a freshened version of the same old male uniform, the (phallus) shirt. Picabia's internal views of hermaphroditic machines may have offered a way out of sex—a way that seems exhilarating in our own gender-bending age—but perhaps it was only a way out of that kind of complicated and demanding sex that happens with real women. The fluid neurasthenic subject was, after all, temporary and unstable, an identity consciously constructed as outside the "real" world. The patient was intended to be cured, and in some of Picabia/*fille*'s constructions, we can see the mechanisms being staged for this reemergence. The new kind of reproduction we have glimpsed, without explicit difference, may have been a reproduction without the political troubles that difference seemed to bring—without, in fact, the actual women that exemplified difference in the social realm. Picabia/*fille*'s poem *Le Germe* suggests this kind of escape, in typically hermetic and elliptical form:

> *Animal-man*
> *Towards nothingness*
> *Envelops his feelings . . .*
> *Of mutual penetration*
> *Mechanism blind and dumb*
> *We will find some wings that live according to Plato*
> *In the appearances of realities.*[67]

The elusive meanings of the poem crystallize in the pairing of the phrases "Of mutual penetration / Mechanism blind and dumb." In this corner of Picabia's Imaginary, the machine replaces the actual female to become a source of *mutual* penetration: both receptive orifice and incisive probe, but above all "blind and dumb." The *fille née sans mère* promised itself (in Picabia's imaginary) to be a machine that borrowed all the androgynous sex appeal and liberated behavior of the New Woman, without her attitude. Setting aside our current desires for hermaphroditic machines, Picabia's early-twentieth-century version may only be, in the end, the ultimate Sadean commodity—a fusion of the neurasthene's electrified Imaginary with the seducer's erotic visions of incorporation and absorption of the Other. In the *FNSM*, the space of difference is narrowed, even collapsed—but this astonishing feat may serve only to produce a polymorphous coupling device of insensate servitude that would, as the *jeune fille Américaine* always promised, keep on going "For-Ever."

CONCLUSION

Picabia's hermaphroditic solution may have addressed his own psychosexual needs, and, in that specificity, left intact the misogynist trajectory of Dada and Surrealism. But the poems and drawings of the *fille* offer present-day viewers a glimpse of the path not taken, an unheeded alternative to the art-historical logjam set up by Marcel Duchamp's frustrated bachelors and isolated brides. The disappearance of the ambiguous, hybrid, and polymorphous sexuality made available by the *FNSM* may be due to the very specificity of sexuality's function within the neurasthenic regimen. What is clear is that the dominant model of machinic sex is still Duchamp's *The Bride Stripped Bare by Her Bachelors, Even (the Large Glass)*, slowly "painted" by Duchamp during almost the entire decade of Picabia's mechanomorphic production (1915–1923). The *Large Glass* stands as a virtual icon of the 1920s' infatuation with the eroticized machine, revised and revisited after World War II by artists as various as Jasper Johns, Robert Smithson, Hannah Wilke, and Rebecca Horn.[68] In the complex vertical composition, Duchamp's bride remains forever isolated on the top, his bachelors ever celibate on the bottom. The only connection between the disrobing bride and impotent bachelors is the "love gas" that the latter spray forth. To be more accurate, what connections exist between the two sexes have the function of *alternations*. In Duchamp's notes for the piece, we hear echoes of the neurasthenic vocabulary of electrical circuitry and "copulatory excess," but Duchamp implicitly accepts the hyper-gendered model of the neurasthenic system that Picabia's project works to complicate:

> there is no discontinuity between the bach. machine and the Bride. But the connections, will be, *electrical*. and will thus express the stripping: an alternating process. Short circuit if necessary. . . . Slow life—Vicious Circle—Onanism.[69]

The promise to "short circuit" the electrical alternation between male and female is an intriguing possibility (one that may have been realized in the later accident that fractured the *Glass*), but as built, the work's separation between bride and bachelors is complete. As William Rubin summarizes its "intricate amatory iconography," "the *Large Glass* constituted . . . an assertion of the impossibility of union, hence, of sexual futility and alienation."[70] While Duchamp posits a female machine (itself a destabilizing move), she remains isolated, her "marriage" unconsummated and her desire unknown—a far cry from the mutual penetration of Picabia's more outgoing, hermaphroditic machines.[71]

As we have seen, technology inhabited Picabia's imaginary in peculiar and particular forms, and his elision into identification with the mechanical *Fille* was more thoroughgoing even than Duchamp's gender-bending presentation of himself as Rrose Selavy. But I have argued that this personal psychological level resonated with more general discourses about sexuality, "modern nervousness," and machines. In fact, although the specifically neurasthenic content of his project seemed to pass largely unremarked, Picabia's enormously evocative name for the *fille née sans mère* proved so productive that it was adopted by his friend Paul Haviland, writing in *291* about the benefits of new technologies such as the camera:

> Man made the machine in his own image. She has limbs which act; lungs which breathe; a heart which beats; a nervous system through which runs electricity. The phonograph is the image of his voice; the camera the image of his eye. The machine is his "daughter born without a mother." That is why he loves her. . . . She submits to his will, but he must direct her activities. . . . Through their mating, they complete one another. She brings forth according to his conceptions.[72]

Haviland's incestuous fantasy is very clear; his "*destinées*" are the cinemagenic females of Fritz Lang's *Metropolis* and Villiers's Eve, who unify voice and image and present a phantasmagoric cross-circuiting (but not short-circuiting) of patriarchal and filial desire. I have hinted all along that the sexed machines of the cultural Imaginary may themselves have veiled anxieties about more practical problematics such as the position of actual women in the world, or fragile subjectivities in the rapidly industrializing urban environment. Although this paper has only hinted at the larger relationships between, for example, the industrial workplace and these motherless machines (and remember that they are not fatherless), or their links to the emergence of a seemingly powerful New Woman, I would like to open a space for further discussion and, potentially, new Imaginaries. Those who have studied the early-twentieth-century fascination with the sexed machine have posited some fairly straightforward interpretations: Andreas Huyssen proposes that the growing fear of technology was displaced onto females who could then be mastered and destroyed.[73] Peter Wollen suggests that Americanism and Fordism became routes to a mechanization of real bodies that had as its goal the control of sexuality, inverted and mirrored, as Rosi Braidotti sees it, in the Sadean dynamic of repetitive mechanistic rituals that ultimately fail to contain the nonproductive energies of *jouissance*.[74] But these formulas seem perhaps too tidy for the complex dynamic of the Picabian hermaphroditic machines, or even for the parallel formation of the phallic woman. With the psychoanalytic depth invited by Picabia's drawings of, and by, the FNSM, we can see the potentially absolute *unfixity* of the machinic phylum—poised, as we are, at what Gilles Deleuze saw clearly as a new episteme.[75] The machine is neither utterly outside nor wholly in the human body, neither male nor entirely female, neither bad nor completely good. The same egg-beater can be male or female, and the mechanomorphic body may turn out to be both bride and groom. Psychoanalysis, that emerging discourse whose origins were not so distant from Picabia's cultural frame (nor, perhaps, entirely unfamiliar to the neurologists to whom he dedicated his *fille*), offered an early vision of the slippage of meanings, the doublings and inversions, the misprisions and parapraxes that refuse reduction to a single "fact" or a fixed identity. Picabia elaborates and extends this view—not as a case of ambiguity, but as an oscillatory shift between what Lacan calls the Imaginary and the Symbolic. In Picabia's hermaphroditic machines, we have both the polymorphous bliss of the self-lubricating system—a pre-Oedipal, pre-linguistic state—and the law of Logos and linguistic *différance*, the insistently productive and patriarchal order of electrical coils, meanings, *mâles*, convalescing narcissists, and even *haricots*.

And if these insights about the sex of machines have pertinence for our understanding of early modernism, they are equally relevant for our charting of the post-World War II configurations seen in Warhol, Stella, Smithson, and their 1960s colleagues, as well as the more recent erotic cyborgs of Donna Haraway or Rebecca Horn, William Gibson or Ridley Scott. Reading through Picabia, we can see much more instability in these discourses of technology, more slippage in those signifiers and more destabilizing effects. Now that the female machine has been disinterred and recreated as Haraway's utopian cyborg, can we rescue her from that old Galatean function as man's daughter without a mother? Far from answering such a question, I can only hope, perhaps mischievously, to open it up for further interrogation. Will it turn out to be a can of worms, or Pandora's box? Probably, like Picabia's *Fille/Fils*, the answer is "both." And if we can make sense out of these kinds of trouble, we may understand the larger troubles that continue to fret our technological dreams.

Notes

1. Francis Picabia, quoted in "French Artists Spur on American Art," *New York Tribune*, October 24, 1915, part IV, p. 2. My thoughts on New York Dada, and the American sources for surrealism's mechanomorphic imagery, were first stimulated by my work in Milton Brown's seminar at the Institute of Fine Arts in 1982; I am grateful to Professor Brown for his comments on my unpublished paper on "New York Dada and the American 'Avant-Garde,' 1910–1925." Versions of the present essay were presented at Princeton University in the Women's Studies Program and the department of art and archaeology, at the Berlin Summer Academy of the Max Planck Institute, at the Program in Science, Technology, and Society at MIT, at the Center for Literary and Cultural Studies at Harvard, and at the Whitney Museum of American Art; I am grateful to each of those audiences for their responses. Helpful readers and sounding boards were found in Rosi Braidotti, Wanda Corn, Rachelle Dermer, Laura Engelstein, Yaakov Garb, Peter Galison, Amelia Jones, Joan Scott, Susan Suleiman, and Robert Wokler.

2. The concept is Michel Foucault's. See his discussion of "the 'micro-physics' of power," in which relations of power are not imposed from above, or from a distance, but "go right down into the depths of society." *Discipline and Punish: The Birth of the Prison*, trans. Alan Sheridan (New York: Vintage, 1979), pp. 26–27.

3. Judy Wajcman, "Technology as Masculine Culture," in *Feminism Confronts Technology* (University Park: Pennsylvania State University Press, 1991), p. 137.

4. The language of Benjamin's generative essay is one of its many complications, for it was published initially in French in the exiled German-language *Zeitschrift für Sozialforschung* V, 1 (1936). For the English translation of the essay and a brief introduction to his work, see Benjamin, *Illuminations*, trans. Harry Zohn (New York: Schocken Books, 1969).

5. See Nancy Ring, *New York Dada and the Crisis of Masculinity: Man Ray, Francis Picabia, and Marcel Duchamp in the United States, 1913–1921* (Ph.D. dissertation, Northwestern University, 1991), pp. 29, 124, 131.

6. The question of who gave the work its title is unsolved. Art historian Francis Naumann ventures this opinion:

   > The photographic image of the eggbeater was first made by Man Ray in 1917. At that time, he titled the picture *Man*. It was the companion piece to his subsequent photograph of lights and clothespins, *Woman*. However, the eggbeater photograph later reappears, retitled and redated *La Femme*, 1920 (Collection Centre National d'Art et de Culture Georges Pompidou). Since the handwriting at the bottom of the photograph does not appear to be Man Ray's, it is tempting to imagine that Tzara, having received *Man* and *Woman* for his magazine *Dadaglobe*, switched their titles when he entered them in the 1921 exhibition at the Théâtre des Champs-Elysées. No copy of *Woman* retitled *L'Homme* exists, however.

   Nauman, "Man Ray 1908–1921: From an Art in Two Dimensions to the Higher Dimension of Ideas," in Merry Foresta et al., *Perpetual Motif: The Art of Man Ray* (Washington and New York: National Museum of American Art, Smithsonian Institution, with Abbeville Press, Publishers, 1988), p. 86, n. 35.

7. While I'm sympathetic to recent discourses about the inseparability and mutability of *both* sex and gender, and to counter-discourses that demand a resituation in actual female bodies (rather than endlessly deconstructed texts), I want here to preserve the useful (if somewhat arbitrary) distinction between sex, as a biological category, and gender, as a social construction. Certainly for the critics of the "*femme nouvelle*" in the early 1920s, there was a clear distinction between gender—that cluster of disturbingly mutable activities such as fashion, profession, and demeanor—and sex, that reproductive role that would be enforced (if necessary) on those whose chromosomes rendered them capable of giving birth.

8. I need both difference and *différance* to denote both the space opened up by the divergent titles (difference), and the complex system of traces that is thereby set into play.

   > *Différance* is the systematic play of differences, of the traces of differences, of the *spacing* by means of which elements are related to each other. [T]he *a* of *différance* also recalls that spacing is temporization, the detour and postponement by means of which intuition, perception, consummation—in a word, the relationship to the present, the reference to a present reality, to a *being*—are always *deferred*. Deferred by virtue of the very principle of difference which holds that an element functions and signifies, takes on or conveys meaning, only by referring to another past or future element in an economy of traces.

   Jacques Derrida, interviewed by Julia Kristeva in 1968, anthologized in Derrida, *Positions*, trans. and annot. Alan Bass (Chicago, 1981), pp. 27–29.

9. Klaus Theweleit, *Male Fantasies*, vols. I and II (Minneapolis: University of Minnesota Press, 1987 and 1989); Jeffrey Schnapp, *Staging Fascism*, work-in-progress, and Hal Foster, *Prosthetic Gods*, work-in-progress. I am grateful to Jeff Schnapp for sharing his work with me, and for alerting me to Hal Foster's similar research into fascism's erotics. For a different view in which technology is neutral, but

ultimately shifted to the woman in order to demonize it, see Andreas Huyssen, "The Vamp and the Machine: Technology and Sexuality in Fritz Lang's *Metropolis*," *New German Critique* 24–25 (Fall–Winter 1982): 221–37.

10. For brief reviews of my take on mechanomorphic art in modernism, see the final chapter of *Machine in the Studio* (Chicago: University of Chicago Press, 1996), my contribution "Artistes et ingénieurs," in *Les Ingénieurs du Siècle*, ed. Antoine Picon (Paris: Musee national d'art moderne, Centre Georges Pompidou, 1997), and my *Painting Machines* (Boston: Boston University Art Gallery and University of Washington Press, 1997).

11. Joseph Stella, quoted in John I. H. Baur, *Joseph Stella* (New York: Shorewood Publishers, 1963), p. 13. The five-panel polyptych was described by Stella as a symphony, but again, it was a curiously *gendered* musical form: "a symphony free in her vast resonances, but firm, mathematically precise in her development . . . highly spiritual and crudely materialistic alike." Ibid., p. 35.

12. Gabrielle Buffet-Picabia, "Some Memories of Pre-Dada: Picabia and Duchamp" (1949), in *The Dada Painters and Poets: An Anthology*, second ed., ed. Robert Motherwell (Cambridge: Harvard University Press, 1989), p. 258.

13. See Estelle B. Freedman, "The New Woman: Changing Views of Women in the 1920s," *Journal of American History*, LXI, 2 (September 1974): 372–93; Carroll Smith-Rosenberg, "The New Woman as Androgyne: Social Disorder and Gender Crisis, 1870–1936," in her *Disorderly Conduct: Visions of Gender in Victorian America* (New York: Alfred A. Knopf, 1985), pp. 245–96; and, for the French case, Mary Louise Roberts, *Civilization without Sexes: Reconstructing Gender in Postwar France, 1917–1927* (Chicago: University of Chicago Press, 1994); and Debora Silverman, "Amazone, Femme Nouvelle, and the Threat to the Bourgeois Family," in *Art Nouveau in Fin de Siècle France: Politics, Psychology, and Style* (Berkeley and Los Angeles: University of California Press, 1989), pp. 63–74.

14. Terry Castle, "The Female Thermometer," *Representations* 17 (Winter 1987): 1–27. Picabia's two drawings are published in his Dadaist journal *391*, 16 (May 1924): 3–4.

15. Castle, "The Female Thermometer," p. 22.

16. Marius Ary Leblond, cited by Silverman, "Amazone," p. 69.

17. Victor José, "La Féminisme et le bon sens," in *La Plume* 154 (September 15, 1895): 391–92, cited in Silverman, "Amazone," p. 72.

18. Joan Wallach Scott, *Gender and the Politics of History* (New York: Columbia University Press, 1988). For a trenchant analysis of the fate of "*la garçonne*" in French postwar fiction, see Roberts, *Civilization without Sexes*.

19. See Roberts, *Civilization without Sexes*, pp. 8–9 and passim.

20. Pierre Drieu La Rochelle, *La Suite des idées* (Paris: Au Sens Pareil, 1927), p. 125, cited in Roberts, *Civilization without Sexes*, p. 2.

21. Nancy Ring, *New York Dada*, n. 15, p. 23.

22. Francis Picabia, "Soldats," *391* 7 (August 1917): 2. Ring translates this as "Hungry madness / desperate attitudes / the sick wall of the feminine sex." *New York Dada*, p. 22.

23. M. Numa Sadoul, writing in *Progrès Civique* (June 13, 1925): 840, cited in Roberts, *Civilization without Sexes*, p. 9.

24. As Siegfried Kracauer wrote of this English dance troupe, which intoxicated Europe in the 1920s: "The hands in the factory correspond to the legs of the Tiller Girls." ("The mass ornament," 1927, translated and published in *New German Critique* 5 [Spring 1975].) Illustrating my point about the European's linkage of things technological with things American, Kracauer was certain the Tiller Girls were American. Writing of them again in 1931 for the *Frankfurter Zeitung* ("Girls und Krise," no. 27, May 1931), he declaimed

the Girls were artificially manufactured in the USA and exported to Europe by the dozen. Not only were they American products; at the same time they demonstrated the greatness of American production. . . . When they formed an undulating snake, they radiantly illustrated the virtues of the conveyor belt; when they tapped their feet in fast tempo, it sounded like *business*,

business; when they kicked their legs with mathematical precision, they joyously affirmed the progress of rationalization; and when they kept repeating the same movements without ever interrupting their routine, one envisioned an uninterrupted chain of autos gliding from the factories into the world.

See Peter Wollen, "Cinema/Americanism/The Robot," *New Formations* 8 (Summer 1989): 24–25.

25. Translated by Robert Martin Adams as *Tomorrow's Eve* (Urbana: University of Illinois Press, 1982).

26. Ibid., p. 98.

27. Duchamp was clearly the most important art-historical and personal influence on Picabia's machine art, but the influence went both ways (as Duchamp's kinship exchange gesture suggests). Léger tells an interesting story about Duchamp that sets the stage for Picabia's importance to the younger painter (the two met in the winter of 1910–11):

> Before the World War I went with Marcel Duchamp and Brancusi to an airplane exhibition. Marcel . . . walked around the motors and propellers without saying a word. Suddenly he turned to Brancusi: "Painting has come to an end. Who can do anything better than this propeller? Can you?"

Fernand Léger, around 1957, quoted in Pontus Hultén, *The Machine as Seen at the End of the Mechanical Age* (New York: The Museum of Modern Art, 1968), p. 140. Picabia's sympathy with Duchamp's perspective, and his pursuit of pure abstraction (earlier and more consistent than Duchamp's), suggests that the synergism of their views was crucially important to them both, and impossible to disentangle in a spurious search for priority.

28. The letterhead on the verso is just visible in the reproduction of the drawing in Figure 4. The Metropolitan Museum of Art, where the drawing is now located as part of the Stieglitz collection, has parsed the name through the paper as "Braevoort House," which I have interpreted as the "l'hôtel Brevoort" referred to by Gabrielle Buffet-Picabia as their lodgings during their first stay. More significantly, Picabia's wife recalls that it was here that Picabia made a suite of watercolors (and drawings?) that served as the basis for his exhibition at Stieglitz's gallery; thus, if the drawing dates from this first trip and was given to Steiglitz with the other works for the exhibition, it would be another explanation for its location in the Steiglitz collection. (On the other hand, Picabia's second wife and executor of his estate, Olga Picabia, dates the work as 1915.) Buffet-Picabia writes of the excitement of this first trip and its effect on Picabia's art:

> Cette ambiance vivifiante ne devait pas tarder à manifester ses effects, c'est-à-dire un irrésistible désir de peindre. Il revint un jour à l'hôtel Brevoort, où nous habitions, avec l'outillage nécessaire à son travail, organisa une installation de fortune et les murs se couvrirent bientôt d'une série d'aquarelles de grandes dimensions qui recréaient aussi un climat inconnu par la richesse de leurs inventions et l'éloquence plastique de leurs "abstractions."

Gabrielle Buffet-Picabia, *Recontres avec Picabia, Apollinaire, Cravan, Duchamp, Arp, Calder* (Paris: Pierre Belfond, 1977), p. 46. Wanda Corn, scholar of these transatlantic exchanges of "Americanisme," informs me that "The Brevoort hotel is where all the French exiles stayed and/or socialized during the war" (email August 15, 1995), which certainly reinforces my inference as to the letterhead's source.

29. The "doctor-mechanic" brings to mind Benjamin's famous comparison, in the "Work of Art" essay, where the painter is like a magician, and the filmmaker like a surgeon who "penetrates deeply into the web of reality." For the importance of X rays in Picabia's ideas, and in the cultural imaginary as a whole, see Linda Dalrymple Henderson, "Francis Picabia, radiometers, and X-rays in 1913," *Art Bulletin*, 71, 1 (March 1989): 114–23. I am grateful to Dr. Henderson for many citations and insights about Picabia's work in this period.

30. William Homer, "Picabia's *Jeune fille américaine dans l'état de nudité* and her friends," *Art Bulletin* LVII (March 1975): 111.

31. Villiers de l'Isle-Adam, *Tomorrow's Eve*, p. 144.

32. William S. Rubin, *Dada and Surrealist Art* (New York: Harry N. Abrams, 1968), p. 56.

33. See, for example, *Voilà la fille née sans mère*, dated 1916–17 (illustrated in *Picabia*, Musée d'ixelles, 1983) and the 1917 *Fille née sans mère* (illustrated in Hultén, *The Machine*, p. 83). The latter image was produced in Barcelona during the first few months of Picabia's rest cure for neurasthenia.

34. "*Je dédie cet ouvrage à tous les docteurs neurologues en général et spécialement aux docteurs: Collins (New-York), Dupre (Paris), Brunnschweiller (Lausanne). F. Picabia.*" Frontispiece, *Poèmes et dessins de la fille née sans mère* (Lausanne, 1918).

35. George M. Beard, *American Nervousness: Its Causes and Consequences, A Supplement to Nervous Exhaustion (Neurasthenia)* (originally published in 1881 by Putnam's, New York; reprint New York: Arno Press, 1972), p. 96. There is a now a large literature on neurasthenia and its relation to modern life. For an excellent history of how neurasthenia related to the emerging modern workplace, see Anson Rabinbach, *The Human Motor: Energy, Fatigue, and the Origins of Modernity* (Berkeley and Los Angeles: University of California Press, 1992). On neurasthenia and manhood, see Mark Seltzer, *Bodies and Machines* (New York: Routledge, 1992). On the medical history, see Francis Gosling, *Before Freud: Neurasthenia and the American Medical Community 1870–1910* (Urbana: University of Illinois Press, 1987). For the French case, see Robert A. Nye, *Crime, Madness, and Politics in Modern France: The Medical Concept of National Decline* (Princeton: Princeton University Press, 1984).

36. Beard, *American Nervousness*, p. 9.

37. Ibid., p. 99.

38. Francis Gosling, *Before Freud*, p. 10, quoting Charles Rosenberg, "The Place of George M. Beard in Nineteenth-Century Psychiatry," *Bulletin of the History of Medicine*, 36 (1962): 249.

39. Joseph Collins, M.D., "The Etiology and Treatment of Neurasthenia. An Analysis of 333 Cases," *Medical Record* 55, 12 (March 25, 1899): 414.

40. See Robert A. Nye, *Crime, Madness, and Politics*, pp. 149–52, passim. Adrien Proust and Gilbert Ballet, *L'hygiène du neurasthénique* (Paris: Masson, 1897). Nye notes (p. 148) that the Proust and Ballet book was "the standard medical text on neurasthenia" in France, and points out that it appeared the same year as Durkheim's speculations on neurasthenia in Emile Durkheim, *Suicide: A Study in Sociology* (1897), trans. John H. Spaulding and George Simpson (Glencoe, Ill.: Free Press, 1951).

41. Collins, "Etiology," pp. 416, 415.

42. Indeed, Picabia's collapse is still described primarily as alcoholism or drug addiction. See *Dada Invades New York* (New York: Whitney Museum of American Art, 1997), p. 79. Whatever contemporary terms are brought to bear on the nature and etiology of Picabia's recurrent illness, it remains the case that his wife at the time, Gabrielle Buffet-Picabia, identified it as neurasthenia, and he was given therapy by three neurologists, to whom he felt grateful enough to dedicate his book of poems and drawings.

43. Collins, "Etiology," p. 419. For Collins's part, he believed that the apparatus using static electrical impulses was more effective than the faradic or galvanic type.

44. Bernard S. Talmey, M.D., *Neurasthenia Sexualis: A Treatise on Sexual Impotence in Men and in Women, for Physicians and Students of Medicine*, (New York: Practitioners' Publishing, 1912), p. 147. Talmey explains the procedure for one particularly gruesome treatment: "When the faradic current is used, one pole is applied to the genitals, the other within the rectum. A sponge electrode may also be placed upon the lumbar spine, while an electric brush is swept over the glans penis, scrotum, hypogastric region, buttocks, perineum and inner surfaces of the thighs" (pp. 153–54).

45. Maurice de Fleury, *Les Grands Symptômes Neurasthéniques (Pathogénie et Traitement)*, 4th ed. (Paris: Félix Alcan, 1910), p. 199.

46. Talmey, *Neurasthenia sexualis*, pp. 66–67.

47. "As a matter of fact veneral excesses are followed by malaise, nervousness, mental depression, lassitude, fatigue, satiety, heaviness in the head, disposition to sleep, dullness of intellect, indisposition to exercise, want of decision, regrets and ill-humor, and the other symptoms of general neurasthenia." Ibid., pp. 74–75.

48. Ibid., pp. 77–78, 83, emphasis added.

49. Ibid., p. 88.

50. As the noted gynecologist Charles Reed put it in his address to the doctors of the Cincinnati Hospital in 1899:

> the genital organs of women, considered in the aggregate, are nothing more or less than a central telegraphic office, from which wires radiate to every nook and corner of the system, and over which are transmitted messages, morbific or otherwise, as the case may be; and it should be remembered right here that telegraphic messages travel both ways over the same wire; that there are both receiving and sending offices at each end of the line.

Cited in Gosling, *Before Freud*, p. 98.

51. The "machinic phylum" is a concept from Gilles Deleuze and Félix Guattari, A *Thousand Plateaus: Capitalism and Schizophrenia*, trans. Brian Massumi (Minneapolis: University of Minnesota Press, 1987), p. 409, passim.

52. The full title of one of Picabia's early paintings, *Edtaonisl (ecclésiastique)*, had included the parenthetical "ecclesiastic"—supposedly a reflection on the artist's experience of observing a priest watching a Russian/American (or Hindu?!) "exotic dancer" named Stacia Napierkowska, to both men's arousal.

53. The writer is a certain P. Rousseau, whose *Histoire des transports* contributed the following: "*c'est le mouvement circulaire, . . . qui règne sur la quasi-totalité de nos mécanismes, . . . depuis le stylo que l'on décapuchonne jusqu'aux pompes d'alimentation des fusées.*" *Trésor de la langue française: Dictionnaire de la langue du XIXe et du XXe siècle* (Paris: Éditions du Centre National de la Recherche Scientifique), vol. 6, p. 802. Compare with the two references in *Le Robert*. Jean Genet: "*Mignon aime l'élégance du geste qui mêle les dés. Il goûte aussi la grâce des doigts qui roulent une cigarette, qui décapuchonnent un stylo*" (*Notre-Dame des fleurs*), and Annie Leclerc: "*Mais voilà: dés qu'ils (les hommes) décapuchonnent leur stylo, ça les prends, ça les reprends, ils n'ont plus qu'un mot à la plume, le Désir*" (*Parole de femme*). In *Le Grand Robert de la langue Française: Dictionnaire Alphabétique et Analogique de la langue Française*, 2nd ed., vol. 3 (Paris: Le Robert, 1987), p. 192.

54. Peter Galison suggested this last association to me.

55. See the definitive essay by Arnauld Pierre, "*Sources inédites pour l'oeuvre machiniste de Francis Picabia: 1918–1922*," *Bulletin de la Société de l'histoire de l'art française* (March 1991): 255–81.

56. . . . *elle se lève et bande la main*
souriant toujours.
Elle gouverne la science d'enchaîner
les degrés de l'eau.

. . .

Je suis le monarque fauvette variété
pudeur de passivité spermatozoïde.
Inesthétique le matelot pâle
près du lac sans soleil.

Picabia, "Zoide," *Poemes et Dessins*, p. 38. Unless otherwise specified, translations are my own.

57. The partial spelling of "*Narcis*" (rather than the proper "*Narcisse*") suggests both "*naquis*," literally "I was born," and "*narcose*," narcosis—in other words, the twin poles of Picabia's awareness.

58. *L'égoïsme conquérant récrée un sot*
Un amant attend le bonheur
Affaires d'apparences
Moi je n'ai jamais vu
Ceux qui les portent
L'inconnu n'a pas de théories
Sur le gaspillage
Le long du fleuve naufragé

Picabia, "Vivre," *Poemes et Dessins*, p. 17.

59. . . . *La vérité de l'âme*
Est la grande lâcheté de l'orgueil académique
Mes yeux dans vos yeux
Je suis content
Dans ma solitude oubliée
     Picabia, "Hélas!" in ibid., p. 18.

60. This image may relate closely to Picabia's additional extramarital affair with the artist Carlos Gregorio (he had already installed both wife and mistress in Zürich when he met Gregorio). The connection is made by William Camfield for *Brilliant Muscles*, reproduced in 1919 under the title *Vagin Brillant*, which "identifies the muscles involved." Other phrases in the work read: "Muscles," "Brillants," "Petit male," "Frottement," "Buche à bouche," and "mécanique de la region sacrée [sic]." See William Camfield, *Francis Picabia: His Art, Life, and Times* (Princeton: Princeton University Press, 1979), p. 116.

61. The original illustration is labeled "*Le Volant du compteur 'Duplex,'*" in *La Science et la Vie* 36 (December 1917–January 1918). See Arnauld Pierre, "*Sources*," p. 258.

62. The editors of the 1992 French reprint have "corrected" the original in this respect, and titled the drawing *Mâle*.

63. I am grateful to Jann Matlock for pointing this out.

64. . . . *C'est le système nerveux*
à l'imagination personnelle
des plaisirs d'éprouver
l'impossibilité.
     Picabia, "Borgne," in *Poèmes et Dessins*, p. 62.

65. Auguste Villiers de L'Isle Adam, *L'Eve Future* (first edition Paris, 1886, second edition, Paris, 1922, English edition, 1982). As I have argued, Adam's book is a crucial one for understanding the complex history of the fantasized cyborg, and how she intersects with the Galatea myth. Peter Wollen provides an excellent brief analysis of this "future Eve," which Wollen compares to E. T. A. Hoffmann's Olympia:

> Caught up in the circulation of desire, the automaton becomes both philosophical toy and sexual fetish or surrogate. Thus Edison . . . is both magus (though American) and marriage-broker (even "idealized" procurer and pimp). His project is the technical realization of the ideal object of masculine desire. The real task of creation is not simply to create a human being, but to create woman *for man*.

"Cinema/Americanism/The Robot," *New Formations* 8 (Summer 1989): 16. Intriguingly, some Picabia scholars have interpreted the *FNSM* in just this way:

> "*The Girl Born without a Mother* . . . refers to the machine as a "creature" made by man for his service—much as God had created Eve, not from woman but from man and for man's use and companionship. The artist was therefore—as Picabia frequently suggested—a god-like figure. But, as God created without the aid of a mother, one eventually encounters concepts of the "unique eunuch" . . . , the "merry widow" . . . and the products of their offspring.

William Camfield, *Francis Picabia* (New York: Solomon R. Guggenheim Museum, 1970), pp. 23–24. I suggest a different reading of these objects here.

66. Breton was probably paraphrasing an epigram attributed to Picabia, that one should "change one's ideas as often as one's shirt." Breton cited in Man Ray's "Photography is not Art," *View* (April 1943): 23, continued in (October 1943): 77–78, 97.

67. *L'homme animal*
Vers le néant
Enveloppe ses sens . . .
De la pénétration mutuelle
Méchanisme aveugle et muet

Nous trouverons des ailes qui vivent selon Platon
Dans les apparences des réalités.
   Francis Picabia, *Poèmes et Dessins*, p. 21.

68. The Smithson work in question is his assemblage titled *Honeymoon Machine* that presents a revision of Duchamp's *célibataires*. See my discussion of the gender dynamics in the Smithson assemblage in *Machine in the Studio* (Chicago: University of Chicago Press, 1996), pl. 8, p. 300ff. For an extended discussion of mechanomorphic art and art machines, see also Caroline A. Jones, *Painting Machines* (Boston: Boston University Art Gallery, 1997).

69. Marcel Duchamp, notes from the *Green Box*, as cited by Amelia Jones in *Postmodernism and the En-Gendering of Marcel Duchamp* (Cambridge: Cambridge University Press, 1994), p. 196. My forced constriction of the meanings of the *Large Glass* is purely instrumental; for a nuanced and sophisticated reading, I refer the reader to Amelia Jones's book.

70. William S. Rubin, *Dada, Surrealism and Their Heritage* (New York: The Museum of Modern Art, 1968), pp. 20, 21.

71. But, as with the counter readings of Picabia's machinic bodies, Duchamp's bride can also be seen as *allowed to exist in her own sphere*, allowed "the possibility of sexual fulfillment as well as her own space of desire." Amelia Jones, paraphrasing François Lyotard in *Postmodernism and the En-Gendering of Marcel Duchamp*, p. 198.

72. Paul Haviland, in *291* 7–8 (September/October 1915): 1. Pontus Hultén suggests that Duchamp "gave" Picabia the name, and the concept, of the FNSM (in his *Machine as Seen at the End of the Machine Age*). Without any supporting evidence, this seems yet another instance of the power of the Duchampian author-function, which Amelia Jones describes in *Postmodernism and the En-Gendering of Marcel Duchamp*. It is clear that Haviland disclaims ownership of the phrase through the quotes he places around it.

73. Andreas Huyssen, "The Vamp and the Machine: Technology and Sexuality in Fritz Lang's *Metropolis*," *New German Critique* 24–25 (Fall and Winter 81–82): 221–37.

74. Peter Wollen, "Cinema/Americanism/The Robot," pp. 7–34. Rosi Braidotti, conversation with the author, April 1995.

75. See Gilles Deleuze on Foucault and the new episteme:

Foucault shows that man, in the classic period, isn't thought of as man, but "in the image" of God, precisely because his forces enter into combination with infinitary forces. It's in the nineteenth century, rather, that human forces confront purely finitary forces—life, production, language—in such a way that the resulting composite is a form of Man. And, just as this form wasn't there previously, there's no reason it should survive once human forces come into play with new forces. . . . What happens when human forces combine with those of silicon, and what new forms begin to appear?

Deleuze, *Negotiations, 1972–1990*, trans. Martin Joughin (New York: Columbia University Press, 1995), pp. 99–100.

DONNA HARAWAY

Deanimations: Maps and Portraits of Life Itself[1]

> *Get a Life!* SimLife, *the genetic playground, allows you to build ecosystems from the ground up and give life to creatures from the depths of your imagination. . . . It's up to you to keep your species off the endangered list! Give life to different species in the Biology Lab and customize their look with the icon editor.*
>
> —ADVERTISEMENT IN *SCIENCE NEWS* 142, 20 (NOVEMBER 14, 1992): 322

CREATION SCIENCE

The user manual for the Maxis computer game *SimLife* opens with the words of Supreme Court Chief Justice Oliver Wendell Holmes, "All life is an experiment."[2] That grounding juridical point is equally the foundation of this essay on the comedic portraiture and cartography of "life itself." My focus is on advertising, joking, and gaming dimensions of genetic portraiture and mapping. These contemporary practices have taproots into the geometric matrices of spatialization and individualization constructed in early modern Europe. The matrices emerged from the instrumental, epistemological, and aesthetic innovations of perspectivism, which became prominent in the narrative time called the Renaissance. "Perspectivism conceives of the world from the standpoint of the 'seeing eye' of the individual. It emphasizes the science of optics and the ability of the individual to represent what he or she sees as in some sense 'truthful,' compared to superimposed truths of mythology or religion."[3] Perspectivism engages types of troping that their practitioners find hard to acknowledge. I want to spelunk through the taproots of spatialization and individualization to see how the carbon-silicon-fused flesh of technoscientific bodies at the end of the second Christian millennium get their semiotic trace nutrients.

In Maxis games, as in life itself, map making is world making. Inside the persistent Cartesian grid conventions of cyber-spatializations, the games encourage their users to see themselves as scientists within narratives of exploration, creation, discovery, imagination, and intervention. Learning data-recording practices, experimental protocols, and world design is seamlessly part of becoming a normal subject in technoscience. Cartographic practice is learning to make projections that shape worlds in particular ways for various purposes.

The Maxis games invite an equation with Christian readings of the creation discourse in Genesis. *The SimEarth Bible* is the title of that game's strategy book. The *Bible*'s introduction tells the reader that SimEarth is "a laboratory on a disk for curious people to experiment with."[4] The author is frankly Christian in his theistic beliefs about evolution, but the game and the strategy manual are deeply enmeshed in "Judeo-Christian" mimesis—i.e., Christian salvation history—even in totally secular interpretations. So too is the perspectivism, which was critical to the history of Western early modern and Renaissance art and map making, enabled by a "Judeo-Christian" point of view. And what was "point of view" before the implosion of biologics and informatics has become, since that impaction in narrative and material spacetime, "pov." Pov is the cyberspace version of secularized creation science's optical practice.

This respectable creation science is not about opposition to biological evolution or promotion of divine special creation. The creation science of the Maxis games, and of much of contemporary technoscience, including molecular biology, genetic engineering, and biotechnology, is resolutely up to the minute in leading-edge science. The secular creationism is intrinsic to the narratives, technologies, epistemologies, controversies, subject positions, and anxieties. "Give life to different species in the Biology Lab and customize their look with the icon editor," urges the *SimLife* advertisement. This is a kind of paint-by-bit game that fills portrait galleries in the cyber-genealogies of life itself. Getting into the spirit, I call the narrative software of my essay "Sim-Renaissance™." I am interested in the official versions of scientific creationism in life worlds after the implosion of informatics and biologics.

My pov in this examination of perspective technologies is that of the chief actor and point of origin in the drama of life itself—the gene. This slant gives me a curious vertigo that I blame on the godlike perspective of any autotelic entity. The gene is the subject of the portraits and maps of life itself in the terminal narrative technology proper to the end of the second millennium. Sociobiologist Richard Dawkins, an inspiration for the Maxis game makers, explained that the body is merely the gene's way to make more copies of itself, in a sense, to contemplate its own image. "Evolution is the external and visible manifestation of the differential survival of alternative *replicators*. Genes are replicators; organisms and groups of organisms . . . are *vehicles* in which replicators travel about."[5] Mere living flesh is derivative; the gene is the alpha and omega of the secular salvation drama of life itself. Faced with this barely secular Christian Platonism, I am consumed with curiosity about the regions where the lively subject becomes the undead thing.

LIFE ITSELF

Following the rules of the game, I mutate the term "life itself" from Sarah Franklin.[6] The instrumentalization of life proceeds by means of cultural practices—sociopolitical, epistemological, and technical. Informed by Foucault on biopower and the history of the concept of life, Franklin analyzes how nature becomes biology, biology becomes genetics, and the whole is instrumentalized in particular forms.[7] "Life," materialized as information and signified by the gene, displaces "Nature," preeminently embodied in and signified by old-fashioned organisms. From the point of view of the Gene, a self-replicating auto-generator, "the whole is not the sum of its parts, [but] the parts summarize the whole."[8] Rather, within the organic and synthetic databases that are the flesh of life itself, genes are not really *parts* at all. They are another *kind* of thing, a thing-in-itself where no trope can be admitted. The genome, the totality of genes in an organism, is not a whole in the traditional, "natural" sense, but a congeries of entities that are themselves autotelic and self-referential. In this view, genes are things-in-themselves, outside the lively economies of troping. To be outside the economy of troping is to be outside finitude, mortality, and difference, to be in the realm of pure being, to be One, where the word is itself.

   In the game of life itself, "[i]t's up to you to keep your species off the endangered list!" Fetishism has never been more fun, as undead substitutes and surrogates proliferate. But fetishism comes in more than one flavor. Nature known and remade as Life through cultural practice figured as technique within specific proprietary circulations is critical to Franklin's and my spliced argument. I hope Marx would recognize his illegitimate daughters, who, in the ongoing comedy of epistemophilia, only mimic their putative father in a pursuit of undead things into their lively matrices. Marx, of course, taught us about the fetishism of commodities. Commodity fetishism is a specific kind of reification of historical human interactions with each other and with an unquiet multitude of nonhumans, which are called nature in Western conventions. In the circulation of commodities within capitalism, these interactions appear in the form of, and are mistaken for, things. In proprietary guise, genes displace not only organisms, but people and nonhumans of many kinds, as generators of liveliness. Ask any biodiversity lawyer whether genes are sources of "value" these days, and the structure of commodity fetishism will come clear.

FETISHISM OF THE MAP

However, I am interested in another, obliquely related flavor of reification that trans-mutes material, contingent, human, and nonhuman liveliness into maps of life itself and then mistakes the map and its reified entities for the bumptious, nonliteral world. I am interested in the kinds of fetishism proper to worlds without tropes, to literal worlds, to genes as autotelic entities. Geographical maps are embodiments of multifac-eted historical practices among specific humans and nonhumans. Those practices con-stitute spatiotemporal worlds; that is, maps are both instruments and signifiers of spatialization. Geographical maps can, but need not, be fetishes in the sense of appear-ing to be non-tropic, metaphor-free representations of previously existing "real" prop-erties of a world that are waiting patiently to be plotted. Instead, maps are models of worlds crafted through and for specific practices of intervening and ways of life.

In Greek, *trópos* is a turn or a swerve; tropes mark the nonliteral quality of being and of language. Fetishes—themselves "substitutes," that is, tropes of a special kind—pro-duce a characteristic "mistake"; fetishes obscure the constitutive tropic nature of themselves and of worlds. Fetishes literalize and so induce an elementary material and cognitive error. Fetishes make things seem clear and under control. Technique and sci-ence appear to be about accuracy, freedom from bias, good faith, and time and money to get on with the job, not about material-semiotic troping and so building certain kinds of worlds rather than others. Fetishized maps appear to be about things-in-them-selves; non-fetishized maps index cartographies of struggle,[9] or more broadly, cartogra-phies of noninnocent practice, where everything does not always have to be a struggle.

The history of cartography can look like a history of figure-free science and tech-nique, not like a history of "troping," in the sense of worlds swerving and mutating through material cultural practice, where all of the actors are not human. Accuracy can appear to be a question of technique, and to have nothing to do with inherently nonliteral tropes. Such a "real" world that preexists practice and discourse seems to be merely a container for the lively activities of humans and nonhumans. Spatialization as a never-ending, power-laced process engaged by a motley array of beings can be fetishized as a series of maps whose grids non-tropically locate naturally bounded bod-ies (land, people, resources—and genes) inside "absolute" dimensions like space and time. The maps are fetishes insofar as they enable a specific kind of mistake that turns process into non-tropic, real, literal things inside containers.

People who work with maps as fetishes do not realize they are troping in a specific way. This "mistake" has powerful effects on the formation of subjects and objects. Such people might well know explicitly that map making is essential to enclosing entities (land, minerals, populations, etc.) and readying them for further exploration, specification, sale, contract, protection, or management. These practices could be understood as potentially controversial and full of desires and purposes, but the maps themselves would seem to be a reliable foundation, free of troping, guaranteed by the purity of number and quantification, outside of yearning and stuttering. Questions of "value," that is, tropes, could be understood to pertain to decisions to learn to make certain kinds of maps and to influence the purposes to which charts would be put. But the map making itself, and the maps themselves, would inhabit a semiotic domain like the high-energy physicists' "culture of no culture,"[10] the world of the non-tropic, the space of clarity and uncontaminated referentiality, the kingdom of rationality. That kind of clarity and referentiality are god tricks. Inside the god trick, the maps could only be better or worse, accurate or not; but they could not be *themselves* instruments for and sediments of troping. From the point of view of fetishists, maps—and scientific objects in general—are purely technical and representational, rooted in processes of potentially bias-free discovery and non-tropic naming. They would say: "Scientific maps could not be fetishes; fetishes are for perverts and primitives. Scientific people are committed to clarity; they are not fetishists mired in error. My gene map is a non-tropic representation of reality, i.e., of genes themselves." Such is the structure of denial in technoscientific fetishism.[11]

That is how the mistake works. Perhaps worst of all, while denying denial in a recursive avoidance of the tropic—and so unconscious—tissue of all knowledge, fetishists mislocate "error." Scientific fetishists place error in the admittedly irreducibly tropic zones of "culture," where primitives, perverts, and other lay people live, and not in the fetishists' constitutional inability to recognize the trope that denies its own status as figure. In my view, contingency, finitude, and difference—but not "error"—inhere in irremediably tropic, secular liveliness. Error and denial inhere in reverent literalness. Error inheres in the literalness of "life itself," rather than in the unapologetic swerving of liveliness and worldly bodies-in-the-making. Life itself is the psychic, cognitive, and material terrain of fetishism. By contrast, liveliness is open to the possibility of situated knowledges, including technoscientific knowledges.

CORPOREALIZATION AND GENETIC FETISHISM

Gene mapping is a particular kind of spatialization of the body, perhaps better called "corporealization." If commodity fetishism is the kind of mistaken self-identity endemic to capital accumulation, and literalization of the categories is the form of self-invisible circulatory sclerosis in important areas of scientific epistemology, what flavor of fetishism is peculiar to the history of corporealization in the material and mythic times of Life Itself? The goal of the question is to ferret out how relations and practices get mistaken for non-tropic things-in-themselves in ways that matter to the chances for liveliness of humans and nonhumans.

To sort out analogies and disanalogies, let us return briefly to commodity fetishism. The Hungarian Marxist philosopher Georg Lukács defined this kind of reification as follows: "Its basis is that a relation between people takes on the character of a thing and thus acquires a 'phantom objectivity,' an autonomy that seems so strictly rational and all-embracing as to conceal every trace of its fundamental nature: the relation between people."[12] Marx defined commodity fetishism as "the objective appearance of the social characteristics of labour."[13] Corporealization, however, is not reducible to capitalization or commodification.

I define corporealization as the interactions of humans and nonhumans in the distributed, heterogeneous work processes of technoscience. The nonhumans are both those made by humans, e.g., machines and other tools, and those occurring independently of human manufacture. The work processes result in specific material-semiotic bodies—or natural-technical objects of knowledge and practice—such as cells, molecules, genes, organisms, viruses, or ecosystems. The work processes make humans into particular kinds of subjects, called scientists. The bodies are "real," and nothing about corporealization is "merely" fiction. But corporealization is tropic and historically specific at every layer of its tissues.

Cells, organisms, and genes are not "discovered" in a vulgar realist sense; but they are not made up. Technoscientific bodies, such as the biomedical organism, are the nodes that congeal from interactions, where all the actors are not human, not self-identical, not "us." The world takes shape in specific ways, and cannot take shape just any way; corporealization is contingent, physical, tropic, historical, interactional. Corporealization involves institutions, narratives, legal structures, power-differentiated human labor, and much more. The processes "inside" bodies—like the cascades of action that constitute an organism or that constitute the play of genes and other entities that make up a cell—are interactions, not frozen things. A word like "gene" specifies a multifaceted set of interactions among people and nonhumans in historically contingent, practical, knowledge-making work. A gene is not a thing, much less a "master molecule" or a self-contained code; instead, the term "gene" signifies a node of durable action where many actors, human and nonhuman, meet.

Commodity fetishism was defined so that only humans were the real actors, whose *social* relationality was obscured in the reified commodity form. But "corporeal fetishism," or more specifically gene fetishism, is about mistaking *heterogeneous* relationality for a fixed, seemingly objective thing. Strong objectivity in Sandra Harding's terms[14] and situated knowledges in my terms are lost in the pseudo-objectivity of gene fetishism, or in any kind of corporeal fetishism that denies the ongoing action and work that it takes to sustain technoscientific material-semiotic bodies in the world. The gene as fetish is a phantom object, like and unlike the commodity. Gene fetishism involves "forgetting" that bodies are nodes in webs of interactions, forgetting the tropic quality of all knowledge claims. My claim about situated knowledges and gene fetishism can itself become fixed and dogmatic and seem to stand for and by itself, outside of the articulations that make the claim sensible. That is, when the stuttering and swerving are left out, a process philosophy can be just as fetishistic as a reductionist one. Both scientists and nonscientists can be gene fetishists; and U.S. culture in and out of laboratories is rife with signs of such fetishism, as well as of resistance to it.

With a little help from Marx, Freud, and Whitehead, let me precipitate from the preceding pages what has been left in solution until now; i.e., the intertwining triple strands—economic, psychoanalytic, and philosophical—in the gene fetishism that corporealizes "life itself" through its symptomatic practices in molecular genetics and biotechnology, for example in the Human Genome Project (medicine), biodiversity gene prospecting (environmentalism and industry), and transgenics (agriculture and pharmaceuticals). I do not mean that scientists or others in these areas necessarily practice gene fetishism. Corporealization need not be fetishized, need not inhabit the culture of no culture and the nature of no nature. Under widespread epistemological, cultural, psychological, and political economic conditions, however, fetishism is a common syndrome in technoscientific practice.

It takes little imagination to trace commodity fetishism in the transnational market circulations where genes, those 24-carat-gold macromolecular things-in-themselves, seem to be themselves the source of value. This kind of gene fetishism rests on the denial of all the natural-social articulations and agentic relationships among researchers, farmers, factory workers, patients, policy makers, molecules, model organisms, machines, forests, seeds, financial instruments, computers, and much else that bring "genes" into material-semiotic being. There is nothing exceptional about genetic commodity fetishism, where focus on the realm of exchange hides the realm of production. The only amendment I made to Marx was to remember all the nonhuman actors too.[15] The gene is objectified in and through all of its naturalsocial (one word) articulations; and there is nothing amiss in that. Such objectification is the stuff of real worlds. But the gene is fetishized when it seems to be itself the source of value; and those kinds of fetish-objects are the stuff of complex mistakes, denials, and disavowals.[16]

The hardest argument for me to make is that there is a psychoanalytic quality to gene fetishism, at least in cultural, if not in personal psychodynamic, terms; but I am driven to this extreme by the evidence. According to Freud, a fetish is an object or part of the body used in achieving libidinal satisfaction. In the classical psychoanalytic story about the fear of castration and masculine subject development, fetishism concerns a special kind of balancing act between knowledge and belief. The fetishist-in-the-making, who must be a boy for the plot to work, at a critical moment sees that the mother has no penis, but cannot face that fact because of the terrible ensuing anxiety about the possibility of his own castration. The youngster has three choices—become a homosexual and have nothing to do with the terrifying castrated beings called women, get over it in the recommended Oedipal way, or provide a usable penis-substitute (a fetish) to stand in as the object of libidinal desire. The fetishist knows and does not know that the fetish is not what it must be to allay the anxiety of the all-too-castratable subject.

For Freud, the penis-substitute is the objectification inherent in a process of disavowal of the mother's (real) castration. The fetish is a defense strategy. "To put it plainly: the fetish is a substitute for the woman's (mother's) phallus which the little boy once believed in and does not wish to forego—we know why."[17] Or, as Laura Mulvey put it, "Fetishism, broadly speaking, involves the attribution of self-sufficiency and autonomous powers to a manifestly 'man' derived object. . . . The fetish, however, is haunted by the fragility of the mechanisms that sustain it. . . . Knowledge hovers implacably in the wings of consciousness."[18] The fetishist is not psychotic; he "knows" that his surrogate is just that. Yet, he is uniquely invested in his power object. The fetishist, aware he has a substitute, still believes in—and experiences—its potency; he is captivated by the reality effect produced by the image, which itself mimes his fear and desire.

Since technoscience is, among other things, about inhabiting stories, Freud's account of fetishism casts light on an aspect of the fixations and disavowals necessary to belief in "life itself." Life itself depends on the erasure of the apparatuses of production and articulatory relationships that make up all objects of attention, including genes; it relies as well as on denial of fears and desires in technoscience. Disavowal and denial seem hard to avoid in the subject formation of successful molecular geneticists, where reality must be seen to endorse the specific practices of intervention built into knowledge claims.

The odd balancing act of belief and knowledge that is diagnostic of fetishism, along with the related cascade of mimetic copying practices that accompany fascination with images, is evident in biotechnological artifacts—including textbooks, advertisements, editorials, research reports, conference titles, and more. Belief in the self-sufficiency of genes as "master molecules," or as the material basis of life itself, or as the code of codes, not only persists, but dominates in libidinal, instrumental-experimental, explanatory, literary, economic, and political behavior in the face of the knowledge that genes are never alone, are always part of an interactional system. That system at a minimum includes the proteinaceous architecture and enzymes of the cell as the unit of structure and function, and also the whole apparatus of knowledge production that concretizes (objectifies) interactions in the historically specific form of "genes" and "genomes." There is no such thing as disarticulated information—in organisms, computers, phone lines, equations, or anywhere else. As the biologist Richard Lewontin put it, "First, DNA is not self-reproducing, second, it makes nothing, and third, organisms are not determined by it."[19] This knowledge is entirely orthodox in biology, a fact that makes "selfish gene" or "master molecule" discourse symptomatic of something amiss at a level that might as well be called "unconscious."

But if I am to invoke Freud's story, I need a particular kind of balancing act between belief and knowledge, one involving a threat to potency and wholeness at critical moments of subject formation. Can gene fetishism be constructed to involve that kind of dynamic? Leaving aside individual psychosexual dynamics and focusing on the social-historical subject of genetic knowledge, I think that such an account makes rough sense, at least analogically. But first, I have to rearrange Freud's account to dispute what he thought was simply true about possession of the "phallus," that signifier of creative wholeness and power. Freud thought women really did not have it; that was the plain fact the fetishist could not face. I rely on feminism to insist on a stronger objective claim, namely that women are whole, potent, and "uncastrated." Wholeness here means inside articulations, never reducing to a thing-in-itself, in sacred, secular, or psychoanalytic terms. Freud got it wrong, even while he got much of the symbolic structure right in male-dominant conditions. Freud, and a few other good men (and women), confused the penis and the phallus after all.

My correction is necessary to make the analogy to gene fetishism. Organisms are "whole" in a specific, nonmystical sense; i.e., organisms are nodes in webs of dynamic articulations. Neither organisms nor their constituents are things-in-themselves. Sacred or secular, all autotelic entities are defenses, alibis, excuses, substitutes— dodges from the complexity of material-semiotic objectifications and apparatuses of corporeal production. In my story, the gene fetishist "knows" that DNA, or life itself, is a surrogate, or at best a simplification that readily degenerates into a false idol. The substitute, life itself, is a defense for the fetishist, who is deeply invested in the switch, against the knowledge of the actual complexity and embeddedness of all objects, including genes. The fetishist ends up believing in the code of codes, the book of life, and even the search for the grail. Only half jokingly, I see the molecular biological fetishist to be enthralled by a phallus-substitute, a mere "penis" called the gene, which defends the cowardly subject from the too-scary sight of the relentless material-semiotic articulations of biological reality, not to mention the sight of the wider horizons leading to the real in technoscience. Perhaps acknowledging that "[f]irst, DNA is not self-reproducing, second, it makes nothing, and third, organisms are not determined by it" is too threatening to all the investments, libidinal and otherwise, at stake in the material-semiotic worlds of molecular genetics these days. So the fetishist sees the gene itself in all the gels, blots, and printouts in the lab, and "forgets" the natural-technical processes that produce the gene and genome as consensus objects in the real world. The fetishist's balancing act of knowledge and belief is still running in the theater of technoscience.

The third strand in my helical spiral of gene fetishism is spun out of what Whitehead called the "fallacy of misplaced concreteness."[20] Growing out of his examination of the still astonishing concatenation of theoretical, mathematical, and experimental developments that mark the European seventeenth century as "The Century of Genius," Whitehead foregrounded the importance to the history of Western natural science of two principles: (1) simple location in space-time, and (2) substances with qualities, especially primary qualities defined by their yielding to numerical, quantitative analysis. These were the fundamental commitments embedded in seventeenth-century and subsequent Western practices of spatialization, including cartography, and the role of these principles in the history of philosophical and scientific mechanism is not news. Whitehead wrote in 1925, when mechanism, the wave-particle duality, the principle of continuity, and simple location had been under fruitful erosion in physics for decades. These dated conventionally from Maxwell's mid-nineteenth-century equations founding electromagnetic field theory and continuing with the developments in quantum physics in the 1920s and 1930s, and were tied to work by both Niels Bohr in wave mechanics and Albert Einstein on the lightquantum, among other critical transformations of physical theory.

Whitehead had no quarrel with the utility of the notion of simple location and the attention to primary qualities of simple substances—unless these abstract logical constructions were mistaken for "the concrete." Albeit expressed in his own arcane terminology, "the concrete" had a precise meaning for Whitehead, related to his approach to "an actual entity as a concrescence of prehensions." Stressing the processual nature of reality, he called actual entities actual occasions. Objectifications had to do with the way "the potentiality of one actual entity is realized in another actual entity."[21] Prehensions could be physical or conceptual, but such articulations, or reachings into each other in the tissues of the world, constituted the most basic processes for Whitehead. I ally with Whitehead's analysis to highlight the ways that gene fetishists mistake the abstraction of the gene for the concrete entities and "occasions" that make up the biological world.

So, gene fetishism is compounded of a political economic *denial* that holds commodities to be sources of their own value, while obscuring the socio-technical relations among humans and between humans and nonhumans that generate both objects and value; a *disavowal*, suggested by psychoanalytic theory, that substitutes the master molecule for a more adequate representation of units or nexuses of biological structure, function, development, evolution, and reproduction; and a philosophical-cognitive *error* that mistakes potent abstractions for concrete entities, which themselves are ongoing events. Fetishists are multiply invested in all of these substitutions. The irony is that gene fetishism involves such elaborate surrogacy, swerving, and substitution, when the gene as the guarantor of life itself is supposed to signify an autotelic thing in itself, the code of codes. Never has avoidance of acknowledging the relentless tropic nature of living and signifying involved such wonderful figuration, where the gene collects up the people in the materialized dream of life itself.

Inside and outside laboratories, genetic fetishism is contested, replicated, ironized, indulged, disrupted, consolidated, examined. Gene fetishists "forget" that the gene and gene maps are ways of enclosing the commons of the body—of corporealizing—in specific ways, which, among other things, often write commodity fetishism into the program of biology. I would like to savor the anxious humor of a series of scientific cartoons and advertisements about the gene in order to see how joking practice works where gene fetishism prevails. We move from Maxis's *SimLife* to maps and portraits of the genome itself.

GENOME

My reading of comic portraiture and cartography—the story of life itself—picks up after the implosion of informatics and biologics, especially in genetics, since the 1970s. Still absent from Webster's 1993 unabridged dictionary, *genome* progressively signifies a historically new entity engendered by the productive identity crisis of nature and culture. The cultural productions of the genome produce a category crisis, a generic conundrum in which proliferating ambiguities and chimeras animate the action in science, entertainment, domestic life, fashion, religion, and business. The pollution works both ways: culture is as mouse-eaten as nature is by the gnawings of the mixed and matched, edited and engineered, programmed and debugged genome.

A 1991 residential seminar at the University of California Humanities Research Center spent considerable time on the Human Genome Project. One philosopher in the seminar put his finger on potent double meanings when he understood the science studies scholars, who suggested the term "the cultural productions of the genome" as the title for a conference, to be referring to musical, artistic, educational, and similar "cultural productions" emerging from popularization of science. The science studies professionals meant, rather, that the genome was radically "culturally" produced, and no less "natural" for all that. The gene was the result of the work of construction at every level of its very real being; it was constitutively artifactual. "Technoscience *is* cultural practice" might be the slogan for mice, scientists, and science analysts.

Attending to how the permeable boundary between science and comedy works in relation to the genome—and at the risk of giving comfort to those who still think the cultural production of the genome means its popularization—I pursue my story literally by reading the comics. My structuring text is a family of images, all cartoon advertisements for lab equipment drawn by Wally Neibart and published in *Science* magazine in the early 1990s (see Figures 1 and 2). I am reminded of David Harvey's observation that advertising is the official art of capitalism.[22] Advertising also captures the paradigmatic qualities of democracy in the narratives of life itself. Finally, advertising and the creation of value are close twins in the New World Order, Inc. The cartoons explicitly play with creation, art, commerce, and democracy.

The Neibart cartoons suggest who "we," reconstituted as subjects in the practices of the Human Genome Project, are called to be in this hyper-humanist discourse: Man™. This is man with property in himself in the historically specific sense proper to the New World Order, Inc. Following an ethical and methodological principle for science studies that I adopted many years ago, I will critically analyze, or "deconstruct," only that which I love and only that in which I am deeply implicated. This

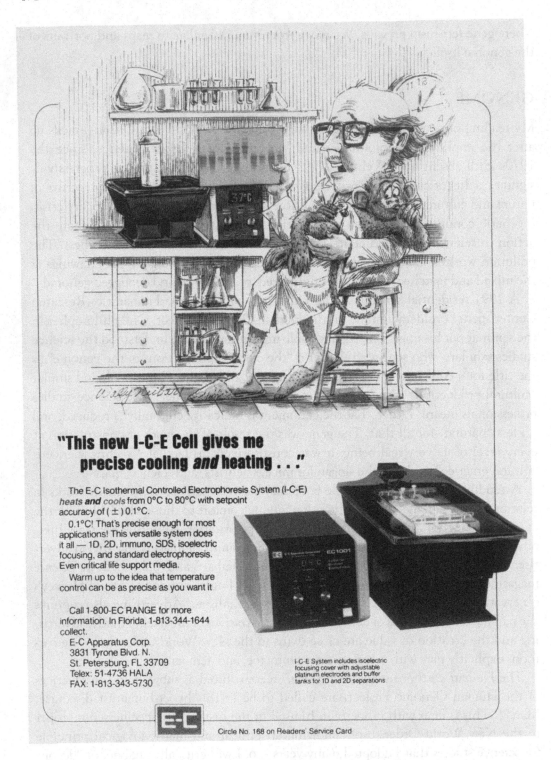

Figure 1. Night Births. Courtesy of E-C Apparatus Corporation.

Figure 2. Portraits of Man. Courtesy of E-C Apparatus Corporation.

commitment is part of a project to excavate something like a technoscientific unconscious, the processes of formation of the technoscientific subject, and the reproduction of this subject's structures of pleasure and anxiety. Those who recognize themselves in these webs of love, implication, and excavation are the "we" who surf the Net in the sacred/secular quest rhetoric of technoscience.

Interpellated into its stories, I am in love with Neibart's comic craft. His cartoons are at least as much interrogations of gene fetishism as they are sales pitches. His cartoons depend on a savvy use of visual and verbal tropes. In his wonderful cartoon image advertising an electrophoresis system, a middle-aged, white, bedroom-slippered and labcoat-clad man cradles a baby monkey wearing a diaper (Figure 1).[23] Addressing an audience outside the frame of the ad, the scientist holds up a gel with nice protein fragment separation, generated by the passage of charged molecules of various sizes through an electrical field. The gel is part of a closely related family of macromolecular inscriptions, which include the DNA polynucleotide separation gels, whose images are familiar icons of the genome project. In my reading of this ad, the protein fragment gel metonymically stands in for the totality of artifacts and practices in molecular biology and molecular genetics. These artifacts and practices are the components of the apparatus of bodily production in biotechnology's materializing narrative. My metonymic substitution is warranted by the dominant molecular genetic story that still overwhelmingly leads unidirectionally from DNA (the genes), through RNA, to protein (the end product). In a serious and persistent joke on themselves, the kind of joke that affirms what it laughs at, molecular biologists early labeled this story the Central Dogma of molecular genetics. The Central Dogma has been amended over the years to accommodate some reverse action, in which information flows from RNA to DNA. "Reverse transcriptase" was the first enzyme identified in the study of this "backward" flow. RNA viruses engage in such shenanigans all the time. HIV is such a virus, and the first (briefly) effective drugs used to treat people with AIDS inhibit the virus's reverse transcriptase, which reads the information in the viral genetic material, made of RNA, into the host cell's DNA. Even while marking other possibilities, the enzyme's very name highlights the normal orientation for control and structural determination in higher life forms. And even in the reverse form, Genes "R" Us. This is the Central Dogma of the story of Life Itself.

In the Neibart cartoon, while the scientist speaks to us, drawing us into the story, the monkey's baby bottle is warming in the well of the electrophoresis apparatus. The temperature monitor for the system reads a reassuringly physiological 37 degrees Celsius, and the clock reads 12:05. The time is five minutes past midnight, the time of strange night births, the time for the undead to wander, and the first minutes after a nuclear holocaust. Remember the clock that the *Bulletin of Atomic Scientists* used to keep time in the Cold War; for many years it seemed that the hands advanced relentlessly toward midnight. As Keller argued persuasively, the bomb and the gene have been choreographed in the last half of the twentieth century in a dance that intertwines physics and biology in their quest to reveal "secrets of life and secrets of death."[24]

In the electrophoresis system ad, Neibart's image suggests a reassuring family drama, not the technowar apocalypse of secular Christian monotheism, nor the Frankenstein story of the unnatural and disowned monster. But I am not reassured: all the conventional rhetorical details of the masculinist, humanist story of man's autonomous self-birthing structure the ad's narrative. The time, the cross-species baby, the scientist father, his age, his race, the absence of women, the appropriation of the maternal function by the equipment and by the scientist: all converge to suggest the conventional tale of the second birth that produces Man. It's not "Three Men and a Baby" here, but "A Scientist, a Machine, and a Monkey." The technoscientific family is a cyborg nuclear unit. As biologist—and parent—Scott Gilbert insisted when he saw the ad, missing from this lab scene are the post-docs and graduate students, with their babies, who might really be there after midnight. Both monkey and molecular inscription stand in for the absent human product issuing from the reproductive practices of the molecular biology laboratory. The furry baby primate and the glossy gel are tropes that work by part-for-whole substitution or by surrogacy. The child produced by this lab's apparatus of bodily production, this knowledge-producing technology, this writing practice for materializing the text of life, is—in fruitful ambiguity—the monkey, the protein gel (metonym for man), and those interpellated into the drama, that is, us, the constituency for E-C Apparatus Corporation's genetic inscription technology.

I over-read, naturally; I joke; I suggest a paranoid reading practice. I mistake a funny cartoon, one I like immensely, for the serious business of real science, which surely has nothing to do with such popular misconceptions. But jokes are my way of working, my nibbling at the edges of the respectable and reassuring in technosciences and in science studies. This nervous, symptomatic, joking method is intended to locate the reader and the argument on an edge. On either side is a lie: on the one hand, the official discourses of technoscience and its apologists; on the other, the fictions of conspiracy fabulated by all those labeled "outsider" to scientific rationality and its marvelous projects, magical messages, and very conventional stories.

My interest is relentlessly in images and stories and in the worlds, actors, inhabitants, and trajectories they make possible. In the biotechnological discourse of the Human Genome Project, the human is produced in a historical form, which enables and constrains certain forms of life rather than others. The technological products of the several genome projects are cultural actors in every sense.

PORTRAIT™

A second Wally Neibart cartoon for a *Science* ad makes an aspect of this point beautifully—literally (Figure 2). Evoking the world of (high) art, this ad puns on science as (high) cultural production. That should not prevent the analyst from conducting another, quasi-ethnographic sort of "cultural" analysis. I think Neibart subtly invites a critical reading; he is laughing *at* gene fetishism, as well as using it. Our same balding, middle-aged, white, male scientist—this time dressed in a double-breasted blue blazer, striped shirt, and slacks—is bragging about his latest acquisition to a rapt, younger, business-suit-clad, white man with a full head of hair. They get as close to power dressing as biologists, still new to the corporate world, seem to manage. The two affluent-looking gentlemen are talking in front of three paintings in an art museum. (We assume they are in an art museum—that is, if the *Mona Lisa* has not been relocated as a result of the accumulated wealth of the truly Big Men in informatics and biologics. After all, in 1994 William H. Gates, chairman and founder of Microsoft, purchased a Leonardo da Vinci notebook, *Codex Hammer*, for a record $30.8 million in a manuscript auction.)[25]

Neibart's three paradigmatic portraits of man on display are not of male human beings, nor should they be. The self-reproducing mimesis in screen projections works through spectacularized difference. One painting in Neibart's ad is da Vinci's *Mona Lisa*; the second is Pablo Picasso's *Woman with Loaves* (1906); the third, gilt-framed like the others, is a superb DNA sequence autoradiograph on a gel. The Italian Renaissance and modernist paintings are signs of the culture of Western humanism, which, in kinship with the Scientific Revolution, is narratively at the foundations of modernity and its sense of rationality, progress, and beauty—not to mention its class location in the rising bourgeoisie, whose fate was tied progressively to science and technology. Like the humanist paintings, the sequence autoradiograph is a self-portrait of man in a particular historical form. Like the humanist paintings, the DNA gel is about instrumentation, framing, angle of vision, lighting, color, new forms of authorship, and new forms of patronage. Preserved in gene banks and catalogued in databases, genetic portraits are collected in institutions that are like art museums in both signifying and effecting specific forms of national, epistemological, aesthetic, moral, and financial power and prestige. The potent ambiguities of biotechnological, genetic, financial, electrical, and career power are explicitly punned in the ad: "I acquired this sequence with my EC650 power supply." The E-C Apparatus Corporation offers "the state-of-the-art in Power Supplies"—in this case, a constant power supply device.

The unique precision and beauty of original art become replicable, everyday experiences through the power of technoscience in proprietary networks. The modernist opposition between copies and originals—played out forcefully in the art market—is erased by the transnational postmodern power of genetic identification and replication in both bodies and labs. Biotechnological mimesis mutates the modernist anxiety about authenticity. "Classic sequence autoradiographs are everyday work for E-C Electrophoresis Power Supplies." No longer oxymoronically, the ad's text promises unlimited choice, classical originality, eighteen unique models, and replicability. At every stage of genome production, in evolutionary and laboratory time, database management and error reduction in replication take the place of anxiety about originality.

But a calmed opposition between copy and original does not for a minute subvert proprietary and authorial relations to the desirable portrait in all its endless versions, although the subjects of authorial discourse have mutated, or at least proliferated. Just as I am careful to credit Neibart and seek copyright releases, E-C is careful to confirm authorial and property relations of the beautiful, framed DNA sequence autoradiograph, which is reproduced in the ad "courtesy of the U.S. Biochemical Corporation using Sequenase™ and an E-C Power Supply."[26] E-C used the molecular portrait of man with permission, just as I must, in the escalating practices of ownership in technoscience, where intellectual and bodily property become synonymous. The "great artist" of the technohumanist portrait is a consortium of human and nonhuman actants: a commercially available enzyme, a biotech corporation, and a power supply device. Like the art portraiture, the scientific portrait of man as gel and database signifies genius, originality, identity, the self, distinction, unity, and biography. In eminently collectible form, the gel displays difference and identity exhaustively and precisely. Human beings are collected up into their paradigmatic portrait. No wonder aesthetic pleasure is the reward. The autoradiograph reveals the secrets of human nature. Intense narrative and visual pleasure are intrinsic to this technoscientific apparatus, as it is to others, which nonetheless try to ensure that their productions can only be officially or "scientifically" discussed in terms of epistemological and technological facticity and non-tropic reality. Genes *are* us, we are told through myriad "cultural" media, from DNA treated with reagents like Sequenase™ and run on gels, to property laws in both publishing and biotechnology. Narrative and visual pleasure can be acknowledged only in the symptomatic practices of jokes and puns. Displayed as "high science," explicit "knowledge" must seem free of story and figure. Such technohumanist portraiture is what guarantees man's second birth into the light and airy regions of mind. This is the structure of pleasure in gene fetishism.

The strong bonding of biotechnology with the Renaissance, and especially with Leonardo da Vinci, demands further dissection. Commenting on the potent mix of technique, ways of seeing, and patronage, a venture capitalist from Kleiner Perkins Caufield & Byers summed up the matter when he observed that biotechnology has been "for human biology what the Italian Renaissance was for art."[27] Leonardo, in particular, has been appropriated for stories of origin, vision and its tools, scientific humanism, technical progress, and universal extension. I am especially interested in the technoscientific preoccupation with Leonardo and his brethren in the "degraded" contexts of business self-representation, advertising inside the scientific community, science news illustration, conference brochure graphics, science popularization, magazine cover art, and comic humor.

Consider Du Pont's remarkable ad that begins, "Smile! Renaissance™ non-rad DNA labeling kits give you reproducible results, not high backgrounds."[28] The text occurs underneath a color reproduction of Andy Warhol's giant (9′ 2″ × 7′ 10½″) 1963 photo-silkscreen, in ink and synthetic polymer paint, that "clones" the Mona Lisa. Filling in a grid of five Mona Lisa's across and six down, Warhol's multiplied version is entitled Thirty Are Better Than One. In Warhol's and Du Pont's versions, the paradigmatic, enigmatically smiling lady is replicated in a potentially endless clone matrix. Without attribution, Du Pont replicates Warhol replicates da Vinci replicates the lady herself. And Renaissance™ gets top billing as the real artist because it facilitates replicability. But how could Warhol, of all artists, object to his work being anonymously appropriated for commodity marketing under the sign of "debased" high art and high science enterprised up? In the Du Pont ad, the only mark of intellectual property is—in a comic, recursive self-parody—Renaissance™. The mythic chronotope itself bears the trademark of the transnational biotechnology corporation. Recursively, the brand marks detection and labeling tools, for the code of codes, for life itself.

IN THE COMPANY OF GENES

The company the gene keeps is definitely upscale. Fetishes come in matched sets. Master molecule of the Central Dogma and its heresies, the gene affiliates with the other power objects of technoscience's knowledge production: neuro-imaging, artificial intelligence, artificial life, high-gloss entertainment, high technology, high expectations. The ten-part series, "Science in the 90s," which ran from January 5 to May 8, 1990, gives a broad sense of what counts as cutting-edge technoscience for the news writers and editors of Science. The excitement came from high tech/high science, including neuroscience, computing and information sciences, and molecular genetics. The boring and discouraging notes came from (very brief) consideration of ongoing racial and sexual "imbalance" in who does technoscience and the troubles that arise when "politics" gets into a scientist's career.

The chief power sharer in the gene's new world community is the nervous system. Even the UNESCO Courier carries the news that links mind and origins, neuron and gene, at the helm of life itself: "No one would deny that, within the highly organized framework of a human being, two 'master elements' account for most of our characteristics—our genes and our neurons. Furthermore, the nature of the dialogue between our genes and our neurons is a central problem of biology."[29]

Every autumn since 1990, Science, the magazine of the American Association for the Advancement of Science, has put out a special issue updating its readers on progress in genome mapping, and especially in the Human Genome Project. The table of contents of the first special issue highlights the tight coupling of genetic and nervous systems in the discourse of millennial science.[30] Citing a recent example of homicidal mania, *Science* editor Daniel Koshland, Jr., introduced the issue with the argument that hope for the mentally ill—and for society—lies in neuroscience and genetics. Necessary to the diagrams of life itself, the tie to informatics is explicit: "The irrational output of a faulty brain is like the faulty wiring of a computer, in which failure is caused not by the information fed into the computer, but by incorrect processing of that information after it enters the black box."[31] In addition to the articles on the genome project and the map insert, the issue contains a research news piece called "The High Culture of Neuroscience" and eight reports from neurobiology, spanning the range from molecular manipulation of ion channels, to a study of primate behavior, to a psychological assessment of human twins reared apart.

Located in the potent zones where molecular genetics and neurobiology ideologically converge, this last study on twins reared apart lists as its first author Thomas Bouchard, a former student of Arthur Jensen. Jensen promoted the idea of the linkage of genetic inheritance, IQ, and race in his famous 1969 *Harvard Educational Review* article. The special gene map issue of *Science* was the first major professional journal to publish Bouchard's controversial work, which ascribes most aspects of personality and behavior to genes. Many of Bouchard's papers had been rejected through peer review, but he brought his message successfully to the popular media. Following *Science's* publication, Bouchard's ideas gained authority and prominence in public debates about genetics and behavior.[32]

Cartography, the high science of the Age of Exploration, tropically organizes the first *Science* gene map issue, from the design of its cover to the content of its prose. Collectively labeled "The Human Map," the cover is a collage of mapping icons—including a Renaissance anatomical human dissection by Vesalius, a Mendelian genetic-cross map superimposed on the great scientist's facial profile, a radioactively labeled region of metaphase chromosomes, a linkage map and bit of a sequence data rendered by the cartographical conventions that have emerged in the genome projects, a flow diagram through the outline of a mouse body, and a computer-generated colored-cell map of an unidentified abstract territory. The cover design is explained inside: "Just as the ancient navigators depended on maps and charts to explore the unknown, investigators today are building maps and charts with which to explore new scientific frontiers."[33]

The reference to the Renaissance cartographers, a common rhetorical device in genome discourse, is not idle. Genomics "globalizes" in specific ways. Species being is materially and semiotically produced in gene mapping practices, just as particular kinds of space and humanity were the fruit of earlier material-semiotic enclosures. Traffic in bodies and meanings is equally at stake. The orthodox stories of the Renaissance and early modern Europe are useful to my narrative of genome mapping as a process of bodily spatialization akin to enclosing the commons in land, through institutions of alienable property, and in authorship, through institutions of copyright. Harvey points out that the introduction of the Ptolemaic map into Florence from Alexandria in 1400 gave Europeans the critical means to see the world as a global unity.[34] The Ptolemaic map and its offspring were the air pumps of scientific geography, embedded in material, literary, and social technologies that made the "global" a mobile European reality. "[M]athematical principles could be applied, as in optics, to the whole problem of representing the globe on a flat surface. As a result it seemed as if space, though infinite, was conquerable and containable for purposes of human occupancy and action."[35] The elaboration of perspective techniques in mid-fifteenth-century Florentine art was entwined with the construction of individualism and perspectivism critical to modern spaces and selves. The sixteenth-century Flemish cartographer, Gerardus Mercator, after whom a biotechnological corporation is named, crafted projections of the globe geared to navigation on the high seas in a period of intense world exploration by Europeans. All of these practices constituted a major reworking of conceptions of space, time, and person. And all of these practices are in the family tree of genetic mapping, which is a distributed, located practice enabling certain sorts of power-charged global unity. No wonder Mercator's grids and projections line the scientific unconscious of biotechnology researchers and advertisers.

Bruno Latour illuminates the mobilization of worlds through mapping practices.[36] Cartography is perhaps the chief tool-metaphor of technoscience. "Mapping Terra Incognita (*Humani Corporis*)," the news story toward the less technical front of *Science*'s first special issue on the genome project, has all of the expected allusions to Vesalius's Renaissance anatomy.[37] This kind of ubiquitous new-world imagery, like the extended propaganda for cybernetics in the United States in the 1950s and 1960s, indicates a "distributed passage point," through which many popular and technical projects get loosely associated with the high gloss of molecular biology and biotechnology.[38] The second article on genome mapping in the special issue, "Mapping the Human Genome: Current Status," charts another kind of intersection, one Latour called an "obligatory passage point."[39] This node represents the fruit of the mobilization of resources and the forging of alliances among machines, people, and other entities that force others to pass through *here*, and nowhere else. The sociotechnical achievements of molecular biology are a node through which many *must* pass: paleoanthropologists who wish to resolve evolutionary arguments, physicians who wish to diagnose and treat disease, developmental biologists who seek resolution of their questions, ideologists who proclaim legitimation for or exemplary condemnation of technoscience. Molecular biology does not just claim to be able to decode the master molecule; it installs the tollbooths for a great deal of collateral traffic through nature.

The human genome map inserted into the special issue of *Science* in 1990 inaugurated the practice of annually giving each subscriber-member of the AAAS a personal copy of the most up-to-date chart available. The practice reverberates with *National Geographic*'s presentation to subscribers of the new Robinson projection map of the globe in its January 1988 issue, which featured on the front cover the holographic portrait of the endangered planet earth at the dawn of the decade to save man's home world. (A holographic ad for McDonald's, with appropriate words from the transnational fast food chain's founder, graced the back cover.) Just as all subscribers to *National Geographic* are automatically members of a scientific society, and so patrons of research, all subscribers to *Science* are members of the AAAS and share symbolically in its ideological and material privileges. As subscribers, "we" are the constituents of technoscience, a mapping practice of the highest order. With more than 150,000 subscribers, *Science* reaches about three times the number as does *Nature*, its British sibling and nearest world-class competitor. *National Geographic* reaches millions.

In a mid-1990s ad for DNA-cutting enzymes, New England Biolabs invokes the imploded global bodies materialized by both *National Geographic* and by the Human Genome Project (Figure 3). The oxymoronic Global Native embodies the Global Gene, literally. Difference is mapped and enclosed; art, science, and business join in the dance. From the left side of the page, against a black background, the body of a beautiful young woman with generically (and oxymoronically) "indigenous" facial features flows forward. Her body is the mapped terrain globe, shaped to her lovely female contours; she is its soul. Of the earth, she moves through it as both its spirit and flesh. Arms raised in a dance gesture, the native woman is clothed with the tissue of the mapped planet, which billows into a semicircle continuous with her figure. Marked off by its geometric coordinates, the projection map shows the bulge of west Africa and the Atlantic Ocean. The seas are dotted with the great sailing cutter ships of Europe's age of exploration and marked with the fabulous Latin names bestowed by the navigators' culture. The map-woman is an animated Mercator projection.

The earth is both the woman's body and her dress, and the color-enhanced regions highlighting the beige tones of the swirling hemispherical corpus/fabric are like style elements in a United Colors of Benetton celebration of global multiculturalism. To remember the slave trade and the middle passage across the region of the world shown on this lovely map seems petty. The woman-earth's body confronts text at the midline of the page: "Mapping the Human Genome." The earth and the genome are one, joined in the trope of the technoscientific map. "Advanced by a diverse range of 8-base Cutters," the new cartography will be enabled by New England Biolab's restriction enzymes. Map, women, earth, goddess, science, body, inscription, technology, life, the native: all are collected in an aestheticized image like a Navaho sand painting that places the holy people inside the four sacred mountains. Who said master narratives, universalism, and holism were dead in the New World Order's extended networks? Advanced by the code-analyzing restriction enzymes given by the globalized history of race and gender, naturalization has never been more florid. I doubt that is what New England Biolabs meant to signify in its ad, which promised "exceptional purity and unmatched value essential for success in your genomic research."

202

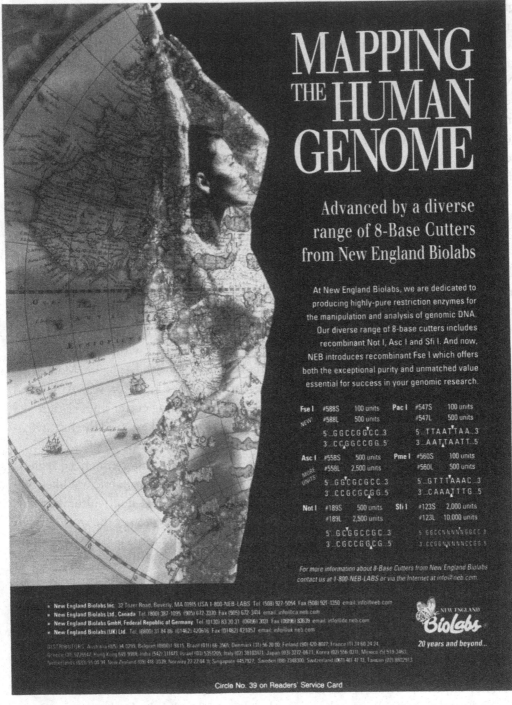

Figure 3.   Global Native. Courtesy of New England BioLabs.

In short, biotechnology, in general, and the Human Genome Project, in particular, aim high. No wonder the Human Genome Project's apologists called it biology's equivalent to putting a man on the moon. Where else could he go with all that thrust? The Human Genome Project is discursively produced as "one small step. . . ." At this origin, this new frontier, man's footprints are radioactive traces in a gel; at the dawn of hominization, the prints were made in volcanic dust at Laetoli in Ethiopia; at the dawn of the space age, a white man, acting as surrogate for mankind, walked in moon dust. All of these technoscientific travel narratives are about freedom, the free world, democracy, and, inevitably, the free market.

REPRESENTATION, RECURSION, AND THE COMIC

Under the signifiers of freedom and democracy, a third Neibart cartoon (not illustrated here) completes this comic essay's catalogue of the savvy artist's potent jokes. Two senior white male scientists in business suits, one the same successful fellow who acquired the techno-humanist portrait of man in the form of a DNA separation gel, stand with their hands clenched above their heads in the sign of victory on a stage above a cheering mob at a political convention. The figures in the crowd wave the red, white, and blue banners inscribed with the names of their constituencies: DNA, protein, AGTC, RNA, PCR, and all the other molecular actors in the genomic drama. "With 90% of the vote already in, it is a landslide" for the E-C Apparatus Corporation's power supply. The joke makes the concretized entities of the biotechnological laboratory into the voters in the democracy of science. The molecules and processes— themselves the feat of the scientists in the productive drama of the laboratory—are the actors with a vengeance. The sedimented feats of technoscientific virtuosity authorize their ventriloquists under the sign of freedom and choice. This is material subject construction, Oedipal and not.

Jokingly ironized in the Neibart cartoon, this scene is also gene fetishism at its most literal. Literary, social, and material technologies converge to make the objects speak, just as Shapin and Schaffer showed us in the story of Robert Boyle's air pump.[40] In the culture of no culture conjugated with the nature of no nature, the objects speak with a withering directness.

It is not new to link the stories of science and democracy, any more than it is new to link science, genius, and art, or to link strange night births and manly scientific creations. But the interlocking family of narratives in the contemporary U.S. technoscientific drama is stunning. The Neibart cartoon must be read in the context of *Science 85*'s cover of a decade ago, "The American Revolution." The magazine cover featured the chip and the gene, figured, as always, as the double helix, against the colors of red, white, and blue, signifying the New World Order, Inc., of nature "enterprised up,"[41] where free trade and freedom implode. This warped field is where, to misquote the

Supreme Court Chief Justice, "Life Itself is always an experiment." It is also a venture in marketing.

What are advertisements in technoscience doing? Do the ads in magazines like *Science* matter, and if so, how? Can I make a case for reading these materials as even gently ironic, rather than celebratory and instrumental in strengthening gene fetishism? Is anxious humor enough to force the trope into the open and disrupt literalism? Who besides me is anxiously laughing or crying at these ads? I do not know enough about how ad designers in technoscience produce their work, how graphic artists' views do and do not converge with scientists' or corporate managers' discourse, or how readers appropriate and rework ad images and text. I do know that the ads are more than pretty designs and helpful information. They are part of the visual culture that makes the gene fetish—and the epistemology of the gene fetishist—so productive.

Although many of the ads contain considerable technical information, I do not think a strong case can be made for seeing these ads principally as sales strategies. The companies that supply the key equipment and products to biological and engineering labs have more effective mechanisms for informing and servicing clients. Company and product name recognition is enhanced, and I would not argue against modest functionalist economic readings of such ads. Urged to learn more about potentially powerful tools, readers get toll-free phone numbers and reader-response cards for ordering catalogues.

More significantly, the readers of these ads taste the pleasures of narrative and figuration, of recognizing stories and images of which one is part. Advertising is not just the official art of capitalism; it is also a master teacher of history and theology in postmodernity. The debates about historical and literary canons should be taking place in graphic artists' studios in corporations, as well as in classrooms. The ads draw from and contribute to a narrative and visual world that activates the unconscious mechanisms that issue in the possibility of a joke. The joke is a sign of successful interpellation, of finding oneself constituted as a subject of knowledge and power in these regions of sociotechnical space. Whoever is inside that joke is inside the materialized narrative fields of technoscience, where, in the words of a recent Du Pont ad, "better things for better living come to life." These ads work by interpellation, by calling an audience into the story, more than by informing instrumentally rational market or laboratory behavior. Such interpellation is the precondition of any subsequent rationality, in epistemology or in other such duplicitous free markets. In the Book of Life Itself, fetishism in all its flavors is comic to the end.

Finally, the Neibart cartoons draw on the comic in quite another sense than "funny." In the literary analysis of the comic mode in drama, "comic" means reconciled, in harmony, secure in the confidence of the restoration of the normal and noncontradictory. Shakespeare's comedies are not funny; rather, their endings restore the normal and harmonious, often through the ceremonies of marriage, through which opposites are brought together. The comic does not recognize any contradictions that cannot be resolved, any tragedy or disaster that cannot be healed. The comic mode in techno-science is reassuring in just this way.[42] For those who would reassure us, the comic is just the right mode for approaching the end of the second Christian millennium.

Edgy and nervous, I must end by jokingly repeating myself in a comic recursion that restores few harmonies. In a Sydney Harris cartoon in *Science*, a white male researcher in a lab-coat reads to a white female scientist, similarly dressed, surrounded by their experimental animals and equipment, "Here it is in Genesis: 'He took one of Adam's ribs and made the rib into a woman.' Cloning, if I ever heard it."[43] Woman™ cultured from the osteoblasts of Man™: this Genesis replicates salvation history compulsively, repeating *in saecula saeculorum* "a few words about reproduction from an acknowledged leader in the field."[44]

Figuring the implosion of informatics and biologics, this bastard scriptural quotation comes from a Logic General Corporation ad for a 1980s software duplication system. In the foreground, under the earth-sun logo of Logic General, a biological white rabbit has her paws on the grid of a computer keyboard. The long-eared rodent is a cultural sign of fecundity, and "breeding like rabbits" is a popular figure of speech. But Logic General's hare, a brand of technoscientific Easter Bunny, evokes the pregnancy-test rodent famous in the history of reproductive medicine. Looking into the screen of a video display terminal, the organic rabbit faces its computer-generated image, who locks its cybergaze with the ad's reader. In her natural electronic habitat, the virtual rabbit is on a grid that insists on the world as a game played on a chess-like board, or Cartesian grid, made up of a square array of floppy disks. The disks constitute a kind of Mercator™ projection at the end of the second Christian millennium. The replication-test bunny is a player in *SimLife*. Remember the game ad's version of the injunction to be fruitful and multiply: "Give life to different species in the lab and customize their look with the icon editor."

Both the pregnancy-test and replication-test rabbits in the Logic General ad are cyborgs—compounds of the organic, technical, mythic, textual, economic, and political. They call us, interpellate us, into a world in which we are reconstituted as technoscientific subjects. Inserted into the matrices of technoscientific maps, we may or may not wish to take shape there. But, literate in the material-semiotic practices proper to the technical-mythic territories of the laboratory, we have little choice. We inhabit these narratives, and they inhabit us. The figures and the stories of these places haunt us, literally. The reproductive stakes in Logic General's text—and in all the tropic, materializing action of the laboratory—are future life forms and ways of life for humans and unhumans. The genome map is about cartographies of struggle—against gene fetishism and for livable technoscientific corporealizations.

Notes

1. Thanks especially to Sarah Franklin, Helen Watson-Verran, Caroline Jones, Peter Galison, and Bruno Latour. The uncut essay appears in my book, Modest_Witness@Second_Millennium. Female-Man©_Meets_OncoMouse™ (New York: Routledge, 1996).

2. Michael Bremer, SimLife User Manual (Orinda, Cal.: Maxis, Bremer, 1992), p. 9.

3. David Harvey, The Condition of Postmodernity (Oxford: Basil Blackwell, 1989), p. 245.

4. Johnny L. Wilson, The SimEarth Bible (Berkeley: McGraw Hill, 1991), p. xviii.

5. Richard Dawkins, The Extended Phenotype (London: Oxford University Press, 1982), p. 82. On the gene as a sacralized object in U.S. culture, see Dorothy Nelkin and M. Susan Lindee, The DNA Mystique (New York: Freeman, 1995), pp. 38–57.

6. Sarah Franklin, "Life Itself," June 9, 1993, Center for Cultural Values, Lancaster University. See also Sarah Franklin, "Life," in Encyclopedia of Bioethics (New York: Macmillan, forthcoming) and "Romancing the Helix," in Romance Revisited, ed. L. Pearce and J. Stacey (London: Falmer Press, 1995), pp. 63–67.

7. Michel Foucault, The Order of Things (New York: Pantheon, 1971); and The History of Sexuality, vol. 1, trans. Robert Hurley (New York: Pantheon, 1978). Barbara Duden, Disembodying Women (Cambridge: Harvard University Press, 1993).

8. Franklin, "Romancing the Helix," p. 67.

9. Chandra Talpade Mohanty, "Cartographies of Struggle," in Third World Women and the Politics of Feminism, ed. C. Mohanty, A. Russo, and L. Torres (Bloomington: Indiana University Press, 1991).

10. Sharon Traweek, Beamtimes and Lifetimes (Cambridge: Harvard University Press, 1988), p. 162.

11. See Helen Watson-Verran, "Re-negotiating What's Natural," Society for Social Studies of Science, October 12–15, 1994, New Orleans, for analysis of how both admitted and denied tropes work within knowledge systems developed by European and aboriginal Australians in contending for land possession. Communication in power-laced practical circumstances makes the work of codification, situating, and mobilization of categories explicit for all parties, changing everybody and everything in the process, including the categories. This kind of articulation precludes fetishism—nothing gets to be self-identical. The maps and the facts turn out to be tropic to the core and therefore part of knowledge practices. This analysis is important for understanding knowledge production in general, including natural science.

12. Georg Lukács, History and Class Consciousness, trans. Rodney Livingstone (Cambridge, Mass.: MIT Press, 1971), p. 83.

13. Karl Marx, Capital, vol. 1, trans. Ben Fowkes (New York: Random House, 1976), p. 176.

14. Sandra Harding, Whose Science? Whose Knowledge? (Ithaca: Cornell University Press, 1991).

15. See Michael Flower and Deborah Heath, "Anatomo-Politics: Mapping the Human Genome Project," Culture, Medicine and Psychiatry 17 (1993): 27–41, for the semiotic-material negotiations solidifying the "consensus DNA sequence" that instantiates "the" human genome.

16. The word "fetish" is rooted in a mistake and disavowal of the colonialist and racist kind, one shared by both Marx and Freud, in which "Westerners" averred that "Primitives" mistook objects to be the real embodiment or habitation of magical spirits and power. Fetishism, these rational observers claimed, was a kind of misplaced concreteness that depended on "Primitives'" lower powers of abstract reasoning and inferior forms of religious faith. "Primitive" fetishes were about "magical thinking"; i.e., they were about the potency of wishes, where the desire was mistaken for the presence of its referent. Anthropologists discarded this doctrine of fetishism, but the racialized meaning, connoting the underdeveloped, irrational, and pathological, persists in many domains. The irony of the doctrine of "primitive" fetishes is that, if one follows Whitehead's explanation of the "fallacy of misplaced concreteness" that comes from the belief in simple location, relation- and observer-free preexisting objects, and a metaphysics of substantives with primary and secondary qualities, then the children of the Scientific Revolution are the world's first and maybe only serious fetishists, whose most extraordinary abstractions are taken to be reality itself. See A. N. Whitehead, Science and the Modern World (New York: Mentor, 1948, orig. 1925), pp. 41–56.

17. Sigmund Freud, "Fetishism," in *Sexuality and the Psychology of Love*, ed. P. Rieff (New York: Collier, 1963, orig. 1927), p. 205.

18. Laura Mulvey, "Some Thoughts on Theories of Fetishism in the Context of Contemporary Culture," *October* 65 (Summer 1993): 3–20, p. 7.

19. Richard Lewontin, "The Dream of the Human Genome," *New York Review of Books* (May 28, 1992): 31–40, p. 33.

20. Whitehead, *Science and the Modern World*, p. 52.

21. A. N. Whitehead, *Process and Reality* (New York: Free Press, [1929] 1969, p. 28.

22. Harvey, *Condition of Postmodernity*, p. 63.

23. *Science* (February 1, 1991): back cover.

24. Evelyn Fox Keller, *Secrets of Life, Secrets of Death* (New York: Routledge, 1992), pp. 39–55.

25. Carol Vogel, "Leonardo Notebook Sells for $30.8 Million," *New York Times* (November 12, 1994): A1, A11.

26. Sequenase™, a DNA polymerase used in sequence analysis, is marketed in versions, for example, Sequenase Version 1.0 or 2.0, like software, such as Microsoft Word 5.0—another signifier of the bond between informatics and genomics.

27. Joan O'C. Hamilton, "Biotech: An Industry Crowded with Players Faces an Ugly Reckoning," *Business Week* (September 26, 1994): 84–90, p. 85.

28. *Science* 18, 1 (1995): 77. A nonradioactive DNA-detection tool from Boehringer Mannheim is called Genius™ System, with the slogan "leaving the limits behind." An ad in *Biotechniques* 17, 3 (1994): 511, links the Genius™ System protocols with the toe pads of a tree frog, "allowing it to perform the most sensitive maneuvers . . . in pursuit of insect prey." The company offers natural design, delicacy, transcendence, and genius. Who could want more?

29. François Gros, "The Changing Face of the Life Sciences," *UNESCO Courier* (1988): 7.

30. *Science* 250 (October 12, 1990).

31. Daniel Koshland, Jr., "The Rational Approach to the Irrational," *Science* 250 (October 12, 1990): 189.

32. Nelkin and Lindee, DNA *Mystique*, pp. 81–82. Thomas J. Bouchard, Jr., D. T. Lykken, M. McGue, N. L. Segal, and A. Tellegen, "Sources of Human Psychological Differences: The Minnesota Studies of Twins Reared Apart," *Science* 250 (October 12, 1990): 223–28. Arthur Jensen, "How Much Can We Boost IQ and Scholastic Achievement?" *Harvard Educational Review* 39 (Winter 1969): 1–123.

33. *Science* 250 (October 12, 1990): 185.

34. Harvey, *Condition of Postmodernity*, pp. 244–52.

35. Ibid., p. 246.

36. Bruno Latour, *Science in Action* (Cambridge: Harvard University Press, 1987), pp. 215–57.

37. Barbara Culliton, "Mapping *Terra Incognita* (*Humani Corporis*)," *Science* 250 (October 12, 1990): 210–12.

38. Geoff Bowker "How to Be Universal: Some Cybernetic Strategies," *Social Studies of Science* 23 (1993): 107–27.

39. Latour, *Science in Action*, p. 245. J. C. Stephens, M. L. Cavanaugh, M. I. Gradie, M. L. Mador, and K. K. Kidd, "Mapping the Human Genome: Current Status," *Science* 250 (October 12, 1990): 237–44.

40. Steven Shapin and Simon Schaffer, *Leviathan and the Air-Pump: Hobbes, Boyle, and the Experimental Life* (Princeton: Princeton University Press, 1985).

41. Marilyn Strathern, *Reproducing the Future* (New York: Routledge, 1992), p. 39.

42. See Sharon Helsel, "The Comic Reason of Herman Kahn" (Ph.D. dissertation, History of Consciousness Board, University of California at Santa Cruz, 1993).

43. *Science* 251 (March 1, 1991): 1050.

44. *Science* (May 1, 1983), Logic General Corporation advertisement.

Fred Jameson says that postmodernism is
what you get when the modernization process
is complete and Nature is gone forever.
This installation makes you feel that explicitly.
It suggests that we no longer have any place
on earth that could serve as the ostensible
ground for pristine, nonhuman being. We
could see Faraday's as marking the disappearance of the religious construction of Nature,
Nature being all that is other than human.

LEO MARX

"Nature" and "Technology" can be seen as the two Others
that are always mobilized to define the human. And if one
of the points of Faraday's Islands is that the technological
is deeply human, then I think we also have to examine how
"Nature"—as a word, as a concept, as a representation—
is equally imbricated in the human.

CAROLINE JONES

The phrase "human nature" is something that I think of as having
a double meaning. "Human nature" is actually everything that
we build around us, which is why I tend to use metaphors such
as gardens and islands for this kind of work. A lot of decisions
about how the islands were built came out of this idea; the
plywood was supposed to suggest boardwalks and pilings;
the exposed wires hanging down were meant to seem like roots,
and so on. I'm not trying to say that "Nature" is gone. It's just
that it's totally problematic. This is not pure technology, and
there is no pure nature. The world is a given. It might preexist us,
but we're stuck with whatever we've done with it. So, for example, if we fix the hole in the ozone layer, then the ozone layer itself
becomes a human construction.

PERRY HOBERMAN

Seeing Wonders

KRZYSZTOF POMIAN

Vision and Cognition

COGNITION AS VISION

For ancient authors, *to know* means *to see*. It is indeed vision that offers them the best example of sensory cognition. It is also vision that is supposed to be an analogue of intellectual cognition. And when the latter is divided into an intuitive and a discursive cognition, this last, according to a general opinion, cannot unfold its sequences of syllogisms so as to produce science without having its foundation in the cognition of principles, which is itself a kind of vision.

Ancient theories of vision differ one from another.[1] But they all agree that to see an object is to establish with it an immediate relation such that nothing qualitatively different from the soul, on the one side, and from the object of vision, on the other, could find itself between them. This is obvious in the case of atomists who reduce vision to some modality of touch, because (in their opinion) the *eidola* or *simulacra*, which are emanations of objects, enter into the eye by the pupil and strike directly the soul, composed of some subtle matter. The idea of contact between the object and the soul is accepted also by Plato: an invisible body formed through some melting of an inner fire going out from the eyes with the daily light, transmits movements of the object "until they reach the soul causing the perception which we call sight."[2] The position of Stoics on this point seems to have been similar to that of Plato.[3]

This Platonic theory is firmly rejected by Aristotle. "It is, to state the matter generally, an irrational notion that the eye sees in virtue of something issuing from it; that the visual ray shall extend itself to the stars or else go out merely to a certain point, and there coalesce, as some say, with rays which proceed from the object."[4] But the idea of contact is nevertheless preserved, as is shown by the role Aristotle ascribes to the transparent medium: "As vision outwardly is impossible without light, so also it is impossible inwardly. There must, therefore, be some transparent medium within the eye, and, as this is not air, it must be water. The soul or its perceptive part is not situated at the external surface of the eye, but obviously somewhere within: whence the necessity of the interior of the eye being transparent, i.e. capable of admitting light."[5] Now the light, through the transparent medium of which it is the entelechy, transmits to the soul the movement coming from the outside and causes therefore the passage of the sensory faculty from potentiality to act, making it identical to the actually visible: "if to perceive by sight is just to see, and what is seen is colour or the coloured, then if we are to see that which sees, that which sees originally must be coloured."[6]

The Aristotelian theory of vision eliminates therefore the qualitative difference between the eye and the air because it makes both of them participate in the transparent medium. This enables it to identify vision with a perception of distant objects in which different intermediaries are involved, and to state at the same time that the form of an illuminated object impresses itself directly upon the soul: "it is better, instead of saying that the sight issues from the eye and is reflected, to say that the air, so long as it remains one, is affected by the shape and colour. On the smooth surface the air possesses unity; hence it is that it in turn sets the sight in motion, just as if the impressions on the wax were transmitted as far as the wax extends."[7] It follows that Aristotelian theory belongs to the same family as that of Plato and the Stoics.[8]

The immediate character of vision, and of sensory perception in general as understood by ancient authors, is manifest in the metaphors they use. This is the case in the Platonic metaphor of a block of wax upon which we impress our sensations and our conceptions "as we might stamp the impression of a seal ring."[9] We just met this metaphor in a quotation from Aristotle who uses it also in another important statement: "the sense is what has the power of receiving into itself the sensible forms of things without matter, in the way in which a piece of wax takes on the impress of a signet-ring without the iron or gold."[10] The same metaphor is used also by Stoics, who compare the gaze to a stick or rod that touches objects.[11]

Theories of intellectual cognition are as diverse as those of vision, which serve them as models. But they all assimilate cognition to a play between three partners: the intellect that perceives, the source of a metaphysical light, and an intelligible object. According to Aristotle the intellect is double: receptive and active, "capable of becoming anything" and "capable of producing them all, similar to some kind of a state like the light." In illuminating its object, this active intellect abstracts an intelligible form, i.e. causes the form's passage from potentiality into actuality as the light causes the passage into actuality of colors that otherwise would remain only potential. And the receptive, passive intellect receives this form, which inscribes itself upon it as upon a "writing table on which as yet nothing actually stands written."[12] The Platonic idea of an intellectual knowledge does not appeal to such a theory of abstraction, because it assumes that intelligible objects are really separated from sensible ones. And it assimilates the intellect to something like an eye, which sees the former in the same way the soul sees the latter, while the light comes from the metaphysical equivalent of the sun. Stoics too patterned intellectual cognition after their theory of vision .[13]

In a phrase, *to know is to see*. And to see is to establish an immediate relation between the soul itself and the object looked upon. In such a conceptual frame, nature can be known only through the naked eye. The use of observational instruments that intervene between the eye and its object, despite their being qualitatively different from both, cannot indeed be recognized as legitimate; hence the secular absence of interest among philosophers for lenses and for glasses.[14] Neither can one conceive, in this frame, the very idea of cognition in regard to the past; indeed, in order for cognition to be possible, its object must be present to the cognitive faculty, and this means that object and cognitive faculty must be strictly contemporaneous one with another. Finally, in such a conceptual frame, there is no place left for epistemology. The immediate relation between the cognitive faculty and its object entitles one to consider knowledge as a reflection of objects themselves; "being" has priority with respect to cognition. As we have seen, then, the theory of vision plays, in fact, the role of a theory of sensory cognition, and it furnishes also a model of intellectual cognition. Epistemology is therefore at one and the same time both inconceivable and superfluous.

The victory of Christianity and the assimilation by its theologians of ancient philosophy modified only in some respects the model of cognition as vision. From the fifth to the twelfth century, Western thought is dominated by St. Augustine. As well as the *Timaeus*, translated by, and with the commentary of, Chalcidius, Augustine's works are vehicles of the Platonic tradition revised and corrected in order to be put in conformity with the teaching of the Church. For instance, the idea of freedom of the soul (with regard to cosmic determinations to which it was obedient according to the pagan tradition) deepened the cut between the soul and the body, and furthered the isolation of the senses, servants of the latter, from the intellect, exclusive property of the former. And the identification of a transcendent God, the God of the Scriptures, with the source of metaphysical light necessary for the occurrence of an act of intellectual cognition, connected such intellectual cognition strongly with the Revelation and stressed therefore its contrast with sensory cognition, considered as incurably profane.

One of the most important legacies of this period resides in the clarification of the concept of faith. For faith is not cognition. It relates to invisible beings and is acquired not by sight but by hearing. Yet the knowledge it provides is superior to any knowledge that may result from cognition with respect to its object, to its origin and to the authority that warrants its content. Cognition must therefore be subservient to faith as the visible is subservient to the invisible, the earth to the heavens, the present to the very distant past, the time of origins (and of the presence of God among men) to the future (which will close time and open eternity). Hence an approach to the natural world and history that treats phenomena and events as expressions of the invisible: symbols, allegories, signs; in other words, instead of being interesting in themselves, they are related immediately to those contents of faith that seem to make them meaningful. One finds examples of such an attitude in *Hexaemerons* and in encyclopedias of the early Middle Ages, which follow the model set by Isidore of Seville.[15]

Since the twelfth century, the West begins to translate from Arabic and Greek. The discovery of Arab optics, in particular those of Alhazen, exerts an essential influence on the evolution of this science.[16] Attempts to put the rediscovered Aristotelian and naturalistic tradition in harmony with the theology molded by St. Augustine renew almost all philosophical problems. This is illustrated by controversies concerning the relation between divine illumination and the natural light of the intellect. Such explorations investigated the degree of autonomy of the latter with respect to the former in an act of cognition patterned after an act of seeing. This is illustrated too by controversies concerning relations between the *credibile* and the *intelligibile*, and by controversies concerning the concept of *species*. *Species*—an image or a semblance of an object (perceived either by a sensory or an intellectual vision) supposed to emanate from that object itself and to arrive at the receptive organ—is that through which the object may be known. A *species intelligibilis* is a form without matter. A *species sensibilis*, material at the point of departure, loses its materiality as it goes through the medium.[17] Therefore it is not qualitatively different either from the object out of which it emanates, or from the soul it is received by.

Nevertheless, knowledge *per speciem* offers an immediate relation only for Aristotelians who identify cognition with the grasping of a form by sensory faculty or by intellect. Such is, for instance, the position of Aquinas. His teaching does not leave therefore any room for human cognition, which would be intuitive. The expression *cognitio intuitiva* seems to be absent from his works. Words like *intueor* or *intuitus* are seldom used, and they apply principally to angels and to God. Aquinas distinguishes not between intuitive and discursive cognition, but between a *cognitio discursiva sive ratiocinativa* and a *cognitio sine discursu seu intellectus*.[18] The position of Duns Scotus is very different. According to him, intellectual cognition reaches its plenitude when it grasps an existent object as existent. Cognition of this kind cannot proceed through the agency of a *species*; it must be an immediate relation between an existent soul and an equally existent object, and it may be attained in the course of our terrestrial life. Such a *cognitio intuitiva* is opposed to a *cognitio abstractiva*, which grasps not an existent object, but only its image. Ockham extends these ideas of Duns Scotus to sensory cognition and concludes that the very concept of *species* is void and must be eliminated. According to Ockham, cognition is certain only when it is immediate.[19]

All of these controversies show that the Middle Ages maintained the model of cognition as vision, the latter being identified with an immediate relation between the soul and the object of its sensory or intellectual gaze. Changes introduced into this model result principally from the fact that henceforth faith superimposed itself upon cognition, and divine illumination superimposed itself upon natural light, be it physical or metaphysical. All consequences of the model rest in force: the cognition of objects that cannot be grasped by sight is as inconceivable as the cognition of the past through the agency of its remains. However, the importance granted to the past by Christian teaching, and Church endeavors to make intelligible the content of a faith that would preside over the transformation of theology into a science, both awakened an interest in the letter of sacred and profane texts. Increasingly, there were attempts at criticism that would eliminate contradictions between different authorities.[20] On the other hand, epistemology is still unthinkable. But the controversies between Thomists, Scotists, and Ockhamists concerning (in this context) the certainty of cognition, the problem of abstraction and of intuition, and the notion of evidence,[21] shifted the attention toward the relation between the soul and its object, and put into question the status of intermediating agencies such as *species*. In the fifteenth century, with Cusanus and Valla, there appeared the first examples of a practice of cognition of the past through the agency of its remains.[22] But only Galileo with his telescope could inaugurate the cognition of objects that are beyond the reach of sight.

FROM COGNITION AS VISION
TO COGNITION AS PRODUCTION

In his *Ad Vitellionem paralipomena* (1604), Kepler, without even being aware of it, gave the first blow to that age-old identification of vision with an immediate relation of the soul to an object it gazes at. He established indeed a qualitative difference between the fixation of the image in the optical field "on the white and reddish wall of the concave surface of the retina" and the grasping of this image by the soul. The first belongs to optics, the second to "physics," which seems to have been for Kepler both a physiology and a psychology, the study of nervous transmission and the study of perception itself. Now optics stops at the moment of fixation of the image: the "luminous representations" cannot propagate themselves through the nerve "because placed among the opaque and therefore dark parts and regulated by spirits which differ under all respects from humors and other transparent things, it is already completely withdrawn from the laws of optics"[23]—hence the criticism of Vitellion and all ancient and medieval traditions that did not take this discontinuity into account.

The common sense or any other faculty of the soul that has to receive the data of vision enters therefore in relation to something for which the similitude to the object itself becomes a new problem. This problem, which Kepler only stated and which he left to "physicists," is approached by Descartes. In his *Dioptrique* (1637), he gives it a negative cast, inspired by the invention of *"ces merveilleuses lunettes qui, n'étant en usage que depuis peu, nous ont déjà découvert de nouveaux astres dans le ciel, et d'autres nouveaux objets dessus la terre, en plus grand nombre que ne sont ceux que nous y avions vus auparavant."*[24] After having noted that *"c'est l'âme qui sent, et non le corps"* and that *"c'est par l'entremise des nerfs que les impressions, que font les objets dans les membres extérieurs parviennent jusqu'à l'âme dans le cerveau,"* Descartes stresses:

> Il faut . . . prendre garde à ne pas supposer que, pour sentir, l'âme ait besoin de contempler quelques images qui soient envoyées par les objets jusqu'à cerveau, ainsi que font communément nos philosophes; ou, du moins, il faut concevoir la nature de ces images tout autrement qu'ils ne font.

Descartes makes this claim because images are not the only stimuli able to excite our thought. The same effect is produced by *"les signes et les paroles qui ne ressemblent en aucune façon aux choses qu'elles signifient."* And images themselves bear only a very partial similitude to their objects like the copperplate engravings that

> n'étant faites que d'un peu d'encre posée çà et là sur du papier, nous représentent des forêts, des villes, des hommes, et même des batailles, bien que, d'une infinité de diverses qualités qu'elles nous font concevoir en ces objets, il n'y en ait aucune que la figure seule dont elles aient proprement la ressemblance; et encore est-ce une ressemblance fort imparfaite.[25]

It follows that it is not the similarity of an ocular image to its object that makes us feel this image *"comme s'il y avait derechef d'autres yeux en notre cerveau, avec lesquels nous la puissions apercevoir."* This is caused rather by *"les mouvements par lesquels elle est composée, qui, agissant immédiatement contre notre âme, d'autant qu'elle est unie à notre corps, sont institués de la Nature pour lui faire avoir de tels sentiments."*[26] Vision is characterized therefore as an indirect relation between the soul and the objects of its gaze, and even between the soul and the image of these objects, because the movements that are transmitted by the nerves and that mediate between the soul and its objects are qualitatively different from both. Thus, a problem arises from the conversion of the visual image into a movement and from the action of the latter upon the soul; it is a particular case of the more general problem of communication of substances.

The consequences derived by Descartes from the new theory of vision upset the traditional philosophical landscape. As the nerves transmit to the soul only movement, and as this is sufficient to awaken in the soul all the diversity of thoughts (in the Cartesian meaning of this term), the movement, provoked in the nerves by bodies, must be diversified at the very starting point. Yet the only factors able to do this are, according to Descartes, the magnitude, the shape, and the position of parts of bodies. In other words, these are the only characteristics of bodies the information on which is encoded in the characteristic of movements transmitted by the nerves to the soul. It follows that

> nous n'apercevons point en aucune façon que tout ce qui est dans les objets, que nous appelons leur lumière, leur couleurs, leurs odeurs, leurs goûts, leurs sons, leur chaleur ou froideur, et leurs autres qualités qui se sentent par l'attouchement, et aussi ce que nous appelons leurs formes substantielles, soit en eux autre chose que les diverses figures, situations, grandeurs et mouvements de leurs parties.[27]

Sensible qualities do not belong therefore to bodies themselves. They are rather akin to passions of a soul—a soul that reacts, in conformity with its nature, to external stimuli.[28]

This being admitted, if the only cognition we may have is a sensory cognition, then the real world (in which are located bodies differentiated by shapes, magnitudes, and positions of their parts), would be completely beyond our reach; so much so that we would even be unable to know that bodies are actually differentiated in that manner. But our situation is not that unhappy. Indeed, according to Descartes, we have a direct access to reality that enables us, so to say, to short-circuit the senses. Thanks to the *cogito*, my intellect puts beyond doubt my own existence as a thinking substance. It demonstrates then the existence of God, who alone may explain the presence in us of the idea of infinity at which we should never arrive by our own forces because of our very finitude. Going further, the intellect, in grasping the existence of God, demonstrates the reality of bodies, for we grasp bodies (as we comprehend a piece of wax as wax, though it be molten or solid), which are reduced only to their extensions, i.e. precisely to the shape, magnitude, situation, and movement of their parts.

These demonstrations take the form of logical inferences only in order to be reproducible. Really each of them consists in grasping by the intellect of clear and distinct ideas, and of the necessary connections that unite them so as to make impossible the admission of one of them without all of the others. For Descartes, to have a certain cognition is precisely to grasp such ideas:

> car la connaissance sur laquelle on peut établir un jugement indubitable doit être non seulement claire mais aussi distincte. J'appelle claire celle qui est présente et manifeste à un esprit attentif, de même que nous disons voir clairement les objets lorsque étant présents ils agissent assez fort, et que nos yeux sont disposés à les

regarder; et distincte celle qui est tellement précise et différente de toutes les autres, qu'elle ne comprend en face de soi que ce qui paraît manifestement à celui qui la considère comme il faut.[29]

The intellect therefore is identified here with the eyes, in front of which are placed strongly illuminated objects whose outlines stand out sharply against the background. But these eyes seem to function in a manner closer to the theory of vision of Plato or of Duns Scotus than to that of Kepler or of Descartes himself.

The belief in intellectual intuition indeed coexists in Cartesian philosophy with the recognition of the indirect caracter of sensory vision. The first grasps the really existent objects—shapes, magnitudes, movements, respective positions—and grasps them with the certitude awarded by divine guarantee. This is why the intellectual intuition may unfold itself in demonstrations of the same type as those of mathematics. The second puts the soul in relation not with objects themselves but with the movement transmitted by nerves. Between objects and ocular images, on the one side, and sensations, on the other, there is an insuperable barrier. Sensory vision may be studied by anatomy with the help of instruments and of mathematics; it is a part of physics. Intellectual vision has a distinctively metaphysical character. And this duality is projected on things themselves. Among the qualities commonly ascribed to them, some, according to Descartes, are real; others are only affections of the soul. Both present themselves as coming from the senses. Only the intellect, because it is endowed with intuition, enables us to separate, among these sensory semblances, primary from secondary qualities, objective data from subjective affections, and information concerning reality from illusions.

The duality I just described is not specific to Cartesian philosophy only, or even only to the rationalist current. Despite his placing himself at the opposite pole, Locke compares understanding to a *camera obscura*.[30] In so doing, he compares it also to the eye, of which the *camera obscura* was itself a model since Leonardo—it is referred to as such by both Kepler[31] and Descartes.[32] In Locke, simple ideas are treated as analogous to ocular images, and the relation between understanding and simple ideas is therefore an immediate one. Hence it can be used as a foundation of the equally immediate relation between the mind and ideas in general. Indeed, Locke compares the mind to the eyes and speaks of intuitive cognition and of a knowledge that imposes itself on the mind as solar light imposes itself on sight.[33]

But between external objects and the ideas of these objects grasped by the understanding is placed the movement that conveys through the nerves primary qualities of things: matter, extension, shape, and movement. The secondary qualities, however (colors, sounds, flavors), are nothing more than our sensations, produced by primary qualities without our knowing how such a thing occurs.[34] The relation between ideas of external things and these things themselves is therefore an indirect one. And if external senses are treated as objects of physics—Locke speaks of eyes and other organs, of nerves, of "animal spirits," of the brain—the understanding and the mind belong to metaphysics. In the theory of cognition, then, the philosophy of Locke is characterized by the same duality as the philosophy of Descartes.

The ninety-one years that separate the first edition of Locke's *An Essay concerning Human Understanding* (1690) from the first edition of Kant's *Kritik der reinen Vernunft* (1781) witnessed several attempts to eliminate the disagreement between physical theory of the ocular vision and metaphysical belief in the reality of intellectual intuition. Thus, George Berkeley, in his *Essay towards a new Theory of Vision* (1709), tries to show that we neither see the distance between objects, nor their magnitude.[35] In other words, the data of sight do not authorize us, according to Berkeley, to introduce between primary and secondary qualities of things a sharp, fundamental distinction such as was admitted by both Descartes and Locke.

In Berkeley's opinion, our ideas of distance are not the data of sight but the products of the synthesis of visual and tactile sensations, and, inside each of these two classes, of multiple sensations that may be called *elementary*. The term is absent from Berkeley's text but his approach is obviously based on the conviction that one can dissociate an idea into its ultimate components without any remainder.[36] For the synthesis of sensations—tantamount to the fabrication of an idea—is not imposed by any necessary connection between these sensations themselves. Such a connection does not exist. The soul produces ideas according to its habit of doing this as it gives names according to its habit of indicating definite things by definite words.[37] But if there is nothing real that would correspond to our ideas of distance and of magnitude, then visual space itself vanishes, and things we believe to be external to ourselves, because of the confidence we grant to sight, in fact exist only in our minds.[38] Vision is not therefore for Berkeley a matter of physical investigations. It is appropriated by a psychology that practices an introspection and analyzes its data.

In the *Treatise concerning the Principles of Human Knowledge* (1710), Berkeley goes a step further. Henceforth all physics is reduced to psychology because things, time, space, and motion are nothing but ideas. And psychology leads to spiritualist metaphysics. As all apparently external objects are actually in our mind, the unique substance is a spiritual one that thinks, desires, acts, and perceives. Ideas are its productions. Hence they do not refer to anything, nor do they exert any influence, even on other ideas. Deprived of the slightest spontaneity, they depend completely in their being and in their mutual relations upon the spirit that perceives them.

The spirit maintains with itself an immediate relation in which the distinction of subject and object is irrelevant. This is the last remnant of cognition as vision: an intellectual intuition deprived of all cognitive virtue because it does not produce ideas but only a feeling of oneself. To know, the spirit has to turn itself to the effects it produces, i.e. to ideas. And to perceive ideas, it must produce them. For the spirit, being purely active, deprived of any receptivity, perceives only when and insofar as it produces. *Cognition as vision is thus replaced by cognition as production.* This does not seem to apply to human beings, who discover easily that some ideas they have do not depend upon their will. As these ideas can come only from some spirit, they must be imposed by a superior spirit, by God. So they are produced by the spirit upon which they are imposed, but under the impact of an even more powerful will. Hence it is absolutely true that to know ideas is to produce them.[39]

In this way, Berkeley gets rid of the problem of communication of substances with all its difficulties, such as the question of impact of external objects upon our sensory organs, or the issue of perception by the soul of data transmitted by nerves. He thus eliminates the duality present in Descartes and in Locke. But he pays for this with a duality of cognition and feeling (needed to justify the opposition between ideas and the spirit); equally, he incurs the need to introduce God as a transcendent guarantor of ideas of sensible things, which differ from the products of imagination only because of Him.

In *A Treatise of Human Nature* (1739) finished at La Flèche a century after the publication of *Dioptrique*, Hume eliminates all incoherency from the position of Berkeley and gives achieved form to the model of cognition as production. Now, it is an exclusively human production:

> We may observe, that 'tis universally allow'd by philosophers, and is besides pretty obvious of itself, that nothing ever is really present with the mind but its perceptions or impressions and ideas, and that external objects become known to us only by those perceptions they occasion. To hate, to love, to think, to feel; all this is nothing but to perceive.
>
> Now since nothing is ever present to the mind but perceptions, and since all ideas are deriv'd from something antecedently present to the mind; it follows that 'tis impossible for us so much as to conceive or form an idea of any thing specifically different from our ideas and impressions. Let us fix our attention out of ourselves as much as possible: Let us chase our imagination to the heavens, or to the outmost limits of the universe; we never really advance a step beyond ourselves, nor can conceive any kind of existence, but those perceptions which have appear'd in that narrow compass. This is the universe of our imagination, nor have we any idea but what is there produc'd.[40]

This quotation, and the entire work of Hume, introduces us to an anthropocentric world antipodal to Cartesian metaphysics and its various continuations. For Hume, it is obvious that all perceptions that intervene between the mind and external objects are affections of the former: no essential difference can be traced between emotions such as love or hatred and, for instance, visual sensations. Able only to establish relations among perceptions, the mind does not possess any power that would put it in contact with an object in such a way that it could experience not a perception but a substance, if this term means something more than a set of qualities, i.e. of perceptions.[41] The mind, therefore, has no resource that would enable it to circumvent perceptions, to attain directly the things themselves. Its relations with the outside are necessarily indirect. But the way outside (through perceptions) is unfit for traffic. Indeed, we cannot, whatever we do, pass from perceptions to the causes they are occasioned by and that remain, for us, unknown forever.[42] It is impossible therefore to compare our ideas of external objects with the objects themselves in order to verify whether the former agree with the latter; the very project of such an operation is a sheer absurdity.

Indeed, the idea that there are external objects at all is not received from the senses. The relation of being external with respect to us is imposed by our imagination on sense data, which in themselves tell us nothing about that. The same is true of any idea of existence that might be independent with respect to our perception of it.[43] In such a situation epistemology is as inconceivable and superfluous as it was when cognition was identified with vision, albeit for completely different reasons. The cognitive faculty was considered then as purely receptive, and so external objects could leave upon it their impressions. For Hume, on the contrary, the cognitive faculty is active and external objects are but its projections. But to ask questions about cognition—its nature, its reliability, the legitimacy of its proceedings—one has at first to assume that an interaction occurs between the external and the internal, the world and the mind, the given and the produced, etc. And that knowledge results from such an interaction.

According to Hume, on the contrary, our knowledge consists only of relations we establish between impressions and ideas. The most important among these relations is that of causality, because it seems to transcend our senses and to inform us about things and existences that cannot be seen or felt.[44] Yet an analysis of our idea of causality shows that what is constitutive of it and without which we cannot think about it, is the idea of the necessary connection between two objects; terms such as *power, energy, force, efficiency, necessity* are only its synonymous names. And if we pursue our analysis further, trying to discover where this idea of necessary connection comes from, we arrive at the conclusion that we are ourselves its only authors:

> Upon the whole, necessity is something that exists in the mind, not in objects; nor is it possible for us ever to form the most distant idea of it, consider'd as a quality in bodies. Either we have no idea of necessity, or necessity is nothing but that determination of the thought to pass from causes to effects and from effects to causes, according to their experienc'd union. . . . The efficacy or energy of causes is neither plac'd in the causes themselves, nor in the deity, nor in the

concurrence of these two principles; but belongs entirely to the soul, which considers the union of two or more objects in all past instances. 'Tis here that the real power of causes is plac'd, along with their connexion and necessity.[45]

We arrive here at the extreme point of Hume's radical anthropocentrism and we measure all the effects of the overthrow of the model of cognition as vision, inaugurated a century earlier in the work of Descartes and now at its end. When Hume eliminates the idea of a necessary connection and substitutes that of a purely factual one, proceeding from the habit created by repetition, he destroys first of all the very possibility of a bridge between our ideas and the outside; it is only now that we understand why the causes of our sensations are unknown to us forever. Hume thus denies any foundation for the belief according to which we can transcend our perceptions, not toward other perceptions—with this he agrees—but toward something different from any perception in its manner of being. Hence, according to Hume, we are enclosed within the limits of our perceptions and of our imagination, which establishes relations between them. But these limits are those of the universe itself (because we cannot even conceive of any other). And this universe, of course, is produced by humans.

This is why the only relevant questions for Hume are concerned with human nature. All problems of being disappear with the reduction of human nature to the set of our ideas, and with the parallel dissolution of metaphysics into psychology and history. Psychology shows the working of the mind. And the knowledge of past events makes possible an understanding of human nature. These events may be known, provided they are registered in written records; a legitimate reasoning enables us to pass from impressions they create in us to the idea of those who were eyewitnesses of recorded events.[46] Thus the past acquires the status of an object of cognition (which it already acquired a long time ago in the practice of historians).[47] But it acquires such a status only after having been reduced to a set of impressions and ideas.

Likewise cognition, through the agency of observational instruments, is perfectly legitimate—as it was already for Descartes. But Descartes assimilated instruments to materializations of theories. Even if the telescope was, according to him, found thanks to an accident, it furnishes valuable results founded on the laws of optics[48] (and on intellectual intuition). As it seems, Hume seldom mentioned observational instruments. Contrary to Berkeley (who was interested in the microscope), he leaves such questions, as he leaves sensations, to practitioners of natural philosophy.[49] It is certain, however, that in the Humean perspective, instruments are only extensions of our senses: they bring us new perceptions but we remain nevertheless in a universe of which we are the center.[50]

Better than anyone else in his time, Kant recognized the importance of Hume's work.[51] And he derived from it a perfectly valid conclusion: that one could no longer practice philosophy as had previously been done, i.e. using the model of cognition as vision (or some fragments of it). But this did not prevent Kant from discovering that the interpretation of the Humean model of cognition as production is not the only possible one. Nor did it stop him from proposing a new interpretation of this model that was opposed in several respects to that of his predecessor.

Their disagreements are rooted, it seems, in the divergent orientations of their curiosity and intellectual practices. Hume was interested all his life in the science of human nature: in an introspective psychology, in economy, in politics, in history; he published in particular a history of England that remains a classic. Kant turned rather toward physics, mathematics, and the natural sciences, and his reflections concentrated on space. By the eighteenth century and even later, disciplines studied by Hume could be reduced to a collection of ideas that might be isolated from each other without being distorted, because they were connected only by extrinsic relations. Post-Newtonian mechanics and mathematics were much more resistant to such a treatment. It is true that Hume approaches them in the same way, because for him all human knowledge is but a collection of ideas, as any complex idea is but a collection of impressions.[52] But this is precisely the path Kant refuses to follow.

According to Kant, mathematics and physics utter judgments that establish between their components a necessary and universal connection (examples: "7 + 5 = 12" or "Between two points, the straight line is the shortest one" or yet again "In all communication of movement, the action and the reaction must always be equal one to another").[53] These judgments, says Kant, are not analytic: the idea of bringing together 7 and 5 does not contain the idea of 12; the idea of a straight line has nothing in common with the idea of the shortest line between two points; the idea of a communication of movement does not entail automatically that of equality of action with reaction. But, on the other hand, these judgments cannot be synthetic *a posteriori*, that is, come from an experience, because the latter can only ascertain some state of things without being able to arrive at a necessary and universal judgment.

In Kantian language, the judgments of mathematics and physics are therefore synthetic a priori judgments. But once we accept this, all the work of Hume must be taken up again on new foundations. For if human knowledge cannot be reduced to a collection of ideas connected by extrinsic relations furnished by experience, a theory of knowledge must be constructed in order to explain how synthetic a priori judgments are possible. Where resides the faculty of an a priori synthesis and what is its nature? Are we enabled by it to go out of the universe of our perceptions? And how can one include such a faculty into the model of cognition as production?

The answer to these questions is *The Critique of Pure Reason*. In certain respects it rests near the work of Hume. Kant focuses his investigations on human cognition; this constrains him to make the capacity to make an a priori synthesis a faculty of the human mind, to place it inside a human being. On the one hand, he presumes, as Hume did, an exhaustive and disjunctive division of human cognition into a sensory and an intellectual cognition, into experience and thought; on the other hand, he divides senses into internal and external ones. Kant admits with Hume that it is only our sensibility that establishes an immediate relation with objects. It follows, and this point must be heavily stressed, that the very possibility of an intellectual intuition is eliminated, which makes still more difficult the question of the nature of the faculty of an a priori synthesis.

A disagreement with Hume appears however, given the need to choose an approach expected to give a satisfying answer to the question concerning the possibility of synthetic a priori judgments. The very statement of such a question is extraneous to Humean introspective psychology, which allegedly analyzes the real functioning of the human mind. For Kant is not interested in that. He is interested instead in the conditions of possibility for an a priori cognition. His approach is not empirical. It is transcendental. "I call transcendental," explains Kant, "all cognition which applies itself in general not so much to objects as to our manner of knowing objects in so far as it is possible in general."[54] And he practices not psychology but epistemology. In the history of philosophy, *The Critique of Pure Reason* seems to have been the first book at the very center of which are placed neither the principles of cognition as in Descartes and Berkeley, nor human understanding as in Spinoza, Locke, Leibniz, and Hume; neither human nature as in Hume, nor yet the origin of knowledge as in Condillac, but cognition itself, its conditions of possibility and its limits.

A disagreement on method produces a disagreement on results. The latter is present for Hume and Kant in their differing analyses of the most fundamental act of cognition—the formation of a phenomenon from sensations or, as Hume would state it, of a complex idea from impressions. For Kant shows that in order to make a phenomenon present, i.e. in order to have a representation of an object as one object, the diversity of sensations must be integrated in a unique form. Yet such a form cannot itself be a sensation. It must therefore be a priori, coming before any experience and standing ready to receive its data; in other words, it must be inbuilt in the perceptual apparatus of a human being. There are two such a priori forms of sensibility: space for external senses and time for the internal sense. Thanks to the latter we may have intuition a priori—and it is nevertheless sensible!—of time and of space. In other words, we may grasp the very forms of our sensibility as we grasp the phenomena. Hence the possibility of judgments that are synthetic, because they establish connections between phenomena, and that are at the same time a priori, because they are not dependent upon experience.[55]

According to Hume, mind builds complexes starting with simple data (impressions) and using relations that are given to it. Kant, on the contrary, ascribes to mind a spontaneity: a capacity to integrate what is given to it in syntheses displayed in a hierarchy of levels going from forms of sensibility to principles of reason through categories of understanding. At the level of sensibility, forms of thought stand ready to receive the data of intuition and to operate the a priori synthesis. The unifying power of thought increases with the widening of distance from sensibility, while the data of intuition are more and more diluted. But at no level—and this is a fundamental point—do we succeed in leaving the world of phenomena in order to establish some contact with things in themselves.

This is striking already at the level of sensory cognition. Kant summarizes his opinion on this topic:

> We wanted to say that all our intuition is nothing other than the representation of phenomena; that things which are objects of intuition are not in themselves such as we grasp them in our intuition and that their relations are not constituted in themselves such as they appear to us; that if we made abstraction of our subjectivity or even only of the subjective constitution of senses in general, the manner of being of objects and all their relations in space and in time, as well as space and time themselves, would disappear; as phenomena they can exist exclusively in us and not in themselves. The nature of objects considered in themselves and abstracted from all this receptivity of our sensibility is completely unknown to us. We do not know anything about these objects but our manner of perceiving them, the manner which is specific to us and which may quite well not be necessary to all beings, although it is necessary for any man.[56]

With this statement of the impossibility of leaving the world of our representations, Kant arrives at conclusions similar to those of Hume. He stresses their importance, for he insists on the reversal of roles between cognition and its objects (objects having henceforth to conform themselves to cognition and not cognition to objects); and he compares his work in this respect to that of Copernicus.[57] Numerous philosophers of the nineteenth century tried to escape from such an anthropocentrism, to discover ways able to conduct us to things themselves that had somehow been neglected by Hume and by Kant. They never succeeded, however, as far as intellectual and sensory cognition were concerned, to go beyond the conceptual frame imposed by the model of cognition as production. They remained unable to free themselves from problems indissolubly connected to it.

INDIRECT COGNITION

Where are we today? Is the model of cognition as production still valid? Are Hume's and Kant's problems still determining the limits of our epistemological reflection? Are their questions still ours? In order to answer without writing a book on epistemology during the last century, I shall try to take a shorter way. It consists in an attempt at verifying, without entering into details, whether in view of all that happened during the last two hundred years we may still accept the assumptions of Hume and of Kant that made their questions relevant.

Among these assumptions the place of primacy belongs, it seems, to the differentiation of sensory and of intellectual cognition, and the status hierarchy established between them. The former is a physical fact: the action of an object on a sense organ triggers impulses transmitted by nerves to the brain. That they were called *movements* or *animal spirits* is for us unimportant. The only relevant point is the absence of any similitude between these impulses and the objects that trigger them. For it follows that senses do not give us imprints of external objects because between the latter and ourselves intervene the nervous impulses. Sensory cognition therefore establishes between us and external objects only an indirect relation. Physical and indirect, it opposes itself in these two respects to intellectual cognition, which is metaphysical and consists in an immediate grasping of sensations (the latter being the metaphysical equivalents of nervous impulses).

If the intellect could in addition directly grasp things themselves, it would be entitled to proceed to the critique of senses founded on its capacity of confronting their data with objects that are their causes. So it was according to Descartes. But Hume and Kant deny all intellectual intuition. The intellect is for them nothing more than the capacity to associate or synthesize sensations and thus to produce representations, ideas, or phenomena. These productions cannot however confront things themselves, because they remain irreparably isolated from them. Yet besides the senses and the intellect, we have no faculty of cognition, and thus we have no immediate relation with things themselves. Conclusion: the belief that we remain enclosed in the world of human representations is inevitable, insofar as we accept the indirect character of sensory cognition, deny the possibility of intellectual intuition, and refuse to admit any cognition that would be neither sensory nor intellectual, as if the division in these two categories was at the same time exhaustive and utterly disjunctive.

Yet this last assumption can no longer be accepted, if it was even acceptable in the times of Hume and Kant, to say nothing of later in the nineteenth century and into the twentieth. It can no longer be accepted not because of a discovery of some extrasensory metaphysical cognition; such an event never occurred. But we practice every day and at an enormous scale a kind of cognition that, despite its being extrasensory, is nevertheless a physical fact. I refer here obviously to the cognition through the agency of instruments of observation and measurement. I shall try now to sketch some characteristic features of this type of cognition, in order to show that the very fact we are practicing it obliges us to abandon the model of cognition as production.

Instruments of observation and measurement are not simple extensions of senses. Such an opinion could probably be accepted with regard to an optical microscope or a telescope, although already these instruments, as far as they enable us to see objects beyond the reach of the naked eye, introduce a cognition qualitatively different from the only one the latter is able to practice; on this point Descartes as well as Kant would agree.[58] But even if one could reduce this difference to a simple widening of the visual field and an increase in the number of objects grasped by sight (without being aware of the fact that sight is transformed by such moves beyond what it was before Galileo), such an attempt would seem simply incongruous with regard to a Geiger counter, a spectrograph, a radio telescope, or a particle accelerator. In all these cases indeed the instruments we deal with function according to principles sharply different from those governing our sensory organs (although their data might take visual or auditory forms). Such instruments enable us to apply cognition to objects that would otherwise be inaccessible through differences in their very manner of being from ordinary objects of our macroscopic world. To characterize such instruments of observation and measurement as extensions of our senses is to erase without any justification the essential difference between two types of cognition.

This does not mean that instruments belong to the sphere of intellectual cognition. For they are not simple materializations of theories. It is true that without theory one could not build them or discuss their results, i.e. establish in what limits their indications express their effective interactions with objects they are applied to. There is however a deep difference between the statement according to which a theory is necessary in order to build and to use instruments and the statement according to which either may be reduced to a theory. The first is obviously true. The second either means that instruments do not bring anything unforeseen by the theory from which they proceed, or it has no definite meaning. It is therefore either manifestly false or obscure. And it is manifestly false because there are countless examples of results of observations and of experiments no theories have foreseen, without even mentioning those that contradicted theories that were apparently very well grounded.

This cognition through the agency of instruments of observation and measurement is therefore an extrasensory but nonetheless physical cognition. And extra-intellectual—but also productive of elements of discourse: of images, of indications displayed on screens, of photographs, of different types of recordings, and so forth. It is a cognition *sui generis*. And its particularly striking character is its being an *indirect cognition*: what we receive as a result of an observation or an experiment is either the image of an interaction between an instrument and the object to which it is applied, or a set of parameters that describe such an interaction. This enables us, thanks to our knowledge of the instrument used, to infer the properties of the object itself, holding off on the theory, within the limits established by laws of physics. The possibility of reproducing an observation or an experiment and of controlling one instrument through its confrontation with others of the same kind gives us good reason to think that we deal indeed with natural objects, and not with artifacts.

In contradistinction to sensory cognition, which seems immutable (although it has its history too), instrumental cognition evolves in a spectacular manner through an enrichment of the panoply of instruments and their improvement. By its very nature, it creates a history. As this history proceeds, objects on which we are informed by instruments become more and more distant from us: distant in space, distant in time, distant because of their dimensions, distant, in the end, because of their strangeness with respect to laws of the macroscopic world in which we live. In order to be able to have a correct representation of these objects and in order to be able to think about them, it was necessary to modify even some of the most unquestionable of our assumptions concerning, in particular, the ideas of space and of time, the idea of identity, the idea of determinism, and so forth.

The history of physics for approximately a century shows clearly, through paradoxes, contradictions, and difficulties (all of which provoke controversies), the incapacity of the usual language and of the stock of images derived from everyday experience to master conceptually the new universe progressively unveiled. It shows also a struggle with the usual language and with intellectual habits rooted in everyday experience, which, in the end, were both completely overthrown. It shows, in a word, that physicists had constantly to learn anew how to imagine and how to think in order to adapt themselves to results furnished by instruments, and to derive from them conclusions able to be translated in the language that may be understood by instruments and thus to be submitted to a test of observation or experiment. The results of the history, provisional to be sure, reveal the world of microphysics and the universe of megaphysics (current cosmology) to be profoundly *un*-anthropocentric. Many things can be said about these worlds, but it would be difficult to contend that humans can reasonably claim to be at their very center.

Hence, if the division of cognition into sensory and intellectual regimes no longer holds, this is because there exists at least one other type of cognition (which I have termed "instrumental cognition") that has no place in such a division. In reality there exists yet another that has no place in this division: the cognition of the past through the agency of its remains. These two paradigmatic examples of indirect knowledge draw nearer and nearer to one another, so as to melt in some cases. On the other hand, as is shown by the contemporary psychology of perception, the time-honored division of cognition into sensory and intellectual modes of knowing cannot today be considered as disjunctive. Moreover, the development of instrumental cognition and its application to the study of the nervous system has resulted in a new idea of sensory cognition itself.

In sum, if we want to avoid paradoxes, we can no longer identify ourselves with a metaphysical subject of cognition separated by a barrier from the senses. We know that intellectual activity is a function of the brain and that the latter builds our image of the world, i.e. our knowledge, using nervous messages that bring, in a coded form, information about external objects. The presence of the code makes sensory cognition itself an indirect cognition, similar in this respect to instrumental cognition. But it does not establish an insuperable barrier that would separate us from objects them-selves and enclose us in the world of our representations. If we speak about the code, it is precisely because we are able to discover its rules and to know (not always, to be sure), what messages correspond to what characteristics of objects. We know this because observations of, and experiments on, the nervous system made it possible for us to understand the principles governing its functioning. There are certainly plenty of things we ignore in this field as well as in any other. In particular we do not know how nervous messages are transformed into signs, how culture and society intervene in the process, and how the human brain succeeds in thinking about itself. Nevertheless, thanks to instruments, we are able to observe from outside our own nervous system and in particular its cognitive activities, and to overcome therefore the limits that were considered insuperable by our ancestors and that for them were insuperable indeed.

One could draw similar conclusions with regard to intellectual knowledge. Suffice it to say that, on the one side, intellectual cognition, like sensory knowledge, is a physi-cal fact, which is at the same time a semiotic and therefore a cultural one. On the other side, it is also an indirect cognition because there is no axiom, no evidence, which would be a primeval datum, nothing like an absolute a priori. As far as space is concerned, the impossibility of direct cognition has been known since the discovery of non-Euclidean geometries; all the posterior history of formal thought has only corrob-orated this linkage of the physical and the intellectual, time and again. It would be illegitimate, however, to infer from this that our thought is exclusively discursive in the old meaning of this term. For the practice of theoretical physicists shows that, pro-vided certain conditions are met, thought succeeds in grasping reality and in making statements about it that instruments translate into interactions with objects.

The model of cognition that is valid today is neither that of cognition as vision nor that of cognition as production. It is a model of indirect cognition, of which I have sketched now a rough outline. We live decidedly in a world qualitatively different from that of Hume and of Kant. Their philosophies preserve nevertheless a partial validity, as do the philosophies of their predecessors, because, under its most recent strata, our world contains also all the ancient ones. We practice indirect knowledge with our instruments, but we also simply contemplate the world around us. And one lesson of Kant is still fully valid. It states that pure reason cannot transgress the limits of a possible experience, without falling into paralogisms and antinomies.

Notes

1. Cf. John I. Beare, *Greek Theories of Elementary Cognition from Alcmaeon to Aristotle* (Oxford, 1906).
2. Plato, *Timaeus*, trans. by Benjamin Jowett, 45b–46a.
3. Cf. Emile Brehier, *Chrysippe et l'ancien stoïcisme* (1910) (Paris, 1971), p. 81ff.

4. Aristotle, *De sensu et sensib*, 438a, 25–27 (All English translations from Aristotle come from: *The Complete Works of Aristotle*, revised Oxford translation, ed. Jonathan Barnes [Princeton University Press, 1984, Bollingen Series LXXI. 2]).

5. Ibid., 438b, 6–12.

6. Aristotle, *De anima*, 425b, 22–24.

7. Ibid., 435a, 4–10.

8. Cf. David C. Lindberg, *Theories of Vision: From Al-Kindi to Kepler* (Chicago and London, 1976), p. 9ff.

9. Plato, *Theaehetus*, trans. F. M. Cornford, 191d.

10. Aristotle, *De anima*, 424a, 18–19.

11. Cf. E. Brehier, *Chrysippe*, pp. 82–83; D. Lindberg, *Theories of Vision*, p. 10. And cf. St Augustine, *De quant. anim*. 44 = PL, t.32, col. 1060: "*oculi tui hoc modo defendi possunt, quorum est quasi virga visus.*"

12. Cf. Aristotle, *De anima*, 430a, 14–17 and 430a, 1–2.

13. Cf. E. Brehier, *Chrysippe*, p. 96ff.

14. Cf. Vasco Ronchi, *Histoire de la lumière* (Paris, 1956), p. 51ff.

15. Cf. Krzysztof Pomian, *Przeszlosc jako przedmiot wiary. Historia i filozofia w mysli sredniowiecza* [The past as object of faith. History and philosophy in medieval thought] (Warsaw, 1968), p. 37ff.

16. Cf. D. C. Lindberg, *Theories of Vision*, pp. 58ff., 104ff.

17. Cf. Pierre Michaud-Quantin,"Le champ semantique de *Species*. Tradition latine et traductions du grec," in *Etudes sur le vocabulaire philosophique du Moyen Age* (Rome, 1970), pp. 113–50. Also, Anneliese Maier, "Das Problem der 'Species sensibiles in medio' und die neue Naturphilosophie des 14. Jahrhunderts," in *Ausgehendes Mittelalter. Gesammelte Aufsätze zur Geistesgeschichte des 14. Jahrhunderts* (Rome, 1967), t. 2, pp. 419–51.

18. Cf. Roy J. Deferrari and M. Inviolata Barry, *A Lexicon of St. Thomas Aquinas based on the Summa Theologica and selected passages of his other works* (Washington, D.C., 1947), under *cognitio, cognoscere, intueor, intuitus*.

19. Cf. Sebastien J. Day, *Intuitive Cognition: A Key to the Significance of the Later Scholastics* (St. Bonaventure, N.Y., 1947).

20. Cf. M.-D. Chenu, *La théologie au XIIe siècle* (Paris, 1957) and *La théologie comme science au XIIIe siècle* (Paris, 1957).

21. Cf. A. Maier, "Das Problem der Evidenz in der Philosophie des 14. Jahrhunderts," in *Ausgehendes Mittelalter*, pp. 367–418.

22. Cf. Nicolaus de Cusa, *De Concordantia catholica*, III, ii, in *Opera omnia*, ed. G. Kallen, t. XIV, part 3, pp. 328ff. Lorenzo Valla, *De falso credita et ementita Constantini donatione declamatio*, ed. Wolfram Setz (Weimar, 1976) [M.G.H. Quellen zur Geitesgeschichte des Mittelalters, t. 10].

23. Ioannes Kepler, *Ad Vitellionem Paralipomena Quibus Astronomiae pars optica traditur*, V, 2; trans. Catherine Chevalley (Paris, 1980), p. 317. Cf. A. C. Crombie "The Mechanistic Hypothesis and the Scientific Study of Vision" (1967) and "Kepler: *De Modo Visionis*" (1964) both reprinted in A. C. Crombie, *Science, Optics and Music in Medieval and Early Modern Thought* (London and Ronceverte, 1990), pp. 175ff., 185ff.

24. René Descartes, *La Dioptrique*, I: *De la lumière*; A.T., VI, p. 81.

25. Ibid., IV: *Des sens en général*; A.T., VI, p. 109f.

26. Ibid., VI: *De la vision*; A.T., VI, p. 130.

27. R. Descartes, *Les Principes de la Philosophie* IV, 198; A.T., IX, II, p. 317.

28. R. Descartes, *Les passions de l'âme*, art. 12 and 23; A.T., XI, pp. 337 and 346.

29. R. Descartes, *Le Principes de la Philosophie* I, 45; A.T., IX, II, p. 44.

30. Cf. John Locke, *An Essay concerning Human Understanding*, II, 11, ed. Peter H. Nidditch (Oxford, 1975), p. 163.

31. Cf. D. Lindberg, *Theories of Vision*, pp. 164ff., 184–85, 205–06.

32. Cf. R. Descartes, *La Dioptrique* V: *Des images qui se forment sur le fond de l'oeil*; A.T., VI, p. 115.

33. Cf. J. Locke, *An Essay*, IV, 2, 1 and 5; pp. 531 and 533.

34. Ibid., II, 8, 9–12; pp. 135–36.
35. George Berkeley, *Essay towards a new Theory of Vision*, ed. A. A. Luce (London, 1948).
36. Cf. for instance ibid., 45, 49, 77, 103; pp. 188, 189, 202, 212.
37. Cf. for instance ibid., 17, 25, 28; pp. 174, 176, 177.
38. Cf. ibid., 41, 111; pp. 186, 215.
39. Cf. G. Berkeley, *Treatise concerning the Principles of Human Knowledge*, edition quoted, t. II (London, 1949).
40. David Hume, *A Treatise of Human Nature*, I, ii, vi; ed. S. A. Selby-Bigge (Oxford, 1955), pp. 67–68.
41. Ibid., I, i, vi, p. 16.
42. Cf. for instance ibid., I, ii, ii; I, ii, v; I, iv, ii; pp. 7, 84, 187ff.
43. Cf. ibid., I, iv, ii, p. 187ff.
44. Cf. ibid., I, ii, ii, p. 74.
45. Ibid., I, iii, xiv, pp. 165–66.
46. Cf. ibid., I, iii, iv; p. 83.
47. Cf. K. Pomian, "Le passé, de la foi à la connaissance," *le débat* 24 (1983): 151–68; and *Przeszlosc jako przedmiot wiedzy* [The past as object of knowledge] (Warsaw, 1992; finished in 1964).
48. Cf. V. Ronchi, *Il cannochiale di Galileo e la Scienza del Seicento* (Turin, 1958), p. 68ff., and R. Descartes, *La Dioptrique* VIII and IX; A.T., VI, pp. 165ff., 196ff.
49. Cf. G. Berkeley, *Essay*, 85, p. 206, and D. Hume, *Treatise*, I, i, ii, p. 8.
50. D. Hume, *Treatise*, I, ii, i, p. 28.
51. Immanuel Kant, *Prolegomena* preface.
52. Cf. for instance D. Hume, *Treatise*, I, i, iv and I, iv, ii, pp. 10ff., 207.
53. I. Kant, *Kritik der reinen Vernunft*, B 16–17.
54. Ibid., B 25.
55. Cf. ibid., A 36ff.
56. Ibid., A 42.
57. Ibid., B XVI.
58. See Joel Snyder's comments on this issue in the present volume, and his discussion of Ian Hacking's position.

LORRAINE DASTON

Nature by Design

INTRODUCTION[1]

I begin with three objects, all made of stone and all at one time or another viewed as straddling the boundary between art and nature. The first is a cameo of probably Hellenistic origins depicting two helmeted figures in profile[2] (Figure 1), in all probability the onyx described by the thirteenth-century natural philosopher Albertus Magnus, which he had once seen at the shrine of the Three Kings in Cologne. After ascertaining that the image was made of stone rather than glass, Albertus concluded "that this picture was made naturally and not artificially," adding that "[m]any others like this are found."[3] No modern eye, however untutored in the techniques of cameo carving or late classical motifs, could mistake this piece for a work of nature rather than of art. Not only the intricacy of the craftmanship but also the content of the image marks it for us immediately and indelibly as made by human hands. How could Albertus have thought otherwise? It will not do to dismiss Albertus as credulous or ignorant. Although he knew little about gem cutting, he knew enough about techniques of incising, engraving, embossing, and carving stones to discuss how similar images might be made artificially. Like Pliny, he warned the unwary against forged natural curiosities of this kind.[4] Moreover, it is not primarily on grounds of what we know about how cameos are made that we base our conviction that the Ptolemy cameo is artificial rather than natural. Nor was Albertus's judgment to the contrary unique: throughout the late Middle Ages and well into the seventeenth century, European scholars classified

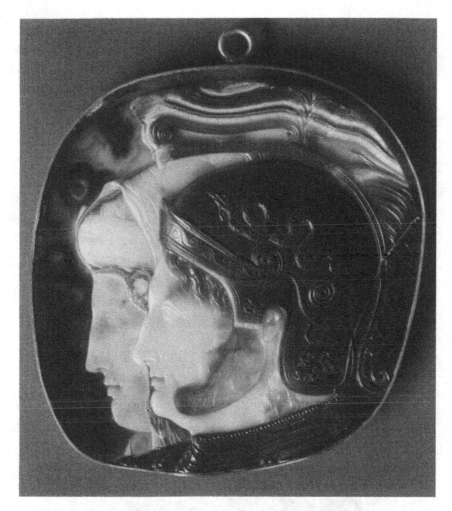

*Figure 1.   "Natural?"—"Ptolemy" Cameo, Hellenistic or Roman.
Courtesy of Kunsthistorisches Museum, Vienna.*

what seem to us to be unmistakably artifacts, such as Etruscan vases, as natural objects.[5] Why and when did the boundary between the natural and the artificial shift so dramatically?

The second object presents the converse problem: an object we take to be natural teeters on the edge of the artificial. It is probably a fossil ammonite, a "stone of Hammon," belonging to the collection of antiquities, portraits, seashells, stuffed animals, "curious petrifications," and exotica assembled in mid-seventeenth-century Paris at the Bibliothèque de Sainte-Geneviève. The first curator, Claude du Molinet, puzzled over this set of five interlocking rocks: "Most of those who have seen it have believed it to be artificial, but the cleverest sculptors in Paris judged it [to be] natural."[6] (Figure 2)

234

Figure 2.   "Artificial?"—Stone of Hammon (fossilized ammonite).
From Claude du Molinet, Le Cabinet de la Bibliotheque de Sainte Genevieve
(Paris: Chez Antoine Dezallier, 1692), Figure XVI.
Courtesy of the University of Chicago Library.

A Paris guidebook published some fifty years later still wavered: "it is difficult to discern whether it is a work of art or a caprice of nature."[7] Molinet was not credulous in matters of natural history; elsewhere in his catalogue he doubted the existence of unicorns, sirens, and footless birds of paradise. But like Albertus, he located the boundary between art and nature in forms that no longer seem in the least ambiguous to us.

The third object is a small limestone panel ornamenting the *Kunstschrank* presented by the Lutheran citizens of Augsburg in 1632 to their ally King Gustavus Adolphus of Sweden. Constructed of oak and ebony, the cabinet was richly inlaid with medallions of enamel, beaten silver, marble, agate, lapis lazuli, and intarsia panels of multicolored woods, and crowned with a mound of crystals, corals, and shells surrounding a goblet fashioned from a seychelles nut chased in gold and ornamented with the figures of Neptune and Thetis.[8] The limestone panel painted by Johann König with a biblical scene (Exodus 14:26) in which Moses leads the Israelites through the miraculously parted Red Sea (Figure 3) is small (430 mm × 354 mm) and easily lost amid all the splendor of amethyst, coral, gold, and silver, not to mention the contents of the cabinet's many drawers. Yet the panel captures in miniature the deliberate juxtaposition and even fusion of art and nature that characterizes the *Kunstschrank* as a

Figure 3.   "Art and Nature Collaborate"—König Panel from Uppsala Kunstschrank (probably early-seventeenth-century; the Kunstschrank was constructed 1625–31). Courtesy of the University of Uppsala.

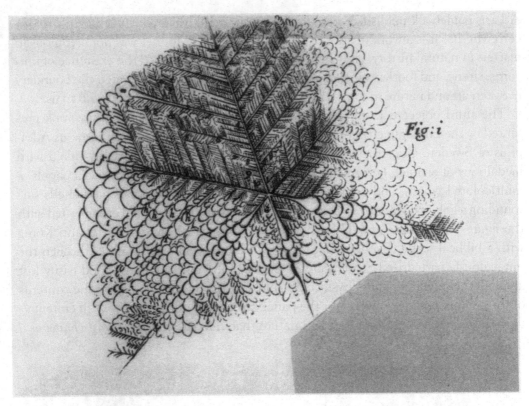

Figure 4.   "The Art of God"—Crystalline Formation in Frozen Urine, Magnified.
From Robert Hooke, Micrographia: Or, Some Physiological Descriptions of Minute
Bodies Made by Magnifying Glasses (London: John Martyn and James Allestry, 1665),
Scheme 8, Fig. 1. Courtesy of the Bayerische Staatsbibliothek.

whole. The contours of the stone are worked into the painting as mountains and
waves, a work painted as much by nature as by the artist. Albertus's onyx and Molinet's
ammonite straddle the boundary between art and nature because their provenance
could only be guessed from their form; König's limestone panel is intentionally ambiguous, a game of forms played across that same boundary.

It is my aim in this paper to retrace this line and to chart its shifting course in the critical period from the mid-sixteenth to the mid-eighteenth century. I will focus on the
borderlands between art and nature, on objects that challenged, toyed with, or outright undermined that ancient but unstable frontier. The literary and literal prominence of such objects—automata, seashells, figured stones, hybrids like the limestone
panel—in both treatises on natural history and in the Wunderkammern testifies to the
fluidity of the boundary during the early modern period. Although my primary concern in this essay will be with the conceptual geodesy of how art and nature were

mapped, it is impossible to ignore the distinctive sensibility of wonder that saturated these objects. Because so much recent anthropological and historical literature emphasizes the anxiety evoked by objects that straddle culturally fundamental boundaries,[9] I shall attempt to redress the balance by dwelling upon the peculiar and peculiarly intense pleasure excited by the art/nature ambiguities of the sixteenth and seventeenth centuries. Even monsters, the prototypical classificatory anomaly and one often associated with horror,[10] could become pleasing wonders if their aberrant forms could be viewed as nature's art rather than nature's errors.

My story is primarily one of forms, their appearances and their causes. Art and nature create form, and the analogies between the ways in which they do so have never ceased to shuttle back and forth between the two poles. But the forms characteristic of art on the one hand and nature on the other, their relative valuation, and their proper causes have changed almost beyond recognition, as the three ambiguous objects indicate. Beyond the history of form lurks a history of intelligence—what is it, who has it, and how much and what kind are needed to produce forms; and also a history of wonder—what evokes it, who merits it, and how it depends on the categories into which experience is parsed. In the course of the early modern period, nature was transformed from artisan into art in natural history and natural philosophy. Despite proclamations from Bacon, Descartes, and other visionaries of the new philosophy that the ancient opposition between art and nature had been dissolved, it not only persisted but actually hardened in the late seventeenth and eighteenth centuries. But by the 1660s, the dividing line between art and nature did not run where it once had. My account of how the boundary between art and nature came to be redrawn during this period centers on the redistribution of intelligence and wonder among three makers of form: nature, God, and the human artisan.

NATURE AS ARTISAN

If Albertus Magnus was able, after some hesitation, to ascribe the Ptolemy cameo to nature rather than to art, it was because he believed that nature and art brought about such forms by many of the same means, and that nature was the superior craftsman of the two. Nature and art often worked in similar ways to imprint form upon matter, but the forms of nature were at once more finished in appearance and more penetrating in essence than those of art: "Art imitates nature, but cannot attain to many of nature's works."[11]

This conviction that nature would always outstrip art still resonated at the turn of the seventeenth century, as when in Shakespeare's *The Winter's Tale* the shepherdess/princess Perdita refuses to include the "carnations and streaked gillyvors, / Which some call nature's bastards" in her winter bouquet: "For I have heard it said / There is an art which in their piedness shares / With great creating nature."[12] But by then humble and imperfect imitation was only one of a list of commonplaces balancing the reciprocal claims of art and nature. Art still aped nature, but it also might extend, assist, complete, contravene, or even surpass nature.[13] Not all Elizabethans, for example, shared Perdita's dim view of creating new varieties of flowers and fruits by grafting: the Oxford-educated philosopher John Case praised the alleged grafting of a pear tree onto a cabbage as "a wonderful fact of art! The bloom of one plant thus grafted changes the whole tree into another species. . . . [W]hat can prevent me from concluding that something natural has really been done by art?"[14] The Italian apothecary Ferrante Imperato in his treatise on natural history claimed that "art conduces to the perfection" of stones and metals;[15] French potter and naturalist Bernard de Palissy spoke of how artificial fountains improved upon natural ones because "one has [here] helped nature, just as to sow grain, to prune and labor in the vineyards is nothing else but helping nature."[16] Although the majority of Renaissance writers might still have sided with Perdita on nature's superiority to art,[17] there were countervailing views among elite artists and artisans and the collectors who patronized them.[18]

Among artisans in particular, an aesthetic of technical virtuosity bespoke a growing sense of the ability of art—here the mechanical and decorative arts, rather than what had already begun to be differentiated as the "fine arts"—to rival and even surpass nature. The *Meisterstück* emerged unevenly in the urban guilds of northern and central Europe in the course of the fifteenth century as a more or less standard piece of work to qualify younger journeymen before a jury of older craftsmen.[19] But by the early sixteenth century, the masterpiece had become a display of extraordinary dexterity, sometimes to the point of ignoring utility altogether, such as a bed too delicate to be slept in.[20] From ancient times there had been a tradition of describing extraordinary technical achievements as wonders, as in the ever-changing list of the seven wonders of the world that originated in Alexandria in the third century B.C.[21] Beginning in the sixteenth century, the vocabulary of *mirabilia* becomes almost interchangeable with that of artisanal masterpieces, a number of which found their way into *Wunderkammern*.[22]

In the sixteenth and seventeenth centuries a common aesthetic of the marvels of art and nature emerged, one that exploited the ancient opposition of art and nature to evoke wonder. Because the opposition was still a conceptual reflex during the early modern period, its violation was startling. Bedrock assumptions quaked, and the intensity of the wonder was correspondingly seismic. These pleasing paradoxes, at once art and nature, aroused wonder most intensely by blurring the line between nature and human artisan. The early modern *Wunderkammer*[23] was the showcase for the aesthetic of the marvels of art and nature, mingled at several levels: juxtaposition, fusion, and imitation.

In the first instance, the rarities of art and nature were displayed side by side within a "cabinet," "studio," "museum," or "repository," as the collections were variously called:[24] a petrified crab cheek-and-jowl by "a patent of the King of China on tissue paper painted with gold flowers like brocade";[25] a two-headed cat by "masterpieces" of lathe turning.[26] Although the contents of collections differed markedly according to means and motivation, the princely cabinet of Rudolf II in Prague from that of the Neapolitan apothecary Imperato, it was the rule rather than the exception for most collections to embrace both artificialia and naturalia. The collection of the French antiquarian Boniface Borilly was for example dominated by Roman medals, but it also boasted "a head of a rat from the Indies," "three well-polished coconuts, garnished with ivory, serving as flasks," and a celebrated "cyclops."[27] The actual physical arrangement of many collections (in contrast to the classifications of catalogues and inventories) was often calculated to highlight this heterogeneity.[28]

The marvels of art and nature might be brought still closer to one another by fusion in a single object. Nautilus shells carved and ornamented in gold to serve as a pitcher, a *Handstein* made of a glittering chunk of Bohemian ore to represent Christ's crucifixion on Calvary, the Johann König panel of the Uppsala *Kunstschrank*—these were all examples of hybrids of art and nature. Hybrids undermined the art/nature opposition not only by transforming natural materials by human craftmanship—the simplest piece of furniture did as much—but also by exploiting analogies of form, e.g., between a mound of ore and a hilltop fortress, between branches of coral and branches of trees, or between the lip of a nautilus shell and the lip of a pitcher. Nature had, as it were, already begun the work of art. Virtuosity created a momentum of embellishment of its own: since all ornamentation was strictly speaking gratuitous, there was no logical stopping point for the heaping of costly materials upon bravura craftsmanship.[29] In the case of the natural marvels of the *Wunderkammer*, the spiral of virtuosity had already begun before the human artisan even touched the object. Naked and natural, the ostrich egg or rhinocerous horn were already wonders, rare and finely wrought. Nature's admirable workmanship was a gauntlet thrown down to the human artisan, who enriched the delicate pearly shell of the nautilus with still more delicate carvings, burnished its luster with gold, outdid its rarity by adding fabulous figures of dragons and satyrs, and finally threw in a branch of coral for good measure. In these hybrids, art and nature competed as well as collaborated with one another, and in both cases nature tended to merge with art—or rather, with the artisan.

Finally, the objects of the *Wunderkammer* brought the poles of art and nature together through outright mimickry. Within an aesthetic of mimesis, of which the endlessly repeated stories of Zeuxis and Parrhasius were emblematic, the illusionary imitation of nature in trompe l'oeil painting or bronze casts made from shells and reptiles represented the peak of artisanal achievement: art finally pulled even with nature, after centuries of competition.[30] There was an unmistakable note of awe in Cornelius de Bie's praise for Johannes de Heem's fruit paintings—"D'Heem paints, nature is astonished"—and of triumph in Palissy's boast that the ceramic casts of shells and animals in his Tuileries grotto would be "so close to nature, that it will be impossible to recount."[31] The mimetic aesthetic of the marvelous was symmetric: if the marvels of art imitated nature, then the marvels of nature also imitated art. The skilled hand of the artist faked the veining of marble on plaster; the skilled hand of nature faked a landscape of ruins and rivers on Florentine marble.[32]

Because the metaphors of nature's workmanship had been a staple of philosophy and poetry since Antiquity, it is important to be specific about the kind of "work" nature allegedly accomplished in the objects displayed in the early modern *Wunderkammer*. As maker of marvels, she was neither Aristotle's fabricator of mundane, functional objects like beds and ships, nor was she the blacksmith of the medieval *Roman de la rose*, nor the semidivine creator exalted by neoplatonic art theory during the Renaissance. Rather, she was akin to the goldsmith, the clock maker, or ivory turner—a maker of luxury items, as elaborate and expensive as they were useless. Naturalia and artificialia testified triumphantly to the difficulties of material and scale effortlessly overcome by skill: the hard, porcelain-like substance of seashells molded into frills, whorls, convolutes, and spirals by nature; dense ivory turned into geometric filigree by art. Freed from the demands of utility, the virtuoso artisan could play with form and matter, just as nature occasionally "sported" with her ordinary species and regularities. Hence nature's finest workmanship came in the late sixteenth and early seventeenth centuries to be associated with some her most bizarre productions, just as mannerist artists rejoiced in the grotesque. The sixteenth-century French surgeon Ambroise Paré saw nature at play not only in the pleasing but useless variety of seashells,[33] but also in the equally dysfunctional construction of an African "monster" with sets of ears, eyes, and paws pointed in all four compass directions: how could "each perform its function [*faire son office*]? . . . [I] can only say that nature is here at play, in order to make us admire the grandeur of her works."[34] Even human monsters could be admired and imitated as nature's art: a Venetian surgeon describing a dissection of female conjoined twins remarked that "the Painter, who was employ'd to draw them, affirm'd, That if they were done in Ivory, he would have paid any money for them."[35]

The three stones of my introduction, all of which were displayed in early modern collections, belonged to a class of objects marvelous even among the other marvels of the *Wunderkammern*.[36] Among the forms of nature that mimicked those of art, none was more surprising and therefore wondrous than figured stones. According to Aristotelian natural philosophy, principles of form inhered within plants and animals, ensuring the integrity of species.[37] But minerals lacked even a "nutritive soul," and therefore usually displayed the most irregular forms in nature. The humblest weed was a masterpiece of symmetry and organization compared to the average rock. How then to explain "a stone naturally sketched with the figure of trees," or an agate "in which nature has painted the hemisphere of the heavens"?[38] In the works of sixteenth- and seventeenth-century naturalists, readers could find illustrations of still more marvelous minerals distinguished by their strange forms: the Swiss humanist Konrad Gesner divided his stones into fourteen classes according to what they resembled, including (Class V) natural fossils (literally, "things dug up") similar to works of art;[39] the Roman Jesuit Athanasius Kircher described stones naturally lettered with the Greek and Roman alphabets, flower-like crystals of topaz, and human figures found in marble;[40] Oxford naturalist Robert Plot produced plate after full-page plate of the star-stones, scrotum stones, shell stones, etc., that could be found in the quarries of Oxfordshire.[41] If nature was a virtuoso artisan, how did she craft her wares?

The explanations for figured stones were almost as varied as the objects themselves. Early modern naturalists added little original to the repertoire of causes to be found in Albertus Magnus and other medieval lapidarists, but they elaborated these in far greater detail, with reference to specific objects often contained in their own collections, as in the case of Cardano, Gesner, Aldrovandi, Plot, and Kircher. Some explained the uncanny resemblance between certain figured stones and living forms by granting minerals life: Italian natural philosopher Girolamo Cardano believed stones "suffer maladies, old age, and death," since all things which are "mixed [in composition] live."[42] Alternatively, the germinating form might be impressed from without upon vaporous exhalations from the bowels of the earth, either by celestial influences (as in the case of Albertus's explanation of the Ptolemy cameo) or by animal or plant seeds borne thither by wind or water.[43] Seeds of ferns or fish, forms of kings or mountain ranges realized their forms more or less perfectly in the soft matter, which eventually hardened into figured stones. Those who, like Palissy, Nicholas Steno, and Robert Hooke, argued that at least some figured stones were organic remains further invoked a "spiritus lapidificus" to explain not only how plants and animals had become petrified, but also how coral and kidney stones were formed.[44] Palissy himself owned "more than a hundred pieces" of petrified wood and had heard of a German prince whose cabinet contained "the body of a man partly petrified"; these examples and his own experience as a potter in making casts of shells and reptiles persuaded him that "some salty and generative substance" caused the stony imprints of plants and animals—and also the occasional rain of frogs.[45]

All of the above explanations assumed that the material of figured stones had originally been soft, fluid, or even vaporous, and that the form that distinguished these stones as remarkable had been impressed upon them either internally or externally during this malleable stage. This mechanism of form imprinting soft matter, as seal stamped soft wax, was ubiquitous in ancient and medieval natural philosophy.[46] In these processes of imprinting soft matter, a spectrum of formal principles were assumed to be at work, from the immanent formal principles contained in seeds to the half-internal/half-external, half-conscious/half-unconscious force of the imagination, to the deliberate designs of celestial intelligences or human artisans. Anchoring this spectrum of gradations of deliberation in design for natural operations was blind chance at one end and the sport of nature at the other. Although these two extreme explanations of figured stones were in many ways poles apart, both ultimately strengthened the analogy between nature and artisan.

The image made by chance was an important trope in Renaissance theory, stemming from Pliny's story of Protogenes and given new impetus in the writings of Leone Battista Alberti and Leonardo da Vinci on how artists might find inspiration in the fortuitous forms of marble or a spotted wall.[47] In the works of sixteenth-century naturalists, chance was one of several standard explanations for figured stones.[48] Cardano for example explained the famous agate of King Pyrhhus, described by Pliny as bearing the image of Apollo with his lyre and all nine muses, as the product of chance and petrification: an artist had painted the scene on marble, which then "by chance, or by industry" was placed somewhere where agates were generated, and the image imprinted itself upon the still-congealing stone.[49] But chance became an ever more suspect explanation for form in general, and for figured stones in particular, during the middle decades of the seventeenth century. If there was one subject upon which most natural philosophers were in absolute agreement, it was that chance or fortune was a vulgar error, "no proper cause at all, but a kind of *ens rationis*."[50] In the context of theological and natural philosophical debates over Epicureanism, chance, especially as an explanation for strikingly regular natural forms, seemed a weapon that atheists might use against divine providence.

An especially artful figured stone might also be explained as a *lusus naturae*, an expression of nature's whimsy and ingenuity.[51] All of nature's organic productions exhibited remarkable intricacy and symmetry, each structure fitting form exquisitely to function. Figured stones did not surpass ordinary organisms in complexity or regularity of form, but they were, in the eyes of early modern naturalists, extravagantly afunctional. What possible aim could nature have had in fabricating the geometric forms of crystals, the outlines of a cat in marble, a jagged mountainous landscape in limestone, or the imprint of a seashell in slate, other than to dazzle the spectator into openmouthed admiration? As Plot remarked of figured stones, they "seem rather to be made for his [man's] *admiration* than use."[52] Like the artisanal masterpieces displayed in *Wunderkammern*, the sports of nature were in part defined by their artistry, and in part by their uselessness. The Flemish physician and naturalist Anselmus Boetius de Boodt abandoned all attempts to explain the hexagonal form of certain crystals, concluding that "nature wishes us to admire these things, not to comprehend them."[53]

When Hooke, John Ray, and other late-seventeenth-century naturalists contested the *lusus naturae* explanation of figured stones, their main target was neither its anthropomorphism nor its lack of mechanism. Rather, they protested with the ancient and equally anthropomorphic commonplace that "Nature does nothing in vain," insisting that it was quite contrary to the infinite prudence of Nature, which is observable in all its works and productions, to design everything to a determinate end, . . . that these prettily shaped Bodies should have all those curious figures and contrivances (which many of them are adorn'd and contriv'd with) generated or wrought by a *Plastic Virtue*, for no higher end than only to exhibit such a form.[54]

Playful nature vied with prudent nature in the late-seventeenth-century debate over figured stones; sober utility trumped admirable extravagance. At first glance, it seems as if one anthropomorphism had simply given way to another, both of antique lineage.[55] Yet the new characterization of nature to be found in the works of natural philosophers in the late seventeenth and early eighteenth centuries was not merely an Aristotelian revival, despite its Aristotelian slogan. In the context of theological, philosophical, and political debates of the period it marked a major turning point in the meanings and distribution of wonder and intelligence.

NATURE AS ART

On June 24, 1678 John Locke visited the apartment of Marie de Lorraine, Duchess of Guise, in Paris, where he admired a small grotto decorated with a fountain and "a very fine artificial rocke of marble, agates, cornelian & fine branches of Corall."[56] Locke's use of the word "artificial" to describe highly regular and/or worked objects, including naturalia, was standard in early modern Latin and several vernaculars: jurists spoke of "natural" (direct) and "artificial" (elaborated) inferences drawn from evidence; naturalists described how certain wheel-like fossils grew "after a very artificial manner"; philosophers praised "the *Orderly, Regular and Artificial Frame* of things in the Universe."[57] This usage permitted near-paradoxical locutions concerning "the Infinite Regularity and Artificialness" of nature.[58] It was also larded with ambiguity: Was nature "artificial" as art or as artisan?

The answer to this question hinged on a division of cosmic labor between God and nature, and on the degree of deliberative intelligence implied by design. Although early modern natural philosophers overwhelmed by the variety, beauty, and, occasionally, whimsicality of flowers, seashells, figured stones, and even monsters sometimes paid homage to God for not only sustaining but also ornamenting his creation, these aesthetic and jocular offices were usually left to nature, as God's "chambermaid" or "quartermaster."[59] When for example Plot upheld the *lusus naturae* explanation of figured stones, he argued that it was

the wisdom and goodness of the *Supreme Nature*, by the *School-men* called *Naturans*, that governs and directs the *Natura naturata* here below, to beautifie the World with these varieties, which I take to be the end of such productions as well as of most *Flowers*, such as *Tulips, Anemones, & c.* of which we know as little use as of formed stones.[60]

There were at least three reasons for assigning nature rather than God the responsibility for such embellishments. The first was quite general, and applied to all natural philosophy, wondrous or no: philosophical (and medical) explanations were by definition confined to the natural, however devout the naturalist.[61] The second concerned the dignity of work, or lack thereof: although some saw nothing demeaning about having God attend to "the most minute and seemingly most trivial and contemptible transactions on this great exchange of the world,"[62] most agreed with Isaac Newton that " 'God' is a relative word and has a respect to servants."[63] Finally, the extravagant and even bizarre character of the objects made it unseemly to attribute them directly to God: nature might sport, but God did not.

However menial or undignified the tasks assigned to nature, the very fact of a division of labor implied some measure of autonomy for nature. It was exactly this autonomy that was at issue in late-seventeenth-century debates about the nature of nature. Although the mechanical philosophy has often been represented by historians of science as a declaration of nature's independence from the meddling interventions of divine providence, some of the foremost mechanists insisted vehemently on nature's absolute dependence on God, and on God's equally absolute perogative to alter his creation at will.[64] England's most prominent mechanical philosopher, the chemist Robert Boyle, was particularly loud in his protests against granting nature the slightest discretion in her operations. Indeed, he went so far as to deny nature even bare existence, suggesting that it was a merely "notional" entity.[65] For Boyle, the central issue was usurpation: those who admired the works of nature stole praise, gratitude, and, above all, wonder from God. It was disrespectful and even idolatrous to suggest that God needed an assistant, "to imagine, as we commonly do, that God has appointed an intelligent and powerful Being, called nature, to be, as his viceregent, continually watchful for the good of the universe in general, and of the particular bodies, that compose it."[66]

Although Boyle took too lofty a view of God's exalted station to be able to countenance too much divine labor, he was also loathe to allow God servants, for this would lead willy-nilly to an ensouled and potentially usurping nature. Boyle's solution was to claim that nature was artifact rather than artisan. Moreover, it was an artifact of a peculiar kind, immediately recognizable from the *Wunderkammern* inventories: an "engine" or "automaton," words Boyle used interchangeably. Appealing over and over again to the Strasbourg clock (itself a fanciful and intricate masterpiece catering to the wondrous sensibility)[67] and to the automata that swam like real ducks or tooted like real flutists, Boyle envisioned the world as nothing but a "great automaton," composed of still smaller automata, in the manner of Chinese nested boxes, and God as the most ingenious of engineers. Boyle suggested that decorum would be best served if the divine artificer arranged for "all things to proceed, according to the artificer's first design, and the motions of the little statues [of the Strasbourg clock], that at such hours perform these or those things, [and] do not require, like those of puppets, the peculiar interposing of the artificer, or any intelligent agent employed by him."[68] Ingenious automata would eliminate the need for uppity servants, in particular "an intelligent and powerful being called nature," and at the same time keep God's hands clean of demeaning labor.[69]

The problem with "intelligent and powerful" servants is that they may rebel. It is not surprising, given the prolonged and bloody attempts by seventeenth-century European monarchs to consolidate and extend their power vis-à-vis ambitious nobles, prelates, and commoners, that the fear of usurpation penetrated not only late-seventeenth-century political but also philosophical, literary, and theological discourse. Boyle's nature *Free Inquiry* recalls Milton's Satan in *Paradise Lost*, both too close to the throne for comfort. Natural philosophical debates about God's dominion over the universe echoed coeval political debates about the king's dominion over his subjects.[70] Yet Boyle's chief concern was not insurrection but idolatry; more specifically, the idolatry of misplaced wonder. Nature the virtuoso artisan might steal the "admiration" (a word in seventeenth-century English still redolent of its Latin root, *admiratio*, "to wonder") due to God. Boyle cautioned that God is "jealous," and that even those Christians who recognize that nature is subordinate to God might give "in practice, their admiration and praises" to nature rather than to God.[71] From this standpoint, the effusive admiration for nature's handiwork found in the writings of naturalists like Cardano and Plot, and at the heart of the *Wunderkammer* sensibility, verged on the worship of false gods. The intensely pleasurable wonder of the ambiguous marvels of art and nature shaded imperceptibly but dangerously into the religious wonder of reverence and awe. For Boyle, God must be acknowledged to monopolize not only agency in the universe, but also the wonder of his rational creatures.[72]

Nature had become art, and God artisan—and an ingenious maker of wondrous objects to boot. Boyle's God pitted one kind of stock *Wunderkammer* object against another, the automata (themselves marvels of art imitating nature) against the figured stones and nautilus shells, in a struggle over who and what properly merited wonder. Had God then simply taken over tasks formerly assigned, as well as the admiration paid to nature? Setting aside for a moment the vexed question of divine labor, the art of God differed markedly from the art of nature—and also from human art. Within the *Wunderkammern*, the awe-inspiring natural objects and artifacts had displayed the art of external forms, of appearances. In contrast, the art of God revealed its finest workmanship only upon closer, internal scrutiny. In late-seventeenth-century natural philosophy, a new opposition opened up between the human art of macroscopic exteriors and the divine art of microscopic interiors. René Descartes thought the main difference between the "machines" of art and those of nature was that nature's were composed of tinier and more perfect wheels and springs.[73] Robert Hooke thought it hardly worthwhile to examine man-made products under the microscope, which revealed them to be "rude, misshapen things." Under magnification the point of the finest needle was as rugged as a mountain range, "whereas in the works of Nature, the deepest Discoveries shew us the greatest Excellencies. An evident Argument, that he that was the Author of these things, was no other than the Omnipotent."[74] (See Figure 4.)

The Cambridge philosophers Henry More and Ralph Cudworth, and later Gottfried Wilhelm Leibniz, worried less about the idolatry of nature than about the indignity of a God without servants, but they also registered the fears of misplaced wonder that had exercised Boyle. None were satisfied that the mechanical philosophy had adequately explained form matched to function, much less the afunctional forms of nature's sports and errors. All appealed to a standard set of counterexamples—sympathetic cures, musical instruments that vibrated in unison, the power of the maternal imagination, the spider's web, geometric crystals—in order to justify their assumption of an ensouled nature, variously described in terms of "plastic powers" or "spirit of nature" or "indwelling active principles." However, they were all at pains to insist on the inferiority of the soul of nature to the rational soul of humans, much less to God. Although ensouled nature was elevated above the stupid matter of the mechanical philosophy, Leibniz warned that a fully anthropomorphized nature would revive "heathen polytheism";[75] Cudworth admitted that human actions may lack the "Constancy, Eaveness and Uniformity" of natural operations, but that we nonetheless surpass nature in acting consciously.[76] It would not do to admire nature excessively.

These philosophers granted nature intelligence, since "artificial" form required it, but intelligence of the very lowest order. Perhaps nothing is more revealing of how the anti-mechanists understood ensouled nature than their recurring analogies to the kind of labor it performs. Far from rivaling God, ensouled nature was his "servant," his "Drudging Executioner," the "manuary Optificer" to God's "Architect." Just what it meant to be a servant in the seventeenth century is made painfully clear by Cudworth's elaboration of the latter analogy: "We account the Architects in every thing more honorable than the Manuary Optificers, because they understand the reason of the things done, whereas the other, as some Inanimate things, only do, not knowing what they do."[77] The labor of nature was drudging labor, and the knowledge of nature was tacit knowledge, akin to the unconscious habits that sustain the art of the musician or dancer. This lowly, brutish servant was a far cry from the virtuoso artisan of the *Wunderkammern*. Even those late-seventeenth-century philosophers still attentive to the problem of natural form could barely grant nature half a soul, the least degree of intelligence consistent with design.

CONCLUSION: IMAGINATION AND THE FALSIFICATION OF FORMS

Half a soul could still serve to imprint forms on soft matter, and not only in the production of figured stones. The closest analogy More could hit upon to the "Spirit of Nature," whose dim intelligence molded the forms of the world "without Sense or Animadiversion," was the maternal imagination unconsciously molding the fetus. Citing Kircher's account of a man with a birthmark in the shape of Pope Gregory XIII seated on a throne with a dragon at his feet, More insisted that here nature imitated art, for what else was the maternal imagination that had impressed the birthmark but the work of *"exorbitating Nature"*?[78] The half soul of nature worked through the half soul of the mother's imagination.

The maternal imagination remained a stock explanation of misshapen offspring well into the eighteenth century, affirmed by Leibniz, Nicholas Malebranche, and many savants. But it was a dark and dubious kind of creativity, as passive as it was unconscious. A pregnant woman who witnessed a criminal broken on the wheel bore a child with bones broken in telltale places; a child born with a monstrous calf-like head was chalked up to the mother's dismay at losing a prized cow a few months prior.[79] Voltaire distinguished sharply between the "active" imagination of invention in mathematics, mechanics, poetry, and art, and the "passive" imagination of the ignorant, the mad, and pregnant women. The passive imagination did not depend on the will; indeed, as the instrument of passion and error, it overcame the will: "it is an internal sense that acts imperially [*avec empire*]."[80] Imagination was thus split in two, between art and nature: the active imagination created new forms in art and technology; the passive imagination enslaved the mind to false forms fabricated by the body.

Among the false forms of the imagination came to be numbered many of the figured stones that early modern naturalists had so admired. Although the hypothesis that some of these stones were petrified organic remains steadily gained adherents in the early eighteenth century, this accounted for only a fraction of the striking objects numbered among figured stones. It could not explain geometric crystals or landscape marble or moss agate: "But though a real petrification were allow'd in some cases, it would not be rational to plead this in all the figur'd stones we see."[81] Explaining these stubborn anomalies became a matter of explaining them away, as projections rather than as products of the imagination. Leibniz ridiculed those naturalists who had found shapes of stars and the moon in marble, or Apollo and the nine muses in agate. The resemblances they saw existed not in the stones but only in their imaginations.[82] The false forms of the human imagination now substituted for the true forms of nature's once-fertile imagination. If figured stones showed signs of genuine artistry, then these were the marks of the forger, not of nature. One nineteenth-century French naturalist went so far in his skepticism as to dismiss all figured stones as human artifacts.[83]

The boundary between art and nature had been redrawn, but it was as sharp and distinct as it had ever been. By the early decades of the eighteenth century, collectors had begun systematically to separate artificialia and naturalia, and by the end of the century a taste for marvels had become synonymous with bad taste. If art and nature had moved closer to one another in natural philosophy, it was at least as much because nature had become artificial, "the Art of God,"[84] as that art had become natural. This was the departure point for the argument from design of the eighteenth-century physico-theologists. Nature was designed, not designer; moreover, its forms commanded admiration for their utility rather than their extravagance. Art continued to imitate nature, albeit ever more hesitantly, but nature could imitate art only in strained metaphors. The provenance of objects like Albertus's onyx or Molinet's ammonite were no longer ambiguous, and the König limestone panel lost its meaning as a collaboration between nature and art.

It is possible to tell this as a story of anthropomorphism vanquished, but this would be deeply misleading. As David Hume pointed out with devastating clarity, to insist that God rather than nature was the artisan was simply to displace, not eliminate, anthropomorphism.[85] Moreover, to further insist that design implied the deliberative intelligence of humans or God, that watch implied watchmaker, was to exacerbate anthropomorphism with vaulting anthropocentrism. Aristotle had staunchly denied that design in nature—the bird's nest or the bee's honeycomb—required *conscious* purpose: "animals other than man . . . make things neither by art nor after inquiry or deliberation."[86] Human (and, anthropomorphically, divine) art is only one special case of designed form, anomalous in its reliance on consciousness and deliberation. At the heart of the debate over nature's forms lurked a new anthropocentrism, which took a peculiarly human brand of intelligence and made it the measure of the distance between design and chaos.

Notes

1. Unless otherwise noted, all translations are my own.
2. Adolf Furtwängler, *Die antiken Gemmen Geschichte der Steinschneidekunst im klassischen Altertum* (1900), 3 vols. (Amsterdam/Osnabrück: Adolf M. Hakkert/Otto Zeller, 1965), vol. 2, pp. 250–51, table LIII.
3. Albertus Magnus, *Book of Minerals* (*De mineralibus*, comp. ca. 1256), trans. Dorothy Wyckoff (Oxford: Clarendon Press, 1967), II.iii.9, pp. 130–31. The identification of the cameo, now at the Kunsthistorisches Museum in Vienna, is Wyckoff's.
4. Ibid, II.iii.9, pp. 131–33; Pliny the Elder, *Historia naturalis*, 10 vols., Loeb Classical Library (Bks. XXXVI–XXXVII), trans. H. Rackham (Cambridge and London: Harvard University Press and William Heinemann, 1962), vol. 10 XXXVII.lxxiv, p. 323.

5. Frank Dawson Adams, *The Birth and Development of the Geological Sciences* (Baltimore: Williams & Wilkins, 1938), pp. 123, 255.

6. Claude du Molinet, *Le Cabinet de la Bibliothèque de Sainte Geneviève* (Paris: Chez Antoine Dezallier, 1692), p. 218.

7. Germain Brice, *Description de la ville de Paris*. Reproduction de la 9e édition (1752), ed. Pierre Codet (Geneva and Paris: Librairie Droz and Librairie Minard, 1971), p. 302.

8. On the *Kunstschrank,* which has been in the possession of the University of Uppsala since 1695, see Thomas Heinemann, *The Uppsala Art Cabinet* (Uppsala: Almqvist and Wiksell, 1982), also Hans-Olof Boström, "Philip Hainhofer and Gustavus Adolphus' *Kunstschrank* in Uppsala," in *The Origins of Museums. The Cabinet of Curiosities in Sixteenth- and Seventeenth-Century Europe,* ed. Oliver Impey and Arthur MacGregor (Oxford: Clarendon Press, 1985), pp. 90–101.

9. The seminal reflections on the cultural meaning of classificatory anomalies are by Mary Douglas, *Purity and Danger: An Analysis of the Concepts of Pollution and Taboo* (London: Routledge & Kegan Paul, 1966), pp. 36–40.

10. Arnold I. Davidson, "The Horror of Monsters," in *The Boundaries of Humanity. Humans, Animals, Machines,* ed. James J. Sheehan and Morton Sosna (Berkeley and Los Angeles: University of California Press, 1991), pp. 36–67.

11. Albertus Magnus, quoted in Gérard Paré, *Les Idées et les lettres au XIIIe siècle: Le Roman de la Rose* (Montréal: Edition du Centre de Psychologie et de Pédagogie, 1947), p. 67.

12. William Shakespeare, *The Winter's Tale* (1610–11), IV.iv, pp. 86–87.

13. George Puttenham, *The Art of English Poesie* (1589), ed. Gladys D. Willcock and Alice Walker (Folcroft: Folcroft Press, 1936, 1969), pp. 303–04.

14. John Case, *Lapis philosophicus* (1599), quoted in Charles B. Schmitt, "John Case on Art and Nature," *Annals of Science* 33 (1976): 543–59, on p. 547; see also Elisabeth B. MacDougall, "A Paradise of Plants: Exotica, Rarities, and Botanical Fantasies," in *The Age of the Marvelous,* ed. Joy Kenseth (Hanover: Dartmouth College, 1991), pp. 145–57, on p. 153, for early modern grafting techniques and their reception.

15. Ferrante Imperato, *Dell'Historia naturale di Ferrante Imperato Napolitano libri XXVIII* (Naples: Constantio Vitale, 1599), preface, n.p.

16. Bernard Palissy, *Discours admirables de la nature des eaux et fontaines* (1580), ed. Anatole France (Paris: Charavay Frères, 1880), p. 212.

17. A. J. Close, "Commonplace Theories of Art and Nature in Classical Antiquity and the Renaissance," *Journal of the History of Ideas* 30 (1969): 467–86.

18. Thomas DaCosta Kaufmann, *The School of Prague: Painting at the Court of Rudolf II* (Chicago: University of Chicago Press, 1988), p. 94.

19. Hans Huth, *Künstler und Werkstatt der Spät Gothik* (1925), 2nd ed. (Darmstadt: Wissenschaftliche Buchgesellschaft, 1967), p. 15ff.

20. Walter Cahn, *Masterpieces Chapters on the History of an Idea* (Princeton: Princeton University Press, 1979), p. 20.

21. Theodor Dombart, *Die sieben Weltwunder des Altertums* (1967), 2nd ed. (Munich: Ernst Heimeran Verlag, 1970), p. 7.

22. Cahn, *Masterpieces,* pp. 17, 47.

23. In addition to classics such as Julius Schlosser, *Die Kunst- und Wunderkammern der Spätrenaissance* (1908), 2nd rev. ed. (Braunschweig: Klinkhardt & Biermann, 1978) and David Murray, *Museums: Their History and their Use* (Glasgow: James MacLehose, 1904), there has been a recent efflorescence of literature on early modern collections: Giuseppe Olmi, *Ulisse Aldrovandi: Scienza e natura nel secondo cinquecento* (Trent: Libera Università degli Studi di Trento, 1976); Adalgisa Lugli, *Naturalia e amirabilia: Il collezionismo enciclopedico nelle Wunderkammern d'Europa* (Milan: Mazzota, 1990); Impey and MacGregor, *Origins*; Krzysztof Pomian, *Collectors and Curiosities: Paris and Venice, 1500–1800*

(1987), trans. Elizabeth Wiles-Portier (Cambridge: Polity Press, 1990); Antoine Schnapper, *Le Géant, la licorne et la tulipe: Collections et collectionneurs dans la France au XVIIe siècle* (Paris: Flammarion, 1988); Kaufmann, *School of Prague*; Kenseth, *Age of the Marvelous*; Elisabeth Scheicher, *Die Kunst- und Wunderkammern der Habsburger* (Vienna, Munich, and Zurich: Molden Edition, 1979); Wolfgang Liebewein, *Studiolo: Die Entstehung eines Raumtyps und seine Entwicklung bis um 1600* (Berlin: Gebr. Mann, 1977); Horst Bredekamp, *Antikensehnsucht und Maschinenglauben: Die Geschichte der Kunstkammern und die Zukunft der Kunstgeschichte* (Berlin: Klaus Wagenbach, 1993); Paula Findlen, *Possessing Nature: Museums, Collecting, and Scientific Culture in Early Modern Italy* (Berkeley and Los Angeles: University of California Press, 1994); Andreas Grote, ed., *Macrocosmus in Microcosmo: Die Welt in der Stube. Zur Geschichte des Sammelns 1450–1800* (Opladen: Leske+ Budrich, 1994).

24. On the rich early modern terminology of collections, see Johann Daniel Major, *Unvorgreiffliches Bedencken von Kunst- und Naturalien-Kammern insgemein* (1674), reprinted in D. Michael Bernhard Valentini, *Museum Museorum, oder Vollständige Schau-Bühne aller Materialien und Specereÿen nebst deren natürlichen Beschreibung* (Frankfurt: Johann David Tunners, 1704), pp. 4–11; also Paula Findlen, "The Museum: Its Classical Etymology and Renaissance Genealogy," *Journal of the History of Collections* 1 (1989): 59–78.

25. Balthasar Monconys, *Voyages de M. de Monconys*, 4 vols. (Paris: Pierre Delaulne, 1695), vol. 2, p. 198. Monconys is here describing the collection of one "gentleman named S. Victor" in Ghent, one of the many he visited on his travels.

26. Pierre Borel, *Les Antiquitez, raretez, plantes, mineraux, & autres choses considerables de la ville, & Comté de Castres d'Albigeois* (Castres: Arnaud Colomiez, 1649), p. 136.

27. Philippe Tamizey de Larroque, ed, *Boniface de Borilly. Lettres inédits écrites à Peiresc (1618–1631)* (Aix-en-Provence: Garcin et Didier, 1891), p. 48; Schnapper, *Géant*, p. 95.

28. Laura Laurenich-Minelli, "Museography and Ethnographic Collections in Bologna during the Sixteenth and Seventeenth Centuries," in Impey and MacGregor, *Origins*, pp. 17–23, on p. 21; cp. Schnapper, *Géant*, p. 11, concerning parallel evidence from guidebooks and visitors' reports.

29. Cp. how very similar objects were depicted in Dutch still lifes of the period: Norman Bryson, *Looking at the Overlooked: Four Essays on Still Life Painting* (Cambridge: Harvard University Press, 1990), p. 126.

30. Pliny, *Historia*, XXXV.xxxvi, vol. 9, pp. 311–23; see Kris and Kurz, *Legend*, pp. 64–79, for Renaissance adaptations of this artistic mythology. On the wonder associated with early modern trompe l'oeil painting, see Arthur K. Wheelock, Jr., "Trompe-l'oeil Painting: Visual Deceptions or Natural Truths?" in Kenseth, *Age of the Marvelous*, pp. 179–91. On casts, see Ernst Kris, "Der Stil 'Rustique'. Die Verwendung des Naturabgusses bei Wenzel Jamnitzer und Bernard Palissy," *Jahrbuch der Kunsthistorischen Sammlungen* N.F. 1 (1926): 137–208.

31. Cornelius de Bie, *Het Gulden Cabinet van de edel vry Schilder Const* (Antwerp: Ian Meysens, 1661), p. 219; Palissy, "Devis d'une grotte pur la royne mère du roy," in *Discours*, pp. 465–71, on p. 469. On the rivalry of art and nature in grottoes see Lugli, *Naturalia*, pp. 103–12.

32. Jurgis Baltrusaitis, *An Essay on the Legend of Forms* (1983), trans. Richard Miller (Cambridge and London: MIT Press, 1989), pp. 63–70.

33. Ambroise Paré, *Les Monstres et prodiges* (1573), ed. Jean Céard (Geneva: Librairie Droz, 1971), p. 117.

34. Ibid., p. 139.

35. "An Extract of an Italian Letter Written from Venice by Signor Jacomo Grandi, to an Acquaintance of his in London, concerning some Anatomical Observations, and two odd Births," *Philosophical Transactions of the Royal Society of London* 5 (1670): 1188–89, on p. 1189.

36. The Ptolemy cameo was displayed by Ferdinand III in the Vienna treasury of the Habsburgs: Rudolf Distelberger, "The Habsburg Collections in Vienna in the Seventeenth Century," in Impey and MacGregor, *Origins*, pp. 39–46, on p. 40.

37. Aristotle, *On the Soul*, II.iv, 415a14–415b27.

38. Imperato, *Historia*, p. 663; Girolamo Cardano, *De la subtilité* (1550), trans. Richard le Blanc (Paris: Guillaume le Noir, 1556), f. 318v.

39. Konrad Gesner, *De rerum fossilium, lapidum et gemmarum* (Zurich: n.p., 1565), pp. 86–96.

40. Athanasius Kircher, *Mundus subterraneus* (1664), 2 vols. (Amsterdam: Janssonio-Waesbergiana, 1678), vol. 2, pp. 22–48.

41. Robert Plot, *The Natural History of Oxfordshire* (Oxford: Theater, 1677), pp. 80–130.

42. Cardano, *Subtilité*, f. 164v.

43. Elaborating a suggestion from Aristotle on the generation of metals: *Meteorology*, III.vi, 378a16–378b4.

44. On the early modern debate over fossils in the modern sense, see Adams, *Birth*, pp. 81–132, 254–59; Martin J. S. Rudwick, *The Meaning of Fossils: Episodes in the History of Palaeontology* [1976], 2nd ed. (Chicago: University of Chicago Press, 1985), pp. 1–84; Paolo Rossi, *The Dark Abyss of Time: The History of the Earth and the History of Nations from Hooke to Vico* (1979), trans. Lydia G. Cochrane (Chicago: University of Chicago Press, 1984), pp. 7–23; Norma E. Emerton, *The Scientific Reinterpretation of Form* (Ithaca: Cornell University Press, 1984), pp. 19–75. On the early modern views about the formation of coral, see Schnapper, *Géant*, p. 22; on kidney stones, see Walter Charleton, *Spiritus Gorgonicus* (Leyden: Ex Officina Elseviviorum, 1650), pp. 2–13.

45. Palissy, *Discours*, pp. 325, 332–38.

46. See Aristotle, *Meteorology*, Iv.ix, 386a18–386b10, concerning impressibility of bodies; also Katharine Park, "Impressed Images: Reproducing Wonders," in this volume.

47. H. W. Janson, "The 'Image Made by Chance' in Renaissance Thought," in *De artibus opuscula XL. Essays in honor of Erwin Panofsky*, ed. Millard Meiss, 2 vols. (New York: New York University Press, 1961), pp. 254–66. The artist Protogenes painted a "marvelously executed [*mire factus*]" picture of a dog, but could not get the foaming at the mouth quite right. Finally, after many unsuccessful attempts, he literally threw in the sponge, and produced exactly the desired naturalistic effect by chance: Pliny, *Historia*, XXXV.xxxvi; vol. 9, pp. 337–39.

48. See, e.g., Kircher, *Mundus*, p. 40, who lists chance as the first cause of such stones.

49. Pliny, *Historia*, XXXVII.iii, vol. 10, p. 167. Pliny attributed the image to nature: *"non arte, sed natura sponte"*; Cardano, *Subtilité*, f. 137r.–v.

50. Robert Boyle, *A Free Inquiry Into the Vulgarly Received Notion of Nature* (1686), in *The Works of the Honourable Robert Boyle* (1772), ed. Thomas Birch, 6 vols. (Hildesheim: Georg Olms, 1966), vol. 5, pp. 158–254, on p. 202. On the philosophical antipathy to chance in the seventeenth century, see Lorraine Daston, "Fortuna and the Passions," in Thomas M. Kavanagh, ed., *Chance, Culture and the Literary Text, Michigan Romance Studies* 14 (1994): 25–48, pp. 25–27.

51. On the *lusus naturae* in early modern natural history see Paula Findlen, "Jokes of Nature and Jokes of Knowledge: The Playfulness of Scientific Discourse in Early Modern Europe," *Renaissance Quarterly* 43 (1990): 292–331, esp. pp. 311–13 concerning figured stones.

52. Plot, *Natural History*, p. 80.

53. Boodt, *Ioallier*, p. 279.

54. John Ray, *Observations: Topographical, Moral, & Physiological* (London: John Martyn, 1673), p. 124. Compare the very similar passages in Robert Hooke, *Micrographia: Or some Physiological Descriptions of Minute Bodies Made by Magnifying Glasses* (London: John Martyn and James Allestry, 1665), pp. 110–11; on the relationships between the Hooke and Ray passages, see C. E. Raven, *John Ray: Naturalist* (1942) (Cambridge: Cambridge University Press, 1986), p. 422.

55. The image of nature at play, particularly in the creation of beautiful and useless variety, is at least as old as Pliny, *Historia*, vol. 6, p. 160; Aristotle repeatedly invokes the principle that "Nature does nothing in vain," e.g., in *Physics*, II.viii, 198b10–199b33.

56. John Lough, ed., *Locke's Travels in France 1675–1679* (Cambridge: Cambridge University Press, 1953), p. 201.

57. Albéric Allard, *Histoire de la justice criminelle au seizième siècle* (1868) (Aalen: Scientia Verlag, 1970), pp. 263–64; John Beaumont, "Two Letters written by Mr. *John Beaumont* Junior of *Stony-Easton* in *Somersetshire*, concerning Rock-Plants and their growth," *Philosophical Transactions of the Royal Society of London* 11 (1676): 724–42; Ralph Cudworth, *The True Intellectual System of the Universe* (1678) (Hildesheim and New York: Georg Olms, 1977), p. 154.

58. Cudworth, *System*, p. 148.

59. E.g., Walter Charleton, *The Darknes [sic] of Atheism dispelled by the Light of Nature: A Physico-Theological Treatise* (London: William Lee, 1652), p. 115.

60. Plot, *Natural History*, p. 121.

61. See, e.g., Fortunio Liceti, *De monstrorum caussis, natura et differentiis libri duo* (1616), 2nd ed. (Padua: Paulum Frambottum, 1634), pp. 51–52, concerning exclusively natural or "philosophical" explanations for monsters.

62. Charleton, *Darknes*, p. 121.

63. Isaac Newton, *Mathematical Principles of Natural Philosophy* (1687/1713), "General Scholium," trans. Andrew Motte, rev. Florian Cajori (1934), 2 vols. (Berkeley and Los Angeles: University of California Press, 1962), vol. 2, pp. 544–45.

64. On the relationship between the doctrine of divine voluntarism and seventeenth-century concepts of natural law, see John R. Milton, "The Origin and Development of the Concept of the 'Laws of Nature,'" *Archives of European Sociology* 22 (1981): 173–95.

65. Boyle, *Free Inquiry*, p. 161. Although Boyle composed this treatise in 1666, it was first published in 1686, and an abridged Latin translation, *Tractatus de ipsa natura*, appeared in 1688.

66. Ibid., p. 164.

67. The clock adorned the cathedral at Strasbourg, and was built (1570–74) by a team of artisans led by the Swiss mathematician Conrad Dasypodius: Cahn, *Masterpieces*, p. 91.

68. Boyle, *Free Inquiry*, p. 163.

69. Ibid., p. 164.

70. Steven Shapin, "Of Gods and Kings: Natural Philosophy and Politics in the Leibniz-Clarke Disputes," *Isis* 72 (1981): 187–215; Francis Oakley, *Omnipotence, Covenant, and Order An Excursion into the History of Ideas from Abelard to Leibniz* (Ithaca: Cornell University Press, 1984), pp. 92–113.

71. Boyle, *Free Inquiry*, pp. 183, 192.

72. Ibid., p. 253; cp. Boyle, *Of the High Veneration Man's Intellect Owes to God* (1685), in *Works*, vol. 5, pp. 130–57, on p. 153.

73. René Descartes, *Principia philosophiae* (1644), in Charles Adam and Paul Tannery, *Oeuvres de Descartes*, 12 vols. (Paris: Léopold Cerf, 1897–1910), IV.cxcvii, vol. 9, pp. 321–22.

74. Hooke, *Micrographia*, pp. 8, 2.

75. Leibniz, Letter to Thomasius (April 20/30, 1709), quoted in Catherine Wilson, "*De Ipsa Natura*: Sources of Leibniz's Doctrine of Force, Activity and Natural Law," *Studia Leibnitiana* 19 (1987): 148–72, on p. 163.

76. Cudworth, *True Intellectual System*, p. 162.

77. Ibid., p. 156.

78. Henry More, *The Immortality of the Soul* (London: James Flesher, 1662), pp. 193, 175.

79. Nicholas Malebranche, *De la recherche de la verité* (1674–75), 6th ed. (1712), in *Oeuvres de Malebranche*, ed. Geneviève Rodis-Lewis, 2 vols. (Paris: J. Vrin, 1963), II.i.vii.3, vol. 1, pp. 238–39; "Extrait d'une lettre de Mr. le Prieur de Lugeris en Champagne, sur un enfantement arrivé au mois de Mai dernier," *Journal des Sçavans* (1690): 41–42.

80. Voltaire, "Imagination, Imaginer (Logique, Métaphys, Litterat. & Beaux-Arts)," in *Encyclopédie, ou Dictionnaire raisonné des sciences, des arts et des métiers*, ed. Denis Diderot, 17 vols. (Neufchastel: Chez Samuel Faulche & Compagnie, 1765), vol. 8, pp. 560–63, on p. 561.

81. Beaumont, "Two Letters," p. 738.

82. Gottfried Wilhelm Leibniz, *Protogaea*, ed. Christiano Ludovico Scheidio (Göttingen: Johann Schmidt, 1749), pp. 44–46.

83. Baltrusaitis, *Essay*, p. 99.

84. See Thomas Browne, *Religio Medici* (1642). Facsimile reprint of 1643 ed. (Menston: Scolar Press, 1970), I.xvi, p. 35. Cp.

85. David Hume, *Dialogues Concerning Natural Religion* (comp. 1779), ed. Norman Kemp Smith (New York: Macmillan, 1947, 1986), pp. 176–81.

86. Aristotle, *Physics*, II.viii, 199a20.

KATHARINE PARK

Impressed Images:
Reproducing Wonders

By the time she died in 1320, Margarita of Città di Castello had acquired a considerable local reputation for sanctity. An acknowledged visionary despite being blind from birth, she had lived for a number of years as a Dominican tertiary. Although illiterate and of humble origin, she had commented on the Psalms with the authority of a master of theology, and eyewitnesses reported that she had levitated during prayer.[1] Impressed by these abilities, the local Dominicans decided to embalm her corpse as a relic. Accordingly, they had her body publicly opened and eviscerated on the high altar of their church, burying her entrails in a vessel in the convent cloister. Some time later, as the miracles associated with her proliferated, the friars decided to exhume the entrails, in order to transfer her heart to a golden reliquary for display. In the words of the author of one of Margarita's two fourteenth-century Latin *vitae*,

When [the vessel] had been taken out, and while brother Niccolò was cutting the reed to which the heart was attached, . . . suddenly three wonderful [*mirabiles*] little stones fell out of the reed, with different images impressed [*ymagines impressas*] on them. On one was seen sculpted the face of a very beautiful woman with a golden crown, which certain people interpreted as a likeness of the glorious blessed virgin Mary, to whom the blessed Margaret was attached with enormous devotion. The second showed a little boy in a cradle, surrounded by cattle, which certain people said signified Christ or the birth of Christ. On the third little stone was sculpted the image of a bald man with a gray beard and a golden cloak on his shoulders; before him knelt a woman dressed in the Dominican habit, and they said that this pictured the blessed Joseph and the blessed Margarita. On the side of the same stone was a white dove, which they said represented the Holy Spirit, by which Mary conceived her Son. And thus it appeared that where the heart of Margaret was, there also was found a wonderful treasure.[2]

The author of Margarita's other *vita* added that "persons worthy of belief" reported that she had said, "Oh, if you knew what I carry in my heart, you would be struck with wonder [*miraremini*]."[3]

Among other things, this is a story about seeing wonders. Margarita had seen wonders; like another Dominican tertiary, Catherine of Siena, with whom her cult was associated, she was one of several notable visionary women in late-thirteenth- and fourteenth-century Tuscany and Umbria (Figures 1 and 2). And after her death, her body revealed its own wonders to all who viewed her opened heart. What was the relationship between the wonders Margarita saw and the wonders generated in her body, between the marvelous images that presented themselves to her internal vision and the marvelous images impressed on the stones contained in her heart?

Figure 1. Venetian altarpiece from the early fifteenth century, Murano, Museo Civico Vetrario. Shows holy women belonging to the Dominican Third Order: from right to left, Giovanna of Florence, Vanna of Orvieto, Catherine of Siena, Margarita of Città di Castello, and Daniella of Orvieto. Courtesy of Osvaldo Böhm.

*Figure 2.   Detail of Figure 1, showing Margarita of Città di Castello holding
her heart with its three stones. Courtesy of Osvaldo Böhm.*

I will approach these questions in the light of ideas about seeing and generation in
the period between about 1250 and 1350. Modern historians tend to analyze images
primarily as representations, but for Margarita's contemporaries, images could have
other, more potent functions, operating also—and perhaps more importantly—as
replicas or reproductions of an original, partaking of or reproducing the original's
power. As art historians, David Freedberg and Hans Belting have explored the nature
and function of what Freedberg calls "secondary images," in the form of reproductions
of cult images such as the "Fair Mary" of Regensberg.[4] This paper extends their work
into the realm of the history of science and medicine, by considering a particular kind
of secondary image or reproduction: the "impressed image," to borrow a phrase from
Margarita's biographer. In addition to being supernatural (as in the case of the figures
on the stones in Margarita's heart) or artificial (as in the case of seal imprinted in warm
wax), images might be impressed by natural means. Indeed, late medieval natural
philosophers considered the production of impressed images to be a fundamental type
of physical causation, linking processes as apparently dissimilar as visual cognition and

the generation of fossils. In discussing the relationship between these various types of images, I will argue that the idea itself was suffused with assumptions about gender that informed not only the Latin treatises of medical theorists and philosophers, but also contemporary accounts of the experiences and responses of Christian laymen and -women, including aspiring female saints.

GENERATION AND VISION: THE PHYSIOLOGY OF IMPRESSED IMAGES

Consider, for example, the personification of nature as Jean de Meung first described her in the *Romance of the Rose* (ca. 1275):

> Nature, who thinks on the things that are enclosed beneath the heavens, was inside her forge, where she put all her attention to forging individual creatures [also coins: *pièces*] to continue the species [*espièces*]. . . . Nature, sweet and compassionate, . . . continues always to hammer and forge and always to renew individuals by means of new generation. When she can bring no other counsel to her work, she cuts impressions [*emprainte*] of such beings, which give them true forms in coins of different moneys.[5]

For Jean de Meung, Nature's primary function was generation, which she performed like a minter, stamping forms on matter, just as impressions were stamped on the metal of coins. Like a minter, she produced similar individuals within species, just as individual coins belonging to a given currency were stamped from a single die—a highly resonant analogy in the context of the later thirteenth century, when the explosion in the number of western European authorities minting money had put enough coins in circulation to fuel the definitive transition to a money economy.[6]

Jean de Meung's image also neatly embodied contemporary natural philosophical theories of generation, which portrayed this as the impression of form on pliant matter. In the case of animals, for example, most philosophical writers of the later thirteenth and fourteenth centuries accepted the model proposed by Aristotle in *The Generation of Animals*. According to Aristotle, "what the male contributes to generation is the form and the efficient cause, while the female contributes the material," for "the female, as female, is passive, and the male, as male, is active, and the principle of the movement comes from him."[7] Thus "the female always provides the material, and the male that which fashions it [into shape]."[8] In more concrete terms, the father contributes seed, which supplies the fetus with form, or soul—that which gives it its identity, shapes it, and makes it grow; in contrast, the mother supplies only the matter, which Aristotle identified with menstrual blood. Thus the end product of generation was ideally a son identical to his father in every respect—a physical as well as a mental reproduction, like the coin that issued perfect from the minter's die. It was in this sense that Aristotle described all daughters as defective births or "mutilated males."[9] Although high and late medieval European scholars debated the details of this process—medical writers, following Galen, tended to attribute a more active role to the mother than stricter Aristotelians did—the general model was nonetheless widely accepted as a description of the biological relationships between parents and child.[10]

The model of the impression of form on matter also underlay medieval natural philosophical theories of the generation of plants and minerals. In his *Book on Minerals* (ca. 1260), for example, the Dominican philosopher Albertus Magnus wrote:

> just as in an animal's seed . . . there comes from the seminal vessels a force capable of forming an animal, which [actually] forms and produces an animal, and is in the seed in the same way that an artisan is in the artifact that he makes by his art; so in material suitable for stones there is a power that forms and produces stones, and develops the form of this stone or that. . . . When dry material that has been acted upon by unctuous moisture, or moist material that has been acted upon by earthy dryness, is made suitable for stones, there is produced in this, too, by the power of the stars and the place, . . . a power capable of forming stone— just like the productive power in the seed from the testicles.[11]

In the case of stones, in other words, the stars (and their derivatives, geographical location) take the male role, supplying the "formative power," which gives to appropriately constituted matter its specific and substantial form.[12]

Although in most cases the form imparted by the heavens was thought to determine only the general appearance and properties of the mineral—sapphire is hard and blue, for example, and is good for disorders of the eyes—Albertus noted that in some unusual cases, such as a cameo he had seen in Cologne (Figure 3), the heavens might additionally impress a recognizable image, which reflected or reproduced their own spatial configuration: snakes, for example, or faces, or a king's head. Like Jean de Meung, Albert envisaged this process as a kind of stamping or sealing. Explaining why naturally impressed images of this sort appear only on gems, he wrote,

> they do not appear in other kinds of stone because the material in them is heavy, gross, and earthy, and does not respond to the moving powers; and therefore heaven cannot move it and make an impression on it. But in precious stones and certain marbles, as we have already said, the material is vaporous, and therefore images of this sort are produced in these [stones]. . . . It is as if a stamp [*sigillum*, more literally, seal] were pressed upon hard earth or stone, leaving no imprint at all; but if pressed upon water, it makes an imprint, and if the water freezes, then the figure persists in the ice.[13]

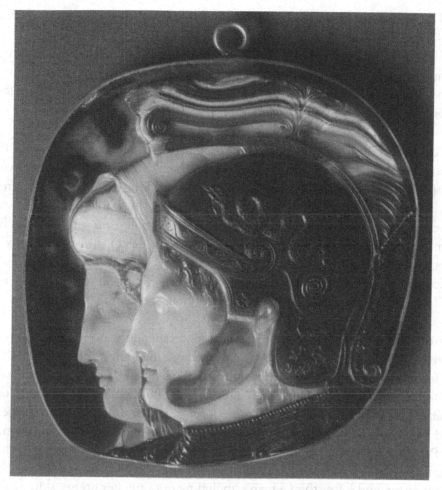

*Figure 3.   "Ptolemy" cameo, seen by Albertus Magnus
in the shrine of the Three Kings in Cologne Cathedral.
Courtesy of the Kunsthistorisches Museum, Vienna.*

Similar principles were thought to govern the process of generation in both humans and animals. Like precious stones, the matter of the female body was presumed to be particularly soft and malleable, which made it especially apt to receive impressions, including the form impressed on the menses by the male. In the analysis medieval philosophers inherited from Aristotle, which underpinned high and late medieval physiology, women's complexion was dominated by the qualities of coldness and wetness, while men were predominantly hot and dry. This difference accounted for many of the characteristics generally attributed to women in thirteenth- and fourteenth-century medical and philosophical writing, from their inability fully to digest their food—resulting in the buildup in their wombs of moist and poisonous waste, which needed to be evacuated each month—to their notorious indecisiveness and change-ability.[14]

In particular, medical and philosophical writers explained many of the psychological characteristics attributed to women using the language of impression on the passive and malleable female body. In his *Questions on Animals*, for example, Albertus Magnus noted that

> the complexion of the female is more humid than that of the male, and the humid receives [impressions] easily but retains them poorly. For the humid is very mobile, so that women are inconstant and always seeking novelties. For this reason, when she is engaged in the act under one man, she would like to be at the same time under another, if it were possible, so that women are without faith. For woman is a mutilated man and has the nature of defect and privation with respect to the male . . . ; and therefore what she cannot get by herself, she endeavors to get by lies and diabolical deceptions. Whence, to put it in a nutshell, one should beware all women like poisonous snakes and horned devils.[15]

Albertus's contemporary Peter of Spain reiterated these ideas, noting that the humidity of the female complexion meant that women retained impressions only with difficulty; for this reason, he argued, they find it hard to believe in promises and are less prone to intractable lovesickness than men: as Peter put it in his commentary on the *Viaticum*, "the impression of any desirable form in the brain of a man is deeper and more difficult to eradicate than the impression of a form in the brain of a woman."[16]

As Peter's words indicate, medical and philosophical writers interpreted such matters in highly literal and material terms. A woman's entire body was thought to be colder, moister, and softer than a man's, including not only her uterus, skin, and muscles, but also her sensory organs and the brain they served. This had important implications for sensory cognition, which, like generation, was also explained in terms of the impression of images on soft or subtle matter, although the matter in this case was not the menses but a substance called *spiritus*, a vapor thought to be distilled from blood and contained in the arteries, veins, nerves, and the cerebral ventricles.[17] *Spiritus* and its organs functioned as the primary instruments of sensation. In the case of vision, for example, an object radiated out forms or images of itself (generally known as *species*) into a transparent medium; these were then transmitted into the eyeball, up the *spiritus*-filled optic nerve, and into the ventricles of the brain. There, the *species* might be further manipulated by the "internal senses" of common sense, imagination, estimation, and cogitation, located in the two anterior cerebral ventricles, or stored in the hinder ventricle of memory for further use (Figure 4).[18] This explains, for example, why people seeking to remember tilt their heads backward and look at the ceiling, so that the impression-laden spirit flows into the organ of memory in the last ventricle,

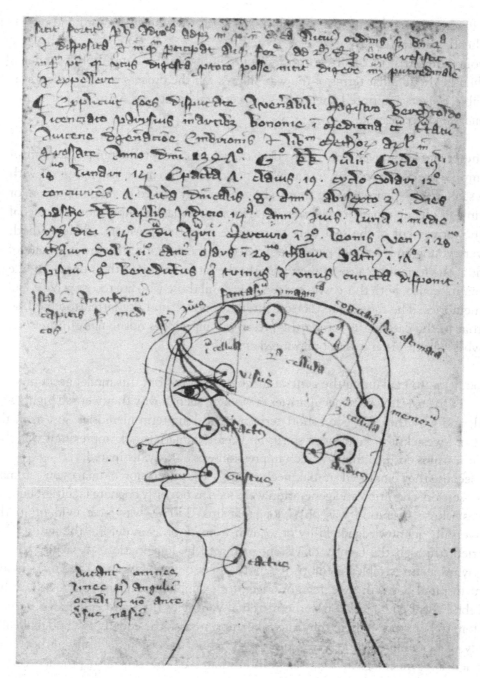

Figure 4. Mid-fourteenth-century illustration from a German medical manuscript compilation, showing the five internal senses located in the three cells of the brain and connected to each other and to the organs of the five external senses by spiritus-filled nerves. Courtesy of the Bayerische Staatsbibliothek München (Clm. 527, fol. 64v).

while when trying to imagine something, they tilt their heads forward and rest their foreheads on their hands.

Following Aristotle, thirteenth- and fourteenth-century natural philosophers described in the familiar terms of the wax impression the process whereby the forms of sensible objects imprinted themselves on cerebral spirit. As Aristotle put it in *On Memory and Recollection,*

> the process or movement involved in the act of perception stamps in a sort of impression of the percept, just as when persons do who make an impression with a seal. This explains why, in those who are strongly moved oweing to passion, or time of life, no mnemonic impression is formed; just as no impression is formed if the movement of the seal were to impinge on running water. . . . Hence both very young and very old persons are defective in memory; they are in a state of flux, the former because of their growth, the latter, owing to their decay. In like manner, also, both those who are too quick and those who are too slow have bad memories. The former are too soft [literally, moist], and the latter, too hard, so that in the case of the former the presented image does not remain in the soul, while on the latter it is not imprinted at all.[19]

In addition to explaining the general mechanism of vision, this model accounted for some of the psychological peculiarities of women, such as why they were so highly sensual, and why they had good short-term but bad long-term memories: just as in the case of lovesickness, their moist spirits and cerebral matter took impressions well, but their softness ensured that any such impression was quickly dissipated.

Because they employed similar mechanisms—the impression of images on soft matter—sensory cognition and generation were seen as not only cognate faculties, but faculties whose operation was physiologically linked. This was most evident in the universally acknowledged ability of women to mark or even deform the fetuses they carried, through the faculty of vision, mediated by imagination—a faculty that in many respects resembled memory. If a mother saw, or even fantasized about, an object that inspired her with love, fear, or anger, for example, the images impressed on her cerebral spirit could travel from her cerebral ventricles through her agitated nervous system to her uterus, where they might be imprinted on the equally soft and malleable body of her unborn child. According to Soranus, the second-century author of an influential gynecological treatise, "various states of the soul also produce certain changes in the mould of the fetus. For instance, some women, seeing monkeys during intercourse, have borne children resembling monkeys. The tyrant of the Cyprians who was misshapen, compelled his wife to look at beautiful statues during intercourse and became the father of well-shaped children."[20] For the same reason, Leon Battista Alberti would later recommend that "wherever man and wife come together, it is advisable only to hang portraits of men of dignity and handsome appearance; for they say that this may have a great influence on the fertility of the mother and the appearance of future offspring."[21] There is ample evidence that contemporaries put these precepts into practice; many of the birth trays and bowls given to Italian women during pregnancy were decorated with images of handsome, healthy baby boys (Figure 5).[22]

Writing on the erotic imagination, Ioan Couliano has argued that in the European philosophical tradition, "body and soul speak two languages, which are not only different, even inconsistent, but also *inaudible* to each other,"[23] and that the soul must as a result create phantasms or representations out of the body before it can understand anything about the sensible world. This may have been true for some of the later Neoplatonists that principally concerned Couliano, but in the Aristotelian tradition that

Figure 5.   Verso of a wooden childbirth tray from Ferrara, ca. 1460.
Collection of the Museum of Fine Arts, Boston, gift of Mrs. W. Scott Fitz.

dominated late medieval and early Renaissance natural philosophy, such a dualism between body and soul was conspicuously lacking. Not only were soul and body fully compatible and tightly linked in the relationship of form and matter, but they acted on each other above all through the mechanism of impressed forms or images. Impressed images were the way in which body spoke to soul (as when a person learned to know the physical world by observing and manipulating the *species* or images impressed by outside objects in the cerebral spirits) and soul spoke to body (as when a woman stamped her fetus with the object of her desire or fear). Such *species* were not representations, but reproductions, impressed by objects on a soft and yielding medium in the manner of a seal in wax. But although such were used to explain all kinds of physical and psychological phenomena, as I have indicated, in many respects the archetype of impressionable matter was the female body in the process of genera-tion—an understanding that not only shaped the experiences and interpretations of late thirteenth- and fourteenth-century university scholars but may also have influ-enced women themselves.

IMPRESSIONABILITY AND FEMALE RELIGIOUS EXPERIENCE

Margarita of Città di Castello was not by any means the only female visionary or holy woman whose body became, or produced, matter on which images could be impressed. In her study of fourteenth-century Italian hagiography, Catherine Mooney has identi-fied a series of consistent gender differences in twenty *vitae* of contemporary holy men and holy women. These include a clear contrast in the role of images and visuality: religious visions played a far larger part in the lives of Christian women than of men.[24] Not only do these works portray women as having more, and more elaborate, internal visions, in which they participated more directly, but they show women as much more likely than men to be moved, inspired, or tempted by looking at external images. Aldobrandesca of Siena was meditating on an image of the crucified Christ when she saw a drop of blood exude from his side, for example, and when Vanna of Orvieto con-templated the crucifix, her "body remained extended in the manner of the cross, rigid, pale, and insensible" (Figure 1).[25]

Furthermore, some women's bodies manifested physical impressions directly trace-able to their intense meditation on images, as when Catherine of Siena received her (invisible) stigmata directly from an image of the Crucifixion (Figure 1). Despite the powerful model of Francis of Siena in the previous century, described in Arnold Davidson's contribution to this volume, external visual objects played a much more subordinate part in thirteenth- and fourteenth-century hagiographic accounts of male religious experience, which rarely described men's bodies as bearing marks impressed in this or any other way. Statistically, stigmata were an overwhelmingly female phenomenon, as was the spontaneous appearance of betrothal rings in the form of colored marks.[26] Such phenomena were often described in the language of impression; for example, when Catherine of Siena received her first wound from Christ, it occurred during a vision in which he took a nail and pressed it into her hand.[27] In the same way, it was only women whose bodies engendered visual objects, and those objects also were described as the products of sealing or imprinting: Margarita's three stones, with their "impressed images" strongly recall the naturally engraved gems and cameos of Albertus Magnus, and Margarita's story had in turn strong affinities with that of Chiara of Montefalco, whose heart, when opened, revealed all the instruments of the passion.[28]

The large role played by religious visions and images in the *vitae* of holy women certainly reflects lower levels of female literacy and lack of female access to religious texts; where holy men were often famed for their learning and inspired by theological or devotional treatises, or were themselves authors of such treatises, women meditated instead on religious images and inscribed their traces on their own bodies. Such visions and marks served to authorize their voices in ways that were not necessarily seen as inferior to the written word.[29] But the prominence of this kind of story also reflects the widely shared view that women's bodies were impressionable, open to the reception of sensible images and apt to conform themselves to forms received from the outside. As I have indicated in the case of the medical and natural philosophical literature, this idea underlay some of the tropes of clerical misogyny—women's sensuality, their mental incapacity—but it could be put to positive uses as well. Caroline Bynum has argued that late medieval religious women accepted the identification of femaleness with the body (as opposed to the identification of maleness with the soul) and refashioned this idea in ways that emphasized not their own subordination and inferiority, but their identification with the humanity of Christ.[30] I would add that women seem to have identified with a particular *type* of body, soft and impressionable, which they further associated with Jesus, whose body was marked by the thorns, the scourges, the nails and lance. In addition to receiving the marks of Christ's passion in their own bodies, late-thirteenth- and fourteenth-century Christian holy women are reported to have described even their more abstract spiritual experiences in the language of impressed images. Thus Mechtild of Hackeborn had a religious vision in which she saw herself "incorporated in Christ and liquefied in divine love," so that she received "the imprint of resemblance [to God] like a seal in wax."[31] In this way, the impressed image became a vehicle for the idea of the imitation of Christ.

Catherine of Siena developed this figure in another direction, using the wax impression to describe not only Mary's identification with her son in his passion, but also her conception of Jesus, in line with contemporary theories of generation. "The Son was struck in his body," reads one of her letters to two nuns in the monastery of Santa Marta of Siena, and his mother likewise, because that flesh came from her. . . . He had the form of flesh, and she, like warm wax, received the imprint of the desire and love of our salvation by the seal and of the seal of the Holy Spirit, by the means of which seal the eternal and divine Word was incarnate. Thus she, like the tree of mercy, received in herself the consummated soul of the son, which soul was struck and wounded by the will of the Father.[32]

In this passage, Catherine made Mary the type or model of female impressionability. Thus she and women like her could accept the fact of being sealed—externalized in Catherine's case in the marks of her stigmata and engagement ring—as a manifestation of their imitation of Christ and their empathy with his suffering. But they also might use it as a means of imitating Mary herself, as the vessel of the incarnation, the soft matter of her body imprinted with the divine and human form that was her son—an event painters often depicted in terms that suggested the impression of an image along a kind of divine ray. This idea seems to inform the figure on Margarita's third stone, in which her own image replaced that of Mary, paired with Joseph her spouse; like Mary's own virgin body, Margarita's, too, engendered miraculous objects through a process of imprinting.

In all of these ways, it would seem, holy women transformed a discourse of female passivity and objectification into a discourse of female authority and spiritual worth, relating the softness and malleability of a woman's body not to her physical and psychological incapacity, but to her ability to engender Jesus in her heart and to conform her entire being to Christ.

AUTHENTICATING FEMALE SANCTITY

With these ideas in mind, we can return to Margarita and her wonderful figured stones. In addition to expressing the general trope of female impressionability, as I have been describing, the images on the stones served a much more specific purpose: to authenticate Margarita's visions and establish their supernatural source. For Margarita, as I have already mentioned, had been blind from birth. Thus there was no way for images of the Holy Spirit and the Holy Family to have penetrated her body and reached her brain except through supernatural means. Rather than entering through the optic nerve, they must have been impressed on her brain and her consciousness directly, in the form of mystical visions (as the stars impressed the image on Albertus's cameo), whence they could travel to the matter of her heart.[33] Thus the authors of the two prose versions of Margarita's *vita* emphasized that her blindness allowed her to see the things of the spirit with particular clarity: "Deprived of corporeal eyes, so that she might not see the world," as one put it, "she lived on divine light, so that standing on earth she saw only heaven."[34] Like a pregnant woman, she then stamped on the soft matter contained inside her body the images of her visions, mobilized by the passion of love.

It was the nature of these images as reproductions rather than mere representations, the product of a contact direct and palpable as coining or sealing, that authenticated Margarita's visions and confirmed her holy status, in the same way that the child's resemblance to its father demonstrated its legitimacy, or the seal on a document guaranteed its authority, or the stamp on a coin ensured its worth. Indeed, the new prominence of the language of sealing and minting reflects not only the general reemergence in the high Middle Ages of a documentary and monetary regime in which authenticity was of central importance and demanded visual expression,[35] but also the newly urgent problem of authenticating women's religious experience in late-thirteenth- and fourteenth-century Italy. The founders of the new mendicant orders had called for a new spiritual ideal, organized around a life of penance, poverty, and urban religious activism, in place of the traditional model of monastic enclosure. To their surprise, and ultimately to their consternation, this ideal took deepest root not among men, as they had clearly intended, but among laywomen. The second half of the thirteenth century saw an exuberant flowering of female religious life, particularly in Umbria and southern Tuscany, largely as a result of the local influence of Francis of Assisi. Women like Margarita, Vanna of Orvieto, and Catherine of Siena embraced this ideal in significant numbers, committing themselves to lives of strenuous poverty and asceticism, and coming together in spontaneous lay communities.[36]

What André Vauchez has called the "feminization" of the penitential ideal in the late thirteenth century posed a challenge for ecclesiastical authorities, who had their doubts about the wisdom and appropriateness of large numbers of laywomen leaving their families to live autonomous and often relatively visible penitential lives.[37] These doubts were greatly magnified by the strong visionary and ecstatic element in this movement, which began to produce charismatic women, known not only for their extreme asceticism but also for their mystical trances and prophetic revelations. Where the holiness of male candidates for sanctity continued to manifest itself primarily in public acts of mercy and of moral and religious leadership, female holiness, as in the case of Margarita, expressed itself in private visions and prophecies and in remarkable ecstatic acts, such as trances and levitations. The only witnesses to these acts were in general other women—it was, for example, three female companions who reported Margarita's levitation during prayer[38]—and their testimony, if not automatically suspect, was at least seen as requiring strict verification. Thus the authors of Margarita's *vitae* were explicit about their probative intent: one began by describing his initial reluctance to record her story, which he described as surrounded by a "cloud of incredulity."[39] Only after he himself had been convinced— "my mind illumined, if I am not mistaken"—could he take on the task of convincing his readers of the sanctity of this ornament of the Dominican order: the marvelous stones in her heart constituted palpable proof.

If the images impressed on the bodies of late-thirteenth- and fourteenth-century Italian holy women served this purpose for their male supporters and publicists, it may be prudent to ask whose views the theme of female impressionability in fact reflected: the women (for the most part illiterate), or the male hagiographers and amanuenses who produced the texts on which modern knowledge of them depends? Was it men or women who transformed the topos from a trope of intellectual misogyny to an expression of the special dignity of female spirituality? While such questions may ultimately prove unanswerable, they serve as reminders of the virtual impossibility of reconstructing female experience in the absence of sources unmediated by a male voice.

But the function of the images impressed on the stones in Margarita's heart was not confined to authentication: the element of contact that underlay the production of the impressed image also marked it as a reservoir of power. In this sense, the image itself served as the vehicle for the transmission of marvelous properties as well as the sign that the impressed object in fact possessed such properties. As Albertus Magnus put it, apropos of astrological sigils, "we must conclude that if a figure is impressed upon matter, either by nature or by art, [with due regard to] the configuration of the heaven, some force of that configuration is poured into the work of nature or of art."[40] In Catherine of Siena's analysis of the Incarnation, the Holy Spirit functioned as seal in these two ways. The images on Margarita's stones were also analogous to those on Albertus's engraved gems: conduits of supernatural healing, they too signified that these objects "worked wonders."

Notes

1. The two earliest sources for Margarita's life are edited in M.-H. Laurent, "La plus ancienne légende de la B. Marguerite de Città di Castello," *Archivum fratrum predicatorum* 10 (1940): 115–28; and [A. Poncelet], "Vita beatae Margaritae virginis de Civitate Castelli," in *Analecta bollandiana*, ed. Carolus De Smedt et al., (Brussels, 1900), vol. 19, pp. 21–36. On the dating and relationship of these two texts, see Enrico Menestò, "La 'legenda' di Margherita da Città di Castello," in *Il movimento religioso femminile in Umbria nei secoli XIII–XIV*, ed. Roberto Rusconi (Scandicci and Florence: La Nuova Italia, 1984), pp. 217–37.
2. Laurent, "Légende," p. 128.
3. Poncelet, "Vita," p. 28.
4. David Freedberg, *The Power of Images: Studies in the History and Theory of Response* (Chicago: University of Chicago Press, 1989), esp. pp. 103–35; Hans Belting, *Likeness and Presence: A History of the Image Before the Era of Art*, trans. Edmund Jephcott (Chicago: University of Chicago Press, 1994), esp. pp. 440–41.
5. Jean de Meung, *Le roman de la rose*, 1589–1601, ed. Daniel Poirion (Paris: Garnier-Flammarion, 1974), pp. 428–30. Translation heavily revised from Guillaume de Lorris and Jean de Meung, *The Romance of the Rose*, trans. Charles Dahlberg (Hanover, N.H.: University Press of New England,

1983), pp. 270–71. On the use of money as a figure for the relationship between form and matter, see Marc Shell, *Art and Money* (Chicago: University of Chicago Press, 1995), esp. pp. 8–11.

6. Peter Spufford, *Money and its Use in Medieval Europe* (Cambridge: Cambridge University Press, 1988), pp. 240–88; Carlo M. Cipolla, *Money, Prices, and Civilization in the Mediterranean World, Fifth to Seventeenth Century* (Princeton: Princeton University Press for the University of Cincinnati, 1956), esp. pp. 20–45.

7. Aristotle, *Generation of Animals*, 1.20, trans. Arthur Platt, in *The Works of Aristotle*, ed. J. A. Smith and W. D. Ross, 12 vols. (Oxford: Clarendon Press, 1912), vol. 5, 729a10–12 and 729b13–14.

8. Ibid., 2.4, 738b20–21. On ancient and medieval theories of generation and sex difference, see Joan Cadden, *The Meanings of Sex Difference in the Middle Ages: Science, Philosophy, Medicine* (Cambridge: Cambridge University Press, 1994). Anthony M. Hewson details the views of a strict thirteenth-century Aristotelian in *Giles of Rome and the Medieval Theory of Conception: A Study of the De formatione corporis humani in utero* (London: Athlone Press, 1975).

9. Aristotle, *Generation of Animals*, 2.3, 737a28.

10. See Jane Fair Bestor, "Ideas about Procreation and their Influence on Ancient and Medieval Views of Kinship," in *The Family in Italy from Antiquity to the Present*, ed. David I. Kertzer and Richard P. Saller (New Haven: Yale University Press, 1991), p. 158. On the debates concerning male and female seed, see Cadden, *Meaning of Sex Difference*, pp. 117–30; Nancy G. Siraisi, *Taddeo Alderotti and his Pupils: Two Generations of Italian Medical Learning* (Princeton: Princeton University Press), pp. 195–201.

11. Albertus Magnus, *Book of Minerals*, 1.1.5, trans. Dorothy Wyckoff (Oxford: Clarendon Press, 1967), p. 22. See also John M. Riddle and James A. Mulholland, "Albert on Stones and Minerals," and Luke Demaitre and Anthony Travill, "Human Embryology and Development in the Works of Albertus Magnus," in *Albertus Magnus and the Sciences: Commemorative Essays 1980*, ed. James A. Weisheipl (Toronto: Pontifical Institute of Toronto, 1980).

12. Albertus, *Book of Minerals*, 1.1.5, p. 23 and 2.1.5, p. 65. On Albertus's theory of place, see Katharine Park, "The Meanings of Natural Diversity: Marco Polo on the 'Division' of the World," in *Texts and Contexts in Ancient and Medieval Science: Studies on the Occasion of John E. Murdoch's Seventeenth Birthday*, ed. Edith Syxna and Michael McVaugh (Leiden: Brill, 1997), pp. 140–42.

13. Albertus, *Book of Minerals*, 2.3.2, p. 134. On Albertus's discussion, see also Lorraine Daston, "Nature by Design," in this volume.

14. See Claude Thomasset, "The Nature of Woman," trans. Arthur Goldhammer, in Georges Duby and Michelle Perrots, ed., *A History of Women in the West*, vol. 2, *Silences of the Middle Ages*, ed. Christiane Klapisch-Zuber (Cambridge, Mass.: Belknap Press and Harvard University Press, 1992), esp. pp. 48–58; Ian Maclean, *The Renaissance Notion of Woman: A Study in the Fortunes of Scholasticism and Medical Science in European Intellectual Life* (Cambridge: Cambridge University Press, 1980), pp. 41–43; Cadden, *Meanings of Sex Difference*, pp. 183–86.

15. Albertus Magnus, *Quaestiones super De animalibus*, 15.11, ed. Ephrem Filthault, in Albertus Magnus, *Opera omnia*, gen. ed. Bernhard Geyer, 40 vols. to date (Münster: Aschendorff, 1951), vol. 12, pp. 265–66.

16. Cited in Mary Frances Wack, *Lovesickness in the Middle Ages: The Viaticum and its Commentaries* (Philadelphia: University of Pennsylvania Press, 1990), p. 115. See in general ibid., pp. 109–25.

17. On *spiritus* in medieval physiology and psychology, see Marielene Putscher, *Pneuma, Spiritus, Geist: Vorstellungen vom Lebensantrieb in ihren geschichtlichen Wandlungen* (Wiesbaden: Franz Steiner, 1973), pp. 38–68; Ioan P. Couliano, *Eros and Magic in the Renaissance* (Chicago: University of Chicago Press, 1987), pp. 9–11; Siraisi, *Taddeo Alderotti, ad indicem*.

18. See David C. Lindberg, *Theories of Vision from Al-Kindi to Kepler* (Chicago: Chicago University Press, 1976), chapters 5–6, esp. pp. 113–16; and Siraisi, *Taddeo Alderotti*, chapter 7. On the internal senses in particular, see E. Ruth Harvey, *The Inward Wits: Psychological Theory in the Middle Ages and Renaissance* (London: The Warburg Institute, 1975); Nicholas Steneck, "Albert the Great on the Classification

and Location of the Internal Senses," *Isis* 65 (1974): 193–211; and Walther Sudhoff, "Die Lehre von den Hirnventrikeln in textlicher und graphischer Tradition des Altertums und Mittelalters," *Archiv für Geschichte der Medizin* 7 (1914): 149–205.

19. Aristotle, *On Memory and Recollection*, 1, trans. J. I. Beare and G. R. T. Ross, in *The Works of Aristotle*, vol. 3, 450a30–b10. For a more detailed discussion of thirteenth-century elaborations of this idea, see Mary J. Carruthers, *The Book of Memory: A Study of Memory in Medieval Culture* (Cambridge: Cambridge University Press, 1990), pp. 47–58.

20. *Soranus' Gynecology*, trans. Owsei Temkin (Baltimore: Johns Hopkins University Press, 1956), pp. 37–38. On the later history of this theory, see, for example, Lorella Mangani, "Il potere dell'immaginazione materna nel Settecento italiano," *Rivista di folosofia* 86 (1995): 477–91.

21. Leon Battista Alberti, *On the Art of Building in Ten Books* (1452), trans. Joseph Rykwert et al. (Cambridge, Mass.: MIT Press, 1988), p. 299.

22. Jacki Musacchio, *The Art and Ritual of Childbirth in Renaissance Italy* (unpublished Ph.D. dissertation, Princeton University, 1995), chapter 5.

23. Couliano, *Eros and Magic*, p. 5.

24. Catherine Mooney, *Women's Visions, Men's Words: The Portrayal of Holy Women and Men in Fourteenth-Century Italian Hagiography* (Ph.D. dissertation, Yale University, 1991), pp. 174–222. Mooney, "The Authorial Role of Brother A. in the Composition of Angela of Foligno's Revelations," in *Creative Women in Medieval and Early Modern Italy: A Religious and Artistic Renaissance* (Philadelphia: University of Pennsylvania Press, 1994), esp. pp. 52–53. See also Chiara Frugoni, "Female Mystics, Visions, and Iconography," in *Women and Religion in Medieval and Renaissance Italy*, ed. Daniel Bornstein and Roberto Rusconi, trans. Margery Schneider (Chicago: University of Chicago Press, 1996), pp. 130–64; Fragoni, "Su un 'immaginario' possibile di Margherita da Città di Castello," in Rusconi, *Movimento religioso*, pp. 203–16; and Belting, *Presence and Likeness*, pp. 411–16.

25. *Vita B. Aldae seu Aldobrandescae*, 2.21, in Johannes Bolland, *Acta sanctorum quotquot toto orbe coluntur . . .*, ed. Godfried Henschen, April III (Antwerp: Michael Cnobarus, 1675), p. 470; Giacomo Scalza, *Leggenda latina della B. Giovanna, detta Vanna d'Orvieto*, cited in Enrico Menestò and Roberto Rusconi, *Umbria sacra e civile* (Turin: Nuova Eri Edizioni Rai, 1989), p. 134.

26. Caroline Bynum, *Holy Feast and Holy Fast: The Religious Significance of Food to Medieval Women* (Berkeley and Los Angeles: University of California Press, 1987), pp. 200–01, 273. See, for example, the list of thirteenth- and fourteenth-century stigmatics in Antoine Imbert-Gourbeyne, *La stigmatisation: l'extase divine et les miracles de Lourdes*, 2 vols. (Clermont-Ferrand: Librairie Catholique, 1984), vol. 1, pp. xxi–xxiv (ten men, forty-four women).

27. Richard Kieckhefer, *Unquiet Souls: Fourteenth-Century Saints and their Religious Milieu* (Chicago: University of Chicago Press, 1984), p. 95.

28. M. Faloci-Pulignani, "La vita di S. Chiara da Montefalco scritta da Berengario di S. Africano," *Archivio storico per le Marche e per L'Umbria* 2 (1885): pp. 231–38. For more detail, see Katharine Park, "The Criminal and the Saintly Body: Autopsy and Dissection in Renaissance Italy," *Renaissance Quarterly* 47 (1994): pp. 1–3.

29. Mooney, *Women's Visions*, p. 214; Frugoni, "Female Mystics," pp. 134–35. According to Frugoni, laymen of modest education tended to follow the "female" pattern.

30. Bynum, *Holy Feast and Holy Fast*; Bynum, "The Female Body and Religious Practice in the Later Middle Ages," in *Fragments for a History of the Body*, ed. Michel Feher et al., 3 vols. (New York: Zone Books, 1989), vol. 1, pp. 161–219.

31. Cited in Bynum, "Women Mystics in the Thirteenth Century: The Case of the Nuns of Helfta," in Bynum, *Jesus as Mother: Studies in the Spirituality of the High Middle Ages* (Berkeley and Los Angeles: University of California Press, 1982), p. 210; see also Bynum, " '. . . And Woman His Humanity: Female Imagery in the Religious Writing of the Later Middle Ages," in Bynum, *Fragmentation and Redemption: Essays on Gender and the Human Body in Medieval Religion* (New York: Zone Books, 1991), esp. pp. 154–55.

32. Catherine of Siena, *Le lettere di S. Caterina da Siena,* 30, eds. Piero Misciattelli and Niccolò Tommaseo, 6 vols. (Siena: Giuntini e Benivoglio, 1913–22), vol. 1, p. 137.

33. Margarita's contemporary Dante described this process in *Purgatory,* 17.13–18; see Edmund G. Gardner, "Imagination and Memory in the Psychology of Dante," in *A Miscellany of Studies in Romance Languages and Literatures Presented to Leon E. Kastner,* ed. Mary Williams and James A. de Rothschild (Cambridge: W. Heffer, 1932), pp. 275–82.

34. Laurent, "Légende," p. 121.

35. See Brigitte Bedos Rezak, "Medieval Seals and the Structure of Chivalric Society," in Howell Chickering and Thomas H. Seiler, *The Study of Chivalry: Resources and Approaches* (Kalamazoo, Mich: Medieval Institute Publications, Western Michigan University, 1988), pp. 313–15; Spufford, *Money and its Use,* esp. pp. 295–301.

36. See in general Roberto Rusconi, "Pietà, povertà e potere: Donne e religione nell'Umbria tardomedievale," in Bornstein and Rusconi, *Mistiche et devote nell'Italia tardomedievale* (Naples: Liguori, 1992), pp. 11–24; Mario Sensi, "Anchoresses and Penitents in Thirteenth- and Fourteenth-Century Umbria," in Bornstein and Rusconi, *Women and Religion,* pp. 56–83; and Anna Benvenuti Papi, "Mendicant Friars and Female Pinzochere in Tuscany: From Social Marginality to Models of Sanctity," in ibid., pp. 84–103.

37. See André Vauchez, *La sainteté en Occident aux derniers siècles du Moyen Age, d'après les procès de canonisation et les documents hagiographiques* (Rome: Ecole Française de Rome, 1981), passim and esp. pp. 472–78; André Vauchez, "La nascita del sospetto," trans. Monica Turi, in *Finzione e santità tra medioevo e età modern,* ed. Gabriella Zarri (Turin: Rosenberg & Sellier, 1991), pp. 41–42.

38. Laurent, "Légende," p. 124.

39. Laurent, "Légende," p. 119.

40. Albertus, *Book of Minerals,* 2.3.3, p. 136.

DAVID FREEDBERG

Iconography between the History of Art and the History of Science: Art, Science, and the Case of the Urban Bee

At the tensest moment of the conclave that would elect the 56-year-old Maffeo Barberini to the papacy as Urban VIII in August 1623, a prophetic event occurred.[1] A swarm of bees entered the Vatican palace from the meadows facing Tuscany, and settled on the wall of Maffeo's cell. It seemed that Divine Providence had sent this portent to announce the imminent accession to the papacy of a member of that Tuscan family whose coat of arms had long since been transformed from one showing three wasps into an emblem of three bees.

Within a few years it was impossible to go anywhere in Rome without encountering the Barberini bees.

None of the beautiful fountains designed during the primacy of the Barberini by their favorite sculptor, Bernini, is without them. The walls of the ancient city itself, restored by Urban, carry this emblem of his papacy. From ceiling to floor, from the highest cornices to the pavements themselves, from triumphal entryways to modest sacristy doors, on tombs and every imaginable piece of church furniture one may still discover the threefold trigon or inverted triangle of bees that formed the main element in the family's coat of arms. All bear lasting witness to the patronage of the Barberini. Giant bees fly above Divine Providence in the center of the great allegorical ceiling that Pietro da Cortona painted for the Gran Salone of their newly rebuilt family palace on the Quirinal. Bees crawl up the twisted solomonic columns supporting the mighty bronze and gold baldacchino that Gianlorenzo Bernini fashioned, swiftly after Urban's elevation to the pontificate, to rise over the main altar of St. Peter's. Even on the tomb Bernini later made for Urban in the apse of St. Peter's, minute bees climb toward the lid of the sarcophagus, not so much to smell the stench of death announced by the hooded skeleton who inscribes the name of the deceased pope on the black page of death, but rather to rise toward the sweet odor of sanctity, the famous *odor sanctitatis*, that issues from the tomb of the Barberini pope.

But there are two larger bees on Urban's tomb as well, which give the impression of having flown free of the sarcophagus in order to settle—but only briefly, it seems—above the grim reminder of mortality. Since they appear to be on an upward course, they must surely be the ancient symbol of the immortality of the soul; for in antiquity the bees that emerged from the bodies of dead animals stood for the spirit's ability to rise to heaven from the bodies of the dead.

But it was not only by means of great works in bronze and stone that the immortality of the Barberini would be ensured. Countless poems were written in praise of the Barberini, and hardly a book printed in Barberini Rome failed to carry the symbol of the papal family. Few medical, technical, scientific, or geographical works published between 1623 and 1644 lacked that distinctive symbol of sweetness, industry, and power.[2] When Galileo's *Assayer* was published just two and a half months after Urban's accession, for example, the papal bees appeared at the top of the title page, as if to encourage the support of the same Maffeo Barberini with whom Galileo had discoursed in friendship in Florence many years earlier. That friendship would become more than strained, but in this early honeyed phase the bees' presence signaled only mellifluous sounds.

The great Flemish painter Rubens, now known throughout Europe, designed a title page for an edition of Maffeo's own beautiful Latin poems. He powerfully illustrated the scene from Judges 14 in which Samson tears open the jaws of the lion to bring forth the sweetness of honey. Bees pour out of the lion's mouth, but in a lightened space just behind the vivid scene, a trio arrange themselves in the formation of the Barberini trigon. "Out of the strong came forth sweetness" is the central paradox of the riddle posed by Samson after his defeat of the young lion; and so the allusion is not just to the antique notion of the immortality of the soul, or to the honeyed beauty of Urban's youthful poetry. It evokes the benevolence and magnanimity of the Latin-loving poet who had attained the rank of supreme pontiff.

But bees, when irritated, also sting; and under the pressures of doctrine, politics, and a cosmos that was itself being transformed, the liberal benignity of Maffeo Barberini changed to sternness, and then into something chilling and unyielding. Before the end of the first decade of Urban's pontificate, Galileo was sent to the Inquisition by the man who had once been his friend. In 1612, in a series of discussions in Florence, Maffeo had supported Galileo with arguments in favor of his anti-Aristotelian views of the suspension of floating bodies in water. In the following year, when Galileo published his letters on the sunspots—thus calling into question the immutability of the heavens—the young cardinal wrote to him not just that he accepted them, but that he would return to them over and over again with much pleasure.[3] When Cardinal Bellarmine issued his injunction to Galileo in 1616 to renounce his view that the sun was the center of the universe and that the earth revolved round it, Maffeo was instrumental in preventing Galileo from actually being condemned for heresy. Even in 1624 Galileo had several warm meetings with the newly elected pope in the vain hope of having the 1616 injunction lifted. But within a few years Urban realized that he could no longer protect the man whose theories were threatening to divide Christendom and overturn the world upon which the Church was built. The hasty and slightly surreptitious publication of Galileo's vigorous and unsparingly critical masterpiece, *The Dialogue on the Two Great World Systems* in Florence in 1632—in its final form it had not been authorized by the Vatican censor—was the last straw; and by June of the following year the Inquisition had threatened Galileo with torture, forced him to recant his views, and sent him into house arrest at his villa outside Florence.

The reasons for Galileo's progressive estrangement from the Barberini family, and from Urban in particular, and of their sudden abandonment of him have been much discussed; but there is one crucial set of documents in this complex personal and scientific drama that have not even begun to receive the attention they deserve. They take the form of panegyrics on the family of bees.

MICROSCOPES AND THE MELISSOGRAPHIA

The Jubilee year 1625 was the perfect moment for the newly elected pope to celebrate the power of the papacy, the triumph of the Catholic Church over the German heresies, and the glory and prosperity of Rome under the Barberini. The poets prepared epigrams, odes, and panegyrics in their honor. Even the mathematicians and scientists made it clear that none of their discoveries could have taken place under any other auspices than those that all of literary and scientific Rome were now united in acclaiming.

Much could be expected from the Roman *Accademia dei Lincei* (Academy of the Lynxes), that eminent association of scientifically minded humanists founded in 1603 by the eighteen-year-old Prince of Acquastarta, Federigo Cesi. Indeed, Galileo was himself a proud member of the Lincei, and desires to honor the new pope would have been extraordinarily compelling for many reasons. But toward the end of Jubilee year, the Lynxes seemed to be running out of time. They had been working on an elaborate celebration of the chief of all the Barberini emblems, the bee. But this celebration was not just antiquarian, philological, and archaeological. It was to be profoundly natural-historical as well. Just as the year was turning, they finally published three separate works, in which archaeology, philology, panegyric, and scientific investigation were combined in a wholly unprecedented way. Appearing in swift and almost breathtaking succession, they bore the titles *Melissographia, Apes Dianiae,* and *Apiarium* (Figures 1, 2, and 3).

Figure 1.   Academy of the Lynxes, Melissographia, 1625,
engraving by Mathias Greuter dedicated to Urban VIII.

Figure 2. *From Justus Riquius,* Apes Dianiae in monumentis veterum
noviter observatae, *Rome: Giacomo Mascardi, 1625, engraved frontispiece.*

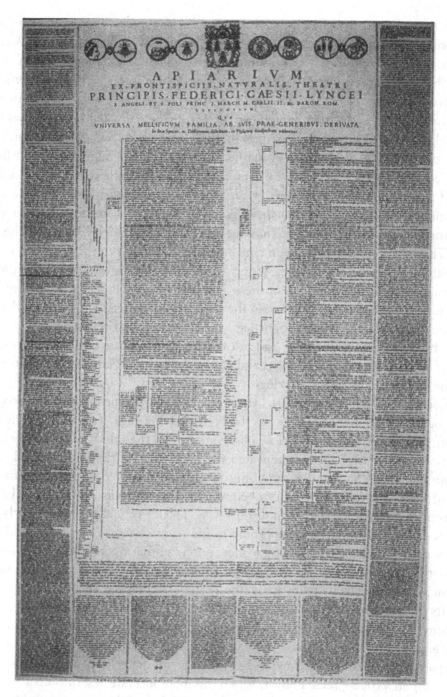

Figure 3. *Federigo Cesi and the Academy of the Lynxes,* Apiarium, *1625,
printed and engraved broadsheet printed on four separate leaves, total
dimensions 107 cm x 69.5 cm. Published by the Lynxes in celebration
of Urban VIII, just as the Jubilee year concluded.*

The first of these three works was an unusually large engraving on a single sheet, measuring 41.6 cm by 30.7 cm. Dedicated to Urban VIII, it carried its title, *Melissographia*, in large Greek capitals, and it was the first printed illustration of a microscopically observed organic being (Figure 1). Signed in the lower right-hand corner by Mathias Greuter, it carries the date of 1625. Cesi was probably referring to this work when he wrote to Galileo on September 26: "The sheet included herewith has been made all the more to show our devotion to our Patrons, and to exercise our particular commitment to the observation of nature."[4] How could the Supreme Pontiff fail to have been impressed? There, in the center, is the family emblem, the trigon of bees, framed by a flourishing pair of bay branches. Above, two putti hold aloft the papal tiara and the keys of St. Peter. But this is no ordinary trigon. These are not bees whose spiky forms have been reduced and mollified by the usual strategies of art. On the contrary. These bees seem to have been *examined* as closely as possible, their forms magnified many times larger than life, each one represented with extraordinary attention to anatomical detail. The precision of these details is remarkable and unprecedented. It is not just that one can easily make out the structure of the head, the tongues, the thorax, the abdomen, the all-important legs, the antennae, and the sting; it is also the astonishing success the engraver had in conveying the texture of the surfaces of the body of the bee, and the diaphanous and flimsy quality of its parts, especially its wings and the delicate extremities of its legs.

"Observed by Francesco Stelluti, Lyncean of Fabriano, by means of a microscope" is the proud inscription that runs across the bottom of the page. This is no simple illustration; it is an examination, a close observation from the life. Whereas other representations of the trigon simply repeated the same view of the bee disposed at the appropriate angles, Stelluti did something quite different. In order to examine the bee, Stelluti viewed it from above, from below, and from the side. And he had his fine engraver present his illustration in just this way. The idea could not more clearly have been to show the bee just as it appeared in life—or rather, as it appeared under the microscope. And then, as if yielding to more purely aesthetic considerations, a number of other details, other parts of the bee, are prettily displayed across the scroll that so elegantly unfolds with its texts across the bottom of the page: on the left, the head with its eyes, tongues, and antennae seen in profile, a frontal view of the head, and the beak; on the right, an antenna, a single eye, cluster of tongues and their casing, and the sting; and finally, in emulation of some graceful printer's vignette terminating a chapter, a pair of hairy posterior legs extending neatly across much of the width of the sheet.

"To Urban VIII, Supreme Pontiff, When this more accurate description of a bee was offered to him by the Academy of Lynxes as a symbol of their perpetual devotion." Thus the inscription at the top of the sheet. On the scroll below, with its illustrations of the parts of the bee, runs a much longer text, at once more complex and more significant.

O great Parent of Things, to whom Nature willingly submits itself, behold the BEE in the BARBERINI escutcheon. Nature has nothing more remarkable than this. Surveying it with a keener gaze, the work of the Lynxes has set it forth in these pictures, and explained it. The genius of the Cesi family has stimulated this sacred labour; the art of Pallas has aided these willing men. Great miracles have emerged as a result of their work with the polished glass, and the eye has learned to have greater faith. Had it not been for the divine discoveries of the new art, who would have known that there are five tongues on the Hyblean body [i.e., the bee's], that the neck is similar to a lion's mane, that the eyes are hirsute, and that there are two sheaths on each lip? Thus it is fitting that while the world looks up to you in wonder, your BEE shows itself even more worthy of wonder.

The sheet may have been intended as a panegyric to Urban VIII—and he could not have failed to be impressed by how much they had seen of the little animal that symbolized his papacy—but it can also be said that Stelluti and his fellow Lynxes could not contain their enthusiasm. They could not help themselves; they were so pleased with the results they had obtained with their new instrument that their panegyrist got carried away by the details of their discoveries. Their pride and their desire to promulgate the results of their achievement were themselves impetus enough for the panegyric mode. Even though fewer than half a dozen prints survive—suggesting that it was distributed to the pope and his family alone—here was the most visible and public statement they had made so far of their commitment to empirical investigation and experiment, and their belief in the power of sight to penetrate the mysteries of nature. In an age when science was torn between the old commitment to the scanning of pure surface and the new drive to theory, hypothesis, and abstraction, this was a crucial step. With the aid of the technology made possible by the perfecting of the telescope, one could begin to reclaim the old hope of arriving at the innermost structures of things.[5]

The eulogistic text of the *Melissographia*, in eight distichs and in more or less immaculate classical form, ostensibly composed in honor of Urban, was written by Justus Riquius (1587–1627). Known as Josse de Rycke in his native tongue,[6] he regularly signed himself as "the Belgian Lynx." His official role in the academy was that of panegyrist, because of his command of ancient rhetoric (which won him the appointment of Professor of Rhetoric at the University of Bologna, the very year he was working on the *Melissographia*). No one could have been in a better position to work on the second of the Lynxes' apiarian offerings of the autumn of 1625, the *Apes Dianiae*.

DIANA'S BEES AND THE ENIGMA OF CHASTITY

The *Apes Dianiae in monumentis veterum noviter observatae*, or *The Bees of Diana recently observed on ancient monuments* is a ninety-line elegiac poem, as Riquius himself described it, in honor of Urban VIII; dated November 1625, it must have been printed hard on the heels of the Lynxes' first astonishing sheet. But the two works could hardly be more different. Where the *Melissographia* points to the future, the *Apes Dianiae* (Figure 2) remains locked in the past. It is an elegiac reflection, replete with recondite scholarly notes, discussing the representation of bees in ancient coinage.

Note D records three coins, illustrated on the frontispiece, in the collections of Urban's brother Antonio. It concludes with the claim that "as far as I know, no one has yet commented on the bees placed under the tutelage of Diana." Indeed, one of the major emphases in the poem and in the notes is the association of bees with Diana and with her coins. What exactly is all this about? The fact that Diana was the goddess of the threefold intersection known as the trivium—"*Diana in Trivio*"—only enhances the connection with the Barberini trigon. But there is much more to the connection than that, and this is what Riquius tries to spell out in his poem.

As everyone then knew, Diana was the stern goddess of chastity, who could strike down Actaeon because he saw her nymphs naked, or turn her favorite Callisto into a bear for having allowed herself to be seduced, albeit unwittingly, by Jupiter. As goddess of the hunt, it was natural enough that on coins she should often be accompanied by a stag. According to Riquius, even the famous many-breasted statue of Diana at Ephesus could not be thought of as in any way unchaste. Her abundant breasts, he asserted, were not for any sexual purpose, but rather for nurture and nourishment.[7] And despite the *horned* stag that so often accompanied her representation on coins, she was also shown with bees, the very model of chastity. As the ancients knew, bees were supposed to be autogenetic; they reproduced without any kind of sexual congress[8] and were therefore particularly pleasing to Diana.[9]

The known chastity of Diana and of bees provides the link to the Lynxes' objectives in the *Apes Dianiae*. As Riquius repeatedly observed throughout his poem, the bees of Diana were precisely suitable as a metaphor; the world, Riquius sang in his introductory "Epigram" to the saintly pope, was all the purer because of the chaste and virginal model of the Supreme Pontiff himself: "*Incorrupta tuos servabunt saecula mores / Virgineo castus Praeside Mundus erit.*"[10] The leitmotif of the poem lies in this parallel between the chastity of bees and that of Urban himself.

But there is more than mere chastity at work. When Riquius writes in his elegy that bees are dear to Diana because of their chastity, because they do not engage in sexual congress and are in fact autogenetic—*ex sese genita*—he adds a note to the following effect:

> As Pliny noted, the way in which bees are generated is a great and subtle dispute amongst scholars. But it is certainly agreed that they produce a foetus without coitus and that they lack either sex. Therefore, since they are virgins, they are consecrated to the Virgin Diana.

And there he leaves it.

But if there was a single most pressing natural historical problem—let us not yet call it biological[11]—that occupied Cesi and his colleagues at the time, it was precisely the problem of generation and reproduction. They were concerned both with the general issue and with particularly difficult and enigmatic cases. Bees were a special crux, for exactly the reason that Riquius spelled out in his note.

In these two documents from the last months of 1625, therefore, pure classical panegyric, in the most learned and traditional form, confronted fundamental and urgent scientific issues—and this on the heels of adopting a wholly new technique for the examination of what we can now surely call biological phenomena. That is, both technological apparatus and scientific curiosity paved the way for determining the bees' reproductive structures and systems, but the inquiry was bound within the elegant constraints of the panegyric mode. Nothing exemplifies this conundrum more strikingly than the way in which this panegyric is used to allude specifically to the problem of the reproductive system of bees. But was there any way in which the strategies of panegyric and of scientific investigation could still come together, or were the two forms already inevitably divided, as they were destined to be in the modern era that was even now in the process of being forged? The answer is complex.

THE STINGLESS KING

The third and final document published by the Lynxes in celebration of Urban VIII as the Jubilee year ended was the most important of all, the huge broadsheet titled the *Apiarium*, or *Apiary*, which Cesi had begun to prepare almost immediately after Urban's accession (Figure 3). As a panegyric, it is one of the most extraordinary examples ever written of that ancient and sycophantic genre. But deeply buried within it, and hidden by a surface that glitters with an immense range of classical learning, is a plea for tolerance, benignity, and restraint, and a foreshadowing of the implications of Galileo's discoveries for the sciences of life.

Cesi's *Apiary* was the most thoroughgoing and most imaginative printed examination of the archaeology, history, literature, and science of bees yet undertaken—but it was also rooted in the concrete reality and minuteness of nature itself. In his letter of September 26, 1625, Cesi announced its forthcoming publication to Galileo. It was intended, he wrote, not only as an expression of the Lynxes' devotion to their patrons, the Barberini, but also as an example of their particular commitment to the observation of nature, *il nostro particolar studio delle naturali osservazioni*.[12]

The *Apiarium* is testimony to an age when the borderlines between science and art, rhetoric and analysis, archaeology and theory, scientific experiment and poetry, were far more fluid than they are now. Printed on four separate sheets joined together, it measures 107 cm × 69.5 cm, very much larger than most other broadsheets published until then. At the top of the sheet is the papal stemma with its trigon of bees, flanked on each side by the obverse and reverse of two ancient coins with bees on them. The first of the coins on the right of the stemma is the Megarensian coin already reproduced and discussed in Riquius's *Apes Dianiae*, while the second of the coins on the left (a bee and a pasturing stag) is a coin from Ephesus, which Riquius describes in his note E as belonging to Cesi himself. The other two coins are from Aptera (showing a profile of Diana and a bee with the inscription *APTA*) and from Metapontum (a bee alongside the two ears of wheat of the Metapontan mint on the obverse, and a bust of Leucippus with the inscription *HERAKLEION* on the reverse).

In the *Apiarium* itself, with its dauntingly dense paragraphs, its all but unreadable "emblems" that surround the central text, its digressive wordplays, and its sheer ambition to form only part of a vast "Theater of Natural History," we notice primarily the stress on papal panegyric, within which is wound an insistence on two things: the fact that the Urban bee does not sting, and that both bee and pope are chastity incarnate. But this chastity is linked inextricably with a manifest fecundity. Indeed, of all the topics within the huge corpus of information the *Apiarium* presents, it is the problem of generation and reproduction, the *GIGNENDI purissima ars* or "purest art of GENERATION," as one of the emblems calls it, that seems to receive the most obsessive attention. The manuscript begins with this issue, but it is framed within a discourse of "stinglessness" that is perhaps crucial to the wider understanding of its purpose.

The *Apiarium* opens with a flurry of classical citations. They serve not only to praise Urban but specifically to testify to the fact that despite his extraordinary power and beauty, the father, king, and supreme lord of bees does not sting. Throughout this great work, ironically enough, the panegyric turns on the parallel between the "king" bee and Urban VIII (these were before the days that it was known that the leader of the hive was in fact the queen).[13] The central section begins with a reference to Columella, who like most recent writers says that the leader of the bees does not have a sting; but Aristotle and Ambrose say that he simply does not *use* his sting; Aelian and Pliny, on the other hand, disagree. And so on, down to modern classical scholars, such as Scaliger and Cardanus.

Why this emphasis on the sting—or rather, the lack of the sting—of the king bee? Because it was precisely the appeal to Urban's benignity and goodwill that motivated this panegyric. Certainly the desire to give evidence of the Lincean researches and the use of a new scientific technique was present here too. Everyone else who panegyrized Urban in that year hoped thereby to win something of his favor; but in the case of Cesi and his fellow Lynxes the need was especially urgent. After all, it was the core group of Lynxes who had encouraged and taken care of the publication of Galileo's *Assayer* of 1623. In it Galileo reemphasized his vigorous and persistent Copernicanism, despite repeated papal warnings not to do so. The work was printed and presented to Urban only two months after his accession to the papacy. It was instantly in demand, and Urban, for all his broad-mindedness, must soon have begun to wonder about the wisdom of having his papacy associated with theories that so patently seemed to threaten the foundations of the Church itself.

"If you should irritate his sting, flee," begins one of the final emblems in the *Apiarium*. But he only stung the wicked and unjust; surely the Lynxes could not be held to occupy such a category. Certainly, they could not have needed Urban's support more desperately. They were in the course of preparing their *magnum opus* on American plants and animals, the so-called *Tesoro Messicano*, in which Galileo's discoveries would be praised openly and even more highly than hitherto. In this book the Lynxes anti-Aristotelianism would be made even plainer, and their own researches depended on the support and protection of both Urban and his nephew Francesco. The Lynxes had to ensure that the Barberini bees would not turn against them or Galileo. "The BENIGNITY of BEES," the *Apiarium* anxiously proclaimed, "wards off both innate and acquired faults."

Rectitude was, of course, also manifested through chastity, on which the *Apiarium* also insists. The Father Pope, or the king bee, procreates and even inseminates without sexual desire. He is not even remotely libidinous, and actually beyond desire. You have to admire him: he knows none of the soft pleasures of sex, none of its mad irritations. He seems to shy away not only from bad odors and from those who are drunk, but particularly from those who have just engaged in sexual intercourse. This is the kind of information that Cesi could have had from ancient writers such as Varro and Columella. The king and father wholly eschews the impurities of lust. He is virginal— yet immensely fecund. But what is all this about, other than the need to panegyrize yet another quality of the pope? It arises from Cesi's central concern with the problem of the generation of bees, and the paradox already noted by Riquius: that whatever their differences about the way in which bees are generated, most authorities, both ancient and modern, agreed that bees—*like the pope*—produce their offspring without coitus. The question of asexual reproduction becomes, then, a question of two great reproductive "miracles" in the kingdom of animals and the kingdom of the Church. Just as science might illuminate questions of the heavens, it promised to cast light on the representative of heaven's rule on earth.[14]

Even without these pressing religious and political considerations, for Cesi the question of generation was one of the central mysteries of nature. The *Apiarium* exhorts: "Behold the admirable work of making offspring—the purest ART OF BEGETTING, the *gignendi ars purissima*, far beyond the gates of desire, the most singular and mysterious spectacle in all the Theatre of Nature." It was just this "spectacle" that Cesi set out to examine in the main section of the *Apiarium*. But these are also poignant words. Could it be that at least part of the intensity with which he pursued the subject had to do with the persistent loss of his own children during childbirth?

Presumably;[15] yet from the very beginning Cesi believed that insight into the order of nature had to begin with the problem of generation and reproduction.

Cesi's immensely complex theory of the reproduction of bees involved his particular interest in the behavior of liquids and solutions at different temperatures (in this case honey, from which future bees were supposed to be generated), his view of the social organization of bees (the king gives rise to all his workers and subjects, who are protective of, and protected by, their king), and his need for order in the midst of a phenomenon so elusive that it seemed to be at odds with the very possibility of verbal representation: *"quod unicum, quod multijungum naturae opus, difficile verbis repraesentaveris"* (how difficult to represent in words this unique and multifarious work of nature!).

What is significant is the realization by Cesi and the other Lynxes that one has to penetrate deep into the anatomy of the bee in order to understand the organs of reproduction. To find the keys to a better classification of the many types of bees, one must go into their interiors, rather than base classification on their exteriors alone.[16] And so, deep within the *Apiarium* are the preliminary results of the Lynxes' work with the microscope, not only by Stelluti but above all by Fabio Colonna. This desire actually to see and observe is enshrouded by the panegyric, but will not be denied by it. It appears most strikingly apparent on those occasions when Cesi expresses his frustration at what cannot be seen with the naked eye: the particles within the liquid honey, the seminal substance of the bees, and the actual place in which they are formed. (Here, in referring to the *gynaeceum apum*, Cesi seems to show some sense of the possibility that female bees may play more of a role than the rest of his discussion may have suggested). For all this, however, only the brand-new instrument, the microscope, could serve; and even it would leave some of that desire unassuaged.

AN EPISTEMOLOGY OF MICROSCOPIC VISION

In April 1624, Galileo went to Rome to try and ingratiate himself with the pope. One of the very first things he did was to confer with Cardinal von Zollern about the use of the microscope—or rather about the two types of microscope that were then available in Rome.[17] It is not entirely clear which of these types Johannes Faber was referring to when he wrote to Cesi on May 11 about Galileo's meeting with von Zollern; but he could barely contain his excitement:

> I spent yesterday evening with our Signor Galileo, who is staying near the Maddalena. He has given a very beautiful *occhialino* to Cardinal von Zollern for the Duke of Bavaria. I examined a fly which Galileo himself showed me; and I was astonished.[18]

By September of the same year Galileo was sending microscopes of his design to a number of people, including one to Cesi and his wife, Isabella. This generous gift was accompanied by a famous letter to Cesi of September 23, 1624, in which Galileo referred to his examination of the horrid flea, and the beautiful mosquito and moth:

> I am sending your Excellency an *occhialino* to view the smallest things as if from nearby. I hope that you will derive no small pleasure and enjoyment from it, just as I did.

And after describing in some detail the correct way of using the new instrument, as well as his own fascinated observation of insects, Galileo concludes:

> But your excellency will have a huge field in which to observe many thousands of specimens. I beg you to notify me of the most interesting things you observe. In sum it gives us the possibility of infinitely contemplating the grandeur of nature, how subtly she works, and with what ineffable diligence.[19]

Of all the senses, the eye was the chief instrument of observation, the microscope its preeminent aid. But what were the limits of vision, and how far could it be aided? This was the problem that continued to plague the Lincei. Even by 1625, when the microscope was clearly ready for use, and the Lincei were ready to tell the world—or at least a part of it—of its possibilities—they remained desperately aware of its limitations. Hence the frustrations expressed in the *Apiarium*, not only about what could not be seen, but also about the fundamental insufficiency of the microscope itself. This, as we shall see, had much less to do with technical matters, such as the quality or the placement of the lenses, than with the very principles upon which the use of the promising new instrument was based.

In the *Apiarium*, there is heroic microscopic detail (see Figure 4). The detail is heroic because of its magnificent intensity, but also because the author proceeds in the knowledge that it cannot be conveyed adequately in words. The clear implication is that one has always to go *beyond* the simple processes of seeing and describing. What *this* entails, however, is the impossibility of ever seeing, or describing accurately, enough. However much one might enlarge the object under examination, one has always to assume that there remains something that is there but is forever beyond the reach of the organs of sight. Sight is essential for good science, but sight alone can never be sufficient. This is already implicit on several occasions in Cesi's account of the generation of bees; but it is made explicit in the final passage on the microscope in the *Apiarium*:

> If you can discern with it many subtly constructed things, you will conclude that there are still other much smaller things yet, which escape and elude even the

Figure 4. Detail of Federigo Cesi's Apiarium, *1625, printed and engraved broadsheet.*

sharpest of instruments constructed by us. This also applies to our telescope: while it draws further things closer to our eyes, you can also judge that there remain other things even further away, which it could never reach. Therefore we accept the fact that there is no small number of very small and very distant things which cannot be seen.

Nothing could be clearer than this acknowledgment of the limitations of the micro-scope and the telescope. They are limitations beyond the limitations of sight itself, and beyond the impossibility of ever seeing everything that goes into the constitution of natural bodies. The mere use of the senses, therefore—however hard Cesi might have insisted on all of them—could never reveal the ultimate basis of things. This was the truly Galilean part of his science. The use of the microscope was only the begin-ning of the real work that still remained to be done.

When one considers these epistemological tensions within the *Apiarium*, one begins to sense a movement away from the raggedness of ordinary visual description toward a more unified picture. What is at stake, finally, is the ordered, mathematically determinable structure of all things—epitomized by the golden, reticulated apian eye revealed by the *occhialino*. Ironically enough, it was the microscope alone that could corroborate such details, and the microscope that could confirm this fundamental perception of the world as mathematically structured. On the face of it the microscope might only have been expected to yield more in the way of descriptive density; yet it had begun to confirm the kinds of geometrical patterns underlying natural forms that Kepler discovered with the naked eye on the bridge in Prague and that Cesi had only recently begun to realize provided the basic clues not only to the secrets of the heavens but to those of earth as well.

PANEGYRIC "SCIENCE"

In so many respects Cesi's account of the bee in *Apiarium* was simply wrong. His view of the crucial role of the king bee, and his insistence on the nonsexual aspects of reproduction, was quite obviously motivated by the convenient coincidence of his reading of classical texts on the bee (notably Pliny the Elder, but many others as well) with his desire to panegyrize the asexual but generative Barberini pope. The texts supporting these views must have seemed especially *ben trovati*. Bounded as well as buoyed by his motivations for such good finds, it was not for Cesi to discover, as Butler did in his *Feminin' Monarchi'* of 1634, that the leader of the hive was in fact the queen bee. For Cesi in the Jubilee year of 1625, the king bee needed to remain a natural analogue for the necessarily chaste ruler of the Church.[20]

But to insist that the *Apiarium* is not "scientific" for having such interests is to use the word naively and ahistorically, and in a way that is uncritically conventional. It is to assume that the modern terms of "science" (whatever they broadly may be) are the only terms by which Cesi's project may be defined. But one does not have to be a committed Kuhnian or Feyerabendian to acknowledge not just that the terms of "science" in the seventeenth century were different from our own but that for Cesi and his fellows their procedures and their descriptions counted as science. The task, therefore, is to determine the extent to which such science differed from preceding approaches to natural history and the nascent sciences of life, or biology, as we now, *pace* Foucault, broadly call it. There *was* a change in seventeenth-century science, and its most exemplary and brilliant figure was indeed Galileo, but to ignore the work of his closest friends in fields other than those that he specifically made his own is to ignore some of the foundational changes in the edifice of what we like to think of—broadly, again—as modern science.

For Foucault, as for Alexandre Koyré, premodern science gave equal—or almost equal—status to mythological, historical, and even what we now call magical explanations as it did to observational ones. Direct observation was not especially valued, and played less of a role than explanations derived from the occult sciences, such as astrology and alchemy. Theory, such as it was, was predicated less on the roles of hypothesis and proof by experiment than on the evidence of ancient and established authority. For Koyré sixteenth-century natural histories went "no further than the stage of a catalogue." In this view, then, pre-Galilean science—and in particular, natural history—lacked "a classificatory theory, the possibility of classifying in a reasonable manner the facts collected."[21] For Foucault, the new *episteme* of the seventeenth century was predicated not on similarity but on difference, not on variety but on identity, on number, and on structure.

Flawed as such views of the distinction between pre- and post-Galilean science may be, they force upon us a reconsideration of the status of the scientific activities of Cesi and his Lincean colleagues. For their work cannot be considered wholly in terms of the old paradigms, or of the old *episteme*, as Foucault would put it. True, Cesi places great store in the old authorities, but he is hardly uncritical. New knowledge, new facts—in vast abundance—demanded from him a critical rethinking of the idea of authority itself, and made him realize "the possibility of classifying in a reasonable manner the facts [he] collected" from the very beginning. Furthermore, the evidence of the old writers as well as the conclusions drawn from the occult sciences could now be tested—*had* to be tested—by means of direct observation and by the possibilities offered by unheard of new techniques of observation. And these new empirical techniques in turn compelled and elicited a whole new theoretical style that focused on the inner operations of things not seen by the eyes. The Lynxes' project was thus wholly divorced from the old theories—such as the many and varied physiognomic ones—that had postulated equivalences and correlations between outer appearance and inner mechanisms and functions.

The *Apiarium* provides critical testimony to these changes. True enough, there is much reliance on ancient sources. The abundant digressions rarely seem to rise above the level of the anecdotal. But not to give this work its due as science would be to omit those aspects that are indeed valuable from a purely empirical point of view (such as the very lengthy and detailed listing of American bees, based largely on firsthand sources) or to misrepresent the beginnings of approaches that would in fact bear lasting fruit. Throughout the *Apiarium*, one realizes the presence of a conscious need for an internally consistent hypothesis that may be in fact—*has been* in fact—ultimately to be proved or falsified by the eventual provision of empirical and experimental data. Still embedded in the antiquarian structure of the panegyrist's theme, this empiricism lay deep in Cesi's labyrynthine Latin text. This is not true in the case of the final Lynxian publication to be addressed in this essay, Francesco Stelluti's magisterial *Persius* of 1630.

THE PERSIO OF STELLUTI

"Lyncean of Fabriano," Stelluti had done the micrographic observations for the *Melissographia*; now he would complete a work, largely unencumbered by the necessity for panegyric, that reunited the seemingly separate paths upon which the *Melissographia* and the *Apiarium* had embarked.

Stelluti's *Persio* was a work whose likes had never been seen before. For many years Stelluti had taken an interest in the literary possibilities of the vernacular Italian of his day, particularly Tuscan. He had written epithalamia on the marriages of Cesi and his brother, and where men like Riquius wrote dedicatory poems in Latin, Stelluti contributed splendidly direct vernacular poems in praise of Galileo to the preliminary pages of both the *Essay on the Sunspots* of 1613 and the *Assayer* of 1623. Barely had he finished collaborating with Cesi, Faber, Colonna, and Riquius on the *Melissographia* and the *Apiarium* when he turned to his next and most important project. This was the translation of that complex and difficult Latin poet, Persius, whose six elegies contain some of the most allusive and obscure lines ever written in all of Latin literature. It was precisely their robustness and colloquial quality—to say nothing of their insistent digressiveness—that Stelluti sought to to emulate in his translation.

The work that finally appeared in 1630 (Figure 5) is an extraordinary performance, remarkable not only for its blunt use of the Tuscan dialect (deliberately reflecting Persius's own Etruscan heritage) and the vigorously free form of what was known as *verso sciolto*—blank verse—but also for its notes. These were not purely literary, or text-critical, or even archaeological, in the manner of someone like Riquius. The attempts at textual criticism were feeble and haphazard, and variant readings were only mentioned when they provided the opportunity for discourse on some other, more interesting subject. Stelluti's real subject in the notes were the items of natural historical and scientific interest, which he added whenever he could find an excuse. The most insignificant reference in the elegies sparked magnificent, if sometimes rambling, excursuses on the work of the Lincei. Thus it came about that this vigorous literary exercise provides us with some of the most crucial information we have on their researches, in the very year in which Federigo Cesi's life and work were so abruptly cut short. The notes of the *Persio* offer eloquent and moving evidence of Stelluti's devotion to the man he admired more than any other, to Cesi's own unfinished projects, and to the kinds of work the Roman nobleman had inspired Stelluti himself to do—most notably with the Lynxes' technology of choice, the microscope.

As examples of the Lincean digressions that adorn Stelluti's *Persio*: Persius mentions a parrot; Stelluti refers to Faber's long examination of the various species of parrots in the *Tesoro Messicano*. Persius scoffs at the soft luxury of poets who scribble on couches made of the wood of the citron; Stelluti comments on the gleaming quality of that same wood when polished—and uses this as a pretext for a long and detailed description of the fossilized wood that Cesi found in the hills around Acquasparta, and which he called *metallophytes*, "of a middle nature between plants and minerals." Persius jokes about poets who write as smoothly as if they were drawing a line with one eye shut; Stelluti adds a long discourse about the advantages of looking with one eye

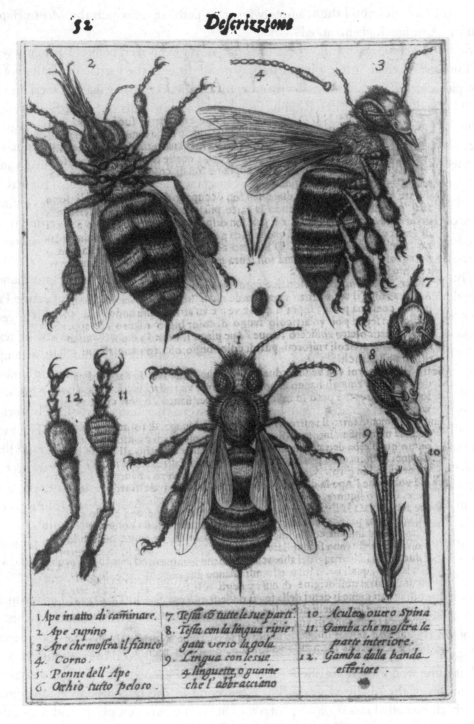

Figure 5.   *Engraving from Francesco Stelluti,*
Persio tradotto in verso sciolto, *Rome: Giacomo Mascardi, 1630.*

only in certain activities, itself a prelude to a detailed discussion of Della Porta's views on binocular and monocular vision in his *De Refractione*, which in turn is followed by a long passage praising Della Porta's work as a whole, insisting on his primacy over the Dutch in the invention of the telescope, and finally hailing the way in which Galileo, *nostro Accademico Linceo*, perfected it. There are hundreds of such digressions, all packed with classical and modern erudition. Desultory, aleatory, seemingly haphazard, they all weave a tapestry of praise to the Lincei and their endeavors.

But let us return to the bee. Stelluti had not yet finished with it; it forms the center of what is justly the most celebrated part of the book. He arrives at it in a typically roundabout way; his pretext for introducing it is the flimsiest. In line 129 of the first satire Persius makes a satirical reference to the town of Arezzo; this provokes what is perhaps the most important note of the book. Stelluti lists several of the many ancient writers who mention Arezzo; but then, observing that some critics have read not "Arreti" for "at Arezzo," but rather "Ereti" for "at Eretum," he points out that the ancient town of Eretum in the Sabine countryside was in fact the present-day Monterotondo, seat of the Barberini country estate. Thus, Stelluti has his pretext. The Barberini were endowed with the greatest of gifts and virtues, he begins. Under the patronage of their symbol, the bee, the arts and sciences had flourished as never before. Having dragged in the bee, it is fitting that the example he chooses is Cesi's *Apiarium*, "full of erudition and original and novel concepts, restricted to one large folio, it is true, but so full of data and theories that it could just as well form a large volume." Stelluti then recalls his own observation of the bee under the microscope, as published in the *Melissographia*. He had discovered things not known to Aristotle or any other ancient naturalist or philospher, he reminds the reader, but now he has examined the bee with still greater diligence; the results of this examination, he says, he has placed at the end of the satire. He does not, after all, wish to interrupt his reader with too long a digression.

The full-page illustration of a bee and its parts (Figure 5) that appears at the end of the satire is perhaps the most unexpected image ever to appear in the edition or translation of a Latin poet. But when compared with the fine and grand engraving that is the *Melissographia*, the much smaller illustration in Stelluti's *Persius* comes as a disappointment. But if the actual contents of this presentation appear to offer no surprises, then the methodology of presentation is strikingly different from its most immediate precedent. In the suppression of the decorative bay leaves and putti surrounding the bees of the *Melissographia*, and the absence of its elegant cartouche and beautifully calligraphic typology, the *Persio* bee is—to put it simply—more factual. However artfully they may still seem to be arranged, the parts of the bee are disposed across the page in a much less contrived way. There is a simple frame. The elements of the illustration are numbered, correlated with an unadorned listing. And the listing itself is not embedded in a poem consisting of fine Latin distichs; instead, the parts are identified in the plainest and most direct vernacular. The facts are presented for what they are, and not (or at least not exclusively) for their antiquarian or panegyric implications.

All this applies equally well to the long description that follows. It is in fact here, rather than in the illustration, that we see the full fruits of Stelluti's observations. There can be no question that it is the verbal description that marks the real break-through, not the illustration of the bee.

It is true that the account by Cesi of the actual microscopic examination of the bee in the *Apiarium* is written in a rather more direct way than the rest of that forbiddingly dense and convoluted work; but still it is embedded in the midst of panegyric and praise, written in a difficult Latin, with the struggle for dispassionate description painfully obvious and the lure of rhetorical devices ever present. Stelluti, on the other hand, separates his description from the rest of his divagations. He assigns it an appendix of its own. It is clearly detached from his rhetorical, digressive, and (muted, but still present) panegyrical agendas; and it is written in a straightforward vernacular, clear and easy to understand.

In the *Persio*, one cannot fail to be struck by the plain directness of the language Stelluti uses. It would be impossible to exaggerate the difference between the complex and insistent density of the *Apiarium* and the lucidity of the description of the bee in Stelluti's *Persio*. Not a single ancient writer is mentioned. It is free of the rhetorical devices that so encumber the *Apiarium*, even in those parts of Cesi's text where panegyric loses its accustomed grip on the exposition. There are light years between the classical Latin of the *Apiarium* and the suppleness and clarity of Stelluti's vernacular.

This distinction, broad though it may be, is crucial. For the Lincei, the vernacular—and by no means the Italian vernaculars alone—played a fundamental role in the new forms and subjects of science. Unlike the older and often occult science the secrets of the new were increasingly intended to be plain and accessible, at least in principle. Just as one could now appeal, at least in the first instance, to the clear evidence of the eyes, rather than to the testimony of ancient books and writers, so too one could rely on the quotidian directness of the vernacular. The vernacular was robust and unfussy, flexible and witty, as Galileo himself showed time after time. It was capable of all the clarity the new science demanded. It could not cloak obscurity in lengthy Ciceronian or worse Latin. Colloquial and invitingly dialogic, it could not appeal to tradition and traditional authority in the way the older scholastic languages of science did. No wonder the major works of Galileo's maturity, the *Assayer* and the *Dialogue on the Two Chief World Systems*, were written in the same robust Italian as Stelluti's *Persio*.

For Galileo as for Cesi, Stelluti, and their Lincean colleagues, tradition and dogma stood in the way of the discovery of scientific truth (although their passionate immersion in classical culture suggests that the ancients could be viewed as a gateway, not merely as a roadblock). Increasingly, Aristotle, Pliny, and the others could go; one had instead to rely on observation and hypothesis. Physics is separated from metaphysics; methodology takes its place. And it was no arcane methodology: even geometry could be learned and grasped without the study of Greek and Latin. Antiquity could no longer be held as the exclusive realm of knowledge; its fruits had to be adapted—and if necessary, rejected—by the pressure of new knowledge from every side.

Latin—and the other classical languages, Greek and Arabic—had been the languages of the Old World; but now the evidence of the New was emerging abundantly every day. The major project of the Lincei in the very years they were editing and supporting Galileo was their great work on the fauna and flora of Mexico. This work is astonishing in its insistence on the importance of vernacular names and vernacular knowledge. Plants and animals had been discovered whose equivalents could not conceivably be found in Aristotle or Dioscorides, and the Lincei knew that it would be vain to seek them there. The new natural history, like the new science, had burst beyond the bounds of the old books.

And just as evidence of the New World pressured old authority, so too did the evidence of the local—once old authority was shown to be incomplete, it became clear that New Worlds could be found on your doorstep too. Cesi and his friends, to an even greater extent than the pioneering botanists of the sixteenth century, found things in the hills and valleys round his home that presented problems no ancient writers could resolve. They were problems of identification, classification, and nomenclature above all, but also of cause and aetiology.

The language of science was thus liberated from its old bondage to authority and tradition; indeed it could only find adequate exposition in the vernacular, whether Meso-American vernaculars or the local languages that alone provide the terms for the local, the lowly, and the unexceptional. By its very nature, the microscope was especially suited to the examination of the lowly, the everyday, and the seemingly insignificant; no wonder that from the very beginning Galileo, Colonna, and Stelluti should have constantly insisted on the scientific interest and relevance of the tiniest louse or flea. In this regard, it comes as no surprise that the second high point of Stelluti's work with the microscope, illustrated in the *Persio*, should be his examination of the lowly curculio beetle that so infested the grain harvests of Tuscany. A more local concern there could not be.

Thus from the global discoveries of the New World came the openness to local and vernacular knowledge, and from the local came the global in turn. One of the most surprising aspects of the work of the Lincei, one soon comes to realize, is its ethnographic range. The region of select knowledge—that is, the region of classical languages and the archaeology, literature, and general culture of the ancient world—is expanded to the wider and more intimate regions of local cultures. When the Lynxes and their friends write about ancient sports they know that the field of modern games has been even less studied,[22] and so they turn their attention to such matters too. Their work is thus deeply concerned with folkways and with the acceptance of popular cultures as much as with the pondering of ancient ones. If ever there was an unnoticed aspect of their work it is this. It offers testimony not to the courtly and exclusive nature of their researches, as has recently been insisted upon,[23] but rather to their belief in the essential humanity of all the sciences. Science can no longer be the exclusive domain it once was; its secrets are essentially accessible, no longer available only to those whose knowledge is predicated on the arcane and elite lessons of the antique. Veiled in classical allusion and wrapped in the shining cloak of panegyric, the Lincei attempted to ensure that whatever threatened to be subversive about their work would never be discovered. But surely the path they charted has profound implications for our understanding of the complex representational strategies and epistemological constraints of early modern science.

Notes

1. Much of this essay derives from my forthcoming book on the study of natural history around Galileo, provisionally titled *The Eye of the Lynx*. More extensive evidence for the essay's arguments will be found there. I am immensely grateful to Caroline Jones for having encouraged and helped me produce this essay in its present form. Here I have deliberately kept the notes to a minimum.

2. See, for example, books such as the Greek and Byzantine scholar Leone Allacci's *Apes Urbanae* (Rome: Grignani, 1633), and the Jesuit Giovanni Battista Ferrari's *De Florum Cultura* (Rome: Paulino, 1633) of the same year. On the latter see especially David Freedberg, "From Hebrew and Gardens to Oranges and Lemons: Giovanni Battista Ferrari and Cassiano dal Pozzo," in *Cassiano dal Pozzo: Atti del Seminario Internazionale di Studi*, ed. F. Solinas (Rome: De Luca, 1989), pp. 37–72; and now, with E. Baldini, G. Continella, and E. Tribulato, *The Paper Museum of Cassiano dal Pozzo, A Catalogue Raisonné, Drawings and Prints in the Royal Library at Windsor Castle, the British Museum, the Institut de France and Other Collections*, series B, Natural History, vol. 1, *Citrus Fruit* (London: Harvey Miller, 1997). Franca Petrucci Nardelli, "Il Cardinal Francesco Berberini Senior e la Stampa a Roma," *Archivio dell Società Romana de Storia Patria* (Rome: 1985), pp. 138–64, provides an excellent survey of the printing history of the books subsidized by Francesco Berberini in particular.

3. Galileo, *Opere*, XI, pp. 495–96.

4. "*Questo è fatto per significar tanto più la nostra divotione a' Padroni et esercitar il nostro particoloar studio delle naturali osservationi,*" [Giuseppe Gabrieli]. *Il Carteggio Linceo*, in *Atti della Reale Accademia Nazionale dei Lincei, Memorie della Classe de Scienze Morali, Storiche e Filologichem* ser. 6, vol. VII (1938–42), p. 1066, no. 866, Cesi to Galileo, September 26, 1625. This reference seems to me to be indubitably to this sheet, rather than to an early draft of the text of the *Apiarium*. Clara Sue Kidwell's valiant dissertation *The Accademia dei Lincei and the Apiarium: A Case Study of the Activities of a Seventeenth Century Scientific Society* (University of Oklahoma, 1970) is the only detailed study of the *Apiarium* ever made, but unfortunately neither the summary of the context of the work nor the translation are reliable.

5. See, in this volume, Lorraine Daston's "Nature by Design" in which she articulates the move by seventeenth-century natural philosophers from "macroscopic" arguments about external resem-

blances visible to all humans, to "microscopic" claims about perfections achieved in the interior of animals by God.

6. Beyond the invaluable articles by Van den Berghe in *Messager des sciences historiques* (Ghent, 1880), pp. 12–32, 189–208, and in the later edition (Ghent, 1881), pp. 160–85, 457–77, see Giuseppe Gabrieli, "Giusto Ricchio Belga: I suoi scritti editi e inediti," in *Rendiconti della Reale Accademia Nazionale dei Lincei. Classe di Scienze Morali, Storiche e Filologiche*, ser. 6, vol. IX (1943), pp. 1–44 (reprinted in Giuseppe Gabrieli, *Contributi alla storia dell'Accademia dei Lincei*, 2 vols. [Rome, 1989], pp. 1133–64).

7. Justus Riquius, *Apes Dianiae in monumentis veterum noviter observatae*, (Rome: Mascardi, 1625) p. 7:
    Hic adyta, & Triviae Virginis esse domum,
    Apparet tota diffusae corpore mammae,
    Nec tamen est ullo foedera passa thoro.
    His alitur mortale genus, vitaeque animantum,
    Vitales succos hinc elementa bibunt.

8. Ibid., p. 6: "*Ex sese genita nullo faedata cubili.*"

9. Ibid.: "*Plus tamen est APIBUS tribuit quod virgo pudicis / Gratior & Castae* [i.e. Diana] *casta volucris adest.*"

10. Ibid., p. 3.

11. See Michel Foucault's strictures and comments on the beginnings of biology in *The Order of Things, An Archaeology of the Human Sciences* (New York: Vintage, 1973), passim but especially pp. xii and 159–62. Published originally in French as *Les mots et les choses* (Paris: Gallimard, 1966).

12. Gabrieli, *Carteggio*, p. 1066.

13. Caroline Jones points out that "this fixation on the *maleness* of the chief reproducing bee, a fixation that served to obscure the gender of the animal's actual sex organs, may have been intimately tied to the patronage of those who studied it. That is, because the natural analogue between Pope and Bee was so compelling, the actual sexuality of the bee could not be perceived by the Barberini-eyed Lincei." Correspondence with the author, July 16, 1996.

14. A point implicit in my argument, which Mario Biagioli has usefully pushed me to foreground more forcefully. Communication with the author and editors, July 16, 1996.

15. It was precisely in his correspondence with Cesi in the first months of 1626—referring frequently to the *Apiarium* and to their work with the microscope—that Fabio Colonna discusses various possible cures for the continued difficulties Cesi's wife had with her miscarriages and in giving birth to healthy male offspring (e.g., Gabrieli, *Carteggio*, letters of February 13, and March 20, 1626, pp. 1100–01 and 1111).

16. Again, see Daston's argument in this volume ("Nature by Design") regarding the significance of the move from macroscopic to microscopic arguments and demonstrations.

17. Galileo's microscope used a convex objective lens and a concave ocular lens, while the "German" one invented by the Dutchman Cornelis Drebbel and brought to Italy by his brothers-in-law Jacob, Abraham, and Egidius Kuffler, was a true compound one using two convex lenses. Galileo, recognizing its potential, was very much taken with the new type.

18. "*Sono stato hier sera col sig. Galilei nostro che habita vicino alla Madalena; ha dato un bellissimo ochialino al Sig. Cardinale de Zollern per il Duca di Baviera. Io ho visto una mosca che il Sig. Galileo stesso mi ha fatto vedere; sono restato attonito*" Gabrieli, *Carteggio*, pp. 942–43.

19. Ibid., pp. 942–43. "*Invio a V.E. un occhialino per veder da vicino le cose minime, del quale spero che ella sia per prendersi gusto e trattenimento non piccolo, chè cosi accade a me. . . . Ma V.E. haverà campo larghissimo di osservar mille e mille particolari, de i quli la prego a darmi avviso delle cose più curiose. In somma ci è da contemplare infinitamente la grandezza della natura, e quanto sottilmente ella lavora, e con quanta indicibil diligenza.*" Gabrieli, *Carteggio*, p. 942.

20. Again, I thank Mario Biagioli for pushing me to emphasize this point. He comments, "Of course, one could also expand on the male homosocial imaginary entailed by Cesi's celebration of [male] bee-

dom—an imaginary that was no doubt shared by many of the boys at the Roman court." Communication with the author and editors, July 16, 1996.

21. Alexandre Koyré, *Etudes d'histoire de la pensée scientifique* (Paris: Gallimard, 1973), p. 53.
22. On this subject, see David Freedberg, "Cassiano on the Jewish Races of Rome," *Quaderni Puteani*, 3, II (Milan, 1992), pp. 41–56.
23. See especially Mario Biagioli, *Galileo Courtier* (Chicago: University of Chicago Press, 1993).

Joseph Leo Koerner

Hieronymus Bosch's World Picture

In the memoirs of his field work in Brazil, Claude Lévi-Strauss describes the moment when alternative realities became a thing of the past.[1] At the end of his travels in the Amazon basin, and after working among native peoples already in contact with the outside world, he got word of an "unknown" tribe living "still savage" in the upland jungles:

> There is no more thrilling prospect for the anthropologist than that of being the first white man to visit a particular native community. . . . I was about to relive the experience of the early travellers and, through it, that crucial moment in modern thought when, thanks to the great voyages of discovery, a human community which believed itself to be complete and in its final form, suddenly learned, as if through the effect of a counter-revelation, that it was not alone, that it was part of a greater whole, and that, in order to achieve self-knowledge, it must first of all contemplate its unrecognizable image in this mirror, of which a fragment, forgotten by the centuries, was now about to cast, for me alone, its first and last reflection.[2]

The viewer is bound to the object in mutual destruction, and, as at an apocalypse, the first shall be last and the last shall be first. Lévi-Strauss, the last white man to thrill to a first encounter, will thereby exhaust the world of possible other worlds. And the last unknown tribe will lose its innocence of other worlds even as it resurrects, in the "counter-revelation" it offers, the white man's own lost belief in a world that is final and complete.

By a coincidence of opposites, this "unknown" was also the historical remnant of modern Europe's original Other. The tribe Lévi-Strauss sought consisted of "the last descendants of the great Tupi communities . . . whom the sixteenth-century travelers saw in their period of splendor." In this déjà-vu of a people without history, Lévi-Strauss observes the cause of Europe's modernizing pluralism and the founding instance of his science of man:

> It was the accounts given by these travellers which began the anthropological awareness of modern times; it was their unintentional influence which set the political and moral philosophy of the Renaissance on the road that was to lead to the French Revolution. To be the first white man to set foot in a still-intact Tupi village would be to bridge a gap of four hundred years and to find oneself on par with . . . Montaigne who, in the chapter on cannibals in his *Essays*, reflected on a conversation he had had with Tupi Indians whom he met at Rouen.[3]

For Lévi-Strauss, the prospect of a belated return to his own historical origins is as thrilling as the promise of a first encounter with the last unknown. Indeed it is quite unclear whether his thrill derives from his anticipated encounter with savages or from his historical transport, through them, back to the originary moment of his own culture.

Lévi-Strauss arrives at the village of people who refer to themselves as Mundé, and sets about studying their "way of thinking and social organization." But since he cannot speak their language and has no interpreter, he must leave empty-handed, concluding: "After an enchanting trip up-river, I had certainly found my savages. Alas! they were only too savage."[4] This sigh, heaved also in the title *Tristes Tropiques*, seems at first merely to express the disappointment of not having been adequately equipped, of having made contact without the tools to make sense. Yet it also describes the condition of mutual indecipherability that the ethnographer anticipates and desires.

Lévi-Strauss retreats, the better to prepare himself for surprise. Yet on his way back into the forest, embarked on a search for yet another "still-savage" tribe, he encounters something truly unexpected. Rounding a bend, he finds himself facing two natives traveling in the opposite direction. They are the leaders of the very tribe that the anthropologist seeks. Having "resolved to leave their village for good and join the civilized world,"[5] they bear with them their most precious possession, a live harpy eagle, as a gift for their future hosts.

The anthropologist arrives too early or too late. Either he encounters innocence and, "alas," cannot penetrate it, or he finds it already on its way to him and therefore no longer pure. Lévi-Strauss bribes the leaders to go back to their village and the eagle, their totem, is "unceremoniously dumped by the side of a stream, where it seemed doomed to die." Even while noting that the jettisoned bird meant the demise of the tribe's identity, Lévi-Strauss returns the natives to their home, where they will play-act as savage informants, as inhabitants of a reality alternative to ours because ignorant of alternatives, forgetting for the while that they had already sought, and thus dwelt within, our now fully ubiquitous world.

For Lévi-Strauss, the forest is not paradisal but tragic. The savage other cannot be observed because it is "alas! too savage," or it will have already discovered us and, measured against modernity, it again is "alas! poor savage." Lévi-Strauss's sadness may be merely a last Romantic yearning for lost innocence combined with the admonishment "we murder to dissect." Yet it has relevance to our present situation at the end of the millennium, in an era of economic globalization, as we turn to wonders not in forests at the world's edge, but in unrecognized historical cultures of a premodern past.

The anthropologist's failed encounter with the unknown locates "alternative realities" in history, defining them as both the founding modern experience and a retrospective fantasy to an earlier time. Even as he laments the passing of indigenous cultures, Lévi-Strauss celebrates his inheritance from the first European explorers, taking as much pleasure in his kinship with Montaigne as in his difference from the savages. More important, he argues that, in its encounter with the New World, the Old World became conscious of its contingency, as a possible but not necessary world, and further that this contingency of worlds gave anthropology its object.

Traditionally defined as that which is but could be otherwise (or, in modal logic, as that which is both not necessary and not impossible), "contingency" is at once settling and unsettling.[6] Europe's unexpected encounter with America made surprises more expectable. Yet it also occasioned a yearning for lost certainty that, in time, fueled the very impulse to explore. Something of this benign, reflective exoticism is present in Lévi-Strauss's encounter with the too-savage savages. Unlike the second tribe that he eventually studies, but that he must drive back to its village in order to do so, the supposedly still-indigenous Mundé are what he really wants to discover, even though their indecipherability leaves him "with a feeling of emptiness." For according to a central Western myth, savages are defined as such by their hermeticism, by their possessing not just a different view of the world, but no proper "view" at all: a reality that so embraces them that it does not admit of, or even give rise to the thought of, alternatives. The New World native, still unaware that it is but one world that he inhabits, becomes indeed a reflection of the European explorer, but of him before he discerns, in his encounter with the native, his own contingency.

This state of dwelling in a "world" without knowing it became a modern ideal. It was summed up in an untranslatable aphorism by Ludwig Feuerbach, composed a few years before the philosopher's death in 1872: "*In der Unwissenheit ist der Mensch bei sich zu Hause, in seiner Heimat; in der Wissenschaft in der Fremde*"[7] (meaning, roughly, "in unknowing man is at home with himself, in his native place; in knowledge, he is in exile"). The sentence surprises by breaking the connection between knowledge and certainty. Certainty, one thinks, depends on knowing, and it is the task of science (*Wissenschaft*) as defined by Rationalist thought since Descartes, to increase certitude, illuminate obscurity, and thus to domesticate the world. In stating, instead, that unknowing fosters belonging, Feuerbach, condensing the Romantic critique of Enlightenment, argues not that science has failed to produce, but rather that it has overproduced knowledge, and of a form that increases uncertainty. Where, for Thales, the first scientist, the world was still "full of Gods" (Plato, *Laws* 10.899B), for the postscientific temperament, the world now is full of theories, of infinite, contingent representations of world. Feuerbach yearns for what the late Hans Blumenberg once termed "the enclaves of unknowing after the triumph of Enlightenment."[8] In late-nineteenth-century Germany, these enclaves were discovered in the vanishing countryside close to home, or, more powerfully, in Ferdinand Tönnies's ideal of the closed, local, natural *Gemeinschaft* of the medieval town as set against the open, global, and constructed urban *Gesellschaft* of the modern world.[9]

In our own century, such imagined enclaves lie further afield, or are discerned, as in Lévi-Strauss, at the point of their extinction. Yet they survive in our thought in various vestigial phantasms of spatial belonging, in which, against the contingency and pluralism of the world, there is set the radically necessary and singular placement of the body. It appears crucially in Edmund Husserl's notion of *Lebenswelt*,[10] which influenced Maurice Merleau-Ponty's "science of pre-science": the utopia of an experience of world before science split "life" and "world." The idea also animates Pierre Bourdieu's term "habitus," defined, with reference to Poincaré, as "a system of axes linked unalterably to our bodies and carried about with us wherever we go."[11] In the writing of history today, it appears most often in idealizing descriptions of premodern spatiality and carnality, and in attempts at describing the medieval conception of the world as the representative alternative historical reality.[12]

Alternative realities, from this point of view, are those that do not know alternative realities. Life-worlds left with their prejudices intact, they are antithetical to our modern consciousness of contingency, which says that "truth is made rather than found,"[13] even as it is only through this consciousness that one recognizes an alternative reality. For as Blumenberg argued early in his career, to speak of realities in the plural makes sense neither in the antique philosophical idea of the reality of instantaneous evidence, nor in the medieval theological doctrine of reality as guaranteed by God. Within the latter view, there may be diabolical deceptions of all kinds, but these are not plural realities but the plurality of falsehood. Only when reality is conceived as the result of the realization of specific contexts—in Blumenberg's terms, when it is regarded as a certainty that constitutes itself only successively, as a never-final and absolute consistency, or as a consistency that refers always to a future in which elements can emerge that might explode the earlier consistency and reveal it to be unreal—only then can one speak of "their" reality, or of "that" society's reality, as being simultaneously real and unique.[14]

Since the late eighteenth century, the historicity of the idea that reality might be plural is most intensely argued with reference to the *Weltanschauung*. Kant coined the word in 1790 to explain why the "world," as a totality, cannot be the object of a "view," except from a transcendent perspective that, when intimated, occasions feelings of the sublime (*Critique of Judgment*).[15] Once launched, the term took on an independent life. At one level, worldview came to indicate the specificity with which each person, culture, or era experiences the world. At another level, it described a subjective relation to the world that was historically specific, and that emerged in Europe during the modern period in the wake of secularization. In this second, narrower definition, worldview implied a particular, self-consciousness that reality is known only through the specific way it is seen. Under such pressures as science's disclosure of plural worlds, the New World's evidence of unknown peoples, and the early modern religious wars' mutually exclusive truths, people—so the story goes—became conscious that their world, its consistency, truth, and purpose, was contingent on their having a specific viewpoint on it. Instead of lamenting this as a loss, the philosophers of *Weltanschauung*, from Christoph M. Wieland, Alexander von Humboldt, and Wilhelm Dilthey to Husserl, Karl Mannheim, and Karl Jaspers, celebrated viewpoint-awareness as a new center of spiritual meaning and as an antidote against the ever-expanding, decentered world being discovered by science. Worldview, in its constitutive acceptance of alternative realities, thus contrasted both to the lost wholeness of the medieval Christian conception of world and to science's dehumanized universe. Its appearance within European thought was believed to mark the hiatus of the modern era by distinguishing the eras "Middle Ages" and "Renaissance."

Because it paired world specifically with "view," because, that is, it articulated the intertwining of object and subject with reference to the faculty of sight, the term *Weltanschauung* had an illustrious career in art history. While normative aesthetics took art's task to be the imitation of reality, and therefore judged individual works against that single standard, the historical study of art, emerging as an academic subject in the nineteenth century, was founded on the belief that different cultures represent reality differently, and that apparently "unrealistic" styles are not to be judged as wrong but to be interpreted as realizations of different contexts. The elaboration of a value-neutral history of style, together with contemporary critical preferences for stylistic uniqueness as the mark of genius, drew attention to the fact that the world, when visualized in art, was contingent on the particularities of person, place, and time. *Weltanschauung*, therefore, was both a consequence of art-historical consciousness and a felicitous motto for the discipline. It announces that art as evidence of the way persons and peoples saw the world ought to be foundational to the understanding of history: pictures are worldviews.

Perhaps the most pivotal use of the term is Erwin Panofsky's in his 1927 essay "Perspective as Symbolic Form." Perspective, equated here with *Weltanschauung*, is both historical and ahistorical. On the one hand, contesting the view that linear perspective as developed in Renaissance art is a categorically truer way of presenting the world, Panofsky asks of his historical material "not whether it has perspective, but which perspective it has."[16] Every artwork has its own perspective corresponding to the particular worldview of the larger culture. On the other hand, linear perspective, in its method of making world contingent on viewpoint, corresponds to the modern *Weltanschauung* in the more narrow sense, as a historically specific consciousness of positionality—what Nietzsche famously termed "perspectivism."

Jan van Eyck's *Madonna of Chancellor Rolin* of around 1435 is amenable to these terms (Figure 1). In its construction of deep space, conveyed by the receding lines of the tiled floor and by the river landscape stretching to the horizon, it locates a sacred scene—the apparition of the Virgin—in a world as if coextensive with our own. Indeed the movement into the picture exerts such a force, and yields so many delights, that the exchange displayed across the picture, between Rolin and the Virgin and Christ, seems eclipsed. The artist employs landscape to mark that exchange: a distant bridge carries Christ's gesture of blessing over to Rolin's praying hands. What Jacob Burckhardt termed the Renaissance discovery of the individual and the world finds its emblem here. The necessary and constitutive connection between viewer and viewed opens a chasm between "medieval man" and his faith. And a newly rehabilitated curiosity about this world,[17] embodied, visually, in the turned figure in the middle ground shown beholding the landscape, replaces ascetic thought directed to an afterlife. Secularization, the process of an increasing worldliness, thus seems the historical condition of the worldview; and the artist Jan van Eyck, probably portrayed as the red-turbaned man standing beside the surrogate viewer, offers his created reality as alternative to God's.

Historians today distrust such apparent modernity. They push van Eyck's picture back into a remoter age, arguing that its mundane world is brimming with symbols, like the Master of Flémalle's famous background fire-screen, which functions visually as the Virgin's halo.[18] They claim that, in the medieval worldview, reality was constituted by signs pointing beyond themselves to God.[19] And they maintain that the hiddenness of these signs in van Eyck indicates not his secular vision but the invisibility of faith to our secular worldview. How then are we moderns expected to see that invisible border between us and the past? Are there pictures of the threshold to an alternative historical reality?

Daniel Boorstin's best-selling history, *The Discoverers*, reproduces a line drawing on its cover, described as an "early 16th century woodcut."[20] In this image, a kneeling

Figure 1.   *Jan van Eyck*, Madonna of Chancellor Nicolas Rolin, *ca. 1435. Louvre, Paris.*

pilgrim gingerly pokes his head, hand, and walking staff through a scrim of stars, to peer from one reality into a host of others. He leaves behind a local *Lebenswelt* of churches, forests, and fields, where the stars are fixed to their spheres, and sun and moon, outfitted with faces, betray an anthropomorphic, pre-Copernican cosmology. The wanderer's head has just breached the boundary of this reality, enabling him to wonder at an infinite succession of worlds arranged as circles placed crosswise to the outline of his sphere. One heavenly body looks mechanical, as if made by Descartes's watchmaker God. What better way to illustrate the historical passage of man, the discoverer, from the closed world of the Middle Ages to the open universe of modernity! While drawn in a quaint medieval style, the woodcut seems to foresee the future.

Yet this quaintness spells trouble. Certain areas of foliage look more William Morris than Dürer; certain hybrids, hard to imagine as sixteenth-century, like the machine-tooled cosmos beside the Mother Goose moon: these indicate a medievalizing print. Indeed, it is an illustration from a popular book on meteorology by Camille Flammarion, published in Paris in 1888.[21] Boorstin's publishers cropped and colored it but forgot to check the source. One might lament the demise of so perfect a picture of breached worldviews. If Boorstin's cover shows the picture one might want, its error raises the question: Can such a picture exist?

Martin Heidegger gave one answer in a lecture delivered in 1938 and published under the title "Die Zeit des Weltbildes."[22] According to Heidegger, the symptoms of the modern age are the hegemony of science, the aesthetization of art as object of experience, the definition of human activity as "culture," and the desacralization of the world. And all these are reduced to the process by which the modern subject constitutes itself as subject by becoming the viewer of a world laid out before it as in a picture. The world picture, in Heidegger's terms, is not a picture of the world but the world as picture. And the "time of the world picture" is the modern era. The argument reiterates the philosophy of *Weltanschauung*, even as its ideological tenor has become more crudely antimodernist. Heidegger laments both the loss of human grounding through perspectivism's abstraction and the functionalization of the world through technology. Nonetheless, his argument is useful, for it states in categorical terms that there can be no transition from medieval to modern world pictures. For according to Heidegger, people in the Middle Ages did not understand the world as a picture, because for them the world, as created by God, places the individual not before it, as its viewer, and therefore as possessor of *Weltanschauung*, but only somewhere within it, as a mere created thing that will be viewed and judged only by an omnivoyant, omnipresent God. I shall attempt to take up Heidegger's challenge by considering some images from around 1500 in which medieval and modern world pictures seem to overlap as in a half-legible palimpsest.

Jheronimus Anthoniszoon van Aken (d. 1516), who signed his works "Hieronymus Bosch," is an art-historical monster. Called in his century "the inventor of devils," the painter of freaks, chimeras, and things, in Lodovico Guicciardini's 1567 account, "fantastiques, & bizares,"[23] Bosch is himself the great unknown of the Northern tradition, the artist who did not, and still does not, seem to fit. Unforeseeable from what came before him, he remains largely un-understood. He is the still-savage major master of the European tradition. His first public defender, the Spaniard Don Felipe de Guevara, reports in 1560 that the people saw Bosch's pictures "as a monstrosity, as something outside the rules of what is taken to be natural."[24] The impossible, in Greek *adynaton*, is contingency's outer limit, and it was there that Bosch was felt to press. Guevara, writing for a courtly audience around 1560, admitted that the artist "painted strange things, but only because he set his theme in Hell, for which, as he wanted to represent devils, he devised compositions of unusual things." While *hoffähig* as portraitist of demons, Bosch occasioned monstrous interpretations. His most famous masterpiece, the triptych sometimes called the *Garden of Earthly Delights*, has been taken to represent, variously, the world before the Flood, life in Eden, the apotheosis of sin, a utopia of a never-fallen humanity, a satanic comedy, a satire on vanity by a Northern Savonarola, a bourgeois parody of courtly love, and a sermon on fantasy. One historian, Wilhelm Fraenger, took Bosch's alterity at face value and read the *Garden* as an actual altarpiece to a non-Christian god.[25] Erwin Panofsky, playing it safe, broke off his monumental account of early Netherlandish painting before discussing Bosch with the learned disclaimer, "This, too high for my wit, / I prefer to omit"[26]—a version, to be sure, of Lévi-Strauss's "Alas! too savage!"

Indeed beyond matters of local interpretation, there is a savagery in Bosch that affects us still today, as the assembled subaltern others of medieval society—the beggars, thieves, witches, and heretics; the quacks and magicians; the Jews, Moslems, and blacks—are all gathered as in some curiosity cabinet of cruelty, there to be vilified, tortured, and damned. In Bosch, Christian culture reveals its barbarism by self-righteously punishing all realities alternative to its own. Collector of stigmatized others, Bosch is himself the quintessential alternative reality, medieval narrow-mindedness on the rampage against competing worldviews. And indeed as soon as one goes to interpret him, his alterity challenges and seduces. To some scholars his art seems encrypted, and demands a specific key, which is often sought in codes he condemns, such as those of alchemy.[27] To others, he pictures the historical loss of any such key.[28] Against the medieval Christian symbolic code, it is argued, Bosch stages the movement to modern semantic uncertainty, in which what something is stands in an unstable, contingent relation to what it means. I shall try to circumvent questions of meaning by concentrating on issues of place and placement. I shall first locate and describe Bosch's pictures of world. Then I shall attempt to place his pictures in the world. Understood as world pictures, Bosch's paintings will help situate the history of knowledge (Feuerbach's *Wissenschaft*) within a history of the emergence of that strange object, art.

The Spanish cleric and erudite Fray Joseph de Sigüença, in his prose masterpiece *The History of the Order of St. Jerome* (1605), defended Bosch against those who term his paintings "non-sense" (*disparates*) and "call him unjustly a heretic."[29] *Disparates* is a hard word to gloss. Derived from the Latin *disparare* ("to separate") and related to the English "disparate," it came to denote, within Spanish art theory of the sixteenth and seventeenth centuries, all that is physically monstrous and deformed, intellectually absurd, aesthetically incongruous, or morally objectionable. Originally a term of disparagement, it soon became a descriptive category naming a specific, popular mode of art, literature, and drama that aimed at grotesque and playfully arbitrary forms: in poetry, for example, nonsense verse; in theater, the farce-intermezzo (*entremés*); and in painting, drolleries or capriccios in the Boschian manner.[30] For such writers as Lope de Vega, Manuel de Melo, and Francesco Quevado, Bosch's pictures defined what *disparates* meant. This makes it difficult, in turn, to understand Bosch through this term, except by noting that, applied to his art, it can both describe and disparage, naming either what Bosch's pictures depict, or what they themselves are. A Spanish satirist in 1600 could vilify his competitors by comparing their farces (or persons) to the *disparates* of Bosch, thereby deliberately confusing satire with satirist.

The ambivalence of "non-sense"—whether it describes Bosch's art or what it depicts—applies also to the more serious accusation dismissed by Sig'uença, that the artist was a heretic. We encounter the notion again in a venomous tract from 1635 attacking Bosch's most famous literary heir, Quevado. In the *Tribunal de la justa venganza*, Quevado appears in league with Bosch, the "ataista."[31] It is possible that seventeenth-century observers in Spain, like some historians in our time, regarded Bosch's various images of apostasy as themselves apostate images; their view might also have been strengthened during the Thirty Years' War, when Bosch's native Low Countries were aligned, as Protestant, against Catholic Spain.[32] Yet the charge of unbelief was at least as slippery in 1635 as it is today for Fraenger's revisionist account. Quevado himself had broached the question in his *El alguacil endemaniado* (1607). Bosch appears as a visitor to hell, who, when asked why he paints his demons so absurdly, answers: "Because I never believed the devils were real." Scholars of Bosch remain uncertain about the artist's faith: whether his monsters are devils or nonsense, and whether, therefore, his *disparates* travesty false religion or reveal religion itself to be a travesty.

Sig'uença's answer is religiously orthodox and seriously intended. Even his strangest pictures—which Sig'uença calls "macaronic," meaning a jumbling of high and low—express the verdict of the prophets on the vanity of the world: "The idea and the art of this manner are based on Isaiah 40:6, where the messenger of God says, 'All flesh is grass.'" Sig'uença understands Bosch within the original Christian idea of contingency. Borrowing from Latinized Aristotelian logic, Christian theologians of the Middle Ages coined the term *contingentia* to express the ontological constitution of the world as that which was created out of nothing, is sustained only through divine Will, and shall pass away. The world, by this definition, is not necessary; it could just as well not have been, or been otherwise, and it owes its existence to God's unconditional being.[33] As I shall suggest, Bosch pictures world in its constitution as that which could be otherwise, and so in order to teach his viewer a proper contempt for this world.

The so-called *Hay Wain* perfectly expresses Bosch's world view (Figure 2). Dating to

Figure 2.   Hieronymus Bosch, The Hay Wain, *open state, ca. 1500–1505,
panel painting. Prado, Madrid. Photograph courtesy Giraudon.*

about 1500, this signed triptych was described by sixteenth-century sources, and exists today in two copies, both of them inferior and in poor condition, in the Prado and the Escorial. Already Sigüenza names the hay at the picture's center as the "grass" referred to in Isaiah. All surrounding matter, from the vagabond on the triptych's outer wings to the interior's spectacles of Paradise, earth, and hell, would thus embellish the central figure of the vanity of the world. World appears here in multiple, overlapping models—what Michel de Certeau, with reference to Bosch, termed "spatial polyglotism."[34] It is present as the subject matter, the hay, which is both a biblical and a vernacular proverbial emblem of world as contingent. Hay fills the pouches of the folk depicted, or sticks to their fingers as their attachment to the world. Bundled on the wagon, it looks like a misshapen globe, or better, like the terrestrial lower half of the spherical "world," such as we see on the outer shutters of Bosch's *Garden of Earthly Delights* (Figure 3). Set against a landscape, the lovers on top of the hay would thus sit on what would be, according to this geographical model, the inhabitable surface of the earth. The hay thereby becomes an allegorical world within the world. Several early compositions after the *Hay Wain* make this valency more apparent. In tapestries in the Royal Palace in Madrid and in the Escorial, the whole scene of the original triptych's central panel is reproduced within a circle that, fitted at the upper right with a cross

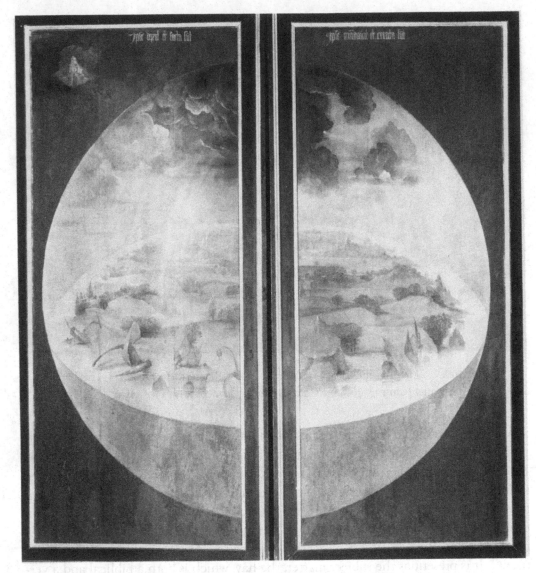

Figure 3.   Hieronymus Bosch, Garden of Earthly Delights,
closed state, ca. 1510, panel painting. Prado, Madrid.

and surrounded at the base by sea monsters and waves, reads like a giant orb or *Reich-sapfel* floating on the deep.[35] Bosch elsewhere superimposes the outlines of the world's orb over an ordinary scene, as in the panel sometimes called the *Stone Operation*.[36] In this panel, now in the Prado, the round format of the image itself, together with the curving border between middle and background that, located halfway up the roundel, could double as the equator of a transparent globe, extend the picture's message of folly to the world as a "whole," represented both as mundane landscape and as outlined globe. The picture thus becomes a macaronic world map.

In the *Hay Wain*, world is most of all present as the triptych's depiction of landscape. Bosch constructs the first genuine *Weltlandschaft*[37] in Western painting. The bird's-eye view unfolds sideways to Paradise and hell, and outward into space, toward infinity at the horizon. Narrow at the sides but expansive to the distance, Bosch imagines a world limited in time but infinite in space. Placing the picture's spectator simultaneously as a pawn in salvation history and as the privileged viewer of an endless universe, Bosch's world picture is both medieval and modern, closed and open, allegory and map.

Bosch offers us a beautiful world view only to anathematize that world as sin. The principle vice is avarice, defined as any positive relation to the world. All other sins—gluttony, anger, lust, etc.—crowd round as versions of love of the world.[38] Bosch depicts sin both by showing sinners and by telling sin's story: fall of the rebel angels and man, exile from Paradise, profusion of sin, and final punishment. World history processes as a false triumph from bad to worse. Bosch shows ephemerality by endowing it with a rigid, necessary structure. Sin might appear chaotic, as bodies grasping helter-skelter at the hay, yet the hay is resolutely at the picture's center. It founds a symmetry that endows the whole with the character of a cosmic diagram. Hell is the negative of Paradise, its black towers being a ruined version of Paradise's curious rocks. And these antitheses surround a composition whose center is maintained both by the hay wagon and by Christ, who, displaying his wounds, appears above a rainbow in the clouds.

Of course, Christ, the hay, and the viewer are only presently aligned. Bosch reminds us of the imminence of this skewing, this future structural dissolution, by suggesting the instantaneous "after" in details like the woman futilely erecting a ladder on the moving mass of hay, and the turbaned man with his already-toppled ladder about to be crushed by the wagon's wheels. In that very next moment, when the hay passes to the right, drawing with it the viewer's gaze, Christ will remain behind at the center now abandoned by the world. True, one might imagine that Christ, peering down from heaven, will keep pace with our movements, as do the sun and moon as we walk the earth, or that the cloud through which Christ peers will cling to the hay, as the perpetual promise of salvation. Such trust in permanence, however, is at odds with the picture's overall message of vanity. It represents that forgetfulness of time, death, and punishment that all actors exemplify and that stands condensed in the motif of the lovers in the *verweile-doch* of lust.

The picture's center is but a momentary alignment of Christ, the world, and the viewer. From any other vantage point in time and space, this relation will be skewed, indeed as it is for all the depicted figures in their rage for the world as center. While the panel's rigid alignment gives the whole the appearance of a necessary structure, it announces that this structure is contingent on the beholding subject. The picture asks the viewer to render a decision on the world here and now. And within the picture's logic, the here and now is the hay itself. It is that shapeless, blank, and mobile mass—equivalent to world—that constitutes the picture's center and principle object, and that appears venerated like a god.

The reference to idolatry—as a general fetishism of things—raises questions about the form and function of Bosch's triptych. We do not know the *Hay Wain*'s original context, whether it was intended as an altarpiece for a Christian altar or an artwork for a secular collection. We know it stood in the church of the royal palace and monastery at El Escorial in the eighteenth century, but it arrived there via secular art collections. Yet whether for a church or *Kunstkammer*, the *Hay Wain*'s triptych format, its symmetrical composition, and even its temporal framework, which places "before" to the left and "after" to the right, derives from altarpieces. More specifically, the scheme whereby side panels representing Paradise and hell flank a central scene of impending damnation recalls the format of Last Judgment altarpieces, which Bosch himself fashioned in numerous versions, including one very large ensemble commissioned by Philip the Fair in 1504 and now lost, but believed to be close to an extant triptych in the Academy of Fine Arts in Vienna.[39]

The retable altarpiece is the model for the *Hay Wain*'s geometry and for its assurance that contingency is framed in a necessary order. In church space this order would extend out from the altarpiece to the altar before it, and beyond that, to a world thought to be oriented around church and altar. For an altar is a sacred place, elevated above ordinary locations not only through the sacrament performed on it, but also through special rites of consecration, which entombed in the altar certain sacred things: martyr's relics, consecrated eucharistic hosts, incense kernels burned during the episcopal rite of the altar's consecration, and documents guaranteeing the authenticity of all these.[40] Endowed with *praesentia*, the altar oriented space around itself as around an absolute center. It directed gazes eastward toward Jerusalem as well as, invariably, toward the miracle performed at it, when the elements of bread and wine were, in Aquinas's term, "transubstantiated" into Christ's real flesh and blood through the agency of the priest. In the late Middle Ages, this miracle became, for the laity, above all a visual spectacle, in which the consecrated host was elevated and placed in special framing tabernacles for prolonged ostentation. The laity received the host in an ocular communion, a *manducatio per visum*, almost as efficacious as gustation proper.[41] Image ensembles erected behind the altar table reiterated in their centralized plan the structured attention of salvific seeing. They functioned variously to glorify, explain, or even bring intercession to the greater spectacle enacted before them, a spectacle that kept all eyes fixed on Christ present, again in scholastic terms, as the substance of the accident of the bread. Bosch himself visualized this mystery in a scene of the Mass of St.

Gregory that adorns the outer panels of his *Epiphany* altarpiece, now in the Prado in Madrid.[42] In the *Hay Wain*, in a gesture that has neither precedent or sequel, the center of this absolute geography is occupied by hay, by the emblem, indeed, of accident without substance.

Bosch's *Hay Wain* probably did not originally stand behind an altar. Perhaps it served as a devotional aid in a place of private worship, such as a privatorium. More likely, however, it was, from the start, a precious work of art within a princely or patrician collection. There it might have functioned to admonish against the enchantment of earthly treasures like itself. The curiosity served by the Renaissance *Kunst-und Wunderkammer* would be repositioned within the medieval catalogue of the vices. A secular context, moreover, would explain the hay's valence as idol. Replacing the cult object at the center of the Christian retable with an image of contingency, the *Hay Wain* would make a moral point about its very status as a worldly thing, and even about the historical passage from sacred to profane that it, as a hybrid art altarpiece, negotiated. Bosch carries over into the new, secular space of art the absolute geography of the sacred, even if only as a ghostly frame. In this space, beginning and end, good and evil, truth and falsehood have fixed and necessary places, structured locations that, in Bosch, are consubstantial with the painted panel itself in its material geometry.

Bosch's portrait of the world's sphere fits snugly in place on the outer panels of the *Garden of Earthly Delights* (Figure 3). The earth's geography conforms perfectly to the picture's geometry because earth was made by God, who appears in the upper left in the position of a divine geometer. The Psalmist's words inscribed at the panels' tops reminds us of this ontological dependency, this relation between a necessary agent and a created, and thus contingent, thing: "For He spoke, and it was; He commanded, and it stood." This providential geography recalls medieval world maps. In the Ebstorf *mappamundi*, dating from around 1235 and destroyed in 1943, geographic and geometric centers—the navel of the world and of the midpoint of the map—converge on Jerusalem and on Christ, shown resurrected from his grave.[43] Beginning in the fourth century and culminating in the crusaders' rallying cry "*ad sepulchrum Domini*," Jesus' empty tomb constituted the place of places around which the world organized itself as around an absolute center. In the Ebstorf example, the world is circumscribed by Christ's body: his head appears in the far east, at Paradise, while his feet and hands mark the points west, north, and south.

According to Horst Appuhn, the map originally served as an Easter Tapestry for the ground before the altar of the nun's choir of the Ebstorf cloister.[44] This further "orients" things, for the map itself would face east, with the altar. The *Hereford Map* of around 1290 similarly inscribes contingent space into the necessary space of God.[45] According to an eighteenth-century source, the map once stood at the central of a triptych backing the Hereford cathedral's high altar. Flanked by shutters depicting (at the left) the annunciating angel Gabriel and (to the right) the Virgin, the map situated God's historical and liturgical entrance in the world, as Christ's incarnation through Mary and as his presence in altar's rite. Read as a version of a *mappamundi*, Bosch's *Garden* superimposes on absolute geography a different space. The landscape of the newly created earth, shown as a disk floating on the waters of the deep, recedes into depth as if observed by a human eye in positional space.[46] Bosch brings together in a single picture two distinct models of world: one contingent on God, the other contingent on viewpoint.

It would be the contingency of perspective that, henceforth, defines the image for European painting until this century. In it became apparent what Immanuel Kant was to state near the opening of the *Critique of Pure Reason*: "It is, therefore, solely from a human standpoint that we can speak of space." Pieter Bruegel the Elder's worldscapes, made in dialogue with Bosch's, already bear eloquent witness to this troubled process. In the Vienna *Carrying of the Cross*, dated 1564, the single, framed, rectangular panel, made to be experienced aesthetically in the *Kunstkammer*, has severed its ties to church space (Figure 4). Bruegel positions the viewer before a vast prospect of the mundane world, and he dramatizes this vertiginous expansion by means of people rushing toward Golgotha in the distance. Christ, the picture's subject, is overlooked by all except the holy figures mourning in the foreground. Bruegel personifies humanity's indifference toward Christ in the figure of Simon of Cyrene at the lower left. According to the Gospel, soldiers compelled the Cyrenian to bear Christ's cross (Matt. 27:32; Mark 15:21); in Bruegel, Simon appears held back by his wife, who, wearing a rosary, stands for false, outer piety. While the multitude march forward with their backs to Christ, Simon, the one called to carry Christ's burden, draws back in the viewer's direction. Christ thus kneels between two immense indifferences, one near the picture's vanishing point, in that empty circle of gawking people on the horizon, the other near the viewpoint, where the beholder's faith is tested. Although tiny in the landscape, Christ appears at the exact center of the panel, and from there looks directly back at us. This vestige of absolute space, of an order located in places themselves (here the painted mark) rather than in positions from which they are observed, is a legacy of Bosch. It appears most momentously in one of Bosch's surviving retable altarpieces.

If measured by its influence, the *Temptation of St. Anthony*, now in Lisbon, and dated to around 1510–15, is Bosch's most important work (Figure 5). More than twenty copies of it exist, and it inspired a huge number of imitations until well into the seventeenth century.[47] Bosch himself made several versions of the theme. Sigüenca, who was close to the Spanish court that zealously collected Bosch, reports that "this painting is seen often; one is in the chapter house of the Order of St. Jerome; another in the cell of the prior; two in the gallery of the Infanta; some in my cell, which I often read

*Figure 4   Pieter Bruegel the Elder, Christ Carrying the Cross, 1564.
Kunsthistorisches Museum, Vienna. Photograph courtesy Bildarchiv Foto Marburg.*

and immerse myself in."[48] I would like to have seen Sigüença's cell, where Bosch's pictures proliferated like the demons they depict. The pious brother seems to have used them for his religious devotions, although by his time the vast majority of Boschian St. Anthony panels were in secular art collections. Originally, though, the Lisbon panels almost certainly functioned as an altarpiece. Contemporary documents inform us that in 1490 Bosch painted the "outer wings" of a retable in the chapel of the Illustre Lieve-Vrouwe Broederschap in the Cathedral of St. John in s'Hertogenbosch; and he seems also to have executed altarpiece wings for the cathedral's High Altar, as well as for an altar dedicated to St. Michael.[49] And we know that altars dedicated to St. Anthony had currency during this period: the retable for the hospital of the Order of St. Anthony in Isenheim, with its sculpted shrine from around 1490 by Nikolas Hagenower and later wings by a painter called Grünewald, is one famous example.

Yet as an altarpiece, Bosch's triptych is certainly unique. For one thing, winged retables ordinarily enclosed a cult image in their shrine, like Hagenower's enthroned St. Anthony, which claimed to make present the power of the saint himself. Or the central image narrated a significant event: a moment in *Heilsgeschichte* or a martyr's death. Bosch's triptych offers no proper cult image, and the specific stories from Anthony's

Figure 5.   Hieronymus Bosch, St. Anthony, open state, 1510–1515.
National Museum, Lisbon. Photograph courtesy Giraudon.

life, as told in such popular, late medieval hagiographies,[50] are exiled to the wings: Anthony's return to his cave, his temptation by the beautiful queen, etc. The central panel extracts the saint from the chain of necessary events and represents him in general attitude of devotion. Bosch's badly preserved *Hermit Saints Triptych,* now in the Doge's Palace in Venice, extends this strategy through all three panels.[51] Saints Anthony, Jerome, and Giles appear there not as objects of devotion but as subjects in devotion. Neither cultic presences nor actors within significant events, they offer, through their inward attitude, a model of subjective piety.

Siguença, always Bosch's best reader, wrote that whereas most artists "paint man as he looks from outside, this artist has the courage to paint him as he is inwardly."[52] This focus on inwardness, congenial to the Counter-Reformation spirituality of Siguença and of the Royal Monastery of San Lorenzo in El Escorial, was also in tune with lay piety in Bosch's time, influenced as it was by the *devotio moderna.*[53] But what does Bosch's inner man look like?

In the Lisbon triptych, we must work to find this inner man, for he is all but lost in the hellish spectacle all around. According to tradition, the temptation of St. Anthony was this kind of spectacle.[54] It was a chaos of phantasms conjured by the devil to tempt and terrify the pious man. At once inner and outer, these abject creatures not only assailed the person but were also *of* the person. In devotional literature through the seventeenth-century, they were calls for both a *contemptus mundi* and a

self-contempt, being at once demonic enticements and projective fantasies, personified sin and sinning person. Describing Bosch's *St. Anthony*, Siguença refers with awe
to the maker of these monstrosities:

> We see . . . the unbounded fantasies and monstrosities that the enemy devises in
> order to confuse his imperturbable soul and distract his fervent love: to this end
> he conjures up living beings, wild animals, chimeras, monsters, fire, death, roar-
> ing, threats, vipers, lions, dragons and fearful birds of all kinds, so that one asks in
> astonishment how it was possible for him to give shape to all his ideas.[55]

The "he" here is ambiguous, referring first to the devil, who conjures demons to cor-
rupt the hermit-saint, but then to Bosch, who pictures demons to edify the viewer.
This prefigures the uncertainty in the Bosch literature about artist's relation to his
work: whether Bosch vilifies apostasy or is himself apostate, whether his *Garden* is a
paradise or a hell, whether he believes or parodies belief. Bosch's pictorial style makes
such distinctions unclear.[56] Refusing to model things in their distinct materiality, blurring the boundaries between mineral, vegetable, animal, human, and spirit and all
into erratic plays of paint, Bosch puts his viewers in an uncertain—one wants to say
"contingent"—relation to everything they see.

In 1604, Carel van Mander wrote that Bosch distinguished himself by his swift,
energetic technique, executing his figures in one go.[57] In contrast to the meticulous
layering of translucent glazes so admired in other Netherlandish painters, in which the
artist's hand is wholly effaced, Bosch's pictures display the temporality of their making.
Their wild outlines, flickering highlights, and textured surfaces announce that they
were created as an act of will. And the many *pentimenti* left visible testify that what is,
in Bosch, could indeed have been otherwise. Bosch's spontaneous forms share features
with those aleatory treasures of the *Kunstkammer*, in which natural objects are worked
to seem other than they are: in the background of the left inner panel, Bosch turns a
mound of earth into a man's buttocks by a few stokes of the brush.

Devilry is an exercise in projective imagination. In the central panel's foreground,
Bosch harnesses a fish like a jousting horse. Spatially estranged, the armor also reads as
a ship's rigging, which, in turn, turns the fish into a decorative ship's prow, and so forth.
What results is the unique creation, the radical singularity that, having no category,
would be classed in the *Kunst- und Wunderkammer* as "error," there to be demonized as
evidence of sin, or celebrated as exemplar of fancy, or (as in Lorraine Daston's 1991
account) naturalized as fact, or indeed all simultaneously, in that ambivalence toward
the world's "curiosities" that Bosch presages for the early modern period.[58]

In Bosch, a palpable sense of contingency extends beyond his individual creatures
to the spatial structure of his scenes. Again contrasting to Netherlandish painting
before him, Bosch refuses to obey the rules of linear perspective. He builds eccentric

architectures that recede chaotically toward an undeterminable distance. Yet even as he refuses the systematic space of perspective, and even as he strews his figures like random blots on the picture surface,[59] he also creates, indeed for the first time in Western art, a coherent, infinite worldscape. And this worldscape, in turn, is subordinated to framing structures, to geometries and necessary placements that diagram an absolute—indeed a non-perspectival—point of view. Bosch's curious penchant for the roundel and the rota, and for eccentric formats that baffle any sense of the image as Albertian open window, work to place the world as it is experienced contingently from within into a fixed and necessary framework opposed from without.

The point about Bosch's St. Anthony is that, as Christian exemplar, Anthony is able to see through the illusions assembled around him, to penetrate beyond the world's accidents to the necessary substance itself. In the central panel of the Lisbon triptych, the saint kneels in prayer before a destroyed chapel. His right hand, doubled by the pointing hand of Christ, directs our gaze toward the cross on the altar. This crucifix, one presumes, both symbolizes Christ's presence in the Mass (again, as substance of the accidents of the bread) and represents an ordinary *corpus Christi* as was usual (and, after Trent, required) for altars. Itself most probably a working altarpiece, Bosch's triptych tells its viewers to look at Christ. In the ritual context of the altar, this means beholding Christ in the *elevatio*. Yet by doubling the scene before the altar, by making altar and altarpiece the subject of an altarpiece, Bosch places Christ in a hall of mirrors. This the viewer must traverse by way of the painting's great temptation.

St. Anthony's temptation consists of a host of parodies. Traditional subjects of religious art and drama appear as if in devilish caricatures: on the far right of the central panel, for example, the scaly tailed tree-woman mounted backward on a giant rat and bearing a swaddled infant, together with the poor, bearded man wearing a blue hat behind, suggest Mary, Christ, and Joseph in their flight into Egypt, while the surrounding three figures hint at images of the Magi.[60] Sacred service appears travestied by devils: just below and to the right of Anthony, three demons in the shape of clerics (a priest and two monks) appear to read prayers from a breviary. And the Eucharistic sacrament is mimicked in a Black Mass performed just behind Anthony's back.[61] An egg born aloft by a frog stands for the elevated Host.

Even the tiny crucifix in the chapel—that last vestige of visual truth and reference of both Anthony's and Christ's deictic gestures—has its own anti-image within the triptych. Just to the right of the crucifix and exactly aligned with it, the ruined column displays, as though in a fresco decoration or in polychromed low-relief, the Golden Calf in a scene of its adoration by idolatrous Israelites. Bosch includes this *mise en abyme*, this painting within a painting, as if it were itself a remnant of an idolatrous culture: just below the picture of the Adoration of the Golden Calf, another ostensible fresco, or relief, exhibits a monkey-demon (or monkey-demon statue) enthroned on a drum and approached by suppliants bearing gifts. And below this is another scene, almost certainly of two Israelites with grapes from the valley of Eschol, and suggestive, perhaps, of a worldly abundance that diverts man from God.[62] Within the triptych's larger picture, then, the Christian chapel would seem to occupy the ruins of an ancient pagan (or, more likely, Jewish) temple, even as it is now threatened by a *reoccupation* by modern demons and idolaters—indeed specifically by Islam, hinted at in the crescent moon on a flag in the left inner shutter.

Bosch's painting of a painting of the Golden Calf, placed beside an altarpiece in Bosch's altarpiece, asks tough questions about the role of images in Christian devotion. The Calf, and with it all the other temptations, enclose the saint like a ruined, eternal envelope, or like the shattered crystal orb of the world. And at the core of Bosch's picture, as the geometric center of his painted panel, the saint looks directly out at us. His eye literally places the contingency of the world into a necessary framework.

Bosch was a master of pictures that see us. His early panel of the Seven Deadly Sins monumentalizes this outward gaze (Figure 6). From Sigüença's account, we know that the panel once hung in the Escorial, in the bedroom of Philip II,[63] the inner windows of which opened, like the fenestration of a private chapel, to San Lorenzo's great domed church. Bosch's roundel takes the form of a giant eye that warns, in the inscription around the pupil, "Beware, beware, God sees." At the pupil's center, as either the image in the eye, or a reflected image of that which the eye sees, stands Christ resurrected from his grave and displaying his wounds. The image recalls the Holy Sepulchre at the world's navel, Jerusalem, in the Ebstorf *mappamundi*, here translated into a veristic image that can capture Christ's reflection as it is cast on the shiny stone of his tomb. Moreover, by turning his painting into an eye, Bosch reverses our usual orientation to images as active viewers to objects passively seen. He makes his work return our glance, indeed hold us in its gaze as we are revealed in our various sins. Read within the figure of the eye, the seven little scenes—representing the sins of anger, vanity, lust, lethargy, gluttony, avarice, and envy—appear as reflections on the eye's white. These scenes, sometimes cited as the first genre paintings of the Netherlandish tradition, together constitute a worldscape of a kind, one wrapped around itself, like the world's orb turned inside out and upside down. The picture, it is implied, visualizes sin as the world-upside-down, here as contingent images on the periphery of God's all-seeing eye. His is a world picture where the *Weltanschauung* is God's.

In the Lisbon *St. Anthony*, Bosch reduces this all-seeing gaze to one spot of paint at the picture's center, yet with it he announces the continued necessity of the center. Centers, as the Golden Calf attests, can be dangerous things, tempting the eye to an interest in the things of the world. Bosch justifies his picture by establishing at its midpoint not an object but a subject, not a thing or curiosity seen but a seer who views us as a curiosity: the inner person with eyes fixed on necessary things. Anthony's outward glaze, which, like the giant eye in the Prado *Seven Deadly Sins*, interpellates and judges

Figure 6.   Hieronymus Bosch, Seven Deadly Sins and Four Last Things,
ca. 1495. Prado, Madrid. Photograph courtesy Giraudon.

us, may stand surrounded by images of our temptations and our misdeeds. Yet painting is not only that wondrous distraction but retains at its geometric center the truth of a holy face.

As fate would have it, the center did not hold. Bosch's imitators ignored the underlying centers, symmetries, and diagrams that locate contingency within a necessary order. Boschian space becomes a surface strewn with clever inventions: demons, arabesques, and saints, all delectable in their variety.[64] Where Bosch labeled the world's contingency as a temptation around a centered inner self, his future followers and fans took the bait and collected Bosch himself as a "curiosity." Their savagery forgotten, his paintings were installed in the space of the art collection. There they would have hung like catalogues of the very exotica that surrounded them: the jokes of nature, the images made by chance, the ethnographic souvenirs, the moralizing prints and pagan gemstones, and the ingenious instruments of art, knowledge, and hygiene. From the diabolical, indecipherable, savage unknown was born the quintessential alternative reality: the modern work of art.

In Bosch, demons remain demons, however obscure their message might be. Only in his reception do they become playacted savages and carnival props. Consider the savages of Bosch's great modernizer, Pieter Bruegel. In his one extant woodcut, dated 1566, Bruegel shows a king and a wildman on a village street (Figure 7). The ruler, it seems, encounters ruleless natural man. Yet the longer we look, the more artificial this difference appears. The wildman's body seems covered by fur, yet the regularized tufts, as well as the gap between these and the wildman's hands and feet, suggest a fur garment. And the wild eyes that peer forth from a shock of hair become, on inspection, eyes of a mask. The king, too, is a masquerade. He is a peasant whose crude artifice Bruegel marks by shading the line between face and beard, and by balancing the crown like a pot on top of a fur cap. Once recognized for what it is—mere rustic entertainment—everything falls in place. The woman to the right is faceless because she too wears a mask; and the crowd in the window locates the play in the street, before a village tavern or brothel. Indeed the scene shows an episode from the popular Flemish

Figure 7.   Pieter Bruegel the Elder, The Masquerade of Ourson and Valentin, *1566, Woodcut on paper.*

play "Ourson and Valentine," in which twins, divided at birth, meet again as knight and wildman.

Bruegel's woodcut exposes the peasants' play. What we took to be natural man was merely a local rustic in carnival clothes. And what therefore seemed like crudeness on Bruegel's part—the unadept treatment of fur, eyes, and crown—turns out to be peasant artifice. This placement of "wildman" in quotes would have been unthinkable in Bosch, who appropriated popular symbolism without ever marking it as popular, which is to say, as other than his own. Bruegel unmasks the wildman by exposing the seams of his outfit, suggesting that savagery is a myth, and that Bruegel's art itself only *seems* strange, foreign, and exotic.

It may be extravagant to discern in a printed line the burden of modernity. The visible gap, in Bruegel's woodcut, between face and mask, which levels wildman and king to rustic players, and declares their art, and indeed culture itself, to be contingent, might simply be a consequence of the graphic mark itself. It might be argued that woodcuts were incapable of achieving, through their heavy lines, the finish demanded for Bruegel's legendary "realism," hence the unique status of this print within the artist's *oeuvre*. Yet it is precisely realism, as the figure of a rejection of artifice,[65] that demands marks to place nature and natural language in quotations.

At 1572, Domenicus Lampsonius, Netherlandish painting's first panegyrist, called Bruegel "this new Jeroon Bos."[66] And Van Mander named Bruegel as the greatest of the sixteenth-century Boschiads—those generally nameless epigones who satisfied the public demand for aestheticized devilry, or *disparates*, during the half century between Bosch's death in 1516 and the Netherlands Iconoclasm of 1566. However, no artist makes Bosch seem more historically remote, and more different from ourselves, than does Bruegel. In Bruegel the devil becomes situated, as the specificity of an artifice or a symbolism that can be viewed with wonder from without, while at the same time evil—as the cruelties of war, punishment, and indifference—derives now relentlessly from the notion "man." The telltale lines in the *Masquerade* woodcut that locate wildness in the practices and beliefs particular to one culture, are unthinkable in Bosch perhaps because he belonged fully to the culture that Bruegel marks as past or primitive, because (I am tempted to say) Bosch still believed in the monsters he painted. The world is contingent in relation to a faith that is not. St. Anthony occupies the absolute center of the painting because the devils around him are not advocates of competing faiths but instigators of apostasy. What Bruegel's markings betray is the Copernican turn, occasioned by the European Reformation that intervened in the half century after Bosch's death, and by the great wars of religion that raged in his own country, that belief itself is contingent on person, time, and place.

Van Mander reports that Bruegel, together with one of his patrons, the merchant Hans Franckert, "went out of town among the peasants . . . to fun-fairs and weddings, dressed in peasants' costume, and they gave presents just like the others, pretending to be family or acquaintances of the bride or the bridegroom."[67] The woodcut wildman has the quality of anthropological field notes. At the same time as the savage becomes familiarized as peasant artifice, the peasant himself becomes unknown. He is not natural man, for he possesses art, and thus he appears to be already embarked on the passage to Bruegel's civility. Yet because his artifice is transparent, unlike Bruegel's, he becomes the native of an alternative reality, with its artifice existing side by side with Bruegel's. Staring out at us not as eyes but as mask, Bruegel's quotidian other bespeaks the modern conditions. World pictures are contingent, not found but made. Henceforth they will be plural.

Notes

1. This essay began as a plenary lecture for the conference "Alternative Realities: Medieval and Renaissance Inquires into the Nature of the World," held at Barnard College in December 1994. My thanks go to Antonella Ansani and her colleagues for their kind invitation to speak. My use of the term "contingency" derives from the workshop "Poetik und Hermeneutik," where I have twice been a grateful participant. Its 1994 meeting, organized by Gerhart von Graevenitz and Odo Marquard, was specifically devoted to "Kontingenz." I also wish to thank Yve-Alain Bois, Susan Buenger, Nick Cahill, Cay Cashman, Jeffrey Hamburger, Serafín Moralejo, and James Marrow for their advice and support. This essay is dedicated to Hans Blumenberg (1920–1996).
2. Lévi-Strauss, *Tristes Tropiques*, trans. John and Doreen Wrightman (New York, 1973), pp. 325–26.
3. Lévi-Strauss, p. 335.
4. Ibid., p. 333.
5. Ibid., p. 344.
6. Hans Blumenberg, "Kontingenz," in *Religion in Geschichte und Gegenwart*, 3rd ed., ed. Kurt Galling (Tübingen, 1959), vol. 3, 1793–1794; Hans Poser, "Kontingenz I. Philosophisch," *Theologische Realenzyklopädie*, ed. Gerhard Müller (Berlin, 1977), pp. 544–58; Erhard Scheibe, "Die Zunahme des Kontingenten in der Wissenschaft," *Neue Hefte für Philosophie 24–25* (1985): 5.
7. Ludwig Feuerbach, *Sämtliche Werke*, ed. Friedrich Jodl (Stuttgart, 1960), p. 310; cited in Hans Blumenberg, *Lebenszeit und Weltzeit* (Frankfurt, 1986), p. 54.
8. Blumenberg, *Lebenszeit*, p. 55.
9. *Gemeinschaft und Gesellschaft* (Leipzig, 1887).
10. Husserl, "Kant und die Idee der Transzendentalphilosophie" (1924), ed. Rudolf Boehm, *Husserliana* (Hague, 1924), vol. 7, p. 232; see Blumenberg, *Lebenszeit*, pp. 10–68.
11. Pierre Bourdieu, *Logic of Practice*, trans. Richard Nice (Stanford, 1990).
12. Most powerfully Henri Lefebvre, *The Production of Space*, trans. Donald Nicholson-Smith (Oxford, 1991).
13. Richard Rorty, *Contingency, Irony, and Solidarity* (Cambridge, 1989), p. 3.
14. Blumenberg, "Wirklichkeitsbegriff und Möglichkeit des Romans," *Nachahmung und Illusion*, ed. H. R. Jauß, Poetik und Hermeneutik, 1 (Munich, 1964), pp. 12–13.
15. Kant, *Critique of Judgment*, trans. Werner S. Pluhar (Indianapolis, 1987), p. 111.
16. Erwin Panofsky, *Perspective as Symbolic Form*, trans. Christopher Wood (New York, 1991), p. 41.
17. On this process, see Hans Blumenberg, *The Legitimacy of the Modern Age*, trans. Robert M. Wallace (Cambridge, 1983), part 3.
18. Erwin Panofsky, *Early Netherlandish Painting* (Cambridge, 1958), vol. 1, pp. 163–64.

19. The classic formulation of this is Johan Huizinga's 1919 *The Autumn of the Middle Ages* (trans. Rodney J. Paynton and Ulrich Mammitzsch [Chicago, 1996]).

20. Daniel Boorstin, *The Discoverers* (New York: Vintage Books and Random House, 1985).

21. Camille Flammarion, *L'Atmosphere: Météorologie populaire* (Paris, 1888); the attribution is Fritz Krafft's in "Die Stellung des Menschen im Universum," *Zur Entwicklung der Geographie*, ed. Manfred Büttner (Paderborn, 1982), pp. 147–81.

22. Martin Heidegger, "Die Zeit des Weltbildes" (1938), *Holzwege* (Frankfurt, 1950), pp. 73–110.

23. Lodovico Guicciardini, *Description de tous les Païs Bas* (Antwerp, 1567), p. 132.

24. Felipe de Guevara, *Comentarios de la Pintura*, ed. Antonio Ponz (Madrid, 1788), p. 44; excerpted and translated in Charles de Tolnay, *Hieronymus Bosch* (New York, 1965), p. 401.

25. Wilhelm Fraenger, *Hieronymus Bosch. Das tausendjährige Reich* (Coburg, 1947).

26. Erwin Panofsky, *Early Netherlandish Painting*, vol. 1, p. 358; quoting Adelphus Müelich, German translation of Ficino's *De vita triplica* (*Medicinarius* [Strasbourg, 1505], fol. 174v).

27. For example, Dirk Bax's aptly titled *Ontcijfering van Jeroen Bosch* (The Hague, 1949).

28. Albert Cook, *Changing the Signs: The Fifteenth-Century Breakthrough* (Lincoln, Nebr., 1985), pp. 81–120.

29. *Tercera parte de la Historia de la Orden de S. Geronimo* (Madrid, 1605), p. 837; the whole passage on Bosch (in the original, pp. 837–41) is translated in De Tolnay, *Bosch*, pp. 401–04.

30. Maxime Chevalair and Robert Jammes, "Supplément aux 'Coplas de disparates'," *Mélanges offert à Marcel Bataillon* (Bordeau, 1962), pp. 358–71.

31. See Helmut Heidenreich, "Hieronymus Bosch in some Literary Contexts," *Journal of the Warburg and Courtauld Institutes* 33 (1970): 171–99.

32. X. de Salas, *El Bosco en la literatura espanola* (Barcelona, 1946).

33. See references in note 6; also Franz Josef Wetz, "Kontingenz der Welt," *Kontingenz*, ed. Gerhart von Graevenitz and Odo Marquard, Poetik und Hermeneutik, 17 (forthcoming).

34. Michel de Certeau, *The Mystic Fable*, trans. Michael B. Smith (Chicago, 1992), p. 66.

35. De Tolnay, *Bosch*, app. pl. 88.

36. Ibid., cat. 1.

37. The term appears first in Eberhard Freiherr von Bodenhausen, *Gerard David und seine Schule* (Munich, 1905), p. 209.

38. Lotte Brand Philip, "The 'Peddler' by Hieronymus Bosch: A Study in Detection," *Nederlands Kunsthistorisch Jaarboek* 9 (1958): 1–81.

39. Hans Belting, *Die Erfindung des Gemäldes* (Munich, 1994), p. 123.

40. Joseph Braun, *Das christliche Altar* (Munich 1924), pp. 525–56.

41. For a recent account, with an updated bibliography, see Miri Rubin, *Corpus Christi: The Eucharist in Late Medieval Culture*, p. 65, passim.

42. De Tolnay, *Bosch*, cat. 31.

43. See, most recently, *Ein Weltbild vor Columbus. Die Ebstorfer Weltkarte*, ed. Hartmut Kugler (Weinheim, 1991).

44. "Datierung und Gebrauch der Ebstorfer Weltkarte und ihre Beziehungen zu den Nachbarklöstern Lüne und Wienhausen," in Kugler, *Weltbild*, pp. 245–59.

45. The relevant sources are given in Marcia Kupfer, "Medieval World Maps: Embedded Images, Interpretive Frames," *Word and Image* 10 (1994): 273–76. Kupfer needlessly rejects the view that the map was originally part of the triptych.

46. This feature has been observed by Klaus Clausberg, "Scheibe, Rad, Zifferblatt," in *Weltbild*, p. 280.

47. Gert Unverfehrt, *Hieronymus Bosch. Die Rezeption seiner Kunst im frühen 16. Jahrhundert* (Berlin, 1980), pp. 151–86.

48. De Tolnay, *Bosch*, p. 403.

49. Bax, *Hieronymus Bosch*, p. 3.

50. Bosch's chief sources are translations of Athanasius' Greek *Vitae Patrum* (the Latin is given in the *Patrologia Latina* 73: 126ff.); on Bosch's vernacular sources, see Bax, *Hieronymus Bosch*, pp. 7–12.

51. De Tolnay, *Bosch*, cat. 24.

52. Ibid., p. 402.

53. On Bosch and the *devotio moderna*, see Paul Vandenbroeck, *Hieronymous Bosch. Tussen volksleven en stadscultuur* (Berchem, 1987), p. 120, passim.

54. Jean Michel Massing, "Sicut erat in diebus Antonii: The Devils Under the Bridge in the *Tribulations of St. Antony* by Hieronymus Bosch in Lisbon," in *Sight and Insight: Essays on Art and Culture in Honor of E. H. Gombrich at 85*, ed. John Onians (London, 1994), pp. 108–27.

55. De Tolnay, *Bosch*, p. 402.

56. Daniela Hammer-Tugendhat, *Hieronymous Bosch. Eine historische Interpretation seiner Gestaltung-sprinzipien* (Munich, 1981), pp. 55–61.

57. *The Lives of the Illustrious Netherlandish and German Painters*, ed. and trans. Hessel Miedema (Doorn-spijk, 1994), vol. 1, p. 125.

58. Lorraine Daston, "Marvelous Facts and Miraculous Evidence in Early Modern Europe," *Critical Inquiry* 18 (1991): 93–124. On the museological category of "error" as historically constitutive of the idea of "art," see Horst Bredekamp, *Antikensehnsucht und Maschinenglauben* (Berlin, 1993), p. 21.

59. Compare Hans Sedlmayr's comments on Bruegel in "Die 'Macchia' Bruegels," *Jahrbuch der kunsthis-torischen Sammlungen in Wien*, n.s. 8 (1934): 137–59.

60. Bax, *Hieronymus Bosch*, p. 113; Ludwig von Baldass, *Hieronymus Bosch* (Vienna, 1943), p. 245.

61. First noted in Enrico Castelli, *Il demoniaco nell' arte* (Milan, 1958), on travestied Eucharists in Bosch, see Jeffrey Hamburger, "Bosch's 'Conjurer': An Attack on Magic and Sacramental Heresy," *Simiolus* 14 (1984): 5–24.

62. Bax, *Hieronymus Bosch*, p. 117.

63. Recorded by Siguença (De Tolnay, *Bosch*, p. 403).

64. On Bosch imitators, see Unverfehrt, *Hieronymus Bosch*, pp. 122–235.

65. On Bruegel's realism as an anti-artifice, see David Freedberg, *The Prints of Pieter Bruegel the Elder* (Tokyo, 1989), pp. 53–65.

66. Lampsonius, *Les effigies des peintres célèbres des Pays-Bas*, ed. Jean Puraye (Liège, 1956), pp. 60–61.

67. *Lives*, vol. 1, 190.

Over time, the thingness of technology has diminished in importance in relation to the embeddedness of technology. In a crude history of technology, as technologies increase in complexity, the "artifactual" part, the tangible mechanical device, plays a smaller part in the totality than the surround of human behaviors, institutions, meanings, know-how, expertise, and so on. By the time you get to really modern technologies they are no longer embodied in a single artifact. This emerges in the installation. Most of the appliances are familiar in ordinary American middle-class households. We think of them as objects. And yet, all of them are linked by a wire to a very complex production-and-use grid, the electrical grid that binds the whole country together as part of a huge technological system. And so one of the things that the installation provokes is the contrast between the "thing," which is what you first think of when "technology" comes to mind, and the invisible or less visible large and encompassing system. It is the system that makes the usual talk about "the impact of technology on society" unrealistic, since the technology is constitutive of such a large part of the society.

LEO MARX

The reason I resist polishing and fetishizing the objects is that I think they've always been part of a system, rather than just reified entities. And this is not only because they plug in to the wall; they're also part of a way of life. I suppose that's why any abandoned technology seems so evocative: because it calls to mind a vanished system and way of life.

PERRY HOBERMAN

Objectivity/Subjectivity

PETER GALISON

Judgment against Objectivity[*]

Judgment against Objectivity[*]

INTRODUCTION: THE BIRTH AND DEATH
OF MECHANICAL OBJECTIVITY

Objectivity is a fighting word. It is lambasted, cherished, hunted, defended; it is realism on Monday, certainty on Wednesday, intersubjectivity on Friday, and truth on Sunday. Claims and counterclaims proliferate: the natural sciences are objective; the social sciences want to be; architecture was in the 1920s. Postmodernism corrodes it, and metaphysics may or may not have captured its essence. Amid the cacophony of these discussions, the term loses its sense, and becomes little more than a contested token in battles from the *Methodenstreit* to the Culture Wars. In the midst of such polemics, a reader can be forgiven for having no conception of what might be meant by claims that objectivity still resides in quantum measurement, democratic politics, and statistical certainty.

Erased in such contemporary debates about objectivity is its genealogy within the conduct of the natural sciences. "Objectivity" is historical. As it is used in the physical, medical, and biological sciences, objectivity is deeply, ineradicably, a nineteenth-century category, one bound up with the process of depicting objects. To get at this visual culture where it most directly intersects notions of objectivity, I will focus on methods for visually classifying the "working objects" of science, rather than, for example, graphical representations of higher theoretical structures. I want to put aside the polemical abstractions about which disciplines have or lack "objectivity." I want to ask, instead: How is objectivity *practiced* at the most rock-bottom level; how is objectivity employed and mobilized by those sorting out the "working objects" of science? In particular, what constituted an *objective pictorial rendition* of the natural world—how did objectivity function, at specific times, for scientists who aimed to represent fossils, clouds, stars, elementary particles, bones, and the electrical activity of the brain?

Nowhere does debate over the classification of such objects come into focus so strikingly as in the spectacular literary genre of the scientific atlas. There are atlases of anatomy, atlases of wounds, atlases of cells, atlases of clouds, atlases of elementary particles, atlases of heads, atlases of peoples, and atlases of stars—in fact, there are atlases of almost any collection of studied objects within science. Many of these collections are explicitly called atlases, others handbooks, guides, or catalogues. But binding them together is the aim of representing the basic species of a field of inquiry, usually addressed to practitioners with the aim of helping to codify existing data and to serve as the basis for further research.

The claim that pictorial objectivity as revealed through atlases is a nineteenth-century concept is *not* to say that there was no notion of getting a true picture of nature long before 1800; of course there was. But elsewhere, Lorraine Daston and I have used the history of scientific atlases and their cognate literary forms to argue that one earlier ideal, that of attaining pictorial "truth to nature," had little to do with objectivity, a notion used in something like its current sense by Coleridge.[1] Truth to nature was associated with a set of practices—practices involving massive artistic and scientific intervention by a natural philosopher whose genius vouchsafed the validity of the move to idealize and correct the unreliable appearances of the given. Individual items misled—this particular skull "erred," only the platonized skeleton would reveal the true form of nature. Individual plants, even individual species contained spurious elements, distorted by the oddities of their history and the circumstances of their observation. Look as we might among the objects of the world, they could only suggest the Goethean *Urpflanz* or the ideal skeleton "behind" the visible. This struggle to get at the hidden true picture was not considered at the time to be "objective"; the term emerged only in the nineteenth century and, when it did, in opposition to the artistry of Genius's intervention.

In the nineteenth century—or, more specifically, after about 1830—both the persona of the natural philosopher and the status of pictorial representations of nature shifted. Instead of a transcendental Genius improving or idealizing nature, the desired character of the natural philosopher inverted to one of self-abnegation. Instead of truth to nature, these scientists aspired to let nature "speak for itself" through a set of instrumentalities that minimized intervention, hamstrung interpretation, and blocked artistic license. More saint-like in self-denial than powerful in genial interpretation, the new scientist of the last two thirds of the nineteenth century set aside the pictorial revelation of metaphysical truth per se, and aimed, happily, at an essentially mechanical registration of natural objects as they came. But this paper is not so much about this displacement of the seventeenth- and eighteenth-century *metaphysical image* (held to be "true to nature") and its replacement by the nineteenth-century *mechanical image* (of "objectivity"). Rather, it concerns a second displacement that occurred as the mechanical image itself increasingly yielded to a third representational strategy predicated on judgment: what I will call the *interpreted image*, emerging in the twentieth century. In both transitions, the changing practices of image making were intertwined with shifts in the moral culture of the scientist-author.

In the case of the mechanical image of the nineteenth century in which "objectivity" first came into prominence, the proclaimed association of automatic practices and moral self-denial are rife. Two examples are illustrative. The first is from Percival Lowell, the American astronomer, as he struggled during the first years of the twentieth century to establish the reality of the "canals" of Mars:

> Each drawing was made as if I had never seen the planet before; only twice did I allow myself even to put in afterward the snow accidentally omitted at the time. About fifteen minutes only was allowed in every instance, so that each drawing does not pretend to represent all that could be seen on that night at the telescope. They were meant to get as nearly as possible impersonal intercomparable representations,—scientific data, not artistic delineations.[2]

After the fact, Lowell could see a great deal that he had not put in the pictures; "snow" at the polar ice caps, for example, was plainly absent from his quick sketches (see Figure 1a). But with pride he reported how he (all but twice) resisted the temptation to reinsert the missing matter and, by so suppressing his impulse to improve, guaranteed the objectivity of his representation. These were "scientific data, not artistic delineations," where artistic correction had previously been precisely the guarantor of Truth. Lowell, in essence, argued that while the artistic delineations might be more complete and accurate, succumbing to the siren call of art would doom the objectivity of the project.

With the collaboration of Carl Otto Lampland, Lowell began, not long after these sketches were made, to begin the photographic exploration of the canals. On May 11, 1905, three days after Mars had been in opposition, Lowell and Lampland were able to capture, on film, the fine lines of the planetary surface. "Thus," Lowell proclaimed, "did the canals at last speak for their own reality themselves." Speak they might, but in whispers: only one-quarter of an inch in diameter, Lowell's photographs of Mars were so blurred, gray, and small that, at the time, they could not even be reproduced.[3] Figure 1b shows the pictures as they appeared in his record book in their original blurry but unretouched form. Though one prominent British astronomer, A. C. D. Crommelin, declaimed that "these photographs did a great deal to strengthen my faith in the

330

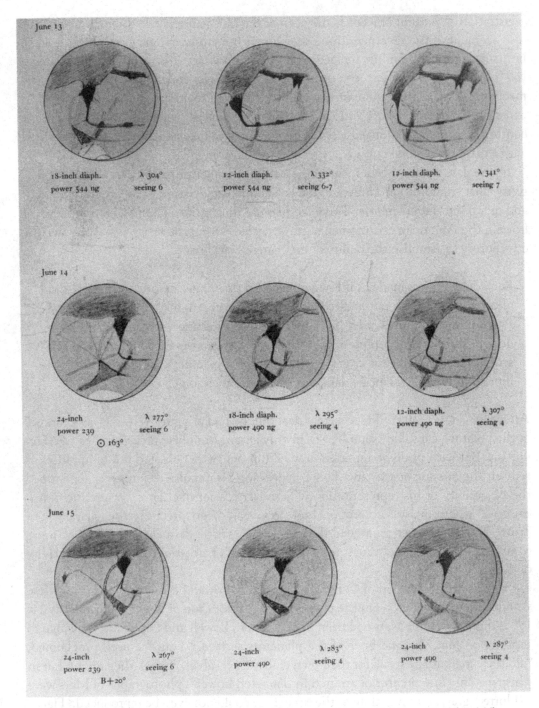

Figure 1a.   Percival Lowell, sketches of Mars showing canals, June 13–15, 1905
(reproduced directly from the record book).

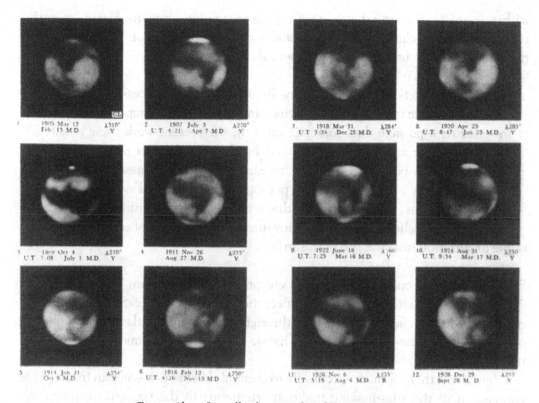

Figure 1b.   Lowell, photographs of Mars, 1905,
courtesy of the Lowell Observatory, Tucson, Arizona.

objective reality of the canals," others looked at the same pictures and were struck by their ambiguity. In response, a desperate Lowell almost succumbed to artistic temptation—he considered having a more neutral party (his friend and fellow Boston scientist, George R. Agassiz) "retouch" the pictures so the canals would be visible in mass reproduction. But Lowell's editors immediately revolted: such retouching would be a "calamity . . . as it would certainly spoil the autographic value of the photographs themselves. There would always be somebody to say that the results were from the brain of the retoucher."[4] This was the classic charge against intervention; Lowell demurred, and in the end, accuracy, completeness, color, sharpness, and even reproducibility was sacrificed. Objectivity would come first.

As Lowell's testimony makes clear, the ideal of removing oneself from the picturing process functioned even in the absence of the photograph. But once photography was available, it served (as Lowell indicated) as a splendid means of breaking that dreaded circle of art, interpretation, and personal predilection. Both before and after Lowell, paeans to the superiority of the photographic over the artistic were commonplaces within scientific discourse. Let one such instance stand in for many. Here is an early-twentieth-century clinical atlas of sectional and topographical anatomy that preserves the nineteenth century culture of mechanical objectivity.

> For the reproduction of their sections Braune and his predecessors were compelled to resort to tracings, hand drawings and engraved lithographic plates, thus introducing a possible dual source of error. In the present work these possible sources of error are entirely eliminated by the introduction of photography throughout. The plates are, therefore, an exact and faithful representation of the original sections. . . . The reason why photography was selected as the medium of recording . . . was . . . due to the fact that it affords the most faithful reflex of the originals, and is altogether free from any intermediate sources of error or of possible idealization from the pen of the artist.[5]

Drawings were erroneous and artists were prone to idealization; photography was faithful, and the photographer would reflect the original in nature. Self-evident criteria of reproduction had altered: where the eighteenth-century atlas maker took it as obvious that idealization was precisely what was called for, by the mid-nineteenth century that very move became anathema.[6]

Objectivity, then, or more specifically *mechanical objectivity,* was not an inextricable component of the atlas-making tradition originating in the sixteenth century, but begins much later in the history of atlas making. Objectivity as it was used at the very center of scientific work had a birth date in the mid-nineteenth century. Moreover, the story of objectivity is a conjoint development, implicating both observational practices and the establishment of a very specific *moral culture* of the scientist. In the first instance, objectivity had nothing to do with truth, and nothing to do with the establishment of certainty. It had, by contrast, everything to do with a machine ideal: the machine as a neutral and transparent operator that would serve both as instrument of registration without intervention *and* as an ideal for the moral discipline of the scientists themselves. Objectivity was that which remained when the earlier values of the subjective, interpretive, and artistic were banished. If the makers of the objective image had a slogan, it might have been: where genius and art were, there self-restraint and procedure shall be.

As we have seen, proceduralism and moral self-abnegation persisted into the early twentieth century, but the pictorial objectivity of atlases soon took another turn. Suddenly one begins to see something only seen in the rarest of late-nineteenth-century atlases—an explicit and repeated call for judgment and interpretation. Within the first third of the twentieth century, both practitioner and practice altered as the self-abnegating scientist and the automatic registration began to yield to judgment, though not at once and not everywhere. This essay explores that turn, beginning with an inquiry into the death of the *mechanical image* and concluding with a sketch of the *interpreted image*. Instead of our imagined nineteenth-century slogan, the twentieth-century atlas writers might say: at the end of procedure begins judgment.

Elements of older strategies for the depiction of nature persist long after new forms emerge. Even after the great efflorescence of atlases espousing nineteenth-century mechanical objectivity, for example, one saw instances of the older, eighteenth-century "truth to nature" that could only be unveiled by genius. Similarly, the death of mechanical objectivity was not sudden. Some atlas writers embraced a vision of mechanical objectivity deep into the twentieth century. This is important: the argument here is *not* that mechanical objectivity suddenly vanishes during the first third of the twentieth century. Rather, it is that during the early twentieth century, the moralized virtue of self-eliminating pictorial practices begins to yield to the moralized virtues associated with active judgment.

To see concretely an instance of the survival of mechanical objectivity, consider, for example, the following excerpt from Henry Alsop Riley's 1960 atlas of the basal ganglia, brain stem, and spinal cord, which perfectly illustrates the goal of mechanical, automatic reproduction safe from interpretation:

> This process [of hand-based illustration], however, makes the illustration a purely selective presentation and therefore the user of the atlas is often uncertain of the exact outline, relations and environs of the structures illustrated. The advantage of a photograph . . . seems to be self-evident. The photograph is the actual section. There is no artist's interpretation in the reproduction of the structures.[7]

For Riley, over-selection was the villain. Allowing the author or artist interpretive autonomy would throw into doubt the reliability of the object depicted. (If, Riley claims, artistic interpretation were to be allowed, then the depiction would become unreliable in its outline, relations, and environs.) Riley contended that hardly anything needed be said to defend the superiority of photographs. So tightly did the photographic image bind itself to the object that he could conclude: "the photograph *is* the actual section." Automaticity welded the image to the object until they stood as one; resemblance became identity.

Still, by this late date in the mid-twentieth century, such an unblinking faith in the photograph could not be sustained completely, and Riley readily conceded that staining was not completely targetable to a specific part of the specimen—his photographs revealed the irregularity of even the best and most technically skilled staining. Alas, even occasional scoring (from cutting) of the samples could be detected. Nonetheless, Riley judged that this photographic procedure ensured that "the accuracy and reliability of the photographs makes up for at times an inartistic appearance,"[8] where being inartistic was a right-handed criticism (rather than a left-handed compliment). Like

Riley, the authors of a 1975 *Hand Atlas* dismissed the artistic in favor of mechanically objective reproduction: "the authors have provided more realistic illustrations by substituting the surgeon's camera for the artist's brush."[9] On the mechanical-objective view, realism, accuracy, and reliability all were identified with the photographic. Nature reproduces itself in the procedurally produced image; objectivity is the automatic, sequenced production of homomorphic images from the object of inquiry to the atlas plate. Photography counted *among* these technologies of homomorphy; its importance was in underwriting the identity of depiction and depicted.

But if mechanical objectivity survived into the twentieth century, it did come to be supplanted across a myriad of scientific fields. My interest is *not* on extra-scientific attacks on objectivity, but rather on the *practices* used *within* laboratory and field inquiry to establish matters of pictorial fact about the basic objects of many scientific fields. For here in the atlases, handbooks, surveys, and guides we are in a central territory of science, far from the speculative frontier of elaborate new theories. In these compendia of pictures the simple (even simplistic) nineteenth-century model of pictures grounded in mechanical objectivity came under the fire of judgment.

JUDGMENT AGAINST OBJECTIVITY

Starting in the early twentieth century, atlas-making scientists began celebrating their use of judgment and interpretation in the production of systematic images of nature. No longer were scientists lionized for their self-abnegation, and their tools celebrated for the ability to present nature "in her own language." Gone too is the ferocious denial of any peculiarly human assessment of evidence. Suddenly, in field after field, atlas makers articulated a new stance toward representation, one that frankly set aside the hard-won objectivist ideals of absolute self-restraint and automaticity. Listen, for example, to Frederic A. and Erna Gibbs, who launched their compendious *Atlas of Encephalography* (1941) with the proclamation that:

> This book has been written in the hope that it will help the reader to see at a glance what it has taken others many hours to find, that it will help to train his eye so that he can arrive at diagnoses *from subjective criteria*.[10]

Surely there are exceptions to every rule, but let us put it this way: in the hundred-odd-year history of late-nineteenth-century scientific atlases one finds scarce evidence of such an utterance, and few that espouse so explicitly the subjective as a necessary part of scientific depiction.

Could it be that the Gibbses simply did not understand the way "objective" and "subjective" had been deployed by the mechanical objectivists of the previous hundred years? Could they be "talking past" those who deplored the subjective? No, the Gibbses understood full well the pictorial practice of mechanical objectivity. And they emphatically rejected it, as is clear from the continuation of their explanation:

> Where complex patterns must be analyzed, such [subjective] criteria are exceedingly serviceable. For example, although it is possible to tell an Eskimo from an Indian by the mathematical relationship between certain body measurements, the trained eye can make a great variety of such measurements at a glance and one can often arrive at a better differentiation than can be obtained from any single quantitative index or even from a group of indices. It would be wrong, however, to disparage the use of indices and objective measurements; they are useful and should be employed wherever possible. But a "seeing eye" which comes from complete familiarity with the material is the most valuable instrument which an electroencephalographer can possess; no one can be truly competent until he has acquired it.[11]

In this context "indices" and "objective measurements" are closely connected. Fourier transformations, auto-correlations, and other attempts to parametrize the complex spikes and wave patterns of the electroencephalogram were positioned precisely as alternatives to the "subjective" criteria. The Gibbses' vaunted subjectivity is not, however, a return to the long-abandoned "truth to nature." Where in the mid-1800s mechanical objectivity was counterposed to the genial intervention in nature to platonize, perfect, average, or derive the *Urpflanz* behind the earthly plant, the procedure accompanying interpreted images was to be far different. Instead of Goethean genius, and in place of proceduralist, bureaucratic self-denial, now the scientist invoked *judgment* based on familiarity and experience. The Genius revealed the true image of nature; the trained expert offered apprentices the means (through the "trained" or "seeing" eye) to classify and manipulate.

Some twenty years later—in a 1950 preface—the Gibbses produced a new edition of their 1941 *Atlas*, expressing the same anti-objectivist sentiment in somewhat different language:

> Experimentation with wave counts . . . and with frequency analysis of the electroencephalogram . . . indicate that no objective index can equal the accuracy of subjective evaluation . . . if the electroencephalographer has learned to make those significant discriminations which distinguish between epileptic and non-epileptic persons. Accuracy should not be sacrificed to objectivity; except for special purposes analysis should be carried on as an intellectual rather than an electromechanical function.[12]

"*Accuracy should not be sacrificed to objectivity.*" In this astonishing statement—astonishing from the perspective of mechanical objectivity—we see the epistemic footprint of the new, mid-twentieth century's regime of the interpreted image. How different this is from the reverse formulation of mechanical objectivity: that objectivity should not be sacrificed to accuracy. One thinks here of Erwin Christeller's insistence in his *Atlas der Histotopographie gesunder und erkrankter Organe* (1927) that "[it] is obvious that drawings and schemata have, in many cases, many virtues over those of photograms. But as means of proof and objective documentation to ground argumentation [*Beweismittel und objektive Belege für Begrunde*] photographs are far superior."[13] In the search for such *objektive Belege*, advocates of mechanical objectivity, roughly from the 1830s to the 1920s, were willing to sacrifice the color, sharpness, and texture of scientific representations for a method that took the brush from the artist's hand and replaced it with instruments. In his time, Lowell's tiny, blurry, black-and-white photographs counted for more than artistic renderings, even if the latter would have been sharp, complete, reproducible, and in color. For advocates of judgment like Gibbs and Gibbs, it was equally obvious that the "autographic" automaticity of machines, however sophisticated, was no longer an acceptable substitute for the professional, practiced eye.

In their radical devotion to mechanical means and their protestation of innocence against the charge of intervention, one senses in nineteenth-century atlas writings a certain defensiveness, a nervousness before the charge that the phenomena were not actually out there, but instead were the mere projections of desires or theories. For Gibbs and Gibbs, that acute anxiety is not present; the idea that the phenomena might be a "mere projection" is simply absent. At one level, this transition from a strident objectivism to a confident culture of scientific judgment should not surprise us. We know from a wide variety of excellent studies that throughout Europe and the United States, the mid to late nineteenth century was precisely the period of maximal scientific institution building, the time when amateur societies coalesced into major state and privately financed fixtures.[14]

These last decades of the nineteenth century were, institutionally, years of transition, during which the persona of the scientist was itself shifting. On the outside, the weighty buildings of the new scientific buildings were hybrids, crossed between neoclassicism and nineteenth-century factory design. Inside the walls, and in the self-image of the investigators themselves, the interior world celebrated the values associated with precision, accuracy, and self-abnegation.[15] In this period of rapid institutional expansion and reformulation of the role and proper comportment of the scientist, it is perhaps not surprising that while these new investigators aim for the durable results of exactness, they were still defensive about their new status. (Should one already call them professionals, trained experts, or following Timothy Lenoir, *Bildungsburger*?) Even as the great brick and stone buildings arose across Berlin, London, Washington, and Paris, laboratory scientists embarked on a nearly fanatical effort to establish their bona fides (in which the epistemology of mechanical objectivity played a part). Only after the institutions themselves had completed the bulk of their construction and the category of the investigator had stabilized does one see the emergence of the more assured ethos that characterized trained experts (with their epistemology of learned judgment).

Reading on in the Gibbses' 1941 *Atlas*, one finds, too, a contrast that would have been unimaginable within the earlier atlas-writing tradition of mechanical objectivity: they oppose an "intellectual" approach to one that is (electro)mechanical. Such a clash again signals a changed vision of *who* the scientist is. No longer most admired for a saint-like (or bureaucratic) self-restraint or an ability to become part of a machinic order transmitting nature undistorted, the scientist now emerged as an intellectual. Neither the eighteenth-century Genius nor the nineteenth-century lay ascetic, the scientist of the twentieth century entered as expert, with a trained eye that could perceive patterns where the novice saw nothing. The "practiced eye" emerges, for example, in geology as well—in atlases, for example, such as Oelsner's 1961 mineralogical study that trained the budding geologist to sort microscopic ore samples. Reflectivity, Oelsner noted, depends crucially on the polishing of the surface, so "beginners using it can often make gross errors." Color too is susceptible to "remarkable misinterpretations" until the neophyte has acquired a "very experienced eye."[16]

Emphasizing the activity demanded of the picture user, the Gibbses went on to liken the development of skills needed to "read" an encephalogram to those required to read a new language bearing an unfamiliar alphabet and a different script. True, they acknowledge, encephalography is not simple to master, but with three months of practice, they promised 98 percent accuracy by an average person.[17] The expert (unlike the Genius) can be trained; and (unlike the machine) the expert is expected to learn—to read, to interpret, to draw salient, significant structures from the morass of uninteresting artifact and background. As another encephalographic atlas (from 1962) put it, "the encephalogram remains more of an empirical art than an exact science."[18] Strikingly, this advocate contrasts empirical art with exact science. Here, the "empirical art" does several things: it first identifies that portion of the wave train that is "regular"—unlike automatic methods that ploddingly must examine each fragment, the eye quickly assesses some portion of the signal as "regular" or "typical." Second, the unaided eye finds "patterns" (which the author inserts into quotation marks). In part, this frank admission of the craft nature of encephalogram reading ties the debate over objectivity to the practice of clinical medicine. But the supplanting of the automatic by the judgmental extends so far beyond the clinical—into the domains of geology, particle physics, and astronomy—that one cannot rely on the specific history of medicine to account for the new emphasis on an active reader.

Before leaving the call of these atlas writers for an "empirical art," return for a moment to the analogy used by the Gibbses in 1941 for their new judgment-based reading: the practice of making a distinction between an Eskimo and an Indian by an un-self-conscious process of totalistic recognition. In this ethnological simile one has a theme that emerges quite widely, not only through Gestalt psychological concern with holistic cognition, but through the wider (and not unrelated) preoccupation with matters of race in the 1930s and 1940s.[19] Judgment as an act of perception and cognition seems to be associated with a picture of reading that is both anti-algorithmic and antimechanistic. Judgment in some sense stands as opposed to a fragmented building-up, to a mechanistic assemblage, and to an automated, protocol-driven set of procedures. Judgment must be acquired laboriously, but it is a labor of a very different sort from that of the nineteenth-century mechanical objectivist. Interpreted images carried force not through the labor behind automation, self-registration, or absolute self-restraint, but through the expert training of the eye.

Consider an atlas located (literally) light years from the human brain, W. W. Morgan, Philip C. Keenan, and Edith Kellman's *An Atlas of Stellar Spectra* from 1943. (See Figures 2a and 2b.) Here the authors set out a classification of stars in the 8–12 magni-

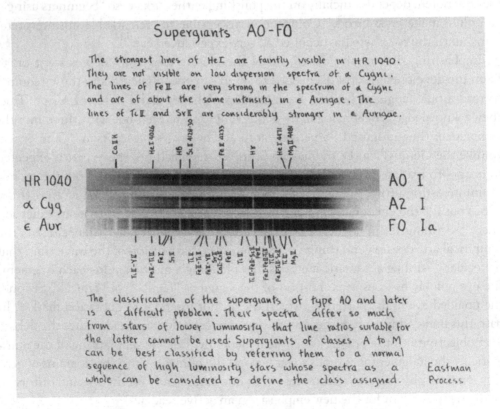

Figure 2a.   "Supergiants A0-F0," plate 20 from Morgan, Keenan, and Kellman,
An Atlas of Stellar Spectra *(Chicago: University of Chicago Press, 1943).*

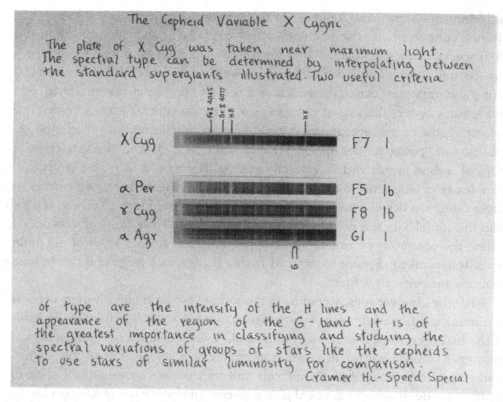

Figure 2b.   "The Cepheid Variable X Cygni," plate 39 from the same atlas.

tude range based on their spectra. The work was carried out with a one-prism spectrograph attached to a 40-inch refracting telescope. Plates were then sorted according to a two-dimensional system: on one axis stood the spectrum (based, for example, on the intensity of the hydrogen lines), yielding the *star type* (O, B, A, F, G, K, M, R, N, S), and on the other axis stood the luminosity (ranked by class I–V, progressing from the dimmest to the brightest). In practical terms, the astronomers first determined a rough type, an "eyeball" estimate of the category of a given spectrum—say B2, a variant of the B-type. Second, using parallax measurements to fix the distance to the star, they found the star luminosity. With the luminosity in hand they could then compare the candidate star spectrum with previously established spectra of similar luminosity. Matching the candidate spectrum against previously sorted spectra for B1, B2, and B3 then fixed the precise classification, which might well not be B2 after all, but rather B1 or B3 (the final classification rarely differed from the rough estimate by more than that).

The process of identifying a star as, say, a B2 class V star might seem purely routine, the kind of sorting that might just as well be effected by an automatic system. Not so, said Morgan, Keenan, and Kellman: "There appears to be, in a sense, a sort of indefiniteness connected with the determination of spectral type and luminosity from a simple inspection of a spectrogram. Nothing is measured; no quantitative value is put on any spectral feature. This indefiniteness is, however, only apparent."[20] Here is an interesting and important claim: *the qualitative is not, for being qualitative, indefinite*. Again and again, one sees this cluster of terms now in the ascendent: what is needed is the subjective, the trained eye, and an empirical art, an "intellectual" approach, the identification of "patterns," the apperception of links "at a glance," the extraction of a "typical" sub-sequence within a wider variation. Reflections like these point to the complexity of judgment, to the variously intertwined criteria that group entities into larger categories defying any simplistic algorithms. But for Morgan, Keenan, and Kellman the complexity and nonmechanical nature of this identificatory process does not vitiate the possibility of arriving at an appropriate and replicable set of discriminations. It may take judgment to sort a B1 from a B2, but such judgments can be unmechanical *and* perfectly definite.

What the observer does, according to the authors, is to combine a variety of considerations: the relative intensity of particular pairs of lines, the extension of the "wings" of the hydrogen lines, the intensity of a band, "even a characteristic irregularity of a number of blended features in a certain spectral region." None of these characteristics could be usefully quantified ("a difficult and unnecessary undertaking"). The root problem is one that has long vexed philosophers: "In essence the process of classification is in recognizing similarities in the spectrogram being classified to certain standard spectra."[21] Of what do these "similarities" consist?

Recognition cannot be grounded in the application of algorithmically fixed procedures; any such attempt would be cumbersome at best, and at worst, would ultimately fail. Our stellar spectroscopists continue with the by-now-familiar appeal to the physiognomic Gestalt:

> It is not necessary to make cephalic measures to identify a human face with certainty or to establish the race to which it belongs; a careful inspection integrates all features in a manner difficult to analyze by measures. The observer himself is not always conscious of all the bases for his conclusion. The operation of spectral classification is similar. The observer must use good judgment as to the definiteness with which the identification can be made from the features available; but good judgment is necessary in any case, whether the decision is made from the general appearance or from more objective measures.[22]

Note that, like the Gibbses, these star atlas authors contrast judgment with objectivity, where objectivity is used quite clearly in the sense of mechanical objectivity: fixed, specifiable criteria of evaluation. But, for both twentieth-century picture classifiers, "mere" objectivity was insufficient. Good judgment could be predicated on no such hard and fast rules of engagement.

Classifying (judging) by luminosity was by no means simple, and illustrates the complex way in which judgment had to be deployed. Certain lines or blends of lines might serve as a basis for calibrating stars relative to a standard in one spectral group; in another it might be useless—the lines might vary hardly at all. Dispersion in the spectrogram—the spreading of spectral lines on the plates—also varied for different spectral types. So long as one used plates of low spectrographic dispersion, hydrogen lines varied with absolute magnitude in stars of type B2 and B3. In high-dispersion plates that separated the "wings" (outlying portions of the broadened spectral line) from the central line, the wings were frequently no longer visible. And since it is these wings that vary with the absolute magnitude, when they are not visible the remaining line looks much the same whether the star it issues from be a dwarf or a giant. Conversely, there are lines visible in the high-dispersion plates that are invisible at lower dispersion. According to the stargazing spectroscopists: "These considerations show that it is impossible to give definite numerical values for line ratios to define luminosity classes. It is not possible even to adopt certain criteria as standard, since different criteria may have to be used with different dispersion." Variations like these made it impossible to specify a one-size-fits-all rule by which to classify: "the investigator must find the features which suit his own dispersion best."[23]

One has here a subtle and interesting confluence of phenomena: on the side of the spectra themselves there is variation that precludes naive rule following. On the side of the observer, there is a peculiarly human ability to seize patterns, and therefore to classify even when our algorithmic forms of reasoning fail. Subjectivity becomes an important feature of classification because the objects do not hold universal essential properties and because it is within our species' nature to be able to classify them univalently.

In sum, Morgan, Keenan, and Kellman draw our attention to four features of judgment: First, they emphasize that classification involves the establishment of similarity relations, and that these similarity relations (e.g., of luminosity) cannot be specified in terms of a fixed set of standard criteria (e.g., line-intensity ratios given for all spectral types). Second, the evaluative process of studying stellar spectra (like the evaluation of "race") is not necessarily a conscious one. With a glance, in a flash of recognition, one sees that a star is "racially" a B-class rather than an F-class entity. Third, the cognitive process at work in interpreted images is represented as holistic, and it is precisely this holism ("decision made from . . . general appearance") that stands in contrast to the "objective measures" of mechanical images (which were piecemeal as well as mechanical). Fourth and finally, nothing in the process of judgment is necessarily vague or indefinite—it is an error, they argued, to suppose that quantitative measures (even were they applicable) are the only way to a determinate classification. All four of these distinguishable features of judgment seem to be captured by the authors' racial-facial simile, and its contrast to quantitative and algorithmic assessment. (Though further discussion would take us too far afield, note that Wittgenstein, too, introduced his version of the racial-facial metaphor, "family resemblance," in the early 1930s precisely to capture a judgment-based, non-mechanical conceptual grouping.)

It might be thought that the atlases that foregrounded judgment differed in subject matter from earlier ones grounded in mechanical objectivity. Perhaps (it might be thought) it was just the twentieth-century material itself that in some way *demanded* judgment, where the subject matter of the nineteenth century required the objectivity of machines. This cannot be the case. There are nineteenth-century X-ray atlases that aspired to mechanical objectivity, and twentieth-century X-ray atlases that relied on judgment; there were anatomical atlases of mechanical objectivity and there were altogether comparable twentieth-century anatomical atlases predicated on judgment. Stellar spectra atlases provide a perfect instance of this continuity of topic and sharp break in the mode of categorical classification. For as we have seen, the Morgan, Keenan, and Kellman atlas argued for judgment over objectivity, root and branch. Strikingly, the atlas that Morgan et al. *explicitly* identified as a direct predecessor was that of the Henry Draper Catalogue of 1918, a volume that quintessentially espoused the image-making goals of mechanical objectivity. To make the contrast as sharp as possible, it is worth pausing for a moment to consider that predecessor volume.

The stunning Henry Draper Catalogue included the classification of some 242,093 spectra from 222,000 stars. It was an opus designed from the outset to last forever: the preface even assured the reader that "various authorities" expected the printing paper itself to be "practically permanent." Edward Pickering (director of the Harvard Observatory) began his preface: "In the development of any department of Astronomy, the first step is to accumulate the facts on which its progress will depend." Nowhere did he expound on judgment as necessary to classify the spectra or on the absence of universal criteria of selection, or on the role of preconscious cognition. On the contrary. Pickering's preface to the Henry Draper Catalogue celebrated the use of scientific management and mechanical objectivity. These were so "automatic" that they were held to be suitable for a replaceable set of hardworking (female) assistants of whom an "average" of five were at work at any given time over four years.[24]

The practice of employing women to do astronomical calculation and classification can be, and has been, read as a labor-historical chapter in workplace history.[25] There are, it seems to me, two further elements that bear on the epistemic status of facticity itself. First, in nineteenth-century mechanical objectivity the very possibility of employing "unskilled" workers served as a tacit guarantee that these data were not the figment of a scientist's imagination, or the results of a preexisting philosophical commitment. In this respect, the workers were identified with the machines, and like the machines in their "emptiness" they offered a transparency through which nature could speak.[26] Second, beyond their supposed "lack of skill" women workers were presumed to offer a "natural" predilection away from the grand speculative tradition. Occasionally, in the context of mechanical objectivity, this presumption conveyed the highest praise. Annie Cannon, who coauthored the great Henry Draper Catalogue with Edward Pickering, was hardly a "mere" computer—it was she who modified and rearranged the older star spectrum classification (A, B, C, etc.) into the long-lived Harvard system of spectral classification. It was also Annie Cannon who showed how these species could be rearranged to display the spectra in a continuous fashion. But it was precisely for her deliberate abstinence from theorizing that she was esteemed by her contemporaries, as is clear from the characterization of her written the year of her death in 1941: "Miss Cannon was not given to theorizing; it is probable that she never published a controversial word or a speculative thought. That was the strength of her scientific work—her classification was dispassionate and unbiased."[27]

Both the Henry Draper Catalogue of 1918 and Morgan et al.'s 1943 atlas on the same subject handled stellar spectra. But where the later authors saw the irreducible need for judgment, Pickering, Cannon, and their epistemically nineteenth-century staff had viewed their ideal atlas as planted on the firm ground of scientific management and mechanical objectivity. So despite Morgan et al.'s use of the Draper catalogue—despite their similarity of subject—the framing of the two projects was quite different. Here and elsewhere, in domain after domain, objectivity, facticity, and scientific management yielded to a new world of sorting nature in which judgment, subjectivity, artisanal practice, and theory were heralded as vital to the scientific project of visual classification.

Atlases of the mid- to late twentieth century, unlike those of the mid-nineteenth, began to be explicit about the need for subjectivity, as in the atlas of *Normal Roentgen Variants that may Simulate Disease* (1973): "The proof of the validity of the material presented is largely subjective, based on personal experience and on the published work of others. It consists largely of having seen the entity many times and of being secure in the knowledge that time has proved the innocence of the lesions."[28] Such a spectrum of the normal required exquisite judgment and extensive clinical training. It built on the famous 1939 treatise of Rudolf Grashey (*Typische Röntgenbilder vom normalen Menschen*), an early call for interpreted images, by means of which the author sought to empart to his readers a sense of the limits of the normal. To Grashey, photographs were *"Steckbriefe"* (wanted posters) that told the radiologist where the territory of the pathological began.[29] Again, one sees interpreted exemplary images analogized to the recognition of a suspect, "other" face.

In particle physics one finds the same kind of argument as that advocated by the X-ray master Grashey: the atlases are there to teach the range of what is known in order to highlight the unusual. In physics, however, the "pathological" becomes the

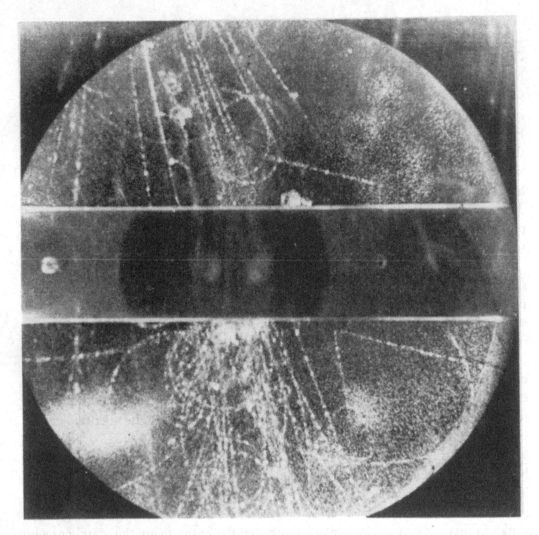

Figure 3.   "The First V-Particle," plate 103 from Rochester and Wilson,
Cloud Chamber Photographs *(New York: Academic Press, 1952).*

rare and unknown species of particles, and the "normal" becomes the known instances of particle production and decay. P. M. S. Blackett, one of the great cloud-chamber physicists of British physics, authored the foreword to George Rochester's 1952 *Cloud Chamber Photographs* (see Figure 3), in which he put it this way:

An important step in any investigation using [the visual techniques] is the interpretation of a photograph, often of a complex photograph, and this involves the ability to recognize quickly many different types of sub-atomic events. To acquire skill in interpretation, a preliminary study must be made of many examples of photographs of the different kinds of known events. Only when all known types of event can be recognized will the hitherto unknown be detected.[30]

Learning to recognize the novel was a matter of training the eye, whether to pick malignant lesions from normal variations, or to extract a kaon from a background of pions.

Whether one was dealing with pions, skulls, hands, lesions, stellar spectra, heartbeats, or brain waves, the problem was the same. Automatically sorted pictures, by themselves, were no longer enough. According to an increasing number of mid-twentieth century atlas makers, more than mechanical images were needed. Only *interpreted* images—interpreted through creative assessment, unconscious pattern recognition, guided experience, and holistic perception—could be made to signify. Only through individual, subjective, creative judgment could pictures transcend the silent obscurity of their raw form. Only the judging eye could pluck the pathological lesion or the previously unseen meson from the tangled pictorial world of "normal variations."

THE ART OF JUDGMENT

Bearing in mind the twentieth-century demand for judgment of images—from electroencephalograms to stellar spectra—one can now come back to our (by now) long-familiar relation of surgeon to medical artist. But where, in the 1800s, our surgeons swore that they policed every line, every dab of color for accuracy, or sought the photographic as an explicit means of avoiding the need for such surveillance, after the 1920s one begins to uncover a very different relation between scalpel and sketch. Here, in a 1968 *Atlas of Precautionary Measures in General Surgery*, Ivan Baronofsky reports, without apology, on the active measures taken by "his" illustrator, Daisy Stilwell, "one of the finest artists in the medical field." He adds: "This accomplishment might be sufficient were it not for the fact that Miss Stilwell is a superb interpreter. It would have been simple for her merely to act as a camera, but instead she brought out the features that justified the picture."[31] In the nineteenth century, being likened to a camera had been the highest praise. The artist's autonomy and interpretive moves were powerful threats to the representational endeavor, threats the camera alone could quell. For Baronofsky, being a "mere" camera carried only opprobrium. To be able to *interpret* was the key; judgment made it possible to sort the significant elements that "justified the picture" from the background. Mere camera-enabled naturalism was too blunt to reveal what the atlas makers and readers wanted to see.[32]

Baronofsky was not alone. John Madden's 1958 *Atlas of Technics in Surgery* did not hesitate to underline just how far representation stood from the surgical theater: "In illustrations, the incisions never bleed and the clamps and ligatures on the cystic and superior thyroid arteries never unlock or slip off. Furthermore, postoperative complications do not occur and there are no fatalities." Bloody incisions and slipping ligatures were the human side of the operating room, hospital-floor pragmatic realism barred from a representational realism founded on judgment:

> In the preparation of the Atlas the importance of having the medical artist present at each operation was stressed. It is only in this way that one may obtain in the illustrations anatomic realism and the creative interpretation of the artist. Only those operations that were witnessed by the medical artist are depicted.[33]

In pursuit of this "anatomic realism," the artist would sometimes witness three or four surgical procedures, with the goal of obtaining a logical visual exposition with no "jumps." To obtain that realism, Madden (like Baronofsky) was perfectly willing to eschew the mechanical objectivity of the camera, and was more than willing—enthusiastic, even—about the adoption of the "medical artist" whose "creative interpretation" offered an accuracy, *a realism*, that more automatic procedures could not match.

No policing of the artist, it seemed, was desirable in these various twentieth-century atlases. (How different Madden and Baronofsky are from Johannes Sabotta, whose famous turn-of-the-century work, *Atlas and Textbook of Human Anatomy*, denounced woodcuts as not "true to life" precisely because they left "entirely too much to the discretion of the wood engraver"—a discretion that photomechanical reproduction would stop cold.)[34] As Wittgenstein, Madden and Baronofsky insisted, it was just the artist's ability to extract the salient that rendered a depiction useful.

The identification of the salient by the self-confident anatomist, surgeon, or scientific illustrator is far from the metaphysical "truth to nature" image extracted by Genius. Goethe, Cruveilhier, Alabinus, and Soemmerring never had as their aim the use of exaggeration or highlighting to facilitate recognition, classification, or diagnosis—they were after a truth obscured by the imperfections of individual appearance. Emphasis in the interest of operational success is a long way from perfection in the interest of metaphysical truth.

One 1954 atlas celebrated the choice to maintain drawings over actual X-ray photographs in pursuit of this operational and diagnostic truth:

> The publisher has done well to retain the original illustrative sketches. A drawing can show so much better the features one is trying to emphasize than the best chosen original roentgenogram. And of course it is such ideal abstractions of sought-for morbid changes that one carries in one's mind as one searches the fluoroscopic screen for diagnostic signs.[35]

Interpolation, highlighting, abstraction—all were subtle interventions needed to elicit meaning from the object or process and to convey that meaning in the representation itself. The images of judgment are neither those of truth to nature nor those of mechanical objectivity.

Even where the object itself is as unchanging as the visible face of the moon, accurate representation was a task of monumental difficulty long after the development of the camera. In 1961, V. A. Firsoff published his *Moon Atlas*, and the difficulties of extracting realism from the vagaries of moment-to-moment astronomical appearances were all too apparent. Judgment, individual judgment, could not be eliminated:

> Nobody who has not himself attempted to map the Moon can appreciate the difficulties involved in such a programme. The lights and shadows shift with the phase and libration and can alter the appearance, even of a clear-cut formation, almost beyond recognition. Thus every region has to be studied under different illuminations and a true picture of the surface relief built up step by step. To some extent the result must needs be one of individual judgment.[36]

Representation need not be homomorphic.[37] That is, the pictures we construct from the world need not correspond in form to something one has seen—or even could see were one to be somewhere else (or even were one to be much bigger or smaller than our given human size). Population density maps, for example, use the visual to express a phenomenon that might otherwise have been presented in tabular form. For the physical sciences such non-ocular representations as tables serve frequently in all branches of theoretical and experimental work, and such illustrations are often the highly processed output of a computer that has not only stored reams of data but manipulated them in controllable ways. When Robert Howard et al. composed their *Atlas of Solar Magnetic Fields* in 1967 (see Figure 4), they had to *choose* how much to "smooth" the data as they grappled with different observations. Even here, in this most physical of the sciences, the role of objectivity is frankly contested by a resurgent subjectivism tied to the twentieth-century emphasis on judgment and interpretation:

> Considerable experience in the handling of the magnetograms has made us cautious in our approach to their interpretation, but for those unfamiliar with the instrument the variation in the quality of the observations can be a great handicap. For this reason we decided that the best way to make the information available was in the form of synoptic charts, which represent a somewhat smoothed form of the data.
>
> Inevitably many decisions had to be made concerning what were or were not real features on the magnetograms. Naturally there is a certain *subjective* quality to these charts.[38]

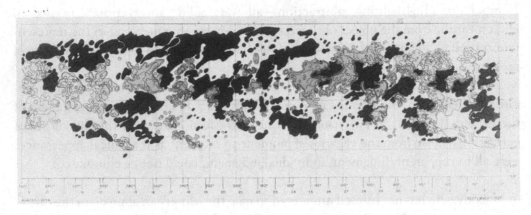

Figure 4. *"Rotation 1417, August-September, 1959," from Howard et al.,*
Atlas of Solar Magnetic Fields, 1967. *Courtesy of the Observatories*
of the Carnegie Institution of Washington, D.C.

Decisions are active, and as such would have had no place within the fundamentally
passive category of nineteenth-century mechanical objectivity.

Gerhart Schwarz (from the Chronic Disease Center of New York Medical College),
collaborating with Charles R. Golthamer (van Nuys, California), also had an active,
artistic conception of pictorial production. Together these two radiologists teamed up
to produce a 1965 Röntgen atlas of the human skull. By this time, the authors argued,
the discipline had advanced to the point where familiarity with normal skull radiology
could be simply assumed as background knowledge: now radiologist, orthopedic surgeon, dental surgeon, neurologist, neurosurgeon, otolaryngologist, and forensic specialist needed not normality but the variants and pseudo lesions that could "vex" even
the expert. Several simultaneous demands made the task complex: first, Golthamer
and Schwarz wanted not a "facsimile" but "a theoretical composite of many different
skulls, containing more than *one hundred* variants and pseudo lesions on each printed
plate." Second, the authors insisted on prints at least of actual size, some even larger
than life. These two constraints, coupled with the "profusion of nature's variants"
promised to overwhelm any possible text. What to do? "It was then that Dr. Golthamer suggested that we might reproduce all radiographs by hand." *Even though the*
X rays already existed, drawings would be created. It was a move unimaginable seventy-
five years earlier: after the hard-won struggle to extract a photograph of Mars, can one
conceive of Lowell reverting to a hand-produced image when he had a photograph
available? Realism (in this mid-twentieth-century context) did not aim at the reflex-
ive correspondence of nature with reproduction, but rather at the half-tone drawing
that *interpreted* particular radiographs.[39]

Golthamer, though he was (on his own account) "an expert painter with many awards to his credit" could not produce a "sufficiently realistic" rendering, nor could Schwarz. Finally, with the aid of the art department director of the College of Physicians and Surgeons, they met with success; the volume represented the combined efforts of two other artists (Helen Erlik Speiden and Harriet E. Phillips). Once the artistic technique (and artist) had been perfected, a more subtle set of concerns arose, issues that get at the very heart of the problem of objectivity in its struggle with judgment:

> The question as to how true to nature the image should be arose for more than one reason. Our initial intention was to make the plates look as "natural" as possible, depicting the normal variant, or pseudo-lesion, as true to its appearance on an actual radiograph as the artist's skill could achieve it. However, after our first plate had been drawn in this manner, we came to realize that painstaking copying of nature was not the purpose of drawings in an anatomic atlas. In many instances, a normal variant, depicted "naturally," remained invisible except to the trained eye of a specialist who was familiar with the lesion to begin with. Reading the completely "natural" plates turned out to be an exercise in "rediscovering" lesions, rather than viewing them. Since a laborious search for lesions in an atlas was surely neither desirable nor practicable, this "natural" manner of graphic presentation would have missed the point altogether. We became convinced that our atlas would gain proportionately in usefulness the more each lesion could be made to look so obvious that a reader would recognize it instantly and without effort.[40]

To bring out the pseudo lesions, the authors depicted *foramen lacerum* "naturally" subdued, and emphasized normal variants and pseudo lesions by "slight optical distortion." "The lesson we learned in preparing the plates for the atlas was that nature may be depicted realistically only by setting off the uncommon and unusual against the background of the 'natural' and common."[41]

If ever one needed evidence that mechanical objectivity had broken down it is here: the enemy of the "natural" (Schwarz and Goldhamer's term) had become the "realistic" (see Figure 5). The real emerged from judgment, and the mechanical transfer of object to representation may well be natural, but the natural was no longer desired. Differing both from the genial improvement of the found object and from the objectivist's mechanical reproduction of the found object, the *interpreted image* is something new. Manipulated to build on the natural, but to bring out features through understanding, the twentieth-century image embodies professional experience; it is the pictorial presentation of the trained eye. A new form of scientific visualization is photographed, painted, and written across this saga of X-rayed lesions.

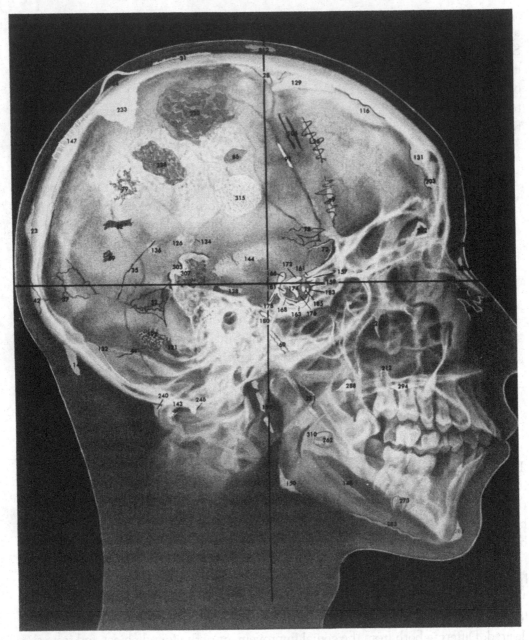

Figure 5. "*Realism vs. Naturalism,*" *plate 1 from Schwarz et al.,* Radiographic Atlas of the Human Skull (*New York and London: Hafner Publishing Company, 1965*).

Here, in the already interpreted image of Figure 5, realism is redefined as one that forcefully takes already existing photographs and replaces them with artwork; a realism explicitly positioned *against* the automaticity of unvarnished photographic naturalism, *against* mechanical objectivity. In making this claim, Schwarz and Golthamer re-situated the nature of depiction; the whole project of nineteenth-century mechanically underwritten naturalism suddenly seemed beside the point. As they arrived at the golden fleece of mechanical objectivity, the purely natural depiction, it proved to be woven of fools' gold. For the image to be purely "natural" was for it to become, *ipso facto*, as obscure as the nature it was supposed to depict. Only by surfacing the oddities against a visual background of the normal could anyone learn anything from the sum of Schwarz and Golthamer's vast labor of compilation.

Golthamer and Schwarz wrote, disarmingly, that it was only after excruciating efforts to depict nature as it was, that they "discovered" the "purpose" of their atlas. Looking back, I would put their concern differently, for what they had discovered was qualitatively unlike the unearthing of a new fossil or the recognition of a never-seen star. Theirs was just as surely a discovery, but one that turned inward to reconstruct not only the kind of evidence they would allow, but the kind of persona that they themselves would need to be. Instead of transparent vehicles for the transport of forms from nature to the reader, the scientist aspired to another ideal, one in which an expert eye counted for more than a mechanical hand. To understand the "discovery" Golthamer and Schwarz had made—to see it repeated over and again as judgment displaced objectivity—is to see just how impossible the interpreted image would have been in the age of mechanical objectivity.

CONCLUSION: PERSONAE AND PRACTICES

The changing ideals of objectivity reformed both pictorial practice and the scientific persona itself. As such, objectivity exists within history and not outside it. And within the history of the natural sciences, the objective image was never a mere synonym for Truth, Certainty, or Consensus. Instead, the objective (mechanical) image stood at a singular moment in the dynamic and contested history of the image, wedged between a pre-nineteenth-century "truth to nature" and a twentieth-century call to judgment. Put otherwise: the scientific image has, historically, been structured to bring forward a variety of often incompatible virtues—mechanical objectivity carried some, but not all of these virtues, and even those it did capture remained primary for a finite time. The pre-nineteenth-century image of a Goethe, Cruveilhier, Albinus, and Soemmerring aimed (in different ways) to depict a world behind the appearances, a *truth to nature*. The resulting tableaux were intended to be better, higher, more universal than anything nature actually made: they revealed a Truth otherwise obscured, and so were truly *metaphysical images*. Not just anyone could pull back the curtain of unstable appearances to reveal these metaphysical images; it was only the Genius who could extract a form more perfect than the best objects we find this side of our sensory limits.

The nineteenth-century machine ideal, by contrast, made pictures into objects of manufacture. Automaticity aimed to secure the identity of the mechanical image with the entity depicted; the mechanical militated against just the kind of intervention that had been celebrated a generation earlier. If the vocabulary of discipline, management, and policing arrived in force during this period, it is precisely because control of the *mechanical image* is factory-like, with an emphasis on the astonishing regularity that such discipline was supposed to produce. Taking place against the background of the mid-nineteenth-century romance with manufactured objects, image technologies instantiated the valued ability to produce identical things.[42] The modernity of manufacture, the dynamics of control, and scientific labor management all figured in the nineteenth-century mechanical image. Self-denial, self-restraint, and supervision were the moral correlates of such production, and they reinforced and affirmed both the social and epistemic rightness of this new way of re-presenting nature. In such a world, Genius necessarily played a distinctly secondary role, entering, if at all, not in the establishment of the ground level "facts" of the matter, but rather in higher-level theoretical constructions out of these facts.

Though judgment, like truth-to-nature, stood in opposition to mechanical objectivity, judgment and truth-to-nature are far from identical. The atlas author of the twentieth century is a more adept version of the *reader*, not a debased echo of the Genius. To the reader-apprentice of the twentieth century, there was no need to rely on the guiding Genius's qualitatively different sensibility. The Gibbses may have been more familiar with the erratic markings of an EEG than the advanced medical student or up-to-date doctor, but the EEG reader is promised 98 percent reading accuracy in twelve short weeks. No part of the self-confidence displayed here is grounded in genius; the self-confident trained experts (doctors, physicists, astronomers) ground their knowledge in guided experience, not special access to reality. (Imagine Goethe promising his readers the ability to construct the *Ur-Formen* of nature after a Gibbs-like high-intensity training course.) Nor are the *interpreted images* that are products of judgment to be likened to the metaphysical images of an earlier age. Explicitly "intellectual," the new depictions not only invited interpretation once they were in place, they built interpretation into the very fabric of the image—but they did so as an epistemic matter. Theirs were exaggerations meant to teach, to communicate, to summarize knowledge, for only through exaggeration (so advocates of the interpreted image argued) could the salient be extracted from the otherwise obscuring "naturalized" representation. The extremism of iconography generated by judgment is there to allow the initiate to learn to see and know, not to display the ideal world behind the real one.

Here, in summary form, is the set of dualisms presented by judgment advocates (in their own terms), ranged against corresponding aspects of mechanical objectivity.

objectivity	judgment
objective	subjective
exact science	empirical art
conscious classification	unconscious classification
reliance on "indices"	seeing eye
(electro)mechanical	intellectual
quantitative	qualitative
universal rules	individual judgments
re-production	interpretation
shared properties	family resemblances

With this set of contrasts in mind, it becomes possible to summarize the three regimes in which pictorial compilations have been embedded. Images—even images as apparently similar as those found in the atlases of science and medicine—turn out to be radically different entities under the three regimes that roughly covered the three periods of pre-mid-nineteenth century, mid-nineteenth century to the early twentieth century, and the last two thirds or so of the twentieth century. The *metaphysical image*, revealing the essence behind the appearance, mediates between the Genius and an audience that learns from the metaphysical images, but will never become the genial author himself.[43] By contrast, the objective, *mechanical image* is produced by scientists committed to the role of a stoic, and, in this resolve, determined to become transparent to nature, a copying mechanism with the affective disengagement of the technical manufacturer. Third and finally, the *interpreted image* is produced not by a moral culture of "towering Geniuses" or neutral, self-abnegating bureaucrats, but by self-confident experts, who trust the trained eye more than master philosophical systems or the automatic conveyance of pictures. While the Genius used the metaphysical image to reveal truth, the technocratic objectivist became a transparent medium for nature to image itself, and the trained expert created images that brought conditioned experience and judgment to the edification of initiates. In the sense used here, "trained expert" designates not so much an initiate into a secret set of skills, but a potential "everyman" who will come in greater or lesser measure to exercise correctly the "experienced eye." "One day," the twentieth-century apprentice could say of the interpreted images of science (as the admirer could never say of the Genius), "I will see like that."

Given the historicity and the contingent nature of these regimes of scientific images, it strikes me as rather doubtful that *the* role of scientific representation can be located. Michael Lynch, for example, maintains that scientific representation is *about* selectivity and mathematization. By contrast, we can see such an assertion as a frequently heard voice of a particular epoch (that of the interpreted image), in which manipulation and restructuring of images was taken not only to be acceptable but praiseworthy. Within the ideal of mechanical objectivity, such intervention was heresy of the worst sort. Recall Henry Alsop Riley's denunciation of drawing as a "purely selective presentation" able to illustrate "anything that the author wishes."[44] Selectivity and mathematization are modes of manipulation that themselves exist within a larger framework of judgment, and judgment within and of images is a *historically specific* form of object classification.

Genius to manufacturer to trained expert; metaphysical image to mechanical image to interpreted image. This epigram, necessarily schematic, joins the epistemological history of the image to the characterological history of the author-scientist. Along with this conjoint history comes a reshaping of the presupposed audience for the image. For different reasons, both the metaphysical and mechanical image presuppose an epistemic passivity on the part of those who see the images: the metaphysical image is self-contained because it is an image of a revealed truth otherwise hidden, and the mechanical image is self-contained because it "speaks for itself" (or for nature). But the interpreted image demands more from its recipient, explicitly. The often-repeated refrain that one needs to learn to *read* the image actively (with all the complexity that reading implies), shifts the assumed *spectator* into an assumed *reader*.

Taken together, these changes in author-artist, reader, and image track a profound shift in the status of the basic low-level objects that make up the disciplinary "facts" of the special sciences. Temporally, the start time for mechanical objectivity appears to sit squarely in the nineteenth century, not in the seventeenth. Spatially, this restructuring of figuration violates national boundaries—our history would only awkwardly separate developments in Germany from those in France, England, or the United States. Should one then speak of an "American-European context" that emerged in the twentieth century, as an explanation that would depict these changes in image making and image understanding as epiphenomenal?

One such approach might involve the invocation of a kind of technological determinism: the shift to objectivity merely reflected the adoption and dissemination of the photographic techniques emerging in the mid-nineteenth century. But this puts the cart before the horse. The ideal of mechanical objectivity could be and indeed was put into practice well before photography became widespread and certainly long before photography entered the atlas-making business. Through policed artistic work, tracing, copying, and the cameras lucida and obscura, film itself entered into an already-existing praxis of mechanical objectivity, and enhanced it.

At the same time, there is a political dimension to the shift from mechanical objectivity and toward judgment but, I suspect, one different from two popular conceptions. The first political reading, articulated frequently in science studies, is that a sociologically glossed Wittgenstein shows science to be now and to have always been a judgment-governed rule-defying activity. Without explicit protocols of action and inference, science is seen as stripped of its authority to make realist claims about the world. In much current work within science studies, judgment comes to stand for the political left, a rebuttal to the ineradicable conservatism of a realistically interpreted science. Against this emerges the second political reading, articulated ever more stridently by conservative critics of science studies, that science is now and has always been an objective reality-reflecting activity. With its strict, universally applicable methods, mechanical objectivity defies the arbitrariness of the subjective, and reflects nature directly. Objectivity in this world is the province of the political right, a refutation of the ineradicable irrationalism of leftist or poststructuralist claims. *Umgekehrt*, as Marx liked to say in his favorite one-word sentence. Both views have their politics ahistorical and backward.

To the extent that mechanical objectivity has a political valence, it would seem to be closer to that of a nineteenth-century, European, technically oriented bureaucrat, a mostly German liberalism that characterized the *Bildungsburger*: cultured in specific ways, above party politics, perhaps, but committed, above all, to a stabilizing program of trade, technological advance, and national unity.[45] To the extent that judgment took on a political valence, it would find its most articulate spokesmen not among firebrand leftists, but among mid-twentieth century *conservatives*: Wittgenstein, as he opposed (and was detested by) the left wing of the Vienna circle, Michael Oakeshott and his followers, who saw judgment as the antidote to an excessive left-wing rationalism, and Michael Polanyi as he struggled to find a place for faith and cultural elitism within modern science. I am afraid, therefore, that I see neither mechanical objectivity nor judgment as salvational moments in a political philosophy read out of science. Of course there are stunning scientific achievements grounded in both the mechanical objectivity of the last century and judgment-emphasizing classifications of this. But in the end, *politically*, I find the opposition between late-nineteenth-century bureaucratic European liberalism and mid-twentieth-century cultural-political conservatism to be a claustrophobic choice indeed. Reading politics out of scientific practice turns out to be a very untransparent affair.

But however one glosses the shifting self-conception of the scientist and the images that accompany it, two points emerge from this story of pictorial practice. First, mechanical objectivity, the nineteenth century's vision of a rock-bottom facticity for the objects with which science works, is a time-specific, hard-won, and contingent category. To depict only what was actually seen meant sacrificing the universalism and truth of the metaphysical image; to rely on photographs often meant abandoning color, accuracy, reproducibility, clarity, even usability on the altar of this mechanical conception of re-production. Second, if the example of the past is any guide, we might do well not to raise twentieth-century judgment as the new standard, or the always present flag of our epistemological continent. Perhaps we should not take it for granted that the metaphysical and mechanical images were "mistaken" and the interpreted image has finally and permanently got it right. We enjoy—both in science and in science studies—re-killing the proceduralism of mechanical objectivity, the way the mechanical objectivists danced on the grave of interventionist genius. But perhaps we are not at the end of the history of image making. Is it too historicist to see the celebration of judgment over mechanical objectivity as historically rooted in the practices of the new academic scientist, philosopher, historian, and sociologist, as its predecessor image techniques were in earlier versions of the natural philosopher?

Notes

*For many helpful comments and suggestions, I would like to thank James Conant, Lorraine Daston, Clifford Geertz, Caroline Jones, Hilary Putnam, and Joan Scott.

1. Lorraine Daston and Peter Galison, "The Image of Objectivity," *Representations* 40 (1992): 81–128.

2. Percival Lowell, *Drawings of Mars* 1905 (Lowell Observatory, 1906), foreword, n.p.

3. Lowell, *Mars and its Canals*, cited in William Hoyt, *Lowell and Mars* (Tucson: University of Arizona Press, 1976), pp. 179, 182–85. As Hoyt notes, the pictures that were reproduced (in the *New York Times, Scientific American, Popular Astronomy, Knowledge and Scientific News*) all failed to show the lines of the canals; one journal's pictures did: the Scottish Review. Lowell was sufficiently concerned by this fiasco that he personally brought the original photographs to show some of the more prominent astronomers.

4. Hoyt, *Lowell and Mars* (1976), pp. 185, 195–96.

5. Richard J. A. Berry, A *Clinical Atlas of Sectional and Topographical Anatomy* (New York: William Wood and Company, 1911), pp. 2–7. Reference is to Wilhelm Braune, *Topographisch-Anatomischer Atlas nach Durchschnitten an gefrorenen Cadavern* (Leipzig: Veit, 1872).

6. The transition from truth-to-nature to mechanical objectivity is discussed in much greater detail in Daston and Galison, "Image of Objectivity," *Representations* 40 (1992): 81–128.

7. Henry Alsop Riley, *An Atlas of the Basal Ganglia, Brain Stem and Spinal Cord* (New York: Hafner Publishing Company, 1960), p. viii.

8. Ibid.

9. Moulton K. Johnson and Myles J. Cohen, *The Hand Atlas* (Springfield, Ill.: Charles C. Thomas, Publisher, 1975), p. vii.

10. Frederick A. Gibbs and Erna L. Gibbs, *Atlas of Encephalography* (Cambridge, Mass.: Lew A. Cummings Co., 1941), preface, n.p., emphasis added.

11. Ibid.

12. Frederic A. Gibbs and Erna L. Gibbs, *Atlas of Electroencephalography*, vol. 1, *Methodology and Controls* (Reading, Mass. : Addison-Wesley Publishing Company, Inc., 1951, 1958), pp. 112–13.

13. Erwin Christeller, *Atlas der Histotopographie gesunder und erkrankter Organe* (Leipzig: Georg Thieme, 1927), cited in Daston and Galison, "Image," *Representations* 40 (1992), fn. 68.

14. In Germany, for example, as the work of David Cahan, Timothy Lenoir, Russell McCormmach, Kathryn Olesko, and others has shown, it was during this period that the university physics laboratory came into existence, along with Ordinarius professorships and the institutionalization of research at the university. From 1830 to 1848, institutes and laboratories multiplied and expanded in the heat of enthusiasm and imperial competition: e.g., Göttingen (1833–34), Leipzig (1837), Berlin (1843), and Königsberg (1847); from the 1840s onward, staffing of the institutes began to include the full complement of *Extraordinarien, Privatdozenten,* and *Assistenten.* Finally, after 1848 the whole of this expanding state-supported infrastructure began to shift in its orientation from teaching toward research, and to expand numerically. From 1870 to 1920 some twenty-three new research institutes launched their operations; Cahan, "Institutional Revolution," *Historical Studies in the Physical Sciences* 15, 2 (1985): 1–65, on 20–21. Laboratory-based medicine burgeoned during these years as well. See,

e.g., Timothy Lenoir, "Laboratories, Medicine and Public Life in Germany, 1830–1849: Ideological Roots of the Institutional Revolution," in *The Laboratory Revolution in Medicine*, ed. Andrew Cunningham and Perry Williams (Cambridge: Cambridge University Press, 1992), "Science for the Clinic: Science Policy and the Formation of Carl Ludwig's Institute in Leipzig," in *The Investigative Enterprise: Experimental Physiology in 19th-Century Medicine*, ed. William Coleman and Frederic L. Holmes (Berkeley and Los Angeles: University of California Press, 1988), and "Social Interests and the Organic Physics of 1847," in *Science in Reflection*, ed. Edna Ullman-Margalit (Dordrecht: Kluwer Academic, 1988).

15. On precision see M. Norton Wise, ed., *The Values of Precision* (Princeton: Princeton University Press, 1995).

16. Finally, most difficult of all, comes paragenesis, the unraveling of the order of growth—and it is to this aim that O. Oelsner, *Atlas of the Most Important Ore Mineral Parageneses Under the Microscope* (Oxford, London, Edinburgh, New York, Paris, and Frankfurt: Pergamon Press, 1961 [German], 1966 [English]) directs his work, see pp. v–vi. The atlas trains the eye by depicting microscopic samples and providing worksheets that the reader superimposes on the picture to provide keys to interpretation.

17. Frederic A. Gibbs and Erna L. Gibbs, *Atlas of Electroencephalography*, vol. 1, p. 113.

18. Hallowell Davis, introduction to William F. Caveness, *Atlas of Electroencephalography in the Developing Monkey* (Reading, Mass., Palo Alto, London: Addison-Wesley Publishing Company, Inc., 1962), p. 2.

19. On holism and politics before and during Nazism, see Anne Harrington, *Hunger for Wholeness: Holism in German Culture, from Wilhelm II to Hitler* (Princeton: Princeton University Press, 1996), and Mitchell Ash, *Gelstalt Psychology in German Culture, 1890–1967: Holism and the Quest for Objectivity* (Cambridge: Cambridge University Press, 1995).

20. W. W. Morgan, Philip C. Keenan, and Edith Kellman, *An Atlas of Stellar Spectra* (Chicago: University of Chicago Press, 1943), p. 4.

21. Ibid., p. 5.

22. W. W. Morgan, Philip C. Keenan, and Edith Kellman, *An Atlas of Stellar Spectra*. In twentieth-century medicine one often sees "clinical judgment" opposed to the view that nature can speak for itself. For example the author of an atlas on electrocardiograms argues that traces cannot replace clinical judgment. See J. Riseman, *P-Q-R-S-T, A Guide to Electrocardiogram Interpretation*, 5th ed. (New York: Macmillan, 1968).

23. Ibid., p. 6.

24. Edward Pickering, *Harvard Observatory Annals* 91 (1918), preface, pp. iii–iv.

25. On women astronomical workers and their often low-paid status within the observatory, see, e.g., Margaret Rossiter, *Women Scientists in America: Struggles and Strategies to 1940* (Baltimore: Johns Hopkins University Press, 1982); also Londa Schiebinger, *The Mind Has No Sex? Women in the Origins of Modern Science* (Cambridge: Harvard University Press, 1989); on the Harvard Observatory see Pam Mack, "Women in Astronomy in the United States, 1875–1920" (unpublished senior thesis, 1977). On women scanners in the high-energy physics laboratory, see Galison, *Image and Logic: A Material Culture of Microphysics* (Chicago: University of Chicago Press, 1997), esp. chapters 3 and 5.

26. On the management and "correction" of workers by use of the personal equation, see Simon Schaffer, "Astronomers Mark Time: Discipline and the Personal Equation," *Science in Context* 2 (1988): 115–45; also Schaffer, "Babbage's Intelligence: Calculating Engines and the Factory System," in *Critical Inquiry* 21, 1 (1994): 201–27. On the notion of the "unskilled" generally and in particular about the mostly women workers who reduced the nuclear emulsion photographs in the 1940s and 1950s see Galison, *Image and Logic*, Chapter 3.

27. Owen Gingerich, "Cannon, Annie Jump," in *Dictionary of Scientific Biography*, ed. Charles C. Gillispie (New York: Charles Scribner's Sons, 1980), vol. 3, p. 50. Katherine Heramundanis, ed., *Cecilia Payne-Gaposchkin: An Autobiography and Other Writings* (Cambridge: Cambridge University Press, 1984).

28. Theodore E. Keats, *Normal Roentgen Variants that may Simulate Disease* (Chicago: Year Book Medical Publishers, Inc., 1973), p. vii.

29. Rudolf Grashey, *Typische Röntgenbilder vom normalen Menschen* (Munich, 1939), cited in Daston and Galison, "Image," *Representations* 40 (1992): 81–128, on 105–07.

30. P. M. S. Blackett, foreword to G. D. Rochester and J. G. Wilson, *Cloud Chamber Photographs of the Cosmic Radiation* (New York: Academic Press, Inc. and London: Pergamon Press Ltd., 1952), p. vii.

31. Ivan Baronofsky, *Atlas of Precautionary Measures in General Surgery* (St. Louis: C. V. Mosby Company, 1968), p. x.

32. The problem of what we "see" raises a fundamental and exceedingly subtle point. Wittgenstein distinguishes between judgments that are in some way so close-lying to our perceptions that the insertion of a distance between what we perceive and what we judge is absurd. Can one really say "I *know* I am in pain" without making a joke? (L. Wittgenstein, trans. G. E. M. Anscomb, *Philosophical Investigations* [New York: The MacMillan Co., 1958], par. 246). This is to be contrasted with the expert judgment, e.g., in algebraic extensions of a series. This distinction between a kind of basic and expert judgment lies deep in Wittgenstein, as Cavell has shown (*The Claim of Reason* [Oxford: Clarendon Press, 1979], esp. chapters 3 and 4.) But in the case of scientific figuration the radical splitting between "seeing" and "seeing as" leads, I believe, to much confusion.

33. John L. Madden, *Atlas of Technics in Surgery* (New York: Appleton-Century-Crofts, Inc., 1958), p. xi.

34. Johannes Sabotta, *Atlas and Text-Book of Human Anatomy* (Philadelphia, 1909), cited in Daston and Galison, "Image of Objectivity," *Representations* 40 (1992): 101.

35. E. A. Zimmer, *Technique and Results of Fluoroscopy of the Chest* (Springfield, Illinois: Charles C. Thomas, Publisher, 1954), pp. v–vi.

36. V. A. Firsoff, *Moon Atlas* (London, Auckland, and Johannesburg: Hutchinson & Co., 1961), introduction, p. 7.

37. Elsewhere (*Image and Logic: A Material Culture of Twentieth-Century Physics*), I have contrasted technologies that are homomorphic (retaining the *form* of that which is represented) with those that are homologous (retaining logical relations within that which is represented).

38. Robert Howard et al., *Atlas of Solar Magnetic Fields* (Washington, D.C.: Carnegie Institution, 1967), pp. 1–2, emphasis added.

39. Gerhart S. Schwarz and Charles R. Golthamer, *Radiographic Atlas of the Human Skull: Normal Variants and Pseudo-Lesions* (New York and London: Hafner Publishing Company, 1965), n.p.

40. Ibid.

41. Ibid.

42. One thinks here of Maxwell's reflections on Herschel's consideration of atoms, manufactured goods, and God. Maxwell identified the three categories of usefulness of identical manufactured objects: cheapness, serviceableness, and quantitative accuracy. "Which of these was present to the mind of Sir J. Herschel we cannot now positively affirm ... though it seems ... probable that he meant to assert that a number of exactly similar things cannot be each of them eternal and self-existent, and must therefore have been made, and that he used the phrase 'manufactured article' to suggest the idea of their being made in great numbers" (W. D. Niven, ed., *The Scientific Papers of James Clerk Maxwell* [New York: Dover, 1965], p. 484). On the culture of manufacture in Victorian England, see Maxine Berg, *The Machinery Question and the Making of Political Economy, 1815–1848* (Cambridge: Cambridge University Press, 1980).

43. Indeed, throughout the late eighteenth century, Genius was a contentious term; and for the older Goethe it was inextricably bound with the rare ability to discover and interpret the hidden laws of nature. There was nothing democratic about it, none of the genius in every man that characterized Fichte's position, for example. English writers, though they differed in certain respects, also emphasized that the Genius could discover laws otherwise hidden: the Genius offered rules to which others conformed. Strikingly, Alexander Gerard's *Essay on Genius* (1774) explicitly contrasted genius with judgment: judgment was a moderating aspect of affect that controlled and directed otherwise ram-

pant imagination. Judgment alone was never genius. Such remarks show two things: first, that the creative action of the genius was not (primarily) the exercise of judgment, and second, that judgment understood in the eighteenth century as capacity for moderation was *not* judgment understood in the twentieth century as non-algorithmic assessment.

On the role of genius, see Simon Schaffer, "Genius in Romantic Natural Philosophy," in A. Cunningham and N. Jardine, *Romanticism and the Sciences* (Cambridge: Cambridge University Press, 1990), pp. 82–92; Myles Jackson, "Genius and the Stages of Life in Eighteenth-Century Britain and Germany," in *Les Ages de la Vie en Grande-Bretagne au XVIIIe Siècle*, ed. Serge Soupel (Paris: Presses de la Sorbonne Nouvelle, 1995), pp. 35–46; Richard Yeo, "Genius, Method, and Morality: Images of Newton in Britain, 1760–1860," *Science in Context* 2 (1988): 257–81. On Gerard see Jackson, "Genius and the Stages of Life," pp. 38–39.

44. Michael Lynch, "The Externalized Retina: Selection and Mathematization in the Visual Documentation of Objects in the Life Sciences," in *Representation in Scientific Practice*, ed. M. Lynch and S. Woolgar (Cambridge, Mass.: MIT Press, 1990), pp. 153–86.

45. In the German case, one thinks here of the older literature by Fritz K. Ringer, *The Decline of the German Mandarins: The German Academic Community, 1890–1933* (Cambridge: Harvard University Press, 1969) and Fritz Stern, *The Politics of Cultural Despair: A Study in the Rise of the Germanic Ideology* (Berkeley and Los Angeles: University of California Press, 1961), both of whom explore the character of the new bourgeois educated elite. More recently, in a series of important articles, Timothy Lenoir has sought to displace the very category of the "professional" (which he finds both post hoc and too narrow) and replace it for the late nineteenth century with the *Bildungsburger*—with its connotations both of a certain kind of career *and* cultural engagements. See, e.g., Lenoir, "Laboratories," "Ideological Roots," and "Science for Clinic," all cited above in note 14.

JAN GOLDSTEIN

Eclectic Subjectivity and the Impossibility of Female Beauty

For much of the nineteenth century, a single theory of psychology exercised near hegemony in France: that associated with the philosophy of Victor Cousin, aptly dubbed by him "eclecticism." Other theories of psychology, to be sure, entered the competitive fray. But by dint of its institutional strongholds, Cousinian psychology captured and long maintained a virtually official, national position.[1]

At the beginning of the July Monarchy (1830–1848), and through the shrewd bureaucratic interventions of Cousin, a subject called "psychology," defined in the Cousinian manner, was added to the standard lycée curriculum in philosophy and made its first substantive part. A decree of 1832 spoke bluntly of the "necessity of commencing the study of philosophy by 'Psichologie' [sic]."[2] Thus, shortly after its Napoleonic inception, the highly centralized state system of secondary education had been enlisted to teach Cousinian psychology to all the male bourgeois adolescents of France. The place of philosophy as the "crown" of lycée instruction, the subject matter that was regarded as most intellectually powerful as well as most expressive of the French national genius, greatly enhanced the strategic importance of this psychology.[3] To staff the educational machine appropriately, Cousin and his "regiment" of loyal disciples used their presence on the critical gatekeeping bodies of the capital (the Ecole normale supérieure, the Sorbonne, the jury of the philosophy *agrégation*) to control the production of philosophy professors. Aspirants who showed heterodox tendencies or failed to toe the Cousinian line were discouraged, if not actively prevented, from completing their degrees.[4]

This institutionally entrenched psychology was furthermore noteworthy for its emphasis on the self, or *moi*. In fact, it seems to have effected the transformation of the word *moi* from an inconspicuous, workaday personal pronoun to a noun bearing both quasi-technical significance and enormous cachet. Almost in anticipation of the journalistically constructed "me generation" of 1980s America, contemporary French observers of the 1830s were fond of pointing out, with only some exaggeration, that the newly regnant Cousinian philosophy substituted the *moi* for God, that it moreover entirely compressed philosophy into the concept of the *moi*.[5]

The purpose of this paper is to examine the kind of subjectivity that this hegemonic, unabashedly self-centered psychology produced and to explore some of the aesthetic implications of that subjectivity. I should make clear that by calling Cousinianism "hegemonic" during the nineteenth century, I do not mean to imply that it was universally disseminated in France. On the contrary, it was inculcated only in those destined to rule—to wit, the male bourgeoisie. Through such institutional arrangements as the protracted exclusion of workers, peasants, and females from the lycées—and the deliberate omission of philosophical psychology from the curriculum of the women's lycées that were finally founded in 1880—Cousinianism became a highly class-specific and gender-specific psychology.[6] It furnished a vocabulary of selfhood to those members of society who were supposed to exert influence and to embody the dominant norms. Much like Latin in early modern Europe, its authority was bound up with the fact that its adepts formed a select company.[7]

I should perhaps also emphasize that my lavishing of so much attention on Cousin should not be taken to mean that I find great intrinsic merit in his philosophy. Cousin interests me as a historical phenomenon. His ideas were derivative (largely from contemporary German idealist sources), frequently garbled, and patched together according to a blatant extra-intellectual agenda ("My political faith conforms in every respect to my philosophical faith,"[8] he announced on more than one occasion). But the successful and long-lasting institutionalization of those ideas, selected from a field of competing possibilities, indicates their basic "fit" with the dominant French cultural values of the period. Conversely, their establishment in the educational system enabled them to reproduce themselves and to become a powerfully articulate cultural force in nineteenth-century France. If only for these reasons, Cousinianism repays close study.

TECHNIQUES OF BOURGEOIS INTROSPECTION

Cousin showcased the *moi* in his philosophy precisely because he believed that the empirical philosophy of the eighteenth century had vitiated that entity, with disastrous sociopolitical consequences. The Cousinian argument on this point ran as follows: by building up consciousness over time through the accumulation of dis-crete, sensory bombardments and their associated ideational forms, *sensualisme* (as the empiricism of Locke and Condillac was rather sneeringly called in France by its oppo-nents) offered no logical ground for psychic unity. It effectively precluded rational belief in an integrated, holistic self. It reduced mind to matter, thus both irrevocably fragmenting mind and rendering it passive. The result of such a philosophy could only be the decay of the belief in personal moral responsibility, in the immortality of the soul, and in divine retributive justice—and all of these erosions had in fact con-tributed to unleashing the frenzied, antisocial behavior that characterized the radical phase of the French Revolution. In the Cousinian vision of things, the stability of post-1789 France as a liberal society and polity (Cousin and his supporters were lib-erals and had no desire to return to the Old Regime) required the restoration of an active, morally responsible self. Taking no chances with the fate of the nation, Cousin therefore postulated this self, or *moi*, a priori.

Since the *moi* had been given to us prior to experience, how could we be sure that we possessed it? This was a question of great practical import to Cousin. It was incum-bent upon him, at least at the beginning, to persuade his audience of the superior veracity of his account of psychology over the empiricist account, to lead them to embrace the axiom that they were innately endowed with a *moi*. Furthermore, the Cousinian *moi* had to be known and articulated in order that society might fully reap its presumed benefits, which included the bourgeoisie's awareness of its dynamic lead-ership role and the deterrence of revolutionary activity through individuals' acute appreciation of themselves as morally accountable beings. For both of these purposes, Cousin's philosophy placed a heavy burden on introspection as the route to firsthand knowledge of the *moi*.

In nineteenth-century France, introspection was called "interior observation." The succinct term introspection was a late-seventeenth-century English coinage, and on the English side of the Channel it enjoyed currency both in ordinary speech and in technical philosophical discourse. Its extremely halting entrance into the French lan-guage occurred in the context of positivist critiques of the scientific legitimacy of the practice.[9] But the earliest of those French critiques, that of Auguste Comte, was stead-fastly French in its vocabulary. As Comte argued in the opening lesson of his public *Cours de philosophie positive* in 1830, the mind was capable of observing all phenomena except its own. Making the observed and the observing organ one and the same resulted in an absurdity: "In order to observe, your intellect must pause from activity; yet it is this very activity that you want to observe. If you cannot effect the pause, you cannot observe; if you do effect it, there is nothing to observe." Far from issuing in sound scientific conclusions, then, "this interior observation gives birth to almost as many theories as there are observers."[10]

As we will see, Cousin at times seemed familiar with this critique of introspection, but it never struck him as definitive or devastating. Rather he took it as indicative of the extreme difficulty of the introspective enterprise, the high-wire virtuosity that it required. Such a challenge in no way dampened his enthusiasm for systematically pursuing the goal of what he often called *repli*, the folding of the mind back on itself.

Introspection was central to Cousin's representation of his own personal itinerary as a philosopher. As he traced that formation in 1826 (when he had attained the ripe age of thirty-four and could afford to muse on his beginnings), its driving force had been twofold: on the one hand, his realization of the destruction wrought by the "analytic spirit" of the eighteenth-century sensualists and hence of the urgent need for reconstruction; on the other hand, his fundamental admiration for the Baconian inductive method that those same sensualists endorsed. His philosophical breakthrough came, he tells us, in 1815 when he decided that the way out of his impasse was to wed Baconian observation to the metaphysics that the sensualists proscribed. Accordingly in that year, he pioneered on himself his "psychological method," a mode of interior observation that, when rigorously conducted, would lead seamlessly to metaphysical speculation. This psychological method, he asserted, "constitute[d] the fundamental unity of my [subsequent] teaching." Hence Cousin's "historical consciousness"—or, perhaps more accurately, his precocious desire for self-memorialization—had spurred him on to "reproduce" faithfully and "in all their weakness" his very first applications of this unusually fecund method and to publish them belatedly under the title *Fragments philosophiques*.[11] That book made available three years' worth of his "obscure and painful labors." The "psychological details" it contained, while "arid and lacking in all apparent grandeur," must nonetheless, he admonished, "never be forgotten since they form the legitimate point of departure for all the future directions that philosophy can and should take."[12]

Cousin's whole system of eclecticism had, in other words, issued from his heroic introspective experience of the years 1815–1817. That experience had both founded the discipline of eclectic philosophy and formed the prototype of all subsequent work in it, much as Descartes's period of systematic doubt while living in solitude in Holland and Freud's interpretation of his own dreams in fin-de-siècle Vienna had served in a joint foundational-prototypical capacity in the disciplines of Cartesian philosophy and psychoanalysis, respectively. Not surprisingly, then, training in Cousinian philosophy—whether in institutions of higher learning or at the more modest level of the lycée—mirrored the master's development by also featuring introspection. An official manual of pre-baccalaureate philosophy instruction written during the July Monarchy by one of Cousin's disciples listed the questions about the *moi* that every student should be able to answer and then specified, "There is but a single way to discover the true responses to these questions, *interior observation*."[13] In another pedagogical handbook from the same period, the same disciple bore witness to the centrality in the philosophy classroom of training in introspection, this time by commenting in a footnote on the difficulties he had encountered in getting his young charges to grasp the "interior reality" (*fait intérieur*).[14] Auguste Comte likewise regarded Cousinianism as fostering introspection among the student population of the late 1830s. One of his harangues against interior observation, and in favor of the positivist method of approaching mental phenomena exclusively through the study of cerebral physiology, mentioned the "deplorable psychological mania that a famous sophist had some years ago momentarily succeeded in inspiring in French youth." His audience would have readily identified the unnamed "sophist" as Cousin and the "psychological mania" as the enthusiastic belief in the scientific efficacy of introspection. The adverb "momentarily" bespoke only Comte's fond wish that the vogue of Cousinianism might quickly pass.[15]

What guidelines did Cousin offer the student embarking on the momentous task of interior observation? The descriptions of the introspective method scattered through the master's writings hardly provide step-by-step instructions, but they do convey some sense of the procedure, as Cousin understood it. Here is the description that figured in Cousin's 1826 account of the strenuous early years of his own career as a philosopher:

> The field of philosophical observation is consciousness. There is no other. But in this field, nothing may be neglected; everything is important because everything hangs together, and if one part is missing, the total unity becomes indiscernible (*insaisissable*). . . . The psychological method consists in isolating oneself from any world other than that of consciousness in order to establish and orient oneself there, where everything is real, but where the reality is extremely diverse and delicate. Psychological talent consists in placing oneself at will within this entirely interior world, in giving oneself the spectacle of oneself and in reproducing distinctly all those phenomena that the circumstances of life bring along only in an accidental and confused fashion. I repeat, long years' practice has revealed to me that the psychological method can be carried out with many different degrees of depth.[16]

Here are descriptions from his celebrated Sorbonne lectures of 1828, when Cousin was allowed to return to his old podium eight years after the Restoration government had banished him from it because of his liberal leanings:

> The psychological method is the conquest of philosophy. That method has already assumed a rank and an uncontested scientific authority that each day increase. . . . What is psychological analysis? It is the slow, patient, and meticulous observation, with the aid of consciousness, of phenomena hidden in the depths of human nature. These phenomena are complicated, fleeting, obscure,

rendered almost undiscernible (*insaissisable*) by their very closeness. The consciousness which applies itself to them is an instrument of extreme delicacy: it is a microscope applied to things infinitely small.[17]

There is, Gentleman, a psychological art, for reflection is, so to speak against nature, and this art is not learned in a day. One does not fold back upon oneself easily without long practice, sustained habit, and a laborious apprenticeship.[18]

Several motifs stand out in these descriptions. First, introspection is depicted as an ascetic, almost monastic discipline tinged with heroism: it requires self-sacrifice and a long and painful tutelage, it goes against the natural grain, and it removes the practitioner from the reassuring world of ordinary social intercourse. Second, in epistemological terms, introspection is decidedly hybrid. At times Cousin stresses its scientificity, as when he metaphorically identifies it with a microscope or, in a passage not quoted above, declares that psychology and physics are on a par as empirical sciences, the first relying on the method of interior observation, "with the aid of that internal light called consciousness," and the second on exterior observation by means of the senses.[19] At other times, however, he designates introspection as an art and asserts that plying it to more than superficial effect requires "talent." In thus suggesting that introspection is a combined art-and-science, Cousin clearly wants to have his cake and eat it too. A zealous anti-positivist, he nonetheless wants his psychology to possess all the authority that the term "science" conferred in the early nineteenth century. At the same time he attempts to fire the imagination of his audience by endowing his key procedure of interior observation with the Romantic ethos of an art that is not equally accessible to all, that does not mechanically chop reality into pieces in order to know it but, instead, apprehends it holistically and organically. The artistic/Romantic aspect of introspection is connected to the third motif of Cousin's descriptions: the extremely elusive nature of consciousness as an object. Fleeting, fugitive, nearly unseizable (to give a literal translation of Cousin's *insaissisable*), it is itself a delicate phenomenon and, to be captured, requires commensurate delicacy on the part of consciousness as instrument.

It is when evoking this elusiveness that, without mentioning Comte, Cousin occasionally addresses the conundrum of the simultaneous employment of consciousness as subject and object—the defining characteristic of introspection and the brunt of Comte's positivist critique of it as a scientific method. In a fragment from his 1817 Sorbonne course, Cousin considers the problem of grasping the spontaneous activity of the *moi*:

All of our researches on ourselves are reflective, and our lot is to seek the spontaneous viewpoint through reflection—which is to say, to destroy it in seeking it. However, when examining oneself in peace, it is not impossible to discern (*saisir*) the spontaneous beneath the reflective. In the very instant of reflection, one senses beneath this activity that turns back on itself an activity that had to be deployed at a prior moment and without reflecting on itself.[20]

Cousin thus recognizes the conundrum and, initially, seems to find it insoluble: we destroy the object of our introspection in the very act of looking for it, he admits. But he will not concede defeat. Peaceful retirement from the hustle and bustle of external reality, he insists, holds out the promise that we can mentally grasp spontaneity in the split second before it is overtaken and deformed by reflection. Introspection, in other words, demands seclusion, vigilance, delicacy, and talent—but it is not, for all that, impossible.

In a slightly different vein, a passage from 1833 (written at a point when Cousin may well have heard Comte's diatribes against introspection) responds in rather poetic language to the problem of the elusiveness of consciousness by emending the definition of interior observation to include subsequent, clarificatory processes of memory and reflection:

> The phenomena of the interior world appear and disappear so quickly that consciousness apprehends them and loses them from view almost at the same time. It is therefore insufficient to observe them fleetingly and during their passage through this mobile theater; they must be retained by the attention as long as possible. One can do still more. One can call back a phenomenon from the bosom of night into which it has vanished, request memory to conjure it up, and reproduce it for consideration at one's leisure. One can recall a certain part of it rather than another part, leave the latter in shadow to accentuate the former, vary the angles so as to cover all of them and to embrace the whole object: that is the task of reflection.[21]

Cousin, it would appear, does not agree with Comte that the introspecting consciousness sees no activity at all because it must cease to function qua subject when it makes itself the object of its investigation. Rather he contends that it sees blurrily and that, by enlisting memory and reflection, it can effectively compensate for its necessarily dim first impressions of the "mobile theater." This formulation of 1833 does, to be sure, represent a further acknowledgment on Cousin's part of the difficulty inherent in introspection. Whereas his metaphor of 1828 construed consciousness itself as a microscope, with extraordinary capacities of sight, in 1833 Cousin depicted a situation of mediation: "reflection is to consciousness what artificial instruments are to our senses."[22] In other words, consciousness acquires sight-enhancing capacities comparable to those of a microscope only by means of the subsequent, reflective reworking in tranquillity of its raw data. But despite the additional level of complexity he acknowledged in 1833, introspection remains for Cousin a valid and eminently feasible undertaking.

FINDINGS OF BOURGEOIS INTROSPECTION: ECLECTIC SUBJECTIVITY

I have thus far combed Cousin's corpus for descriptions of the introspective process. But I have not yet inquired what the introspecting consciousness finds when it encounters the *moi*—what, in other words, Cousin took subjectivity to be and what was drilled into generations of French bourgeois adolescent boys as the very stuff of their subjective being.

Before turning to Cousin's position on this matter, we ought to look at the vocabulary available to him. If "introspection" was an English coinage that the French tended to hold at arm's length during the nineteenth century, so too during the same period "subjectivity" hesitantly entered the French language from a foreign source, in this case the German of Immanuel Kant. It first surfaced in a French dictionary in 1803, where it was attributed to "K" (that is, Kant, according to the dictionary's "Table of Most Frequently Employed Abbreviations"),[23] having made its inaugural appearance in French prose just two years before in Charles Villiers's account of Kant's transcendental philosophy.[24] It had sufficient currency in French academic circles for Balzac to mention it in an 1830 spoof on linguistic fashions:

> Do we speak of philosophy? Oh, whoever you may be, consider the fact that if you fail to follow fashion attentively, you will be written off forever for using words ending in *-ty*, like *objectivity, subjectivity, identity, . . . spontaneity, fugitivity*, when the master has pronounced in favor of *-ism* by employing the words *sensualism, idealism, dogmatism, criticism, Buddhism*, etc. Or if you embrace the *-ism* when he declares the *-ion* fashionable, as in *affection, sensation, . . . argumentation*, you will pass for a fool.[25]

But "subjectivity" was not a term that Cousin favored in his own writing or chose to popularize in France. He seems to have regarded it as smacking unpleasantly of Kantian skepticism, of the German philosopher's insistence that, imprisoned in our mental categories, we are forever precluded from knowing the thing-in-itself or drawing firm conclusions about the nature of the absolute.[26]

But while Cousin never endorsed the term "subjectivity," he offered, on the basis of his scrutiny of consciousness, an unequivocal characterization of the locus of subjectivity—that is to say, of the entity that he designated as the *moi*. Given his moral crusade against sensationalist philosophy, it is hardly surprising that the very first result of Cousinian introspection turned out be a refutation of the fundamental sensationalist tenet. According to Cousin, the unbiased introspecting consciousness quickly discovers that, lo and behold, all is not sensation or a derivative of sensation in the world of consciousness. There are instead three different classes of conscious *faits*: the sensible, the volitional (also called the active), and the rational.[27] Like the Holy Trinity, they are three in one, irreducibly distinct from one another yet all necessarily operating together to form that unity called consciousness.[28] Cousin considered this tripartite classification of the elements of consciousness as one of his signal contributions. "It has really caught on, for I see it reproduced in practically every recently published work of psychology," he commented cheerfully in 1833.[29] That Cousin supplemented sensation by two other elements of consciousness, and hence subsumed the doctrine of sensualism into a larger, variegated structure rather than merely overthrowing it, was the move that led him proudly to name his philosophy "eclecticism."

Cousin was also crystal clear about where, within this "triplicity" (his neologism), the *moi* was situated. "The will alone is the person or the *moi.*" "Our personality is the will, and nothing more."[30] By contrast, the products of sensibility—that is, the data about the external world that came in through our sense organs—were, in the Fichtean language that Cousin liked to adopt, the *non-moi*. And though not technically relegated to the *non-moi* (whose very name seems to connote a pariah status), reason was in Cousin's scheme also utterly foreign to the *moi*. It was "impersonal" by nature, imposing its necessary and universal dictates on the self from some external position. What made volition personal to its possessor (and what in turn made Cousin opt in his prose for the resolutely affirmative "personality" over the skeptically tinged "subjectivity") was that it took a particular self as its source or cause, and took the form of a welling or bubbling up of a particular initiative to action.

Translating these Cousinian conceptions into the language of subjectivity and objectivity, we could say that the operations of reason are objective by definition, that an "impersonal reason" (this was Cousin's preferred expression) merely uses our individual thinking apparatuses as its neutral vehicles. As Cousin instructed the students in his 1828 course, "Your intelligence is not free, Gentlemen. . . . You do not constitute your reason and it does not belong to you." Because the dictates of reason were "universal and absolute," it was *not* "despotic" to "declare entirely crazy (*en délire*) those who do not accept the truths of arithmetic or the difference between beauty and ugliness, justice and injustice."[31] What is, by contrast, subjective in human consciousness is the individual's inclination to activity. Cousin never defines this key concept of activity explicitly or rigorously. But he does imply that activity is a kind of switch point that controls the border between an individual's spirituality and materiality, that it is the mental impetus within that individual whose "mysterious" property is to cause that individual's muscles to move. "When I push one billiard ball into another, it is not the billiard ball that truly causes the movement that it imparts, for that movement has been imparted to it by the hand, by the muscles that, in the mystery of our physiological organization, are at the service of the will."[32] Activity then is an immaterial, voluntary impetus to a highly material, muscular movement. A subjectivity identified with volition or activity would probably be experienced by its possessor as a surge or a push, a pure assertion.

Cousin gives us enough information to infer such a general description but, beyond that, he attributes very little content to subjectivity. He is not, apparently, interested in capturing nuances of emotion or in charting the ebbs, flows, and clashes of delicate currents of thought and feeling. Interiority is not for him a preferred site of experience or place of refuge, although he liked to depict himself to his students as if it were.[33] Instead, interiority attracts him as a polemical resource: he will marshal the evidence of introspection to establish once and for all that the human subject is not passive (and hence perilously exempt from moral responsibility) but active.

His polemical investment in mental activity crops up throughout his writings, coloring his evaluations of earlier philosophers. Thus Maine de Biran earned the accolade of the most important of Cousin's French mentors because of his special sensitivity to matters of volition:

> With Monsieur Maine de Biran, I especially studied the phenomena of the will. That admirable observer taught me to tease out (démêler) in all our knowledge (connaissances), and even in the most simple events of consciousness, the role of voluntary activity, of that activity in which our personality bursts forth (éclate) and reveals itself.[34]

On the other hand, Cousin demoted Kant, finding him guilty of a "psychological error that put him en route to the abyss." By failing to bestow on "voluntary and free activity" the same care and analytic scrutiny that he bestowed on reason, Kant never realized that the "personality is particularly attached to [the] class of [active, voluntary] phenomena, and that reason, although linked to the personality, is profoundly distinct from it."[35] Even in the ministerial decree of the 1832 decree that made psychology the foundation of the philosophy curriculum, Cousin's priorities were evident. Students are first to be taught about the "certitude proper to consciousness" or, in other words, the unassailable scientific validity of the evidence of introspection. But among the three component faculties of consciousness, they will actually verbalize their introspective experience of only one: a unit on "voluntary and free activity" will require them to "*describe* the phenomenon of the will."[36]

Given the specific polemical purpose of Cousinian introspection, it is hardly surprising that its fruits were minimalist and monochromatic rather than richly textured, delicately patterned, and polychromatic: it is only, I think, a slight exaggeration to say that what had to be grasped was the presence within oneself of a self-generated assertiveness or pushiness. Not surprisingly, then, for all its reliance and insistence on introspection, the Cousinian movement did not produce an introspective literature. Cousin and his regiment wrote profusely, but they did not ordinarily cultivate the genres of the diary, the confession, or even the autobiographical reminiscence.[37] Cousin may have learned many things from Maine de Biran, but he evidently had no inclination to adopt his mentor's practice of keeping a daily journal.[38] The bits of autobiography that Cousin wove into the prefaces of his *Fragments philosophiques* are much stiffer, less revelatory, more exclusively confined to professional career and public face than even what Descartes saw fit to share with his readers in the *Discourse on Method*. In sum, Cousin was not himself introspective—in the way that we use that term today, to mean tending to investigate one's motives and feelings and to find initially hidden complexities in them—but he was, in technical philosophical terms, a zealous champion of introspection.

If the chief discovery that Cousin made by means of introspection was that human subjectivity could be identified with activity or will, he also made a second major discovery. Introspection revealed to him the axis on which the fundamental difference between human beings could be plotted. For some people but not for others, the so-called triplicity of consciousness (the operational fusion of the three analytically distinct elements of sensibility, rationality, and will) was an immutable condition, serving as a permanent barrier to their teasing out the *moi* and to forcefully appropriating it for—and as—themselves. At his 1828 course—which was so closely watched by journalists and other commentators, as well as transcribed by stenographers for purposes of immediate publication, that it qualifies as a Restoration-style media event—Cousin expounded this theme to his students:

> The identity of consciousness constitutes the identity of human knowledge. It is on this common base that time sketches all the differences that distinguish one man from another. The three terms of consciousness form a primitive synthesis, in which they exist in a more or less confused state. Often a man stops at that point—that is, in fact the case with the majority of men. Sometimes a man goes further and succeeds in exiting: he adds analysis to the primitive synthesis, develops it by reflection, disaggregates the complex phenomenon by submitting it to a light which, spreading successively over each of the three terms of consciousness, illuminates them reciprocally. What happens then? The man knows better what he knew already. All the possible differences between one man and another reside there.[39]

This is a slippery passage, oscillating as it does between a democratic insistence on our common humanity, save for some cosmetic distinctions, and an elitist insistence on the significant intellectual superiority of a minority of the population. During subsequent decades, Cousin and his followers opted for both of those readings, stressing one or the other according to the circumstances. The first, democratic reading bolstered their claim that all the people, however humble, had (in the manner of Rousseau's Savoyard vicar) spontaneous, unreflective knowledge of certain metaphysical truths—the existence of God, the mind-body distinction, the immortality of the soul—and hence knew, without instruction in the subtleties of Cousinian philosophy, that they were morally accountable beings. This was, for example, the message that Cousin sought to hammer home in a pamphlet written in the wake of the working-class uprising of June 1848, when he and his colleagues feared that the materialist teachings of the socialists had encouraged laboring men to place themselves outside any moral order.[40] On the other hand, the elitist reading of the passage above rationalized the class-specific and gender-specific nature of Cousin's psychology as he established it in practice through policy decisions about who would and would not be taught philosophy in the state schools. The popular classes as well as women and children (these last two were always coupled in his rhetoric) were, he repeatedly asserted, confined to spontaneously acquired knowledge and inherently debarred from reflection.[41] Hence it was useless to try to confer on them the benefits of the philosophy classroom or, what is the same thing, to encourage them to cultivate their subjectivity, to construe themselves as selves. From the eclectic standpoint, subjectivity, like the right to vote under the *monarchie censitaire* (a regime avidly supported by the Cousinians), was not a universal human attribute. It was rather one for which bourgeois males alone had aptitude. The strong class and gender bias of eclectic psychology left its marks on eclectic ruminations about aesthetics.

COUSINIAN AESTHETICS: SOME PRELIMINARY NOTES

Cousin's psychology formed the foundation of a totalizing philosophical system, one that not only expanded upward and outward to embrace ontology and metaphysics but, from its *moi*-centered position, also inflected more circumscribed branches of knowledge such as political theory, the philosophy of history, political economy, and aesthetics. Thus, for example, Cousin construed political economy as the study of human industriousness, which he characterized as "man metamorphos[ing] things" and "put[ting] on them the imprint of his personality, elevat[ing] them into simulacra of liberty and intelligence."[42] The same *moi*-centered perspective is evident in the most sustained work of Cousinian aesthetics, the *Cours d'esthétique* of Cousin's student Théodore Jouffroy—the text that I will use here for a preliminary probe into the aesthetic implications of eclectic subjectivity.[43]

In posing the question of the definition of the beautiful, Jouffroy is quick to dismiss the opinion of the "French school," which finds beauty residing in external attributes of the object, especially its order and symmetry. Rather, in keeping with the Cousinian position on the primacy of psychology, Jouffroy is intent upon making aesthetics a basically psychological affair. "It is within ourselves that we must direct our gaze. It is inside ourselves and by means of consciousness that the question must be attacked."[44] Accordingly, Jouffroy looks to the make-up of consciousness, not to the formal features of the aesthetic object, to find out what produces aesthetic pleasure. He comes up with a variant on the familiar Cousinian refrain about the *moi* and activity, although he substitutes the term "force" for Cousin's "activity":

> The more that these objects resemble man and participate in his nature, the more do they possess the gift of pleasing him; that fact is a constant. Now the nature of man, his intimate, primary and fundamental nature, is force. His law is to act everywhere, continually, endlessly, without rest; it is to live fully and amply, it is to dominate, it is to conquer.[45]

Further developing these views, Jouffroy notes that aesthetic pleasure must also be disinterested, admixed with no attraction to the object on the grounds of its utility or use value. The sentiments of beauty and utility are, he asserts, mutually annihilating. In the course of advancing this argument, he offhandedly transfers to the realm of aesthetics the familiar Cousinian distinction between the cognitive capacities of the bourgeoisie and working class. "Are more examples necessary? Here is one: The upper, rich classes of society, who are in general little burdened by considerations of usefulness, have a greater aptitude for appreciating beauty than do the poor classes, who are occupied with utility every day of the year."[46]

So, too—and again in passing—does Jouffroy transfer to the domain of aesthetics the Cousinian distinction between the psychological capacities of men and women. At least in the examples he adduces, that supposed psychological difference affects the value of men and women as objects, rather than subjects, of aesthetic contemplation; and hence the tenor of his discussion leads him to produce a highly unusual claim about the impossibility of female beauty.

Objects move us aesthetically, he repeats toward the end of his *Cours*, by "their invisible element," by the force in them similar to and therefore able to address the "force that animates us—that is to say, [human consciousness] endowed with the three principal attributes of sensibility, intelligence, and freedom." It is by reference to this trio that we distinguish the merely agreeable from the beautiful and the beautiful from the sublime. For example, the spontaneous movements of a woman are childlike and obey the impulsions of the passions, but they "do not give us the idea of a free force that understands its goal and heads toward it." Such movements therefore strike us as agreeable and nothing more. To acquire either beauty or sublimity, movements must express psychological attributes other than mere sensibility. "Only in face of the spectacle of a man who develops himself with intelligence and freedom, who pursues with his freedom the goal that he identifies with his intelligence, . . . can the beautiful and the sublime appear." The fundamental difference between the latter two lies in their relationship to struggle and, hence, in the quality of the sentiment of personality that they disclose. Sublimity attaches to the "idea of a free, intelligent force struggling against obstacles that impede its development," beauty to the idea of that same force "arriving at its goal easily and without effort." In other words, Jouffroy continues, what we label sublime evokes its characteristically intense aesthetic response because it provides an especially pure, strong, and concentrated expression of the sentiment of personality. By contrast, "there is in the development of a force operating with ease"— and that we consequently experience only as beautiful—"a self-forgetfulness (*oubli de soi-même*) entirely contrary to the sentiment of personality which dominates us when we develop ourselves painfully."[47]

In the hierarchy of aesthetic responses according to Jouffroy, then, the peak is attainable only in the presence of a distilled manifestation of the *moi*. And, given the Cousinian assumption about the inability of women to engage in reflection and thus to detach the *moi* from that primitive synthesis of consciousness in which unreflecting beings are mired, the female body apparently cannot exude that free selfhood experienced by the beholder as beauty. Jouffroy does not make this point in so many words but he insinuates it unmistakably through his choice of examples: the agreeableness of a *woman's* infantile spontaneity; the beauty of a *man* pursuing his free and rational development with ease; the sublimity of a *man* struggling to pursue that same free and rational development against obstacles.

The *moi*-centered aesthetics derived from Cousin's *moi*-centered psychology is evident in many of Jouffroy's discrete aesthetic judgments. Thus he declares that plays with complicated plots (*pièces d'intrigue*) require less talent on the part of the dramatist than plays that depict the unfolding of a single character (*pièces à caractère*).[48] We can take this to mean that a superior artistic talent is needed to seize and render concrete the supreme aesthetic object: the self and its activity. Similarly, he contrasts Fénelon's *Télémaque* with Madame de Staël's *L'Allemagne*, regarding the former as an instance of a unified, ordered, and perfectly coherent artistic work in which all the components are marshaled toward a single end, and the latter as an instance of a brilliant, sparkling, and capriciously variegated work, in which each chapter is animated by a different emotion. Though Jouffroy does not say so, these exemplary works are directly analogous to the types of movements that he characterized earlier in the *Cours*. "There is more pleasure to be had from reading *L'Allemagne* than from reading *Télémaque*," he acknowledges. But, lacking in coherence, *L'Allemagne* is, like the spectacle of a spontaneous woman, agreeable and nothing more. *Télémaque* is, Jouffroy opines, beautiful: "reason recognizes in it a free will and a project conceived in liberty." Apparently, Fénelon's celebrated work qualifies as beautiful because it is manly— which is to say, marked by the activity of that *moi* that men alone enjoy. The gendered attributes of the two works in question are replicated by the sexes of their authors.

Surely it is no coincidence that, to exemplify a work of the sparkling but incoherent sort, Jouffroy selected one of the few well-known works of his day written by a woman.

BY WAY OF CONCLUSION: AN ANTI-PICTORIAL IMPULSE

Using the state-wide pedagogical institutionalization of a particular brand of psychology as its basis, this paper has sought to pin down the features of a subjectivity that was, for much of the nineteenth century, deliberately imparted to nearly all the young bourgeois males of France. That this subjectivity was said to be experienced as pure, unnuanced assertion or propulsion—and that women's nonpossession of it was treated as a self-evident fact—certainly suggests that the conception drew heavily, if implicitly, on phallic analogies. It is not, of course, very surprising to learn that the nineteenth-century French bourgeoisie wanted to differentiate and distance itself from women and from the working class, or that it did so under the cover of a putatively universal human subject, or that these biases affected the canonical modes of reasoning about aesthetic matters. What is, I think, illuminating about an investigation of both the content and the state-mandated dissemination of Cousinian psychology is the sheer literalness, as well as the painstakingly detailed elaboration, of the whole affair.

The "bourgeois subject," we learn from this investigation, was no mere abstraction. It was rather an entity deliberately constructed by a new human science of psychology, one that was, at the propitious political moment, added to and given pride of place in the philosophy curriculum of the nation's secondary schools. Introspection, also taught in the philosophy classroom, was offered up as the key method for fashioning such a subject. The postulate of a triplicity of consciousness—in which the three elements were primitively fused and could be disaggregated only by persons with a talent for reflection—served to separate the sociocultural elite (bourgeois men) from the marginal players (women and workers). Once the *moi*-centered psychology was imported into the realm of aesthetics, the hegemonic "greed" of the male bourgeoisie to reserve to itself all positive attributes became strikingly apparent. Workers could not appreciate beauty, and the movements of female persons could not be characterized as beautiful.

The argument of the paper is, furthermore, epitomized in an absence: Unlike most of the contributions to the present volume, this one lacks pictures. Cousin and his disciples would no doubt be pleased by such austerity, for they regarded immateriality as the essence of the *moi*. That the *moi* stood in radical opposition to brute matter and hence was insusceptible to direct visual representation guaranteed for them its most precious attributes: freedom, immortality, moral accountability. Indeed the Cousinians would probably have assented to the proposition that the very materiality of workers and women—that is, their productive and reproductive roles respectively—explained their cognitive inability to achieve selfhood.

The Cousinians would also be gladdened by the non-pictorial nature of this account because an absence of supporting visual material would, in their view, emphatically distinguish them from their archenemies of the 1830s and 1840s, the phrenologists. By reducing mind to brain and distributing mental qualities among a series of discrete brain organs that they declared visible and palpable in the form of cranial bumps, the advocates of phrenology exemplified just the sort of psychological materialism that the Cousinians deemed inimical to the moral health and political stability of France. Moreover, the phrenologists, who usually subscribed to left-republican or socialist politics, proselytized among the working classes, arguing that phrenology offered an alternate psychology to humble folk who could not go to the lycée to learn Cousin's arcane terminology and subtle introspective techniques. To make a case for the accessibility of their teachings, they had only to brandish their easy-to-read maps of the brain and their plaster heads with the cerebral organs boldly delineated and labeled.[49] By contrast, the Cousinians' anti-pictorial impulse signaled a psychological doctrine that was abstruse, addressed to an elite, and politically cautious in its implications. No, there are no pictures here.

Notes

1. This is a basic argument of my book in progress, tentatively titled *The Post-Revolutionary Self: Competing Psychologies in Nineteenth-Century France*. Material from the book project relevant to this paper can be found in several of my recently published articles: "Foucault and the Post-Revolutionary Self: The Uses of Cousinian Pedagogy in Nineteenth-Century France," in *Foucault and the Writing of History*, ed. Jan Goldstein (Oxford: Blackwell, 1994), pp. 99–115, 276–80; "The Advent of Psychological Modernism in France: An Alternate Narrative," in *Modernist Impulses in the Human Sciences, 1870–1930*, ed. Dorothy Ross (Baltimore: Johns Hopkins University Press, 1994), pp. 190–209, 342–46; and "Saying 'I': Victor Cousin, Caroline Angebert, and the Politics of Selfhood in Nineteenth-Century France," in *Rediscovering History: Culture, Politics, and the Psyche*, ed. Michael S. Roth (Stanford: Stanford University Press, 1994), pp. 321–35, 496–99.

2. See Archives nationales, Paris (henceforth AN): F17* 1795, fol 435, Procès-verbaux des déliberations du Conseil royal de l'Instruction publique, September 28, 1832. Cousin was a member of this council and was present as the meeting.

3. See Jean-Louis Fabiani, *Les philosophes de la république* (Paris: Minuit, 1988), pp 32–33, 60; and my further discussion of this point in "Saying 'I,' " p. 331.

4. Such was, for example, the lot of Hippolyte Taine; see my discussion of his anti-Cousinian psychological views and his failing of the philosophy *agrégation*, in "Advent of Psychological Modernism," pp. 200–01.

5. See, for example, A-Jacques Matter, article "Moi," in *Dictionnaire de la conversation et de la lecture*, 52 vols. (Paris: Belin-Mondar, 1832–1839), vol. 38, pp. 259–61, esp. p. 259; this *Dictionnaire* was intended as a general reference tool for the bourgeois household. See also Louis Peisse, review of Cousin, *Fragmens philosophiques*, 2nd ed., in the daily newspaper *Le National*, September 25, and October 29, 1833, reprinted as an appendix in William Hamilton, *Fragments de philosophie*, trans.

L. Peisse (Paris: Ladrange, 1840), esp. p. 377. Noting that materialist pantheism asserts that "Everything is God," spiritualist pantheism that "God is everything," and idealism that "The *moi* is everything," Peisse puts Cousin squarely in the last camp.

6. See my discussion of the watered-down philosophy curriculum of the female lycées founded by the early Third Republic in "Saying 'I,'" pp 331–34.

7. Interestingly enough, the virtual abolition of the use of Latin in philosophy instruction and the establishment of psychology as the first section of the philosophy curriculum were part of the same series of reforms made by the Conseil royal de l'Instruction publique in the immediate aftermath of the July Revolution and under the aegis of Cousin The first occurred on September 11, 1830, the second on September 28, 1832, in fulfillment of a decree of September 11, 1830. Among the justifications for the first measure was that many ideas of modern philosophy could be rendered "only obscurely and imperfectly" in Latin and that it was important "to maintain the preeminence of the national, popular language in philosophical matters." See AN: F17* 1788, fol. 77; F17* 1795, fols. 434–35. It is almost as if psychological sophistication were being substituted for Latinity as a marker of sociocultural elitism.

8. Victor Cousin, "Préface de la deuxième édition" (1833), *Fragmens philosophiques,* 2nd ed (Paris: Ladrange, 1833), p. lx.

9. On these terminological matters, see the article "Introspection," in André Lalande, *Vocabulaire technique et critique de la philosophie* (lst ed., 1926), 16th ed. (Paris: Presses universitaires de France, 1988). Lalande notes, presumably from the vantage point of ca. 1926, that the term is still rare in French and confined to technical usage. See also entry "Introspection," in Paul Foulquié, *Dictionnaire de la langue philosophique,* 2nd ed. (Paris: Presses universaitres de France, 1969), which includes a quotation from Théodule Ribot's article "Psychologie" in the multi-author collection *De la méthode dans les sciences,* vol. 1 (Paris: F. Alcan, 1909) equating the terms "observation intérieure" and "introspection."

10. Auguste Comte, *Cours de philosophie positive,* 2nd ed., 5 vols. (Paris: Baillière, 1864), vol. 1, pp. 32–33, Lesson 1. The English translation, which I have taken from Gertrud Lenzer, ed., *Auguste Comte and Positivism: The Essential Writings* (New York: Harper Torchbook, 1975), p. 80, is not entirely literal, but it accurately captures the sense of Comte's text. Comte first articulated this argument against the mind's observation of itself in a letter of September 24, 1819, to Valat; see Auguste Comte, *Correspondance générale et Confessions,* ed. P. E. de Berrêdo Carneiro and P. Arnaud, vol. 1 (Paris and the Hague: Mouton, 1973), esp. pp. 58–59.

11. Victor Cousin, "Préface à la première édition" (1826), *Fragmens philosophiques,* 2nd ed. (Paris: Ladrange, 1833), pp. 11–13.

12. Ibid., pp. 49–50.

13. A.-F. Gatien-Arnoult, *Programme d'un cours de philosophie,* 4th ed. (Toulouse: Bon & Privat, 1841), p. 22, italics in the original. The title page reads *"approuvé par le Conseil royal de l'instruction publique."*

14. A.-F. Gatien-Arnoult. *Cours de lectures philosophiques, ou dissertations et fragmens sur les principales questions de philosophie élémentaire* (Paris and Toulouse: J. B. Paya, 1838), p. 81, note 1.

15. See Comte, *Cours de philosophie positive,* vol. 3, lesson 45, esp. pp. 538–39. The sentence before the thinly veiled reference to Cousin begins, *"Quant à leur vain principe fondamental de l'observation intérieure,"* emphasis in the original.

16. Cousin, "Préface à la première édition," p. 12.

17. Cousin, *Introduction à l'histoire de la philosophie* (Paris: Pichon & Didier, 1828), Lesson 2, pp. 5–6. Each lesson is separately paginated in this edition.

18. Ibid., Lesson 5, p. 35.

19. Cousin, "Préface de la deuxième édition," p. viii.

20. Victor Cousin, "Du Fait de conscience," *Fragmens philosophiques,* 2nd ed., pp. 242–52, quotation on p. 246; as the text indicates, Cousin was avowedly in the thrall of Fichte when he wrote it. The provenance of the text is identified as Cousin's 1817 Sorbonne course in the reprinted version of it in Cousin, *Premiers essais de philosophie,* 3rd ed. (Paris: Librairie nouvelle, 1855).

21. Cousin, "Préface de la deuxième édition," pp. viii-ix.

22. Ibid.

23. P. C. V. Boiste, *Dictionnaire universel de la langue française*, 2nd ed. (Paris: Desray, year 11/1803). The first edition of this dictionary, dated 1800, does not contain *subjectivité* and its entry for *subjectif* mentions only the older (and now outmoded) meaning, "placed below"; the second edition contains that older meaning for *subjectif* after the new Kantian one, "belonging to the subject."

24. "Subjectivity" was for Villiers a quintessentially Kantian term, expressing in compressed form Kant's critique of empiricist epistemology, which lodged the laws governing human knowledge in the *objects* of our perception. Against this view Kant insisted that our experience was an amalgam of both objective and subjective components, the latter derived from the structure of our perceptual apparatuses as knowing subjects. As Villiers explained the Kantian principle, "When a hypocondriac, for example, sees the world in shades of black, we say that this blackness is subjective, that it has only a subjective reality for the hypocondriac and no objective reality at all." He went on to say that he was "deliberately multiplying examples in order to clarify for the reader this necessary distinction between *subjectivité* and *objectivité* in the knowledge that we acquire of things." Charles de Villiers, *Philosophie de Kant, ou Principes fondamentaux de la philosophie transcendante* (Metz: Collignon, 1801, year 9), p. 113n.

25. H. de B. [Honoré de Balzac], "Des Mots à la mode," *La Mode 2* (May 1830): 189–94, quotation on 194. This source is cited in Patrice Vermeren, *Victor Cousin: Le jeu de la philosophie et de l'état* (Paris: L'Harmattan, 1995), pp. 18–19. Vermeren believes that Cousin and his disciples are the butt of Balzac's humor.

26. See Victor Cousin, *Cours de philosophie professé à la Faculté des lettres pendant l'année 1818 sur le fondement des idées absolues du vrai, du beau et du bien* (Paris: Hachette, 1836), pp. 25–26. Cousin is here criticizing the Kantian *moi* for its inability "to generate the absolute." In the course of presenting the Kantian position in this passage, he uses the adjective *subjectif* but not the noun *subjectivité*.

27. Cousin, "Préface à la première édition," pp. 13–14.

28. Ibid., p. 38: "*Le triplicité de conscience, dont les élémens sont distincts et irréductibles l'un à l'autre, se résout donc dans un fait unique, comme l'unité de la conscience n'existe qu'à la condition de cette triplicité.*"

29. Cousin, "Préface de la deuxième édition," p. xiv.

30. Cousin, "Préface à la première édition," p. 17.

31. Cousin, *Introduction à l'histoire de la philosophie*, Lesson 5, pp. 9–10.

32. Cousin, "Préface à la première édition," p. 26. "Mysterious" is Cousin's word, and one that he frequently employs in this context.

33. See, for example, the opening lecture of his 1828 course, where he underscores the inwardness that he had cultivated during his years of political banishment from his teaching post at the Sorbonne: "Separated from the public for eight years, I have lost the habit of speaking before an assembly such as this. Accustomed, while in retreat, to those forms of thought that serve well to explain ourselves to ourselves but not always to communicate to others, I fear no longer knowing how to find the words suitable for a large audience and, hence, bringing to this chair the monologues of a solitary man." *Introduction à l'histoire de la philosophie*, Lesson 1, p. 4.

34. Cousin, "Préface de la deuxième édition," p. xxxv. As the pages leading up to this passage indicate, even the fact of Cousin's having had French mentors was of polemical significance because he was regularly accused of being excessively Germanophile.

35. Ibid., p. xv.

36. AN: F17* 1795, fols. 435–36, my italics.

37. One exception is A. Charma, whose *Du sommeil* (Paris: Hachette, 1851) draws on the record of his own dreams that he kept over a dozen-year period. But tellingly, Charma's exercise in this kind of introspection caused him to question some of the tenets of orthodox Cousinianism; see my discussion in "Advent of Psychological Modernism," pp. 199–200.

38. See *Journal intime de Maine de Biran*, ed. A. de la Valette-Monbrun, 2 vols. (Paris: Plon, 1927). The journal, which Maine de Biran kept in a variety of notebooks, pocket agendas, and even on loose sheets of paper, covered the years 1792–1824 and filled some thirteen hundred pages. Much of it was occupied with describing inner states of mind and emotional malaise.

39. Cousin, *Introduction à l'histoire de la philosophie*, lesson 5, pp. 39–40. Cousin made much the same general point—though without specific reference to the contents of consciousness—two years earlier in the preface to the first edition of the *Fragmens philosophiques*: "There is in reflection nothing that is not in the [mental] operation that precedes it, spontaneity. Reflection is, to be sure, a degree of intelligence rarer and higher than spontaneity, but it is bound by the condition that it summarize spontaneity faithfully and develop it without destroying it. Now, in my opinion, the mass of humanity is spontaneous and not reflective." "Préface à la première édition," p. 45. On the 1828 course as media event, see Anne Martin-Fugier, *La vie élégante, ou la formation du Tout-Paris, 1815–1848* (Paris: Fayard, 1990), p. 244, which indicates that this was the first time that stenographers were employed in a nongovernmental setting.

40. See Cousin, *Philosophie populaire, suivie de la première partie de la profession de foi du vicaire savoyard* (Paris: Pagnerre, 1848), pp. 1–14.

41. With respect to the lack of capacity for reflection in the popular classes, see the passage from the "Préface à la première édition" cited in note 35, above, and *Introduction à l'histoire de la philosophie*, lesson 8, p. 15: "I have the deepest respect for good sense, for good sense is nothing but reason itself at its lowest level, in its most popular [i.e., pertaining to *le peuple*] aspect." With respect to the lack of capacity for reflection in that compound being, women-and-children, see *Introduction à l'histoire de la philosophie*, Lesson 9, p. 15: "There [i.e., in the domain of metaphysics], all is obscure for the senses and the imagination, for children and for women; but there, too, all is light for reflection, for he who demands of himself a manly (*viril*) accounting of his own thought." See also Ibid., Lesson 8, p. 19, a passage angrily quoted back to Cousin by Caroline Angebert and discussed in my "Saying 'I,'" pp. 326–29.

42. Cousin, *Introduction à l'histoire de la philosophie*, Lesson 1, pp. 11–12.

43. Théodore Jouffroy, *Cours d'esthétique*, ed. Ph. Damiron (Paris: Hachette, 1845). Given as a set of private lessons in 1826, during the period that the Restoration monarchy had shut down the Ecole normale and deprived Jouffroy of his teaching post, this text was published posthumously; see "Préface de l'éditeur."

44. Ibid., p. 11.

45. Ibid., p. 14.

46. Ibid., p. 27; see also p. 73, where Jouffroy further develops this class-based aesthetics by noting that "*le peuple*" have only a customary sense of order and proportion and lack the reflective power to formulate a general definition of those aesthetic qualities; hence all changes in prevailing aesthetic values must come from "*les hommes éclairés.*"

47. Ibid., Lesson 14, esp. pp. 315–18.

48. Ibid., p. 114.

49. These aspects of phrenology are briefly discussed in my "Advent of Psychological Modernism in France," esp. pp. 192–93, 196–97, and will be more thoroughly explored in my forthcoming book.

JOEL SNYDER

Visualization and Visibility[1]

At the outset of an elegant and persuasive essay exposing "the moralization of objectivity in the late nineteenth and early twentieth centuries" Lorraine Daston and Peter Galison argue that the nineteenth-century physiologist, Etienne-Jules Marey, "and his contemporaries turned to mechanically produced images to eliminate suspect mediation" as a means of abolishing "human intervention between nature and representation."[2]

My interest in this topic was provoked by Daston and Galison's essay, but is tangential to theirs. I am not concerned, as they are, with nineteenth-century compilations of scientific data produced by mechanical means, but with the question of how we should think about what is shown in Marey's graphs and chronophotographs. Much of Marey's work does not fit the mold that Daston and Galison make for it, however right they may be about other nineteenth-century scientists who invented machines for collecting scientific data. For the most part, Marey did not conceive of his precision instruments as impartial mediators substituting for and improving upon an observer's eye or an illustrator's hand. His mechanically originated graphs and photographically generated pictures are visualizations of displacements charted against precisely determined units of time. These movements fall outside the scope of human detection and accordingly, their inscriptions cannot be characterized as especially accurate visualizations of what might otherwise have been registered by an illustrator or scientist. To put this in slightly different form: in most of Marey's experimental work there is no place (literally or figuratively) for human intervention, nothing for a mediator to mediate, no conceptual room into which a scientist might enter and intervene, not because instruments substitute accurate, mechanically produced data for the unreliable, humanly generated variety, but because the displacements registered by mechanical monitors and traced by clockwork-driven inscribers fall outside the scope of human sensibility. Consequently, they do not permit even the possibility of human intervention.

Daston and Galison view the adoption of what they call "mechanical, or noninterventionist objectivity" as an expression of what they claim was the growing fear (beginning in the 1830s) of subjectivity in science and of the moral necessity of "censuring some aspects of the personal."[3] I am uneasy with the notion that the objective/subjective opposition (in which objectivity, in any of its specific guises, is characterized as a constraint of the personal) fits Marey's program of graphic and photographic visualization. It is misleading to view Marey's enterprise as one in which mechanized monitoring and inscribing devices function as more tireless workers than human observers, or as disinterested observers, or to think that his detectors provide better observations than humans can on their own. This view suggests that the instruments are somehow in competition with observers—providing a constraint on the fallibility of the all-too-human scientist. But Marey's mechanically generated records were not typically produced as a remedy for a generalized and febrile anxiety about human perceptual (or moral) *fallibility*, although Marey, among other scientists, was keenly aware of the ineradicable possibility of such failures.[4] In much of his work there is no question of substituting mechanical instruments for a fallible human mediator and of correcting thereby what might otherwise have been falsified (whether by guile or inadvertence). The graphic data show what otherwise cannot be found in the realm of events and processes detectable by human beings and accordingly, questions concerning the reliability or accuracy of machine-generated visualizations cannot be answered by recourse to a human arbitrator, no matter how exquisitely sensitive or impartial. Questions about the accuracy of these data can be resolved only by appealing to other, perhaps even more refined mechanical instruments.

Observers disappear from much of Marey's work and are replaced by graphic records charting relations that cannot be observed, or describing movements that, in Marey's words, entirely "elude" or "escape" detection (and not in the way we might say criminals elude or escape detection). Writing in La Méthode graphique in 1878 about the machinery he constructed for his investigations, Marey makes this claim:

> Not only are these instruments sometimes designed to replace the observer, and in such circumstances to carry out their role with an incontestable superiority, but they also have their own domain where nothing can replace them. When the eye ceases to see, the ear to hear, the touch to feel, or indeed when our senses give deceptive appearances, these instruments are like new senses of astonishing precision.[5]

There are two ideas here about the role of instruments in relation to physiological research: mechanical devices can substitute for and improve upon the performance of an observer; and far more intriguing, machines can be constitutive of their own field of investigation—one in which substitution is not at issue. These tools can provide access to an unknown world—to a new province of study generated by the instruments themselves. But Marey does not go far enough—it is not just that his instruments can usefully be likened to new senses, they are also unlike the senses in one important respect: they not only detect, but simultaneously chart what they register. The detected displacements are properties of the subject under investigation, but one part of the equipment is also in motion: the inscriber. The visualizations, or graphic data, are a function of the imperceptible movement of the experimental subject and of the precisely regulated, revolving motion of the inscriber. Although the detectors *are* like new senses, the data cannot be likened to sensations. The inscriber plots movements, but is in motion itself and so the data are indices both of the displacements of the subject of study and of the machinery tracing the movements. The data owe their existence to the instruments that make them and have no existence apart from the graphing procedures.[6]

Marey's justification for the use of precision machinery in his work reads like a Cartesian denunciation of human sensory defectiveness and inadequacy and like Descartes, his primary concern is with the urgent need to block the errors that follow from putting faith in the depositions of the senses. In the introduction to *La Méthode graphique*, he explains: "When we speak of the defectiveness of the senses, we do not wish only to state their inadequacy for the discovery of certain truths; but primarily, to indicate the errors they lead us to commit."[7] As his prefatory remarks proceed, Marey's rhetoric turns both juridical and self-defensive, and he adopts a language of protection, constant vigilance, and correction. Fearfully, it turns out that the self must be protected from the testimony of its own senses—must be protected from itself:

> Nobody today doubts the need to protect himself from the testimony of sight, hearing, and touch. The sphericity of the earth, its diurnal rotation, the distance of the stars and their huge volumes are determined by correcting what is given to us by our senses.[8]

This vision of a self guarding against itself by recourse to the protection offered by an armory of machines is fundamental to Marey's conception of the compensatory and corrective role played by mechanical instruments in "the conquest of truth." But the insistence on the use of mechanical devices for the production of scientific data is not driven by moral necessity, by the need to eliminate the possibility of human bias, or by other forms of subjective interference in the search for truth. The realm of the graphic and chronophotographic methods is not the domain of the senses and the displacements detected by machines cannot be registered by sight, hearing, or touch. The data produced by his machines are, to use Marey's language, revelations. Our senses provide us with straggling and confused perceptions, giving testimony of a world in chaos. The instruments devised for the graphic and chronophotographic methods "penetrate" this apparent chaos of ceaseless motion and "reveal" an unknown world in the very place the senses testify only to discord and anarchy. Self-protection requires the invention and deployment of an arsenal of precision machinery. The advancement of science itself is conditioned upon the production of data about a world that is knowable only by way of machine-generated documentation. Marey puts it this way:

> Free of the prejudice of the infallibility of our senses and kept on continuous guard against the information they give, science searches for other means in the conquest of truth; it finds them in precision instruments. [Science] has had, for a long time, the means of taking the exact measurement of dimensions, mass, composition, in a word, of the static state of natural bodies; now it begins to study forces in their dynamic state. Motions, electrical currents, variations of gravity or temperature—such is the field of exploration. Our senses, with their too slow and too confused perceptions cannot guide us in this new enterprise, but the graphic method compensates for [*supplée*] their inadequacies; from within this chaos it reveals an unknown world. Inscribing devices measure infinitely small lapses of time; the most rapid and the feeblest movements, the least variations of forces, cannot escape them. These devices penetrate the intimate functions of organs where life seems to exist in ceaseless motion.[9]

I am inclined to say, though an expert in Marey's work might disagree with me, that the results of the graphic method rarely substitute, in any straightforward sense, for data gathered by researchers. In other words, I regard most of Marey's work as falling into a domain in which the mechanical detectors, transmitters, and inscribers of the graphic method, or the high-speed cameras of the chronophotographic method produce data that do not substitute for the kind an investigator might uncover without the aid of these instruments. There is no issue of substitution here—these data replace nothing and "nothing can replace them"; they constitute the raw material of scientific investigation. Marey's procedures direct study away from what we might think of as the phenomenon needing explanation (e.g., the rapid movement of the legs of a horse in full gallop), turning it to the analysis of graphs or pictures that he conceived of as automatic inscriptions of otherwise undetectable displacements. The primary data subjected to analysis and interpretation are mechanically contrived visualizations (either graphic or pictorial) of motions that are unknowable apart from their mechanical realization. When I say that observers drop out of Marey's program, what I mean to suggest is that while the goal of explanation is an understanding of the forces at work in say, the beating of a bee's wings, the subject of investigation—the materials studied by the scientist—are the graphic data themselves. The role of the scientist during the collection of data is, as Marey notes, to make certain the instruments are functioning correctly and the subject, say a runner, knows how to use them, and does, in fact, use them correctly. Intervention ceases to be an issue once the investigator makes certain the machinery is functioning properly. Scientific investigation centers on the graphic records of the registrations inscribed by the machines, or portrayed by the photographic apparatus.

It might seem that there is little difference, in principle, between results obtained by older instruments like microscopes and telescopes and Marey's machinery—since in each of the cases, things that cannot be detected by the "unaided eye" are made visible by instruments functioning as aids to the eye. This is a question deserving separate discussion and is beyond the scope of this essay, but it is worth noting in passing that microscopes and telescopes function as aids to vision in a way that neither the graphic nor the chronophotographic methods can. A scientist looking through a telescope sees Io or Ganymede and not pictures of them. In Marey's procedures, the data are realized by the machinery in the form of visualizations—inscriptions, graphs, pictures. There is no moment during the operation of the instruments in which a scientist can take a peek at anything that might remotely approximate the results of the mechanical, data gathering operation. In Marey's program, the visualized data produced by the inscribing mechanisms have no existence apart from their realization.[10]

In "The Image of Objectivity," Daston and Galison, and in this volume, Galison, in "Judgment against Objectivity," oppose the ingrained tendency of some historians of science to invoke a transhistoric conception of objectivity, by seeking rather to identify and describe the various historically locatable practices that, at any given moment in its history, give specifiable meaning to the term. Yet, despite their insistence on the shifts in the meaning of "objectivity" resulting from determinable changes in specific scientific practices, they nonetheless find an underlying pattern uniting the disparate conceptions of objectivity. In "The Image of Objectivity," they say this:

> We address the history of only one component of objectivity, but we believe this component reveals a common pattern, namely the negative character *of all forms of objectivity*. Objectivity is related to subjectivity as wax to seal, as hollow imprint to the bolder and more solid features of subjectivity. Each of the several components of objectivity opposes a distinct form of subjectivity; each is defined by censuring some (by no means all) aspects of the personal.[11]

Galison puts the matter this way:

> Objectivity as it was used in the very center of scientific work had a birthdate in the mid-nineteenth century. Moreover, the story of objectivity is a conjoint development, implicating both observational practices and the establishment of a very specific *moral culture* of the scientist. In the first instance, objectivity had nothing to do with truth, nothing to do with the establishment of certainty. It had, by contrast, everything to do with a machine ideal: the machine as a neutral and transparent operator that would serve as instrument of registration without intervention *and* as an ideal for the moral discipline of the scientists themselves. Objectivity was that which remained when the earlier values of the subjective, interpretive, and artistic were banished.[12]

This is an appealing approach to understanding claims about "objectivity" in nineteenth-century scientific discourse, but it clashes, or so it seems to me, with Marey's repeated assertion that at least some of his instruments have their "own domain"—a zone of inquiry constituted by instruments possessing inhuman capacities of registration. If Marey thought of his mechanical devices as being capable of producing objective data, it could not possibly have been because they *substituted* mechanical for human registration. Objectivity in these instances would have been achieved by changing venue, by transferring domains from the sensible to the supersensible, by shifting from the perceptible to the imperceptible. The venue shift however, cannot be conceived as effecting a subordination of subjectivity—it is the total elimination of it. I am aware of the implications of this claim. What follows from Marey's rhetoric of penetration and revelation (penetration of the most intimate functions of organs and revelation of a world to which we have access only through the use of precision instruments) is the brain-numbing, perhaps analytically incoherent notion of a freestanding objectivity that cannot take its meaning from being opposed to subjectivity. The objectivity of machine-generated data is not achieved by constraining the personal; it is, for Marey, an emancipation from subjectivity.[13]

Marey's first successful instrument was the sphygmograph (1860), a device used to graph pressure changes in the heart as it goes through repeated stages of expansion and contraction (see Figure 1). It consisted of a free-moving lever with one end resting on the pulse point of a subject's wrist and the other end fitted with a steel stylus resting against a slip of steadily moving, smoke-blackened paper. The rhythmical increase and decrease of the caliber of the artery at the wrist produced a bobbing movement of the lever that in turn moved the stylus up and down against the uniformly moving blackened paper. The stylus moved upward with the distention of the artery, rising to a peak and then dropping with the elastic contraction, and since the paper was moving constantly with the rising and falling of the stylus, the result was a series of inscribed

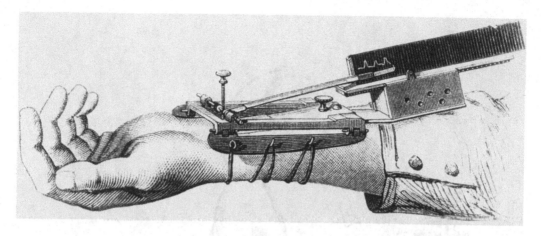

Figure 1.   Etienne-Jules Marey, attachment of the sphygmograph to the wrist for inscription of the trace of the pulse, 1860 (Marey, La Méthode graphique).

upward- and downward-sloping lines, separated by a baseline marking periods of arterial stasis. A finely tuned clockwork mechanism moved the paper at a uniform speed that, in turn, allowed the "reader" of the graph to measure sloped lines and baselines against precisely calibrated lengths of paper, which denoted units of time.

Is the sphygmograph a substitute for an observer, or does it bring us into the domain of something like Marey's "new senses"? Presumably, Marey devised the sphygmograph (the name means "pulse writer") to replace the traditional means of taking a pulse reading by pressing two fingers against a throbbing artery. There is no doubt that the mechanism removes the vagaries of tactile pulse readings by shifting the detection of the pulse from the experimenter's fingers to the mechanical registration of the artery's displacement—initiated and maintained by the movement of the artery itself. The automatic inscription yields a graphic record that is publicly available, quantifiable, and retrievable, even in the absence of the subject and no doubt, the autographic character of the charting procedure is crucial to claims about the accuracy and objectivity of the results. But should we think of this as a substitution or a replacement for observations formerly made by the tips of an experimenter's fingers?

I am uncertain about the answer to this question, but it does seem incorrect to think of the sphygmograph as being a replacement—at least in an unproblematic way—for the older method of pulse taking. The kind of information that can be read off a sphygmographic inscription was not available by means of the digital taking of a pulse.

Let me take another, less problematic example—Marey's graphic procedure for charting what he called "modes of progression" or "locomotion"—walking, running, galloping, and leaping (see Figure 2). The procedure required an appliance called "an

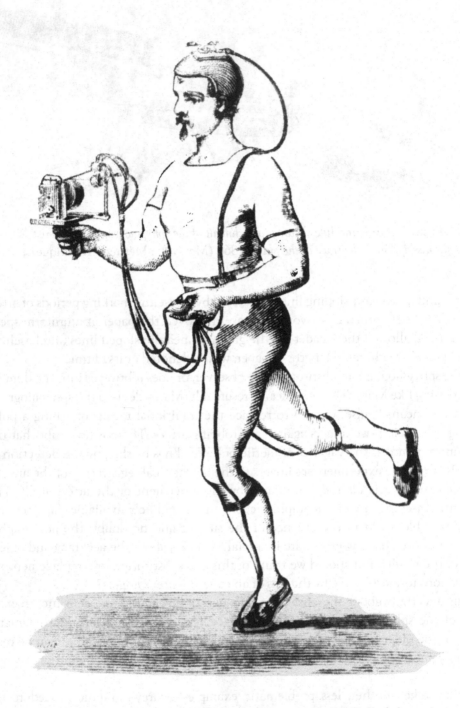

Figure 2. Etienne-Jules Marey, runner wearing experimental shoes and inscribing apparatus (Marey, Animal Mechanism).

experimental shoe," which contained an air-filled chamber in the sole attached to a rubber tube. When a foot, to which the shoe was attached, exerted pressure on the ground, compressed air shot through the tube into a drum with an attached lever, which in turn moved a stylus across a moving sheet of paper. In the least complicated of Marey's locomotion experiments, a human subject wore an experimental shoe on each foot and walked across a level, uniformly hard floor—in more complex experiments, horses were fitted with four experimental shoes and measurements were taken as the animal trotted or galloped across a level floor. The resulting graphs chart the duration, phases, and intensity of the pressure exerted against a uniformly resisting surface by each of the feet and permitted Marey to determine how the feet, driven by the motive force of the leg muscles, do the work of impelling the body forward.[14]

The graphic method, in the case of the sphygmograph, allowed Marey to devise a visual counterpart to readings formerly registered by a physician's fingertips placed on a patient's wrist pulse, and here the issue of restraining the physician's subjectivity is clearly important. But in the animal locomotion experiments, Marey was not looking for a means of restraining an observer's personal bias. His review of a hundred years of patient, but conflicting observations by equestrians, illustrators, and scientists of the gallop of a horse—observations not only of horses in motion, but also of the imprints of their hooves left on carefully prepared sand floors in riding academies—demonstrated to him that these rapid displacements were incapable of being analyzed correctly by traditional methods. The inability to achieve consensus among specialists about the gallop of a horse was not a mark of observational failure, but rather of observational incapacity. The problem was not to censure personal impulse, but to eliminate reliance on observational schemes that were incapable, in the first instance, of resolving the details of swift displacements.

Marey's method might seem to introduce, in the case of rapid animal locomotion, the possibility of finding novel graphic expressions for phenomena that had always been conceived in visual terms, but he was not, in fact, interested in determining, for example, what a horse at full gallop *looks like*; he wanted, rather, an accurate analysis of the mechanics of animal locomotion. I wonder if "phenomena" here isn't being used equivocally. I am inclined to say that Marey's machines produce data about phenomena—about the highly qualified subject of investigation (e.g., the relation of forces at specific instants of a full gallop), but that these data are entirely artifactual, the products of machinery and a conceptual scheme—mechanics—that give intelligibility to the inscribed curves. Whatever the charts may be, they are not illustrations of the movements of horses' legs; when properly deciphered, they are records of work performed by them.

Again, what I want to emphasize—perhaps it is more apparent in the locomotion case than in the example of the sphygmograph—is the character of the graphic results.

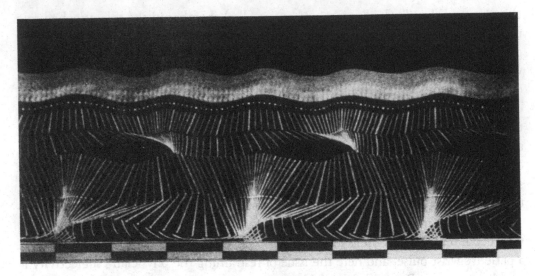

Figure 3.   Etienne-Jules Marey, geometric chronophotograph showing oscillations of the leg during running, 1883 (Marey, Collège de France).

The graphs do not stand in for processes that could have been detected by an investigator bereft of properly designed mechanical instruments. What is shown is realizable only on a machine-made graph. The goal of these experiments—the measurement of duration, intensity, and phases of pressure exerted by each foot or hoof as it performs its role in locomotion is part of a larger project of determining the *work* performed by a body in terms of its constituent mechanical elements, which in turn depends upon having information not incorporated into the graph, for example, the mass of the body and the character of the resistance it encounters in moving forward.

This seems to be an accurate characterization of the graphic method. Marey refers indifferently to the records produced by the method as "curves," "notations," and "tracings," though he sometimes lapses into saying they "represent" certain kinds of movements. I do not want to be overly fastidious about the use of "represent," but I do want to insist that these records, if they are to be thought of as representations, are not to be confused with pictures. If representation is a matter of re-presenting, it is difficult to imagine what the "re" might apply to in cases like the inscriptions produced on a sheet of paper by pressurized rubber balls glued to the hooves of an ambling horse.[15]

The detectors used by Marey in his graphic methodology required some form of physical contact with the subject of his experiments. This was a frustrating requirement built into the procedure. Moving objects that were inaccessible, or that could not be fastened to a recording apparatus by mechanical means (either because of the limitations of the instruments or because the mechanical connection changed the character of the movement to be recorded) could not be charted by the graphic method. In response, Marey devised a chronophotographic apparatus dispensing with mechanical detectors and transmitters. The problem in making use of photography for the study of movement is that a means had to be found that would allow a photograph to show "the relationship existing at any moment between the distance traversed and the time occupied."[16] Marey's solution to the problem began with perfecting extremely high-speed rotary shutters; since his interest was in charting movement against time, the much more difficult problem was to find a means of making serially related images for which the exact shutter speed was known and, crucially, for which the intervals between exposures were precisely determinable. The solution to these problems involved the fabrication of a disk with accurately spaced apertures that was mounted inside the lens of the camera.[17] The disk revolved at high speeds, permitting multiple exposures of brightly lit objects moving across a black background (see Figures 3 and 4). The first chronophotographs were made on fixed plates, but these gave way to moving plates and finally to moving film.

The shift from the graphic to the chronophotographic method introduces conceptual problems relating to photography that were, by and large, overlooked in the late nineteenth and early twentieth centuries. I have argued that Marey's graphic method is primarily a means of making charts, graphs, or diagrams, but chronophotographs are most often pictures. In *Movement*, Marey provides a brief statement of what photography does:

Figure 4.   Etienne-Jules Marey, chronophotograph of fencer showing
two positions of visibility, 1890 (Marey, Collège de France).

Photography, [by contrast with the work of an artist] gives an instantaneous picture of the most diverse objects, and that, too, with the prevailing conditions of light, and all in correct perspective. The appearance of natural objects, as seen by looking with one eye only, is thus reproduced by photography.[18]

He also characterized chronophotography as "a method that demands no material link between the visible point and the sensitized plate on which its movement, from moment to moment, is recorded.[19] Held together, Marey's portrayal of chronophotography, with its emphasis on "the visible point," and his understanding of photography as closely related to monocular vision, would seem to inject an ideal observer back into his graphic program. That is, "the visible point" to which Marey refers is determinable solely in terms of what an observer can see and, as Marey has it, photography reproduces the appearance of natural objects *to the eye*. It would seem to follow that chronophotographs would have to represent what any competent viewer might see from a given perspective.

Marey is a kind of ideal commentator on photography, one who holds nearly all of the contradictory views maintained by late-nineteenth-century writers on photography. If he comes close to suggesting that photographs can only reproduce the appearance of natural objects as seen by one eye, he is also capable of exulting in the production of photographs showing things that could not possibly be seen because they do not exist. In describing a chronophotograph showing a globe, but made by photographing, on one plate and in rapid succession, a single revolving band of curved acetate, he says: "In reality we are dealing with a hypothetical figure, which finds no counterpart in Nature."[20] Again, if he can assert that photographs depict natural appearances, he can also claim: "Although chronophotography represents the successive attitudes of a moving object, it affords a very different picture [image] from that which is actually seen by the eye when looking at the object itself."[21] And finally, if the older graphic procedures fail, "photography comes to the rescue, and affords accurate measurements of time events which elude the naked eye."[22]

Marey's chronophotographic enterprise began in 1882, at the moment when routine instantaneous photography was becoming a practical reality through the commercial production of high-speed gelatin dry plates. As I have tried to indicate, photography had not been coherently formulated when Marey began his chronophotographic experiments, and notions of what a photograph is, or how it relates to what it represents were contradictory and often fantastic.

For reasons that I would like to understand much better than I do, photographs slipped into daily life in Europe and America very discreetly, almost silently, in the years between the early 1840s and the 1880s. For all their pervasiveness, for all the massive production—the selling and buying of portraits, travel views, stereographic cards, cartes-de-visites, and the like—photographs and photographers were rarely thematized in novels, poems, popular literature, or general-interest magazines of the time. It seems, for example, that Charles Peirce was the only nineteenth-century philosopher of note who wrote anything interesting about photographs, and what he said took up all of twelve lines (and has been repeated ad nauseam by contemporary critics who know next to nothing about the rest of his massive and difficult writings). Part of the reason for the silence about photography grows out of its very ordinariness, its ubiquity, and most interestingly for my purposes, its vulgarity and supposed superficiality. By the 1850s, photographic portraits were figured as integumental likenesses, incapable of displaying anything deeper than the sitter's skin and clothing—incapable of evoking the sitter's personality, or the photographer's. Charles Baudelaire condemns photography in the "Salon of 1859" because it is, he says, destructive of the values stemming only from the activity of the imagination. He criticizes photography because it cannot represent anything but "external reality," the immediate, the material, and therefore, the trivial. This is a critique firmly embedded in the fear of popular culture and "the mob," and of what he finally calls "industrial madness."

Baudelaire's critique was indirectly dependent on the insensitivity of photographic materials and the consequent long exposures, which required that every photograph be conceived as a still life, or at best a *tableau vivant*. Baudelaire's notion of "external reality" was modeled on seeing, and he understood seeing itself in terms of looking at a static world: "More and More, as each day goes by, art is losing in self-respect, is prostrating itself before external reality, and the painter is becoming more and more inclined to paint, not what he dreams, but what he sees."[23] What could be seen in a photograph, as Baudelaire has it, was what would have been seen from behind the camera at the time of exposure. The authors of the earliest descriptions of daguerreotypes dwelled on the profusion of detail carried on the surface of the mirrored plate, noting, for example, puddles or stray reflections that had gone unnoticed at the time the plate was made. But with such descriptions, the supposition was always that the bystander or photographer could have seen what came to be represented photographically, if only he or she had carefully attended to the scene in front of the camera. Photographs seemed to corroborate human vision. This assurance came at a price—photography preserved and perhaps exalted the primacy of vision by calling into question the capacity of the human hand to sketch with fidelity what the human eye could see, if only the eye bothered to notice what was in front of it. And so, photographs were characterized as having been "drawn with inhuman precision"—even while the subject of precise delineation was visible to anyone willing to look. Writing in 1840, Edgar Allen Poe struggled to find the words by which he could describe what could be seen in a daguerreotype. He wrote:

All language must fall short of the truth. . . . Perhaps if we imagine the distinctness with which an object is reflected in a positively perfect mirror, we come as near the reality as by any other means. For, in truth, the Daguerreotyped plate is infinitely (we use the term advisedly) is *infinitely* more accurate in its representation than any painting by human hands.[24]

This strict likening of the daguerreotype picture to a reflection in a mirror is a common analogy in the early years of photographic practice. The attempt to identify photographic pictures with camera imagery or with mirror reflections is really a way of aligning photographic imagery with human vision, but the price of this alignment is purchased by disaligning the eye from the hand. On this view, vision is superabundant and skills of hand cannot keep up with it—but a machine can, and accordingly, photography was often addressed in terms of its "inhuman" fidelity.

I want to make this point a little more emphatically. One of the most common ways of describing what photography is, or better, what it does—during the period of its initial publication and for a number of years thereafter (I am talking roughly about the period between 1838 and 1855 in Europe and America)—was to claim that a photograph is a "fixed" camera image. All the inventors of photography—Niepce, Daguerre, Talbot, and Bayard—talk this way. Most of us today want to think of this as a metaphorical fixation, but Daguerre, for example, took it literally. The single most important condition for understanding it literally is that the photograph be taken as the measure of what the image in the camera looked like.

Think of it this way. Daguerre sets up a still life of bas-reliefs and plaster casts in his studio. He photographs it and obtains what looks to him like a one-to-one correspondence between the image on the plate and the image he brought to focus in the camera—and he believes, additionally, that what he sees in the camera is what he sees when he looks at the still life. But now he sets up a camera across the Seine from the Tuilleries, as he did in August of 1839 in his demonstration for the Academies of Science and Beaux Arts. The exposure runs half an hour. Boats pass by, horse-drawn wagons move across the field of the lens. People mill about in the foreground. But the developed plate shows only vacant streets, an empty river, and no figures on the quay. And Daguerre exclaims as he passes the plate around, "Thus I fix the image of the camera." Of course, Daguerre and his audience of believers had to suppress everything they had seen move through the view for half an hour in order to believe that the daguerreotype fixed the image of the camera. Nothing answers to the notion of *the* image of the camera. What appeared in front of the camera was a scene of constant movement. The scene must first be conceived of as a still life before the photograph can be formulated as the fixed camera image. Upon analysis, it turns out that the measure of what constituted the image in the camera is determined by what appears in the photograph and not the other way around. And this is part of what I meant by saying that photography functioned to reassure or authorize vision in the years leading up to the invention of what Marey sometimes thought of as a means of reproducing the appearance of natural objects as seen by the eye.

I need now to return to the question of the observer in chronophotography. Marey delights in discussing chronophotographs showing aspects of motion that "elude vision," or "escape observation"; such pictures work against the conception of photographs as reproducing the appearances of objects. And so here we enter another new domain of mechanical sensibility, which permits us to see, though only in pictorial form, what happens in front of us—before our eyes. The graphic method produced results that were, in a sense, neutral in terms of what we can see. We cannot see arterial pressure rise and fall, nor can we see the alternating pressure of feet pushing against the ground while doing the work of walking. With photography, however, we were supposed to be able to represent what we see and only what we can see. But we cannot see just about everything shown in a chronophotograph of a man running. There is a challenge here to the primacy of vision, to its adequacy. Marey never gave up the conception of photography as representing what we see, but neither did he give up the notion of chronophotography as a form of revelation of the imperceptible, as the registrar of sights undetectable by the human eye. The observer is left wobbling between what is visible to the naked eye and photographically depictable and what is unseeable by the eye but nonetheless reproducible by chronophotographic means.

In the course of discussing an awkward engraving of a horse in mid-stride—one based on his graphs and chronophotographs—Marey provides this excuse for the illustrator: "When considering this ungraceful figure, we are tempted to say with De Curnieu, 'the province of painting is what one sees, and not what really exists.' "[25] Here Marey throws up his hands and performs the wobble. It might seem that the division between "what one sees" and "what really exists" establishes an unbridgeable gulf between vision and the world, but the chronophotographic enterprise is not neutral in respect to what we can see. There is something of a competition here between vision and chronophotography that is elaborated in the course of Marey's discussion of the value of his serial photographs for the practice of painting and sculpting.

In representing a movement, for instance, one of a man, an artist rightly attempts to reproduce a phase which is visible to the eye. It is usually the preliminary or final phase which can be best appreciated. When a machine is in motion, there are certain parts of it which are only visible when they reach their dead points, that is to say, for the brief moment when the direction of movement is changed. And this is also the case with certain movements in man. Some attitudes are maintained longer than others. Now, chronophotograpy on fixed plates could be used to determine these positions. They are recognizable as the ones which have left the most intense impressions on the sensitized plate—in fact, those which have had the longest exposure. . . . In all possible actions . . . there are attitudes which last longer than others and which may be called "positions of visibility." Chronophotography would determine these with the greatest precision.[26]

As between relying on our own eyesight to determine what we see when looking at the most familiar movements, or learning what we actually see by looking at chronophotographic pictures of the same displacements, the issue, for Marey, is settled in favor of the photographs. It is not only that these pictures may represent what we cannot see, but finally, they provide testimony for us, as we cannot for ourselves, of what we actually do see—they authorize us to make claims about what we see right in front of our eyes—and finally educate us about how we ought to judge what we see. Chronophotographs then, can bring us into a domain we cannot see; yet at the same time, they can also show us what we do see, though we cannot warrant having seen apart from the pictorial evidence produced by precision instruments.

Inevitably, an instructional program joined the chronophotographic enterprise as instantaneous, serially related photographs were reproduced, first in Eadweard Muy-

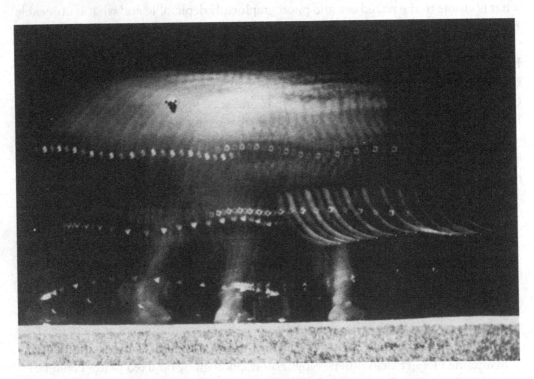

Figure 5.   Etienne-Jules Marey, chronophotograph of elephant in full trot, 1887 (Marey, Collège de France).

bridge's *Animal Locomotion,* then in the book Marey produced especially for artists, and finally in mass circulation magazines (among them the very popular *L'illustra-tion).*[27] When some adventurous illustrators and artists allowed these pictures to inflect their drawings and paintings, the initial reaction was hostile. Marey was fasci-nated by the change in the public response to paintings informed by high-speed pho-tographs representing unaccustomed phases of animal locomotion:

> These positions, as revealed by Muybridge, at first appeared unnatural, and the painters who first dared to imitate them astonished rather than charmed the pub-lic. But by degrees, as they became more familiar, the world became reconciled to them, and they have taught us to discover attitudes in Nature which we are unable to see for ourselves, and we begin almost to resent a slight mistake in the delineation of a horse in motion. How will this education of sight end, and what will be the effect on Art? The future alone can show.[28]

The education of sight, as elaborated by Marey, involves a process of reconciliation between what we see and what we can only come to know through chronophotogra-phy. We cannot be taught to see what we cannot see, but we can learn to expect artists to represent "Nature" without mistakes, and the standard of correctness is set by chronophotographic discoveries. Here, Marey resists the impulse to declare what at other times he is "tempted" to say. He resists claiming: "the province of painting is what one sees, and not what really exists." To the contrary, correctness in art must now, it seems, be determined by insights into nonvisible aspects of motion collected by technological means. Marey struggles to establish the boundaries separating visualiza-tion and visibility—establishing where one ends and the other begins—but he cannot find the means of marking them off from one another and with good reason, for finally it turns out that his photographs, contrary to Baudelaire's urgent denunciation, do not appeal to the eye but rather to the imagination:

> Although chronophotography represents the successive attitudes of a moving object, it affords a very different picture from what is actually seen by the eye when looking at the object itself. In each attitude the object appears to be motionless, and movements, which are successively executed, are associated in a series of images, as if they were all being executed at the same moment. The images, therefore, appeal rather to the imagination than to the senses. They teach us, it is true, to observe Nature more carefully, and perhaps to seek in a moving animal for positions hitherto unseen. This education of the eye, how-ever, may be rendered more complete if the impression of movement be con-veyed to the eye under conditions to which it is accustomed.[29]

This passage comes in the beginning of the final chapter of *Movement*, when Marey turns to the production of moving pictures. Instantaneous pictures of objects in motion, he says, "appeal" to the imagination, while pictures in motion address the eye directly, without further appeal. To comprehend the still photographs, we need to do something with them—associate or integrate all of the represented attitudes appearing in a series—that is, we need to imagine seeing the motion of which each depicted element is but a single transitory and transitional phase. I take Marey to be using the term "appeal" not as a synonym for "attract," but in its legal sense—as the transfer of a case from one court to a court of review. In attesting directly to the eye, motion pictures dispense with any need for interpretation; for him, their testimony is self-evident. Thought of in this way, motion pictures provide a mechanical synthesis of the data of mechanical analysis, through the animation of serially related, static representations of objects in motion. To be "more complete," the education of sight in regard to motion requires seeing something move—requires looking at moving pictures of objects in motion. Marey's work has come full circle: the project that began as an analytic investigation of motion by means of instruments he likened to "new senses of astonishing precision," situating research beyond the competence and reach of the senses, comes to an end with the construction of a mechanical imagination—ends, that is, as a synthetic enterprise capable of educating sight by visualizing both the visible and the nonvisible—by recreating the conditions of visibility.

Notes

1. This would be a far less interesting essay but for the contributions, often in the form of sharp questions, from some good friends: Marta Braun, Lorraine Daston, Arnold Davidson, Michael Fried, Peter Galison, Nancy Henry, Caroline Jones, Nancy Maull, and W. J. T. Mitchell. T. E. D. Cohen saved me from making some insupportable claims about substitution, forcing me to rethink my initial conjectures on the subject. James Conant's comments on an early draft of this paper (presented at the conference "Histories of Science, Histories of Art") were incisive, very useful, and delivered with exceptional good humor.
2. Lorraine Daston and Peter Galison, "The Image of Objectivity," *Representations* 40 (Fall, 1992): 81.
3. Daston and Galison, "The Image of Objectivity," p. 82.
4. There is no issue of human fallibility here. Imagine a person attempting to fly by doing the best he can at imitating, with his hands and arms, the beating of a gull's wings. Is his inability to fly best described as a failure? Perhaps we might say it was a failure, but of what? Where there is no possibility of success, there is equally no possibility of failure.
5. E. J. Marey, *La Méthode graphique dans les sciences expérimentales et particulièrement en physiologie et en médecine* (Paris, 1878), p. 108.
6. In saying the data are manufactured by machines, I am not suggesting there is anything fanciful about them. What I mean to emphasize is the completely factured character of *data* produced by monitoring the imperceptible motions under investigation and plotting them on paper. I would be distraught if someone took me to be saying that the motions under investigation are themselves manufactured, though I am claiming (following Marey) the only way we can know anything about them is by way of machine-generated visualizations.

7. *La Méthode graphique*, p. ii.

8. Ibid., pp. i–ii.

9. Ibid., pp. ii–iii.

10. I do not think it makes sense to think of what can be seen in a microscope or telescope as a visualization or a realization: it seems more appropriate to describe what we see as an enlarged object. Looking through a microscope could be said to reveal a previously unknown world teeming with life. But the issue I am addressing here does not center on revelation. The question, it seems to me, turns on the way we characterize the subject of study. In the case of the microscope, we look at minute objects when we peer into the eyepiece, but Marey's data are diagrams, graphs, pictures—records of processes or events that can be realized only through mechanical visualization. There is nothing comparable to the eyepiece in Marey's work, no way of getting a view of the data apart from looking at the paper on which they are marked. Ian Hacking has a very different position from mine regarding microscopes, claiming, "we do not see, in any ordinary sense of the word, with a microscope." See the delightful discussion of microscopes in chapter 11 of his *Representing and Intervening* (Cambridge, 1983).

11. Daston and Galison, "The Image of Objectivity," p. 82, my emphasis.

12. Peter Galison, "Judgment against Objectivity," this volume, p. 332.

13. See Stanley Cavell, *The World Viewed* (New York, 1979), pp. 21–23. Cavell argues that photography and film answer a need, one beginning in the Reformation, to "escape subjectivity and metaphysical isolation." According to Cavell, the mechanical genesis of photographs allows them to do just this by making the world present to us while we are absent to it.

14. E.-J. Marey, *Animal Mechanism* (New York, 1901), p. 110. The stylus "registers the duration and the phases of the pressure of the foot."

15. I am not suggesting that we obtain a better grasp on "representation" by emphasizing the "re," though I remain convinced that representation is different from presentation.

16. E.-J. Marey, *Movement*, trans. Eric Pritchard (London, 1895), p. 33. See also p. 54: "Since the object of chronophotography is to determine with exactitude the characters of movement, such a method ought to represent the different positions in space occupied by a moving object, i.e., its trajectory, as well as define the various positions of this body on the trajectory at any particular moment."

17. For an excellent discussion of the history of Marey's studies and for a detailed and clear exposition of his instrumentation, see: Marta Braun, *Picturing Time: The Work of Etienne-Jules Marey* (Chicago, 1992).

18. *Movement*, p. 19.

19. Ibid., p. 35.

20. Ibid., p. 31.

21. Ibid., p. 304.

22. Ibid., p. 3.

23. Charles Baudelaire, "The Salon of 1859," in *Baudelaire: Selected Writings on Art and Literature*, trans. P. E. Charvet (London, 1972), p. 297.

24. E. A. Poe, "The Daguerreotype," *Alexander's Weekly Messenger*, January 15, 1840, p. 2.

25. E. J. Marey, *Animal Mechanism*, p. 170.

26. E. J. Marey, *Movement*, p. 179.

27. See E.-J. Marey [with G. Demeny], *Études de physiologie artistique faites au moyen de la chronophotographie*, première série, vol. I, *De mouvement de l'homme* (Paris, 1893).

28. E. J. Marey, *Movement*, p. 183.

29. Ibid., p. 304.

In the emerging world of technology and art, there's a whole rhetoric of interactivity that is often used to make us feel as if we actually have some control over things that are in fact being done to us. In the case of Faraday's Islands, on some level you have absolute control, and on another level you have absolutely no control. And I do think that there is something upsetting about that.

PERRY HOBERMAN

In the media there's a very powerful fiction: we can choose. But the mechanism is fixed, and we can't choose the images. We can switch the channel, we can watch this movie or that movie, we can rent that video. The experience of Faraday's, with its repeating film loops, found slides, and stuttering radio broadcasts, reinforces the fact that these media, when activated, are going to have whatever's been programmed to be there.

CAROLINE JONES

You have a totally different relationship to an image if you actually have to work to see it.

PERRY HOBERMAN

Cultures of Vision

SVETLANA ALPERS

The Studio, the Laboratory, and the Vexations of Art[1]

Seeing that the nature of things betrays itself more readily under the vexations of art than in its natural freedom.

—FRANCIS BACON, *THE GREAT INSTAURATION*

Beginning in the seventeenth century and continuing well into the twentieth, roughly from Vermeer's *Art of Painting* (Figure 1) to Picasso's many self-portraits,[2] a succession of European painters has taken the studio as the world. Or, put differently, the studio is where the world as it gets into painting is experienced. This was not true of European art before and is not true in other pictorial traditions, such as those of Asia. The oddity of this assumption and the pictorial concerns and constraints it entails have not been specifically recognized or defined as such. For one thing, the artist's immediate experience in the workplace is rarely represented in a pure state. Many other factors—conventions of realistic representation, relationship to clients and the marketplace, the nature of display—go into the making and viewing of paintings that are informed by it. More importantly, while practices in it exhibit consistent traits, the studio ambience has not lent itself to the kind of discursive claims that have developed, for example, about the pictorial constructions known as perspective. The realities of the studio are elusive, but they are also determining and astonishingly long-lived in European painting. By looking at paintings made both of

Figure 1. Jan Vermeer, The Art of Painting, *ca. 1665, oil on canvas, 120 cm x 100 cm. Vienna, Kunsthistorisches Museum.*

and in the studio, in the seventeenth century and after, my aim, admittedly a contrary one, is to offer a reasoned account of studio realities.

My subject is neither the studio as an iconographic theme, nor motifs or models worked after life (though both play a part), but rather the situation in which the two overlap: when the relation of the artist to reality is seen in the frame of the workplace. The topic has a certain edge to it today. In some circles, the figure of the painter in the studio is suspect on ideological grounds. Art produced for or in public venues—from earthworks, public sculpture, and assemblages to photography and graffiti—is in, while painting made by someone alone in the studio is out. Caroline Jones, for example, explained what Andy Warhol's factory was an alternative to: "The dominant topos of the American artist was that of a solitary (male) genius, alone in his studio, sole witness to the miraculous creation of art."[3] A way to counter this is to show that painting in the studio—for I agree with its antagonists that this is at the heart of the pictorial medium as we know it—has a narrower and a more humane basis.

In recent times historians and sociologists of science have been considering the role that the workplace for experimental science plays in the nature and the status of scientific knowledge.[4] This is part of a change in emphasis from theory to practice, or from science considered primarily as the building up of natural laws to science as the making of experiments. When I began to think about the realities of the studio I was encouraged for several reasons by the example of studies of laboratory work and life. First, these serve as a corrective. In the study of art not unlike the study of science in this respect, the precepts of a loose assemblage called art theory are too often privileged at the expense of artistic practice itself.

Further, the relationship between the practice of art and the practice of science is particularly striking in the seventeenth century when the studio, in the sense in which I am proposing it here, comes into its own. English (Baconian) experimentation and Dutch descriptive painting share a perceptual or visual metaphor of knowledge of the world. Though they both represent the world, neither is transparent to it. In one case it is represented by a technology such as that of the lens, in the other by painting itself. Experimenting and painting are comparable as conventional crafts. The comparison works in two directions: it grants a seriousness to painters that is appropriately visual in nature (treating them as skillful observers rather than as moral preachers), while it brings experimenters in natural knowledge out of their privileged isolation.[5]

The model of the laboratory has, however, not worked out for painting, at least not in the way that I had hoped it would. There are, basically, two reasons for this.

I began intending to pursue the social construction of the artist's workplace. The questions were to follow on studies of the seventeenth-century house of experiment.[6] Was the site where experiments were conducted a private space or a public one? Was the work done by individuals working alone, with assistants, or in a group? Who were witnesses to the findings? But these laboratory questions—which have led to distinctions between private and public workplaces, places of discovery and demonstration, and have uncovered the essential role of socially inferior assistants—while perhaps of interest about the actual workspace, were not so productive when directed to paintings themselves. Part of the problem is that painters represent the studio experience as seen and experienced by an individual.

We know of course that painters often were not alone when at work. There are Dutch paintings that show a painter sharing his space with students, or assistants, or even a serving maid. But whatever the conditions behind the production of art, the phenomena we are concerned with resist the implications of this kind of social construction: compare Rembrandt's etchings of a model, or his great picture of Hendrickje posing for him as *Bathsheba*, to workshop renderings of him among his students drawing a model. Equally, compare Vermeer's *Art of Painting* (Figure 1) to Mieris's studio painting (formerly in Berlin), which shows a maid coming in the door of the room where the artist is painting his model. The studio may, on occasion, have been teeming with people. But what is represented as the studio experience is a solitary's view. There is something obvious about the point—how else *but* as an individual does one see and experience one's ambience? This is not, of course, necessarily the concern of image making. But the fiction sustained by studio painting is that a painter works alone.

This brings me to the second reason why the studio does not fit the laboratory case. From its beginning in the seventeenth century, the experience of the painter/observer was part of the pictorially represented experiment or experience (in the seventeenth-century sense of the word). The realities of the studio are not only what is observed there (how the world is put together) but the artist's visual and, often, bodily or phenomenological experience of it (how it is experienced). What the painter in his/her painting makes of the world thus experienced is central to the studio as an experimental site. What I am invoking is not a personal matter. It has to do with how every individual establishes a relationship with the world. One of the vexations of art is man. In the laboratory, by contrast, the impact of the interference of the human observer in an account of natural phenomena was neither acknowledged nor taken into account until modern times, and then with a different effect (see Galison in this volume). It is possible to argue that the practice of painting was ahead of the practice of science in regard to the observer. The truth of this might account, at least in part, for the studio's enduring life.

But the link studio/laboratory does focus attention on the enabling constraints under which pictorial investigations in the tradition have been conducted. Indeed, it lets one look on painting as an investigation. What follows are preliminary considerations of the problem. I shall first consider what sort of realities the painter can address in the studio. Turning to the maker, what is the relation of the painter to reality as seen in the frame of the workplace? Then, most importantly because it reflects back on the studio itself, what does not find a place in the studio, or what does the studio exclude? And what have painters done about it? Finally, a brief consideration of the depiction of landscape, outside and in, leads to the notion that a basic reality addressed in the studio is the nature of the studio itself as an instrument of art. With the reference to an instrument, we return, in a unanticipated sense, to the analogy offered by laboratory practice.

Let us begin with what I take as the baseline case. A person is alone in an empty room (Figure 2). It is somewhat darker in the room than in the world outside. Light is let in through several holes in the walls. Curious though it seems, the person has withdrawn from the world for the purpose of attending better to it. In this case, as the interior is empty, he withdraws to attend to the effect of the play of light. This print from Athanasius Kircher's *Ars magna lucis et umbrae* (1646) is of a camera obscura enlarged to the size of a room and set in a landscape. The light conveys some trees as images on the wall.

The painter at work in a picture by Pieter Ellinga Janssens had depicted himself in such a space (Figure 3). The artist, head silhouetted before a window, stands at an easel in the further room (the Dutch painter normally used a room in a house as a studio). The interior is dark. Light coming through large windows reflects off the floor, flooding the far wall and a chair. In the foreground room it puts on a wild show.

Figure 2.   Althanasius Kircher, engraved illustration of a camera
obscura from Ars magna lucis et umbrae (Rome, 1646).

Figure 3.   Peter Ellinga Janssens (1623–82), Interior with Painter,
Lady Reading and Maidservant, *oil on canvas, 83.7 cm x 100 cm.*
Städelsches Kunstinstitut, Frankfurt.

Sunlight angles through the bright upper part of large windows, casts shadows of the
leading on the lower frame, strikes the floor and the wall, and, impossibly, casts the
shadow of a chair and table back onto the window wall where reflected light glistens
off the glass of the shuttered lower window. A number of pictures (nonreflectors), and
a mirror with gilt frame (light reflectors both) are on the walls. A woman reads by the
secondary light and a maid sweeps. Treetops make a faint pattern on the leaded glass.
Again, an optical matter.

What is it like to be in a particular light?

Other Dutch painters are less flamboyant about it than Janssens. Though he never
depicts himself, the best paintings of Pieter de Hooch are in large part lingering investigations of light seen as if while at work. In his paintings, light enters through windows and, often, a distant door, illuminating and reflecting differently off earthen tiles
and metal, fur and gilt. Its interruption is marked by shadows cast by window frames,

chairs, and other domestic appurtenances within the dusky atmosphere of the interior of a house. One understands the recommendation, already made by Leonardo, that light in the studio be from the north. Northern light, as he says, does not vary. It is never—except occasionally in de Hooch's interiors—direct sunlight. The minimally interesting figures in these interiors appear to be stilled by the activity of this sunlight. But de Hooch frequently depicts a child poised to receive the light that floods in through an open doorway. The painter identifies his experience with that of a child. Withdrawing from the world into the studio is a regressive act, in that it rehearses how we come into an experience of the world.

The studio serves as a place to conduct experiments with light not possible in the diffused universal light or the direct solar light of the world outside. Studio light is light constrained in diverse ways. And it is also the light the artist is in; he situates himself in it. This state of affairs is codified in the paintings of Vermeer (Figure 1). The painter, according to Vermeer in *The Art of Painting*, is the man alone in the room with which we began. But his working space or ambience (a better, because a less limited, term) is restricted to a corner. A single window, here hidden by a tapestry, lets light in. It comes, as is usual, from the left, assuring that the painter's right hand does not cast a shadow on his painting as he works. The painter is in the light he paints. When, exceptionally, the light in a painting by Vermeer is from the right (from the left of the painted figure), it is likely that Vermeer imagines himself in her place. See, for example, his *Lacemaker*. Beside her hands is an improbable (for a lacemaker) red tangle of stuff. Seen at the close distance that the diminutive scale of this painting proposes for the eye, the red pigment is a deposit of *his* paint as much as it is a loop of *her* thread.

What is it like when objects are introduced into the painter's lit room? The essential case in the pictorial tradition is the setup known as still life, for example, one of Pieter Claesz's spare arrangements featuring a wine glass, a silver bowl, an olive, a nut. So accustomed are we to tables with objects on them that the oddity of the format has been lost. They had been depicted earlier, but the still life is first situated in the studio in seventeenth-century Dutch painting. Attention to light is an indication of this transformation. There had been displays of collectibles in the works of Frans Francken and Pieter Aertsen, but now the objects on the table, generally domestic in nature, are lit by light from the (painter's) left. Objects are related to each other by reflections off neighboring things on the table—a lemon reflected onto a pewter dish, lit pewter back onto the lemon. The leaded panes of an unseen (studio) window are often reflected on and even through the surface of a goblet. It is curious, though, that though light remains an abiding concern for still-life painters from Heda to Chardin to Cézanne, the window does not figure in the still-life genre itself. Once it was assumed, the light needed no explanation.

Still lifes of flowers, however, continued as pictures of collectibles. In the seventeenth century, flowers were not represented in studio light because the blooms of different kinds and seasons—though represented together in a vase—could never be seen together. In the nineteenth century, rarity goes by the board and flowers in the vase on the studio table are firmly in place. It then becomes disconcerting to come upon a painting of flowers (there is one in Hamburg by the young Renoir) that are, instead, growing in a sunlit flower bed. The diffused light, but also the implied size, position, and distance are disorienting *because* we have become used to flowers in the studio.

The situation with people is a bit different. A person studied by studio light cannot be put outside in the light of a day. The nude is the most aggravated example of this constraint. An interest is that we can trace its history. Giorgione, with Titian following close after, is the presumed inventor of what became the genre of the nude figure in the landscape. In his Dresden *Venus* there is no disparity between the light on the flesh and that on the terrain in which she lies. Light is diffused and universal. This is prior to the studio in the sense in which we are speaking of it here. Once a nude body is painted (as distinguished from being drawn) in studio light and at studio proximity, it is incongruent when set outside. The problem surfaces in the nineteenth century, most egregiously with the Impressionists. Frédéric Bazille bares it in his paintings that insert a nude man, studio-lit, who appears not on, but silhouetted against, a sunlit river bank.[7] Some accommodation must be made. Degas recommended representing nude figures outside at twilight when they appear as silhouettes.

The still-life assemblage has come to be understood as a bourgeois phenomenon. This is to consider its objects as subordinate to man, for our use, manipulation, and enjoyment, conveying our sense of power over things. To paraphrase Meyer Schapiro, still life in this interpretation is the construction of a portable possession.[8] But taking the empty room once more as the baseline, there is a prior sense in which the objects on the table—one can expand this to any objects or people in the studio—are an intrusion on the painter who has retired there. They register something the painter deals with or, perhaps better, plays with in paint. Still-life objects are depicted as if at arm's length and approximately life-size. The distance from what the artist paints is a reality of the studio (studio-size would be perhaps a more accurate description than life-size). The studio is, in an almost primordial sense, a place where things are introduced in the interest of being experienced.

The representation of the painter's hand in Vermeer's *Art of Painting* offers a heightened instance of this studio experience. A shaded blob is where the hand of the painter painting the blue and tawny leaves on the model's wreath should be. Why such an ill-defined blob? One could say that the painter has not yet realized his hand, in the double sense of not yet having fully perceived it as a hand and not yet having fashioned it as a hand. But there is a further twist: the shaded blob also appears to be part of the painting the artist is painting within the painting. Taken together with the

leaves he is putting on the canvas, it has the appearance of the face of the model posing topped by a wreath. (A bit of white at the wrist is the white collar at her neck, the
dark cuff standing out to its right and down is the cloth behind her neck.)

We have caught the painter in the studio playing with his hand in paint. The hand,
if you think of it, is *the* object that a painter always has at arm's length immediately
before him while working. (Velázquez, in his studio in *Las Meninas*, and Rembrandt in
his in the Kenwood *Self-Portrait*, also register this fact: Velázquez's hand mimicking
the color and directionality of worked pigments, Rembrandt's hand mimicking the
instruments with which he paints.) On Vermeer's account its representational status is
unresolved. Though it is part of his body, the hand is represented neither as belonging
entirely to him, nor entirely as an object in the world, nor entirely as a painted image.
We might dub it transitional—involving the relationship, perceptual but also psychological, between an individual and the world.

An experience of ambiguity is a part of the process of perceiving. Our mind works,
albeit quickly, from multiple and conflicting visual clues to work out the place, shape,
and identity of what it attends to. Pictorial equivocation had been entertained by
painters before the seventeenth century. By equivocation I refer to the possibility of
the painter representing the perception of a thing, and representing it for viewers, in
such a way as to encourage the mind to dwell on perceiving it as a process: the painter's
experience of an object as coming into its own, distinguishing itself from others, taking shape. He looks at a modeled blob not yet recognizable as a hand and also sees it as
the flesh of a painted face. The difference the studio site makes is that it frames the
ambiguity as an originating one. In the studio, the individual's experience of the world
is staged as if it (experience, that is) is at its beginning. This gives to studio painting a
forward, probing lean. It is a matter of discovering, not demonstrating.

To return to how experience begins is regressive. And it is in this connection that a
certain line of studio paintings dwell on objects whose status or nature remains unresolved., I think of Degas's disturbing *Portrait of an Artist in his Studio*, now in Lisbon.
Beyond a huge foregrounded palette, the painter, a Degas look-alike, leans against a
wall. In a painting to one side a woman is lounging against a tree; her twin, similarly
dressed, is grotesquely slumped against the wall to the other side. Is she grotesque
because she is only the artist's lay figure? The Courtauld's famous Cézanne painting,
Still Life with Amor, again suggests a vertiginous experiencing of the studio as the
world: the green top of a still-life onion on a table top turning up into a painting, a
plaster sculpture (the "amor") echoed in the truncated drawing of the statue of a flayed
man. What Vermeer kept largely under representational wraps is exposed. One senses
a resistance on the part of the objects in view, which are being bent, against their will,
to be part of the painter's experience. The game can be destructive.

I can't resist introducing into the discussion a squiggle drawn by the British psy-
choanalyst D. W. Winnicott, a drawing of an object on a pedestal that is inscribed

"Frustrated Sculpture (wanted to be an ordinary thing)."[9] Its relevance is pictorial (it looks rather like Cézanne's flayed man) and also psychological. Winnicott's observation of the infant's pre-linguistic experiencing of the world is suggestive, as is his focusing on the one-on-one relationship of person to world. The painter in the studio engages in creative play. And even what this formulation strikingly leaves out or cannot deal with—things beyond an individual's reach or long-term memory or the repressed desires of Freudian dreams—is relevant to the constraints on the studio painter's practice.

The *retreat* to the studio, which is one way to understand it, was celebrated in the seventeenth century by some remarkable self-portraits by artists as painters: Rembrandt at Kenwood and in Paris, Poussin in Paris, the so-called *Las Meninas* of Velázquez and, though not openly a self-portrait, Vermeer's *Art of Painting*. The circumstances of these various canvases are diverse. They were made in different places, on different occasions, for different clients. But each painter, in his own context, dealt with the newly ambiguous role of the artist. For each of them—whether at court (Velázquez), in the home (Vermeer), or away from either (Rembrandt, Poussin)—the working space offers a provisional solution. It is a way to define ground which is the painter's own.

Attention to studio realities marks the end of European history painting as it had been. Painters withdraw from depicting significant (text-related) actions (religious, mythological, and historical) taking place in a greater world. With the striking exception of Rubens, who might be described as an old believer, the imaginary theater is over. (Poussin and the French who follow constitute history painting in new and different terms.)

One can distinguish two different paths into the studio, both of which have had a long life. After his *Christ with Mary and Martha*, *Diana*, and (the newly discovered) *St. Praxedis*, Vermeer turned to objects and people appropriate to the domestic situation of his workplace. The imaginary theater survives only in pictures on the wall. Rembrandt and Caravaggio, by contrast, try to bring the themes of history painting into studios that are not specifically domestic. They deploy models performing in the studio to replace the dramatic figures of history painting. Late in his career, Rembrandt repeatedly represented people who sat in his studio in such a way as to suggest historical personages. A favorite model poses as Aristotle, Hendrickje poses as Bathsheba. Rembrandt attaches historical themes to portraiture, an essential studio genre.

Caravaggio, by contrast, persists in depicting dramatic enactments that, however, he stages with models in the studio. He signals this by a darkened interior and by the lighting, scale, and proximity to him of figures, in particular their flesh and accoutrements. One might describe Caravaggio as configuring historical themes as a sort of still life, the other essential studio genre. He was accused at the time of killing off painting by following nature too closely. Behind the accusation was a notion that art is more or less either after nature (the real) or out of the mind (the ideal). But studio practice—in which nature is vexed by art—confounds these terms. It is painting in the studio, not nature, that poses the threat to painting as it had been.

But the studio presents problems for the artist. Call them strains or call them vexations of art. Prominent among these is the fact that it leaves outside so much of the world and that its condition is that of isolation. Both lead to trying to make connections.

Poussin, as has been recently shown, conceived of his paintings not as sold, but as circulating among friends. He defined the market for his pictures not by an exchange of money (though that did take place) but by an exchange of friendship, which is represented by embracing arms in his Paris *Self-portrait*.[10] One response to the solitude of the studio is to reach out for company. Poussin read and admired Montaigne. It could be that like Montaigne alone in his tower, Poussin valued friendship all the more for the solitude of the studio, depicted here by a stack of empty frames. The court visitations to his studio that Velázquez arranged, like the busy workplaces of his so-called *Spinners* or of Vulcan and his helpers that he also devised, do not disguise the fact that the painter of *Las Meninas* was an elusive loner. It is surprising that Velázquez painted his own portrait at all.

At one level painters have simply tried to buck the studio limits. Surely you can't bring a horse in to model! Well, Horace Vernet did and commemorated the event by painting it. Courbet, choosing the lower road, elected instead a cow. The cow modeling in the studio was the butt of a contemporary print that gets at Courbet's realism, but also of course at his misuse of the studio. What is a cow doing there as the object of serious painterly attention? A similar question could be raised about Courbet's own painting of his studio. Why is the studio so very large and crowded with so many people? Courbet gave it a long title and described in a letter and it has been the subject of much interpretive analysis. Without going into the many particulars, one could say that his is a grandiose attempt to make the studio large enough (the painting is 12 feet × 20 feet or 3.6 meters × 6 meters) to encompass the world—at least, that is, the world of friends and models.

It also takes up two other problems in the studio: it deals with time past and provides the artist with company. *The Painter's Studio, a real allegory determining a phase of a seven-years of my artistic life* is a fair translation of the title Courbet gave his picture. Studio experience is by its nature in and of the present. Remember the empty, light-filled room with which we began. Like the perception/experience that it sustains and extends, it leads forward, not back to the past. Courbet counters this forward lean by bringing in figures and models from history—his own history and paintings past. This convocation of people also saves the artist from himself, from his isolation in the studio. But about this Courbet is clearly ambivalent. Invoking the world and the past, he simultaneously turns his back on it all, with a flourish, to paint a landscape. It seems another piece of the world he is trying to get in.

I want to end with the problematic case of the relationship between landscape and the studio. Already in the sixteenth century Pieter Bruegel made drawings of mountainous valleys outside. Later, Constable did oil sketches of clouds. But as a rule, before the nineteenth century they executed their landscape paintings in the studio, removed from seeing the thing itself. An accepted account of the history of landscape painting turns on locating the moment when painters free themselves from the studio and its conventions and paint in the real landscape outside. This account, a nineteenth-century landscape one, posits the world as the alternative to the studio—the painter went out to do empirical studies and came in to compose. The story is told of Monet digging a ditch in which to lower his large canvas when he went out to capture the "real" play of light and shadow playing on the dresses of women in a garden. A number of painters came to accept the inconvenience, the glare and changing light, decomposition, and challenge to the integrity of the figure encountered in the studio that painting in the world outside the studio entailed.

But there was another departure in taking up landscape, in which the studio remained a determining factor. Thomas Jones, an Englishman working in Italy in the 1780s, is celebrated as one of the first Europeans to work with oil paint outdoors.[11] Liv-

Figure 4.   Thomas Jones, The Bay of Naples, *ca. 1782,*
oil on canvas, 91.4 cm x 152.4 cm. Private collection.

Figure 5.   Thomas Jones, Rooftops, Naples, *April 1782,*
oil on paper, 14.3 cm x 35 cm. Ashmolean Museum, Oxford.

ing in Naples in 1782, he turned from painting panoramic views of bay and hills complete with conventional repoussoir trees (Figure 4) to making, instead, marvelous oil
sketches on paper of rooftops close-by seen in a brilliant sun (Figure 5). But are Jones's
tiny sketches really studio-free? Obviously, they are different from his other landscapes.
They seem more immediately experienced. Jones viewed what he painted from his
window or roof. And what we admire in his rooftops is something experienced and
represented as if on a tabletop in the studio. The small scale (5 in. × 8–13 in. or 14
cm. × 21–35 cm.), close handling, and the steady light are that of a still life. Something about this is very odd indeed. How can one put the landscape on a table like the
objects of a still life? Or to put it differently, how can one experience something with
the expanse of a landscape—its mountains, hills, trees—and its movement—the
clouds—as if it had the visual and bodily presence and the immobility of objects on a
table?

The answer is that one can't, at least not without changing the nature of the realities addressed in the studio. Jones's juxtaposed, lit walls are easily accommodated to
the tabletop. But they are buildings, not the land and sky itself. What would happen if
the painter proposed experiencing something that cannot be experienced in the studio? That is to say, proposed experiencing something without a proper body in a studio
way? It is like proposing Vermeer or Chardin painting a landscape. (Vermeer of course
painted a *View of Delft.* But its status is not like that of objects in the studio, but rather
of the images cast by light onto the wall in the camera obscura box with which we
began. Delft is present, but only to the eye.)

The experiment of experiencing landscape as a still-life motif was one devised by
Cézanne. Despite his treks outside to paint, Cézanne's landscapes share with his still
lifes the studio terrain. To treat landscape in the studio in this way is to alter the nature

Figure 6.   Paul Cézanne, Rocks and Hills in Provence, *ca. 1886–90,
oil on canvas, 65 cm x 81 cm. Tate Gallery, London.*

of painting. It is one thing to display the buildup of still-life objects with pigment and brush, quite something different to do that, as does Cézanne, painting a landscape (Figure 6). Cézanne's often-remarked sacrifice of light in the studio (light is in fact accounted for within the brush strokes themselves) is part of this accommodation. Indeed as landscape approaches the nature of still life, so his still lifes, often depicted before the woven leaves of a tapestry, approach the nature of landscape. The difference between things—between landscape as a motif and still life, or between rocks and trees fields before mountains and bowls or a vase on table tops backed by tapestried leaves—is minimized in the process of painting. Studio representation is construed differently, pitched at a different level than it had been before.

I have some concerns about the account I have offered of the realities of the studio.

First, once these are in place in the seventeenth century, they are too persistent, too much out of time, lacking in historical context. An explanation might be, and though

I think it holds others might well not agree, that something constant in the human perception of the world and its representation was put in place in the art of the studio.

Second, are studio practice and laboratory practice, finally, incompatible? The landscape turn we have just looked at marks a real and addressable change. Is there any account of laboratory practice that can help us better describe and hence understand this studio change? I think there might be. It is the account offered by Galison and Assmus of the transformation in the use of the experimental chamber devised and set up in the Cavendish laboratory in Cambridge, England, in the late 1890s: an instrument designed in order to reproduce clouds came to be used instead to detect subatomic particles.[12]

Put too simply, the physical meteorologist C. T. R. Wilson built his cloud chamber in order to reproduce atmospheric phenomena of the real world in the laboratory: his hypothesis was that condensation nuclei were electrical ions. In its condensation of artificial clouds, his experimental apparatus imitated nature. But working beside Wilson at the Cavendish laboratory were men who were not interested in studying the world that was being imitated within the chamber (clouds) but rather the real things—small, as yet unobserved subatomic particles—that were made visible in the condensation produced there.

The transformation of the meteorologist's dust chamber into the physicist's cloud chamber is felicitously described by Galison and Assmus as a change from mimetic to analytic experimentation. Their title puts it succinctly: artificial clouds, real particles. Couldn't we describe Cézanne's innovation in similar terms? From artificial landscape to real constituents of things as we perceive them? Perhaps this is a more appropriate way of saying (to quote myself above) that with Cézanne, "studio representation is construed differently, pitched at a different level than it had been before." And the paradox in the scientists' changing use of the chamber in Cambridge might also be said to have obtained in the studios of painters. To take up the terms offered by Galison and Assmus again, the mimetic researcher attended to artifice, while the analytic researcher pursued real things.

But it is not only the terms—mimetic to analytic—that I want to note. There is a further and more general interest in this tale of experimental Cambridge. The account emphasizes the activity of experimenters in establishing the character of an instrument (here the chamber) and the effects produced with it. Every instrument produces artifacts/effects that are intrinsic to its construction. But the nature of an instrument and the interpretation of the artifacts produced by it are also subject to human manipulation and interpretation. The lens or glass and the camera obscura are among the most familiar instruments that were used by experimenters in natural knowledge and artists alike. But encouraged by the evocative account of the chamber in Cambridge (and admittedly charmed by the metaphoric affinity chamber-studio), we might consider the studio itself as such an instrument. Perhaps it is here, in the uses of an experimental instrument, that there is a fruitful analogy between the artist and the experimenter, between painting and experimenting.

Starting in the seventeenth century, so an account might go, European artists began to consider the studio as a basic instrument of their art. For some, it was not simply the site *where* they worked, but the very condition of working. The studio as instrument is an invention that has had a long life—from Pieter Janssens's studio as a light box, to Cézanne's studio as a state, or frame, of mind. The historical project, not unlike that which engages historians of science, is to track the changing character of the studio-as-instrument while resisting the tendency to consider it and its products as either simply conventional (the default of art) or simply transparent to the world (the default of science).

A coda. The art/science link to which I was educated privileged the notion of development and progress. It was argued that the problem-solving nature of Italian Renaissance art made it the model for the progress in human knowledge that later came to be associated with science. Here is Gombrich: "The artist works like a scientist. His works exist not only for their own sake but also to demonstrate certain problem-solutions."[13] The immediate reference was what he and others took to be the scientific and demonstrable character of perspective. The studio-laboratory link focuses instead on the constraints under which knowledge is achieved. What is being tracked are the changing conditions of pictorial knowledge, rather than its progress.

Notes

1. This is part of work in progress. I wish to thank to Peter Galison for his persistent encouragement to publish at this stage. The unresolved note on which the paper ends is also, in a sense, due to him. For that I owe an apology. It was only in the course of reshaping the paper for publication that I came to see the broader implications that Galison's account of the cloud chamber might have for an account of the artist's studio.
2. See, in particular, Picasso's *Self-Portrait in the Boulevard Raspail Studio*, 1913, Musée Picasso, Paris.
3. Caroline A. Jones, "Andy Warhol's 'Factory': The Production Site, Its Context and Its Impact on the Work of Art," *Science in Context* 4, 1 (1990): 101.
4. See David Gooding, Trevor Pinch, and Simon Schaffer, eds., *The Uses of Experiment: Studies in the Natural Sciences* (Cambridge: Cambridge University Press, 1989) and Brian Peppard's shrewd account of its implications in the *Times Literary Supplement*, September 29–October 5, 1989, p. 1057.
5. For a substantiated account from the point of view of the practice of art, see Svetlana Alpers, *The Art of Describing: Dutch Art in the Seventeenth Century* (Chicago: University of Chicago Press, 1983).
6. Prominent among studies of early laboratory spaces are Owen Hannaway, "Laboratory Design and the Aim of Science," *Isis* 77 (1986): 585–610 and Steven Shapin, "The House of Experiment in Seventeenth Century England," *Isis* 79 (1988): 373–404.
7. See, for example, Bazille's *Fisherman with Net*, oil on canvas, 134 cm × 83 cm, Foundation Rau, Zürich.
8. Meyer Schapiro, "The Apples of Cézanne: An Essay on the Meaning of Still-life," *Selected Papers: Modern Art, 19th and 20th Centuries* (New York: George Braziller, 1978), pp. 18–19.

9. Illustrated as the frontispiece to *Winnicott and Paradox: From Birth to Creation* (London: Tavistock Publications, 1987).

10. Elizabeth Cropper and Charles Dempsey, *Nicholas Poussin: Friendship and the Love of Painting* (Princeton: Princeton University Press, 1996), pp. 177–215.

11. For Thomas Jones, see Lawrence Gowing, *The Originality of Thomas Jones* (London: Thames and Hudson, 1985). Since this publication, it has been shown that working outdoors was not such an isolated phenomenon. See Peter Galassi, *Corot in Italy: Open-Air Painting and the Classical Landscape Tradition* (New Haven and London: Yale University Press, 1991).

12. Peter Galison and Alexi Assmus, "Artificial Clouds, Real Particles," in *The Uses of Experiment*, ed. David Gooding, Trevor Pinch, and Simon Schaffer (Cambridge: Cambridge University Press, 1989), pp. 225–74. See also Galison, *Image and Logic: A Material Culture of Microphysics* (Chicago: University of Chicago Press, 1997).

13. E. H. Gombrich, "The Renaissance Conception of Artistic Progress," *Norm and Form* (London: Phaidon Press, 1966), p. 7.

Bruno Latour

How to Be Iconophilic in Art, Science, and Religion?

La Vérité est image, mais il n'y a pas d'image de la vérité.

—Marie-José Mondzain

PROLOGUE: TWO WAYS OF POINTING AT ABSENCE

In the first image, soil scientists in the Amazon are gathered around a table in the little restaurant where they house their equipment. They discuss a map, or rather several superimposed types of visual traces: an aerial photograph and a satellite map of this tiny portion of the Amazon to which their expedition is heading (Figure 1).[1] The botanist is pointing with her left index finger at a spot on the map, which is also visible on the photograph although it has a different shape—nuances of grays instead of colored and sharpened boundaries. While her first colleague is coordinating his action by zooming in on the same spot with his hands and eyes, the other one, on the left, makes sure that the documents neither fold nor lose their superposition.

Those scientists are inside a landscape, but they are also *dominating* this landscape through the mediation of the map. They are designating a spot, the site of the botanist's field study, where they hope to go the next morning, and that is supposed to correspond, through a set of more or less predictable transformations, to the blur on the map and the gray area on the photograph to which the botanist is pointing with

*Figure 1.   Soil scientists studying maps of the Amazon rain forest
in a restaurant in Brazil (photograph by B. Latour).*

her index finger. Although the reality of the place she wants to reach is absent, she
points at "it" as firmly as if she wanted to refer to the table on which the documents are
spread. "Here it is," she says, and her colleagues nod approvingly: "*I see.*" Since there
are so many intermediary steps to reach the destination, the student of visual culture
could doubt that this scientist refers to anything, and yet she has collapsed those steps
into one, to the point where a deictic gesture can be used unproblematically to refer to
the site. So do these scientists see something? No, since what is designated is absent;
yes, since they can relate to their field site through a long series of intermediary steps.
If we were to take the inscription spread on the table literally either by denying that it
really refers to something or by claiming that what is referred to is present here, we
would miss what makes it interesting for the student of visual culture. More dramatically, we would have shifted from an "iconophilic" understanding of science to an
"idolatry" of science.[2]

In cell number eight of the San Marco convent in Florence, Fra Angelico (or his
collaborators) also painted deictic hands, those of the angel waiting for the holy
women on Easter morning (Figure 2). The right angel's hand points at the empty
tomb, and the left hand to an apparition of the resurrected Christ, behind the back of

Figure 2. Fra Angelico, Pious Women at the Tomb, *ca. 1440.*
Museo di San Marco, Florence, Italy.

the surprised women. A kneeling monk, Dominicus (on the left side), sees the angel, the empty tomb and the bewildered women, but he watches the scene laterally and his inferior position—almost merged with the frame and the wall—transforms his character into the figuration of a transition between the absent real monk whose cell has been illustrated and the fresco itself whose meaning the poor lonely soul has to recover. Like those of the scientist in the first picture, these hands point at absences—but a different type of absences. The tomb is empty and the whole message of the angel is to convince the holy women that this is *not* what they should look at: "Why do you look for the living among the dead? He is not here, He has risen!" (Luke 24:6.) The apparition of Christ, designated by the angel's left hand, is not more visible and cannot be substituted for the empty tomb, since the women do not see the glorious body of Christ, only the monk in the flesh—and now the visitor—can see it. But what is there to see? More absence, since the same angel's warning applies to the flat painted surface of the fresco: "do not look here, this is not what is in question, beware." If we were expecting to see the apparition *instead* of the empty tomb, here too we would have shifted from faith to idolatry, from "*dulie*" to "*latrie*." We would look for the living among the dead, for the presence among the absence, for what is really alive among the dust, pigment, and dried eggs of a fresco.

The two deictic gestures in the two images point at remote phenomena and absent features; both of them designate a reality; both of them force us to transcend the setting in which we are immersed (the Amazonian restaurant or the San Marco cell); both gestures help us see things that are invisible, and yet they are completely different in their definition of absence, presence, reality, phenomena, transcendence, visibility, invisibility, opacity, and transparence. In this little meditation on mediation, I would like to use art history in order to guide us along the paths of iconophilia. Iconophilia is respect not for the image itself but for the movement of the image. It is what teaches us that there is *nothing to see* when we do a freeze-frame of scientific and religious practices and focus on the visual itself instead of the movement, the passage, the transition from one form of image to another. By contrast, idolatry would be defined by attention to the visual per se. Thus iconoclasm may be defined either as what attacks idolatry or as what destroys iconophilia, two very different goals. Because it seems so difficult to resist the temptation inherent in all images, that is, to freeze-frame them, the iconoclast dreams of an unmediated access to truth, of a complete absence of images. But if we follow the path of iconophilia, we should, on the contrary, pay even more respect to the series of transformations for which each image is only a provisional frame. In other words, we should be iconophilic in all domains at once, in art, in science, and in religion.

ON THE USEFULNESS OF ART HISTORY
TO MAKE SENSE OF SCIENTIFIC PRACTICE

The study of scientific practice has provided us, in the last twenty years, with more and more insights into the fabrication and transportation of information (see the essays by Lorraine Daston, Peter Galison, and Simon Schaffer in this volume). If we could summarize the change of emphasis of such a diverse body of work, we could reuse, with a different meaning, Marshall McLuhan's motto: "the medium is the message." The active locus of science, portrayed in the past by stressing its two extremities, the Mind and the World, has shifted to the middle, to the humble instruments, tools, visualization skills, writing practices, focusing techniques, and what has been called "re-representation."[3] Through all these efforts, the mediation has eaten up the two extremities: the representing Mind and the represented World.

This shift has had the enormous advantage of multiplying the connecting points between art history and the history of science. When science was obsessed by what happened in the Mind or what was the case in the World, the distance with arts, especially the visual arts, was at its maximum.[4] But when science began to be seen as a mediating visual activity, then the visual arts offered a fabulous resource; they had always thought of themselves in terms of mediation and never bothered enormously about the representing Mind nor the represented World, which they took as useful but not substantial vanishing points. To be sure, it was much more difficult to extirpate scientific activity from its epistemological past than to free art history from aesthetics, but once the two moves were completed, a vast common ground was opened and, in recent years, a flurry of studies have "vascularized" the connection between visualization in science and the visual arts.[5]

The social history of the visual arts could teach historians of scientific activity quite a lot in the matter of mediations, since the beauty of a Rembrandt, for instance, could be accounted for by *multiplying* the mediators—going from the quality of the varnish, the type of market force, the name of all the successive buyers and sellers, the critical accounts evaluating the painting throughout history, the narrative of the theme and its successive transformations, the competition among painters, the slow invention of a taste, the laws of composition and the ways they were taught, the type of studio life, and so on in a bewildering gamut of heterogeneous elements that, *together*, composed the quality of a Rembrandt.[6] In art history the more mediators the better, and even now, it is my impression that there is very little in the cultural studies of science at the level of details, heterogeneity, and instability of the best social history of art. Deploying mediations without threatening the work itself—*l'oeuvre*—remains an art history specialty.

If the social history of the visual arts is good at teaching quite a few tricks to the history of science in terms of multiplying mediations, it is also true that it has an *easier* job since it can bracket more easily the question of *what* is carried over through all these mediations and of *who* is doing the carrying over. Once the aestheticians and their ahistorical Beauty have been pushed aside, it is slightly easier to recompose the quality of a Rembrandt, out of a motley crowd of small mediators, than, say, the second law of thermodynamics. In other words, the constructivist character is built into the arts in a different way than into a scientific fact. The more I read about the intermediary steps that make up the picture of the *Night Watch*, the more I may like it. Constructivism adds to the pleasure, going, so to speak, *in the same direction*, toward the multiplication of mediators. In some deep sense, constructivism flatters some essential feature of the arts.

This is not, however, the case with scientific facts. Constructivism, when it multiplies intermediary steps, seems always to *weaken* the claims to truth, to destroy the object under scrutiny. Instruments should be black-boxed, history forgotten, erratic moves erased, local and social circumstances eradicated.[7] Of course, and this confirms the point nicely, it is possible to take much greater pleasure in learning the laws of thermodynamics after having read the social historians on the construction of the first or the second law, but this reading, precisely, takes on some aesthetic character.[8] The same mediators that should have been black-boxed to produce scientific certainty, now that they are deployed by the historian, generate a type of pleasure that we rightly associate with the arts. Even if I exaggerate the differences, it remains fair to say that Beauty is more easily seen as a construction than is Truth.

There is thus an *additional* problem in science studies that should help the commerce between art history and science studies to go both ways.[9] Art history offers extraordinary skills in multiplying the mediators, but history of science insists on a question with which art historians can do away with too easily. The question is not that of Truth, since this epistemological question is no more answerable than that of Beauty in aesthetics, but another question, related to it, that *qualifies* the type of mediators. If you hear the screeching noise of the violin cords, this mediation adds to the quality of your pleasure, exactly as much as the plot you have just read on the program, or the envious glance that you got from a friend who learned that you had been able to get a seat for this most vaunted performance. More exactly, you do not have to build a *stable hierarchy* of those mediators to find it acceptable to consider that Bach's work is indeed "made up" of all these elements associated and combined, no matter how. Extracting from it the "real core" of Bach's work will not necessarily strike you as an important task. Bach, you will say, is made of "all that," snobbery and execution included, scores and lighting, ticketing and pietism, CDs and numerology. In art it remains slightly easier than in science to be constructivist and realist at the same time.

On the other hand, if you sit through a dark and freezing night inside an observatory, there is something you do not want to hear, and that is the screeching noise of your spectrometer that would ruin your data. It is crucially important that, out of the local and temporary array of instruments and set up, all of the mediators function as so many transparent intermediaries. It is essential that a tiny core of information escape from the setting and let you ignore the rest. This is the condition of felicity you want fulfilled. You cannot, as in a performance, be constructivist and realist at once, even though you know pretty well—and scores of science students will remind you of it in case you forget!—that you have "constructed" your data.[10] Once aligned, scientific mediators have to have a way of escaping their origin in a manner that is *not* required from art mediators, where their continuing presence and vibration remain essential. If I am right, what history of science has learned from art history, it should repay by insisting on its own different questions, addressed to *both* fields. I think it would be fair to say that most of science studies (that is not denunciatory) can be defined as an aesthetization of science. This is not meant as a criticism, on the contrary, it was done with the worthwhile intention to "elevate" the study of science to the level and quality of art history.

It seems to me that, as a scholarly community, we begin to know pretty well how to multiply the mediators (especially in the work on visualization in science), but we have no clear idea yet of how to account for the various ways in which the same mediators are telescoped, unfolded, embedded into one another. We cannot simply say that "all of them" count in the making of an observation. If we were stopping at that, something would be missing from the mere deployment of heterogeneous associations. Thus the same care that has been invested in multiplying mediators in art history and then in the history of science, should be now engaged in *specifying* the types of mediations. The notion of mediation itself is much too weak and hazy to define the whole middle range between the bygone representing Mind and the represented World. This is true even if one is careful not to define mediation as what is "in between" (for this I reserve the word intermediary) but rather as that which produces, in part, the elements that come in and out of meditation. If the medium is the message, slightly different types of media (and mediation) will produce enormous differences in types of messages.

THE TRANSCENDENCE OF SCIENCE: THE TRANS-FORMATION OF IN-FORMATION

What would be the simplest way to characterize the type of mediation that renders a visualizing activity scientific? The notion of in-formation captures a first trait, provided we understand the word in a very practical sense, as what put something into a *form*, in its most material aspect of inscription. To travel over distance, matters have to be changed into forms. If there is no trans-formation in the sense of encoding or inscribing into a form, then there is no travel nor transportation and the only way to know something is "to be there" and to point at features silently with the index. The scientist in the first picture could still use her index in this simple commonsense way but only to designate the table or her companions. If she starts to direct her index toward an absent feature, like the field site, then it has to be aiming at an inscription of some sort (at what I am calling in-formation). As soon as one is at a distance from features one wishes to refer to, some *vehicle* has to be invented to carry the reference in a state completely different from the one it had when it was locally and materially present.

Disembodiment and re-embodiment is essential to the task of transformation. It is essential to remember that visualization is only one of the many vehicles that help in this encryption; numbering, tagging, counting, and stuffing are some of the many others.[11] The mass of work now available on visualization in science has been extremely productive in describing a type of "formal matter," so to speak, that is neither in the mind nor in things, but is at once material and symbolic. Actually, the very meaning of symbolic has been completely dislodged out of its mentalist or structuralist past.[12] It is now an activity as empirically loaded—and as observable—as that of child care, fencing, gardening, or baking. This is why more and more work on visualization in science, which had started with an interest in text, then in diagrams, figures, and charts, is now devoted to theory and mathematics, following, by the way, the very path of re-representation.[13]

A second trait of these displacements is as important as the first: the maintaining of constant features through the shifts in representations. Since, by definition, the local matter has been abandoned, how could a form refer to it, if some of the relations were not kept constant? This maintaining of a constant through transformation has nothing to do with the carrying over of the things themselves, as in the naive scenography of realism, since the things have to be abandoned so that we have, at a distance, an information "about" them. But it has a lot to do with conserving a constant through successive transformations of the medium. Information is never simply transferred, it is always radically transformed from one medium to the next. More accurately, it pays for its transport through a heavy price in transformations.

In his major book, Edwin Hutchins has described many examples, for instance, of the ways in which angles are conserved and redescribed from the pelorus operator on the deck of a U.S. Navy ship, all the way to the chart of the plotter, in the cabin, carrying hoeys, rulers, and compasses.[14] Half a dozen media are used to redescribe the information, but they are aligned in such a way that something essential is conserved, and that is the measure of the angles. Each scientific discipline, whether completely "abstract" like topology or completely "concrete" like the building of natural history collections, can be described by the choices it makes in what should be kept constant through which sort of transformations into different media.

As it has been noted, many years ago, by the art historian William Ivins, perspective and relativity theory are two related examples that bring art history and science studies even more closely together in terms, this time, of content.[15] I have proposed to call this common obsession "immutable mobiles" and I stick to this term because it seems to capture pretty well, in my eyes at least, what is scientific in an array of mediators that, otherwise, are very similar to those found in countless other visualizing activities.[16] What they have in common is not the visual itself but the constants carried intact through the transformation of the media. Immutable does not mean that information is transferred unproblematically but that some features have to be maintained *in spite* of the mobility provided to them. In describing the visual practice of scientists, we can be attentive to the textual quality of the document, to the layout of the paperwork, to the intensity of the contrasts, to the enhancement of the features, to the local interpretive traditions, to the relative efficiency of graphs and tables, to the skilled work of rewriting the equations. In that sense, we do exactly the same work as an art historian or a sociologist of art like Howard Becker, but we can also *add the ways* these mediators align with one another, what they choose to keep constant through transformations, and what they determine to discard.[17]

We have thus at least two reasons for being fascinated, for instance, by the work of preparing an electron microscopy image; one will come from the hundred or so intermediary steps going into the construction of the artifactual image, but the second will be the gradual *disappearance* of those hundred steps into one shape that will be kept as a reference through those re-representations.[18] Of course, the quality of the shape can never escape the series of these transformations—and this is what naive realism will always miss so badly and constructivism will always deploy so beautifully—but these transformations are aligned in a way that *justifies* the claim of realism (if not the modus operandi it had imagined). They end up summarizing one another in a way that differs from the deployment of mediators in art.

The *longue durée* history of this type of alignment has been made only in part, but two things are already clear: first, the twin ideas of a calculating Mind and of a mathematizable World are the projection, at its two extremities, of this very specific type of mediation.[19] Wherever this type of network expands, there will be, as a correlate, a certain type of subject and a certain type of object invented at both ends, to sustain the transformations of the forms. This is the strong metaphysical meaning of the expression that the "medium is the message," and this is what justifies the daring and somewhat adventurous move taken, many years ago, by science students who bracketed out the individual internal Mind as well as the World "out there." We were right to extract ourselves from Cartesianism by refusing at once the *ego cogito* and the *res extensa*, in order to focus our attention on the middle ground, since this middle ground—practice, loci, inscription, instrument, writing, groupware—was the active part, and not simply, as we are told, "the means for a Mind to gain access to the World."[20] By holding the mediation we do not miss the essential parts: what happens in the mind of scientists "in there" and what is the real stuff of the world "out there."

On the contrary, by concentrating on the trivial aspects of the cooking of science, we may also end up accounting for its two vanishing points, *res* and *cogito*.

The second important point is that the definition of what counts as an essence or a substance has a lot to do with this question of maintaining a constant through transformations. It is fair to say that, in our scientific cultures, we cannot entertain any alternative notion of what is a substance, except as what is maintained through successive transformations.[21] Of course, in popular metaphysics, we project this sub-stance as what lies "under" the shifting and passing attributes, but this too, like the twin notions of a calculating Mind and a calculable World, is a projection, an extension, an effect of the development of immutable mobiles. What is kept constant from one representation to the next is morphed, quite naturally, into the *thing itself* to which, thanks to "accurate information," we gain access. But in this little shift from information by transformation to information as a mere transfer without any transformation, the word "reference" changes its meaning and instead of being what is carried through the media and the successive inscriptions, it becomes what the thing is, unaltered, unmediated, uncorrupted, inaccessible.[22] This, then, is the ultimate paradox of a historical mediation that provides access to what is then seen as an inaccessible, ahistorical, and unmediated essence.

If the work of mediations that is responsible for these products is erased further, a powerful scenography is then generated: a calculating Mind, a calculable World, a substance that lies under its passing attributes, and the medium of language to circulate in between. All the other types of mediation will now be evaluated according to whether or not they are able to provide an accurate "access to the world." And of course, by comparison, all of the other forms of mediations will be found wanting, and will be condemned as so many fantasies or so many outright lies. Information transfer will be used as the standard, although it does not even do justice to the risky business of producing scientific information. To be sure, the proliferating mediators of art will be able to escape from this indictment, only by accepting their destiny as "forms of art," that is, by abandoning any durable access to the World and any objectivity in the Mind. Hence, this rather terrifying definition of art epitomized in Kant where the free play of subjectivity "mediates" between science and morality. The relative freedom enjoyed by art historians has been obtained at a heavy price, since they had been let loose only because art did not count "seriously" compared to "what we can know" and "what we should do." As to historians of science, according to this scenography, they had nothing to do except reconstruct, in the most whiggish ways, the vagaries of reason slowly ascending to truth through the purification of its concepts.

On the other hand, if we stick to the mediators and join the forces of art history (good at deploying mediators because none of them can be ordered in a stable hierarchy) with the forces of history of science (good at tackling the question of reality and objectivity), then the hideous scenography of mind/world/substance/language disappears and we have to consider heterogeneous associations of mediations *plus* the types of mediations that group or gather the entities in completely different aggregates. Science becomes rich in visualizing skills and art regains many entries into the object. Fiction is no longer free under the pretext that it would be subjective or impotent, and science is no longer merely "accurate," because to be so it would also need to be unmediated, unsituated, and ahistorical. This, then, is the juncture that makes this volume such an interesting venture.

THE ART OF PERSON MAKING
THROUGH BROKEN IMAGES

In order to elucidate this "new deal" between art and science—once the attention to the number of mediators and to their types has been clarified—I want to introduce a third type of mediation. What happens when it is not information that is transported through immutable mobiles, but *persons*? You do not ask a lover "do you love me?" with the same expectation as when you ask "what is the present bearing of Point Loma?" If your lover answers "I have already told you three times! Why do you ask again?" you can deduce, with a pretty good margin of error, that there is no love anymore between you two. The question and the expected answer are not supposed to transport information with the minimum of deformation by propagating through many different representational states, as when you align your statement, your watch, and the clock of Big Ben (itself relying on atomic clocks in Greenwich, and those on the Bureau International du Temps at the Paris Observatory). It is not alignment and re-representation that you are expecting. The question and the answer are supposed to create *persons* who are present to one another in the very act of speaking.

To be sure, it is also a question of presentation, and even of re-presentation, but the meanings of these terms are entirely different. "Present" has first the meaning of a gift that is not due as a payment of any sort of debt or that is not the return half of any sort of barter. The conditions of felicity of the little sentence "I love you" implies that it is given as a gift and that this gift generates in those who give as well as those who receive it a form of personhood: "I am the one who is loved by that one," "*parce que c'était lui, parce que c'était moi,*" as Montaigne said of La Boétie. The word "present" also means that both are present to one another, or in the presence of one another—instead of being "absent" as in many other interactions where we are foreign to one another. More interestingly, it does not just mean being present, since the little sentence "I love you," when uttered rightly, has the other virtue of putting both speaker and listener in the presence of one another *again* and *anew*. Hence the different meaning of re-presentation, as what is presented again, or what provides another chance of being in the presence of someone or something (instead, for instance, of being "in the absence," that is, dead!). Although the conditions of felicity of this "speech act" are difficult to detail, every one of us seems to have an uncanny ability to detect its infelicities: "you don't really mean it," "you say that to please me," "you said that too fast," "you did not say that like the first time." Although it is *un petit je ne sais quoi*," as we say in French, we seem to know a lot about it!

For the purpose of this meditation on mediators, it is very important not to oppose information transfer and person making as objectivity and subjectivity. Information is never transported without being deeply transformed, this, as I said, is the paradox captured in the notion of "immutable mobiles." Thus, as we learned from science studies, there is nothing especially objective about science; this type of mediation simply generates a *form* of transfer, that is, reference, while it projects, at the two vanishing points of its networks, a certain type of subject—the calculating Mind—and a certain type of object—the calculable World. Person making is no more subjective than information transfer is objective. It is simply the case that this new mediation generates, at the two extremities, completely different types of subjects—a person receiving the gift of presence—and a completely different type of object—presence giving. This, however, does not mean subjectivity. It is a full-blown mediation, a form of life, with its own form of judgment, its canon, its empirical world, its own taste and skills. Truth and falsity, faithfulness and infidelity are carefully detected, measured, proved, demonstrated, elicited. Nothing is less unmediated, affective, evanescent than this sturdy, careful, accurate mechanism to evaluate love. A large part of our life is spent—and well spent!—in developing those skills and honing those forms of judgment.

Once the false binaries of objectivity and subjectivity, coldness and heat, nonhumanity and humanity, visibility and invisibility have been put aside, it is much easier to see the differences in the types of mediation. Both are pointing at features absent from the scene of the action; both are transcendent since they designate features beyond the inscription; both are non-realistic since they work by transforming deeply the representation; both define, with a sure skill, truth and falsity, faithfulness and betrayal. Yet, they differ completely on what they point at. In information transfer, everything is sacrificed to the maintenance of a constant that undergoes transformations and, so to speak, jumps from one medium to the next. Einstein's famous "mollusk of reference" is an extreme example of this obsession. Every commonsense definition is modified, but the transfer of information from one accelerated frame of reference to another is saved through the generalization and reconceptualization of the Lorentz *transformations*—a very apt term.[23] In person making what counts above all, what requires the utmost sacrifice, is the designation, here and now, of the person at hand, being presented with the gift of presence. But there is no way to produce this effect by directing attention *away* from the scene. On the contrary, the only way is to *redirect* attention by pointing, through cracks into the discourse, to the character in the flesh listening to the story or watching the scene.

Let me restate this essential point. Redirecting attention away from the wrong direction toward the gift of presence that produces persons here and now, is the obsession of this type of delegation. This differs a lot from transporting information with the minimum of deformation through the maximum of transformations. It differs a lot, but *not* as objectivity used to differ from subjectivity, and especially *not* as an access to a "natural" world used to differ from an access to a "world beyond." Access is precisely *not* what is in question in this person-making form of life, no matter if it is beyond or beneath, before or after. Moreover, "access to a world beyond the present one," paradoxically, is a much better definition of information transfer and of its specific type of transcendence. In person making, there is no interest whatsoever for substances.

In the first picture (see Figure 1), the botanist was directing her index to a world beyond, which was obtained by what remained stable through the transformation of the chart, the aerial photograph, the trip to the field site, the visual evidence they will gain tomorrow, the metrology of cartography and the surveying that held the bearings together. A substance, the locus of the field site, exists through its attributes. In that sense, scientific visualization offers a much more transcendental, immaterial, spiritual vehicle than anything we could think of. The angel, in the second picture (see Figure 2), is doing another job that we can now consider much more clearly: the two hands are *redirecting attention* to something else, something much more important: "He is not here; He has risen! Why do you look for the living among the dead? Go away to Galilea!" But the second hand is not directing attention to the image of the risen Christ *instead*, as if it were displacing something intangible from one representational medium to another, the way our lady scientist could go from an inaccurate map to a more accurate aerial photograph to produce a better reference. The holy women cannot see the apparition at their back![24] So, who can see the apparition? Nobody, and that is exactly the point designated by the finger, that is exactly the angel's warning, what explains the "opacity of painting," to use Louis Marin's magnificent expression.[25] According to him, perspective allows exactly the opposite of what Samuel Edgerton describes as rationalization and transparence.[26] The new coding of perspective does not eliminate the repertoire of the ancient pre-perspectival icons; on the contrary, by stabilizing most of the conventional reading of the rest of the picture, perspective allows Quattrocento painters to *highlight* in a most dramatic fashion the discrepancies, the cracks, that allow the paintings to make the presence real. In Piero della Francesca's *Annunciation*, it is essential to understand that the angel cannot see the Virgin hidden behind the pillar—but this understanding is made possible only by the geometrical convention of the pavement. Standardization makes re-representation possible.

This opacity of painting does not mean, however, as it is often construed, that painting designates something that is always beyond, something above, something spiritual and immaterial, a substance beneath its attributes, or that it designates the indefinite vacuity of belief. On the contrary, it *can* be seen, but not *here* in the empty tomb and not either by *replacing* the empty tomb by the invisible apparition. For the viewer, seeing is no longer the accessing of a substance beyond the present setting, but *being designated* now, here, in the flesh, as someone receiving freely the gift of life anew. What is missing in the picture, to get its meaning, is the monk in the cell, to whom the index of the angel, and the whole painting as an index, is addressing its warning. Beware, your life is in question.

The kneeling painted monk on the left, Dominicus, who sees everything but the apparition (which can only be seen from another point of view not within the picture), marks a perfect transition between what is missing from the picture and how the picture should be read, that is, the "legend" of the fresco. All the more so since, as a father of the Dominican order, he is also the one better suited to teach the friar in the cell how to read, see, behave, and pray. What is designated by the fingers of the angel is visible only by kneeling in prayer and looking not among the dead but among the living, not at the past but at the present. The index here is not about others but about you, not about absent belief but about present persons. Whereas, in information transfer, all the arrows are pointing at entities that are absent, in person making all the arrows, through the cracks, discrepancies, visual puzzles, absurdities of the scene, are pointing at the kneeling monk in the cell putting himself in the presence of what is a present person and not a dead belief, in the presence of what "has risen" and is understood now, in a flash of recognition, by this designated monk who, at last, grasps the gesture of the angel redirecting attention to the text of the Gospel that this painting reenacts, making the text as luminous and as simple as Dominicus's teachings. The fresco is the empty tomb out of which life has risen again.

I have moved surreptitiously from the ordinary sentence "I love you" to the theme of the San Marco fresco as if they were dealing with the *same* type of mediation. The reason for this move is that the only form of talk that we are still good at and that remains close enough to person making, is to be found, nowadays, in what is called "interpersonal relations." The situation is similar to that of Maussian gift giving that used to link whole economies in the past, but remains now visible only in the domestic realm of friendship and family relations, the rest being taken over by market relations (according to the economists at least).[27] This has not always been the case however, and there used to be a time where the most common, public, and collective form of life was not information transfer but person making. I hesitate to use the word "religion" to describe this form of mediation, since religion has been turned, *because* of the contamination of the model offered by information transfer, into something exactly *opposite*: a belief in the existence of a distant substance beyond the realm of experience to which we have access only through the intermediary of special vehicles—a definition that, funnily enough, is a good description of science production, but *not* of person making as defined above.

Even stranger, when considering pictures that have a religious theme, art historians, even sophisticated ones, take the theme, the narrative, as being *about* some scene of the New Testament, or some stories of the Golden Legend. They know very well of course that no picture represents, in any realistic sense, a genuine scene that would exist somewhere, and yet they analyze the picture as if there existed a *body of beliefs* on which the picture would draw or to which it would refer. In so doing, they forget one crucial element of those pictures: not only do they *not* refer to a specific landscape or to a genuine event in any scientific sense, but they do *not* relate in any sort of referential way to a body of beliefs *about* the Virgin Mary or the history of Christ. In other words, although art historians are bona fide constructivists for every aspect of the painting—market forces, varnishes, perspective, programs ordered by the sponsors, and so on in a bewildering display of scholarship—they talk about what the scene represents by using a definition of representation that is utterly scientific, in the sense that it should "refer to" a scene of the Bible and not in the sense that it presents it anew. The visual puzzles to solve—for the painters, their patrons, and their customers—are entirely different if one or the other meaning of representation is chosen.

My contention is that those pictures that are engaged in person making are not at all about reference and access, *not even* in the somewhat innocuous sense that they would allude, refer, or be "about" scenes of the Bible. They do something utterly different, they re-present, in the other sense of the expression, what these stories and scenes really meant—meanings that had been lost by those who read them, but which can be re-understood now again, because of the picture (as in the case of stigmata traced by Arnold Davidson in this volume). Like the fingers of the angel, they redirect attention to what is important and ask, through the discrepancies of the visual display, that we do not look away among the dead, but here among the living, for what is meant. Hence the subtitle of Georges Didi-Huberman's book, *Dissemblance and Figuration*, is itself based on an interpretation of Denys the Aeropagite's theory of images, which uses for what I called "cracks" and "discrepancies," words like *dissimiles, inconsequentes, inconvenientes, deformes, confusae, mixtae.*[28] All the meditations of Christianity are about bringing *real presence*, not about illustrating themes. But compared to this obsession, the historical differences between iconoclasm and iconophilia appear to be very small, even if one takes into account Reformation image breaking, Lutheran search for "mental images," the sorting out of Catholic images after the Counter-Reformation, the Byzantine "economy" of icons, or even the iconoclasm of Modern art.[29] Avoiding information transfer is what all these visual cultures have in common. Iconoclasm, in that sense, might be only one aspect of this long history, another way of multiplying the discrepancies, the cracks, and redirecting attention to what really counts. But what can be done literally by destroying the image itself, can be done figuratively *in* the image. In that sense, all Christian images are born broken.

When they are not, they are broken in yet another meaning of the word: they are so bad that they are ready for the bonfire! The simple demonstration of this is to live the rather horrifying trial of walking by mistake into the rooms of Modern Sacred Art in the Vatican Museum. Thousands of *croûtes* allude to biblical scenes, refer to beliefs, represent biblical stories; not one of them, *not one*, has any sort of pretense at re-understanding anew what those stories meant. Absolutely devoid of theological values, they are also, interestingly enough, devoid of artistic ones—as if mediators of different sorts supported one another and the art ones refused to sit where their person-making brethren had been excluded.

Theology, unfortunately, has been for a long time in the same dire state where epistemology and aesthetics were before the onslaught of constructivism. It is thus of very little help at this juncture, since it has absorbed the language of science to the point where it really believes it has to defend certain "beliefs" in the "real existence" of "real substances" that would reside "beyond" the reach of natural and empirical grasps and that would be accessible only through the successive ladders of more and more immaterial intermediaries. Thus, its most essential phenomenon, its own original type of mediation, its very core, is defined in the exact terms of another one that goes in completely different directions and produces utterly different objects and subjects. An invisible world of belief is mistakenly built beyond the visible world of science, whereas it is almost the opposite: science gives access to a form of invisibility and religion to a form of visibility: *ego, hic, nunc.*

THE "LEGEND" OF SCIENTIFIC AND RELIGIOUS IMAGES

It is difficult for us to pay equal attention to different forms of iconophilia because we tend to confuse the conditions of felicity of visual cultures. This is nowhere clearer than in the notion of belief. Religion, or, to be more precise, person-making mediation, has no more to do with belief than science with a visible world. The notion of belief is the projection on religious mediators of the trajectory of information-transfer ones. We start to talk of belief when we try to grasp the content of a person-making statement by using the reading cue offered by science and then, finding it empty, realizing that no immutable mobile is at work, that no information is carried over, and thus, since this statement is "empty," decide that it should correspond nonetheless to something nonexisting that we call the "content of a belief." Belief is a charitable interpretation but an ill-applied charity. Instead of modifying the reading key, it tries to save the interpretation by offering it a content *it never had* in the lived world of those who uttered those sentences. In other words, no one, absolutely no one, ever believed in anything according to the manner imagined by science.[30] To put it more polemically, the only believers are the ones, immersed in scientific networks, who believed that the others believed *in* something.[31]

Several consequences can be drawn from this. The first is that the information content of religious mediators is nil. Angels, like the one painted on the San Marco cell, are very good at redirecting attention, but if one asks what they did say, what sort of constant they maintained through shifts in messages or inscriptions, the answer is tragically void. In this sense the information content of the whole Bible is nonexistent. This is the case in spite of the Creationists' hilarious attempts to read it as if the geologist Georg Lyell had written it, thus proving that they misconstrue religion talk even *more* than they misunderstand scientific discourse. Creationists are an excellent demonstration that some Christians can be rationalized to the marrow, unable even to retrieve a shred of the kind of talk that would not carry information but transport persons.

What the angels carry, however, is something crucially important: another mediation, "He is not here, He has risen." But who is this He? Not a substance to which one would have access, but another mediation, a life-giving person, a mediator of God. And to this God, in turn, do we finally have access through successive ladders of mediations, like the ones deployed by the scientist in the first picture, accessing one spot in the Amazonian forest? No, since God is another mediation, another way of saying what is present, what is presented again and anew, what is, has been, and will be. But Presence is in no way construed as a substance beneath, everlasting under its attributes. Hence its definition by John in terms of ways of talking, of enunciation: "In the beginning was the Word (*Verbum*)" (John 1:1). To shift from person to substance is to change the reading key and to replace the meaning—what gives life presently—to a completely different one: what has always existed. The second one, contrary to the belief in belief, is not religious; it is, through and through, a scientific way of accessing and managing the transit of immutable mobiles.

The second consequence is that this tension between substance and persons, information transfer and person making, is inherently unstable—as the whole Patristic theology can show. That is, there is no way, no *direct* way (this would go against the very notion of mediation) to put oneself in the presence of presence and to understand, once and for all, the meaning of this message. This is what I meant in the little anecdote above. If, to the question, "Do you love me?" you answer "I have *already* told you," you shift to a temporality where a message could be capitalized once and for all. This is not possible with love talk, since what is required from it is a renewal, on the spot, of what you had indeed said thousands of times to the same person, but that has to be remade anew for the two to be again in the presence of one another, even, for instance, after a quarter of a century of common life. This is why the angels, in religious forms of life, have so much work to do, although they carry no information: they ceaselessly have to redirect attention to the presence, which, by the very passage of time, is always lost. This temporality is well-known and easily experienced in love, but we should use this tiny cue to understand that it has always been the same with religion.

Another way of putting this point is to say that there is no such thing as a pure or direct religious expression, not because it would be ineffable, subjective, spiritual—this is, as we now understand clearly, a charitable but utterly scientistic way of talking—but simply because the meaning of the message is in a presence that becomes absent through the displacement of the mediation itself. Meaning is always missed if it is not renewed now. This is what Rudolf Bultmann indicated long ago by making mythologization and rationalization two synonyms.[32] Rationalization is not a *defect* of religious talk that we could eliminate. In the Gospels themselves, the evangelists are busy rationalizing, inventing scenes, adding anecdotes, making the story smoother, making it look more reasonable. Conversely, the most bizarre rationalization can suddenly retrieve meaning through the cracks or discrepancies of its construction when it is seized again by what is meant, persons being made alive again in the presence of what is person-giving. If we dislike the theme of the Assumption, for instance, we may displace it, break it, or shake it, but we cannot replace it by a purer, less corrupted meaning that would access more directly what is in question *through* that theme. Contrary to many of the Reformation and Counter-Reformation theories of images, directedness, transparence, purity, and access are important properties of information-transfer mediators, but destroy the person-making ones.

A final consequence is that the best way to respect images is certainly not to save them through a "symbolic interpretation" that would lie hidden beneath the popular and metaphoric usage. Contrary to appearances, this way of reading would be thoroughly scientistic, since it would take the relations between the empty tomb (first message), the apparition beneath the holy women (second message), the admonition of Dominicus as a kneeling painted monk (third message), the passage of the Gospel (fourth message), and the understanding of the message by the praying monk in the flesh (fifth message), as *substitutes* for one another *in the same way* as the ones used by the scientist (Figure 1) who could go from a chart, to an aerial photograph, to visual evidence. Such is not at all the way messages order themselves along religious paths. They are not about messages at all, no matter how abstract, pious, reasonable, or gnostic, they are about *messengers*.[33] And they are not about having access to a superior reality beyond, but about designating the speaker as the one who receives the gift of life anew, and suddenly, starts understanding what those messages finally—but always provisionally—meant. Messengers, not messages; persons here and now, not substance there and above or below.

To use again love talk as a template, it is the trembling of the voice, the tone that gives truth-value to the otherwise repetitive sentence "I love you." In the same way, it is the trail through all the discrepancies, the puzzles, and the breaches—in the visual constructions, in the uses of the themes, in the interpretation of the programs—that provides meaning, or not, to the religious icons. The unerring skill with which we define proofs and pass judgments on this tone that sounds, on the face of it, so unseiz-able, is what should be retrieved, in my view, from those paintings by following their "shaking" and their "trepidation."

What we have to retrieve is the carrying of a *movement* that uses the message to pro-duce the *enunciators* of this message. Since this is obviously impossible, the only way it can be done is to render the message unable to do the job of information transfer, in order to force attention away. But this "away" is not the "beyond" of belief, and here again, the path falsely indicated by the message has to be broken, shattered, and inter-rupted, so as to redirect the sight away from the invisible and unalterable substance of the spiritual world. *Always away, but not beyond.* Yet neither is the direction the depth of an individual soul moved by the beauty of a message that is addressed to the *ego, hic, nunc.* Again, to cancel out this third possible reading and escape in the right direction, the message has to be split, cracked, shaken, and redirected away. *Away, but not down* toward the feelings of psychology or even deeper in the dark unconscious. And so on, in a circulation, a spiral, which provides meaning not only for a painting but for the whole setting—theological, institutional, cultural—in which the mediators are gath-ered, reshuffled, and assembled (see Joseph Koerner in this volume). This spiral going from one index redirecting the sight away to another resembles slightly, but only slightly, the way art mediators, as I said above, are unable to be listed in a stable hierar-chy around a core which would be the "real worth of a work." The resemblance is enough to understand the *resonance* of the two, but the difference is still easy to detect, as in the beautiful case of Bach's "blasphemy."[34] Alliances between types of mediators are always provisional.

CONCLUSION: AGAINST ICONOCLASM

What can we conclude in this meditation on mediations? That it is no easier to be iconophilic in science than in religion. On the face of it, the intensity of visual inscriptions seemed to make scientific practice an ideal case for the study of visualiza-tion. But we saw above that this was not the case, since what is visible is only the freeze-frame of a process of transformation that remains extremely difficult to grasp, a proper form of invisibility. Conversely, if we turn our attention toward religious images, there is nothing ineffable in those conditions of felicity. It is perfectly possible to find the empirical grasp that would allow us to write down their specifications, but the grasp should be adjusted to the type of fragile mediators, to the frail feathers of angels.

One of the great interests, to me at least, of the emphasis on visualization in science studies, is to offer us a *vision* of the practical production of facts, which is not very much easier, direct, transparent, or unmediated than that of the religious movement I tried to outline. The quality of the scientific reference continues only if you add yourself to it and push it one step further, or deteriorates if you stop carrying it over; exactly in the same way, albeit through a radically different movement, the meaning of religious mediators can be retrieved only if you add yourself to the list of relays that vibrates through the whole tradition of interpretation. To put it bluntly, Nature is now seen as no less transcendent than God; or to put it even more strangely: angels' work is not very much more difficult to grasp than the work of scientific instruments. To be sure, the ways the mediators circulate is entirely different, but the ways in which they have to be deployed is not so different: visual cultures are immensely complex in both cases, and it is as difficult to claim that we would have a better knowledge "without all that" (meaning the instruments, inscriptions, graphs, and laboratories), as it is to say that we could have better ways of producing persons if we could get rid of "all *that*" (meaning angels, icons, and love talks).

The difficulty is to learn how to be iconophilic for one form of visual culture without being iconoclastic for the others (as has been so often the case in the past). To come back one last time to the example of the San Marco fresco, we can now see how to write down the "specifications" or the "conditions of felicity" of other regimes of mediations, once the tyranny of information transfer is lightened. A tyranny, we should remember, that makes the practice of science incomprehensible as well. To transcribe these conditions, we have to build two lists. The first one provides a message, or a series of successive messages, and the second one is made up of all the breaches that make those messages unfit for normal consumption, and whose succession indicates, as so many relays, the circulation of a meaning that would cut transversely through all these stories, none of which is believed (although each is used for a little time). The first list gives us several layers of meaning: the empty-tomb story of the Gospel, the apparition of Christ, Dominicus kneeling down, and so on. If you displace these inscriptions to align them in the same way as the chart and the photograph in Figure 1, then they become an extraordinarily clumsy and uninformative way of relating you (now in the mind set of a scientist) to events happening in year 30 B.C. in Jerusalem. You cannot push your finger on it as does the scientist on the table at the restaurant and have any sort of access to Jerusalem, as she has to her field site in the Amazonian forest.[35] Harnessed for this use, the mediators appear as a sympathetic and possibly beautiful tissue of lies. This remains so if, shifting this time to art history, the fresco is seen as one of the many examples of a theme inside the program of Dominican visual culture. This time they are no longer outright lies, but mere realizations of a prototype localized in the mental and visual culture (what Michael Baxendall calls the "period eye") of Quattrocento Florence.[36] From bad information they have been turned into good symptoms. Still, they *do* nothing.

But, if you now add the many little elements that are used to redirect attention: the mute angel, the non-visible apparition of Christ, the skewed position of Dominicus who sees nothing, the empty tomb itself, and so on in a list that grows constantly with your knowledge of theology and art history, then you begin to make the whole tissue of messages vibrate all the way to your understanding (now in the mind-set of someone designated by the picture) of what is carried out: this is what was meant by the *invention* of the empty tomb, by the *innovation* of the angel, by the complete *lie* of Galilean apparitions, by this *way of talking* that makes the Evangelist say "He has risen. . . ."[37] On the other hand, if you paint a scene from the Bible, but without shaking its construction by inventing new indices that redirect attention away from it; or worse, if you imitate inventions made at another time and place by other painters for other patrons and customers, then your painting will be much more devoid of religious meaning than an oyster with lemon on a napkin even though you fill it with two thousand angels, hundreds of halos, and countless kneeling worshipers. This is what explains the nightmarish quality of a visit to the Vatican Modern Sacred Art Museum. These *croûtes* simply forgot to renew the "putting into presence" vocabulary and to reinvent new tricks, breaks, and cracks. Their painters, patrons, and viewers have become, literally, *absent-minded*.

The conditions of felicity for such a movement seem, to us, very strange, and become even queerer when we take religious icons as our example, instead of love talk, for which we all have reasonable competence. We should not be surprised by this difficulty, however. About three centuries of forms of life are missing to make the junction between these two ways of talking, since theologians and priests, overinfluenced by the example of immutable mobiles, stopped reinventing new modes of speech and *began to believe*, occupying the antirationalist position that rationalists had devised for them. Christians even went so far as to invent another world beyond the "natural" one, a world to which we could have access only through prayers, discipline, and series of aligned intermediaries! An invisible world beyond the visible one! For my part, a civilization where we can have angels and immutable mobiles circulating, each in their own way, seems a much better place to live than the one in which science is supposed to access the World directly. It also seems better than the rather horrendous culture in which the poor angels are harnessed to do the work of instruments, accessing a world beyond and carrying blank messages back on their return.

Notes

1. Bruno Latour, "The "Pédofil" of Boa Vista: A Photo-Philosophical Montage," *Common Knowledge* 4, 1 (1995): 144–87.
2. Bruno Latour, *Petite réflexion sur le culte moderne des dieux Faitiches* (Paris: Les Empêcheurs de penser en rond, 1996).
3. Susan Leigh Star and Jim Griesemer, "Institutional Ecology, 'Translations' and Boundary Objects: Amateurs and Professionals in Berkeley's Museum of Vertebrate Zoology, 1907–1939," *Social Studies of Science* 19 (1989): 387–420.
4. But cf. Svetlana Alpers, "The Studio, the Laboratory, and the Vexations of Art," in this volume.

5. See, for example, Antoine Hennion, *La passion musicale: Une sociologie de la médiation* (Paris: A.-M. Métailié, 1993).

6. Svetlana Alpers, *Rembrandt's Enterprise: The Studio and the Market* (Chicago: University of Chicago Press, 1988).

7. See Peter Galison, "Judgment against Objectivity," in this volume.

8. Crosbie Smith and Norton Wise, *Energy and Empire: A Biographical Study of Lord Kelvin* (Cambridge: Cambridge University Press, 1989).

9. Antoine Hennion and Bruno Latour, "Objet d'art, objet de science. Note sur les limites de l'anti-fétichisme," *Sociologie de l'art* 6 (1993): 7–24.

10. See Simon Schaffer, "On Astronomical Drawing," in this volume.

11. Adèle Clarke and Joan H. Fujimura, eds., *The Right Tools for the Job: At Work in Twentieth Century Life Sciences* (Princeton: Princeton University Press, 1992).

12. Claude Rosental, "L'émergence d'un théorème logique. Une approche sociologique des pratiques contemporaines de démonstration" (Ecole nationale supérieure des mines de Paris. Thèse de sociologie, 1996).

13. Bryan Rotman, *Ad Infinitum: The Ghost in Turing's Machine: Taking God out of Mathematics and Putting the Body Back In* (Stanford: Stanford University Press, 1993).

14. Edwin Hutchins, *Cognition in the Wild* (Cambridge, Mass.: MIT Press, 1995).

15. William M. Ivins, "On the Rationalization of Sight," *De Capo Press and Plenum Press* (1930, reprint, New York, 1973).

16. Bruno Latour, "Drawing Things Together," *Representation in Scientific Practice*, ed. Mike Lynch and Steve Woolgar (Cambridge, Mass.: MIT Press, 1990), pp. 19–68.

17. Howard Becker, *Art Worlds* (Berkeley and Los Angeles: University of California Press, 1982).

18. Michel Mercier, "Recherches sur l'image scientifique: genèse du sens et signification en microscopie electronique" (Thèse de doctorat, Bordeux I, 1987).

19. Lorraine Daston, "The Factual Sensibility—an Essay Review on Artifact and Experiment," *Isis* 79 (1988): 452–70. See also Daston and Peter Galison, "The Image of Objectivity," *Representation* 40 (1992): 81–128, and Galison, "Judgment against Objectivity."

20. Charles Goodwin, "Seeing in Depth," *Social Studies of Science* 25, 2 (1995): 237–84.

21. Alfred North Whitehead, *Process and Reality: An Essay in Cosmology* (1929, reprint, New York: Free Press, 1978).

22. See Bruno Latour, "Do Scientific Objects Have a History? Pasteur and Whitehead in a Bath of Lactic Acid," *Common Knowledge* 5, 1 (1996): 76–91.

23. Albert Einstein, *Relativity, the Special and the General Theory* (London: Methuen And Co., 1920), p. 99.

24. Georges Didi-Huberman, *Fra Angelico: Dissemblance and Figuration* (Paris: Flammarion, 1990).

25. Louis Marin, *Opacité de la peinture: Essais sur la représentation* (Paris: Usher, 1989).

26. Samuel Y. Edgerton, *The Heritage of Giotto's Geometry. Art and Science on the Eve of the Scientific Revolution* (Ithaca: Cornell University Press, 1991).

27. Marcel Mauss, *The Gift*, trans. Ian Cunnison (New York: Norton, 1967).

28. Didi-Huberman, *Fra Angelico*, p. 56.

29. For this range of iconoclasms and iconophilias, see Olivier Christin, *Une révolution symbolique* (Paris: Minuit, 1991); Joseph Leo Koerner, "The Image in Quotations: Cranach's Portraits of Luther Preaching," *Shop Talk: Studies in Honor of Seymour Slive* (Cambridge: Harvard University Art Museum, 1995), pp. 143–46; Marie-José Mondzain, *Image, icône, économie: Les sources byzantines de l'imaginaire contemporain* (Paris: Le Seuil, 1996); Dario Gamboni, *The Destruction of Art: Iconoclasm and Vandalism since the French Revolution* (London: Reaktion Books, 1996).

30. Elizabeth Claverie, "La Vierge, le désordre, la critique," *Terrain* 14 (1990): 60–75, and "Voir apparaître, regarder voir," *Raisons Pratiques* 2 (1991): 1–19.

31. Bruno Latour, *Petite réflexion*.

32. Rudolf Bultmann, *L'histoire de la tradition synoptique* (Paris: Le Seuil, 1971).

33. Bruno Latour, "Quand les anges deviennent de bien mauvais messagers," *Terrain* 14 (1990): 76–91.

34. Denis Laborde, " 'Vous avez-tous entendu son blasphème? Qu'en pensez-vous?' Dire la Passion selon St Matthieu selon Bach," *Ethnologie française* 22 (1992): 320–33.

35. See Ellen Spolsky, "Doubting Thomas," *Common Knowledge* 3 (2 1994): 110–29, and Malcolm Ashmore, Derek Edwards, and Jonathan Potter, "The Bottom Line: The Rhetoric of Reality Demonstrations," *Configurations* 2, 2 (1994): 1–14.

36. Michael Baxandall, *Painting and Experience in Fifteenth Century Italy: A Primer in the Social History of Style* (Oxford: Oxford University Press, 1972).

37. Bultmann, *L'histoire de la tradition synoptique*.

SIMON SCHAFFER

On Astronomical Drawing

Our object is, or ought to be, not the mere imitation, but the rivalship of nature.
—CHARLES PIAZZI SMYTH, *ON ASTRONOMICAL DRAWING*, 1841

"THE APPEARANCE OF DISTANT REGIONS"

Worries about the adequacy of scientific pictures are part of commonplace work that separates genuine objects from parasites and artifacts. This work is not applied instantly to singular images but to prolonged series of pictures, techniques, and personnel. The psychologist Richard Gregory has used Victorian drawings of nebulae as an example: "it may be doubted if it is possible to make a single observation of anything," he suggests, because observational judgments "take time and knowledge to develop."[1] Images supposed to convey their meaning at a glance were thus made as a result of astronomers' laborious gaze. The gaze and the glance in Victorian nebular astronomy are the themes of this chapter. Many astronomers and draftsmen worked hard between the 1830s and the 1860s with a range of instruments, drawings, and engravings to make an adequate image of a supposedly stellar object, the Great Nebula in Orion. The labor expended in taking such pictures was often publicly described to make images' adequacy telling. Yet when this artful work became visible, the image might be seen rather as an artifact. Mid-nineteenth-century astronomy could be a hard case for the examination of this kind of labor, for it was the norm for a newly positivist theory of science that saw astronomy as a science of angles and times, and the observatory as a kind of institutionalized retina where celestial sights were immediately, almost effortlessly, recorded without prior judgment or experiment. But the norm neglected the systematic disciplines designed in the 1830s and 1840s to produce reliable observers and recorders; new experimental technologies, such as photography and spectroscopy; and the work of drawing and engraving, which mattered especially in stellar and nebular astronomy since images of these bodies needed sketches of mobile and granular shapes rather than merely a record of times and angles. Such sketches were often judged using conventions of physiognomy and caricature, ways of seeing common in the worlds of cheap graphic journalism. So this chapter uses the labors of nebular astronomy in the mid-nineteenth-century British Empire to see how astronomical picturing worked, the shared conventions by which it was governed, and the various milieux in which it mattered.

Nebular drawing was part of the project to establish, or undermine, a Nebular Hypothesis, a story about the evolution of solar systems through the gradual condensation of a spinning cloud of gaseous fluid told by several early-nineteenth-century astronomers and philosophers. They mainly relied on the claims by the Regency's "natural historian of the heavens," William Herschel, that his big reflecting telescopes showed clouds of nebular fluid in deep space whence stars were produced by gravitational condensation. Chief among Herschel's specimens of truly nebulous fluid was the Great Nebula in Orion, "the most wonderful object in the heavens," where he saw evidence of slow development from the "strata" of vast luminous clouds. By arranging well-crafted pictures of nebular types the senior Herschel was able to urge that "the heavens . . . are now seen to resemble a luxuriant garden, which contains the greatest variety of productions, in different flourishing beds. . . .we can, as it were, extend the range of our experience to an immense duration."[2]

Herschel reckoned that in his natural history of the heavens "seeing is in some respect an art, which must be learnt." His art was often compared with contemporary imagery brought from the South Seas by imperial navigators. "European vision" of remote tropical and celestial zones was developed by the voyaging artists sponsored by Herschel's patron, the aristocratic naturalist Joseph Banks, and in Alexander von Humboldt's widely read physical geographies, which mapped contours and vegetation types across huge tracts of exotic territory.[3] Humboldt proposed an aestheticized "physiognomy of plants": "in determining those forms [on which] the physiognomy of a country's vegetation depends, we . . . must be guided solely by those elements of magnitude and mass from which the total impression of a district receives its character of individuality." Whereas in the tropics "the Earth reveals a spectacle just as varied as the starry vault of heaven, which hides none of its constellations," isolated European physiognomists must use representational techniques "to enjoy in thought the appearance of distant regions," celestial and natural historical.[4] Colonial naturalists and astronomers joined in projects of what Michael Dettelbach has acutely named "aesthetic empire." Like Herschel, whom he judged "astronomer, physicist and poetical cosmologist all at once," Humboldt reckoned that condensing nebulae were the sources of stars, there was a close analogy between nebular and organic history, so the physiognomy of natural forms could be applied to astronomical objects.[5]

In the 1830s and 1840s William Herschel's son John, doyen of British astronomy and protagonist of a range of new ways of making images, helped interpret these programs for early Victorian culture by fitting the natural history of the heavens into Humboldt's "applied astronomy" and the physiognomy of nature. In the British empire, astronomers such as the military engineer Edward Sabine joined John Herschel in surveys of magnetic, botanical, and physical variables, and sought to extend their maps to the heavens. In the 1840s Sabine helped translate Humboldt's *Views of Nature* and his *Cosmos*, which Herschel then reviewed. He summarized "the peculiar physiognomy of natural scenes," and, like his close friend Humboldt, Herschel saw that the evolution of stars from overlapping "strata" of nebular fluid was a view to be "expected from one especially conversant with organic forms."[6] On the basis of the Herschels' pictures, the Glasgow astronomer John Pringle Nichol invented a "nebular hypothesis" for stellar and planetary origins. A radical economic journalist, Nichol made this hypothesis a keystone of his reformist account of the law of progress at work in nature and society. As Nichol explained, if powerful instruments like the Herschels' telescopes could not resolve such objects as the Orion nebula into stars, and if its shape kept changing, then there was good evidence of true nebulosity, thus support for his evolutionary story and grounds for an organic physiognomy of the heavens. Nichol asked whether such "void, formless and diffuse" objects were just masses of very distant stars, or instead "monuments of bygone worlds—the fossil relics which mark the early progress of our own planet," containing "the germs, the producing powers of that LIFE, which in coming ages will bud and blossom, and effloresce, into manifold and growing forms, until it becomes fit harbourage and nourishment to every varying degree of intelligence and every shade of moral sensibility and greatness!"[7]

Science journalists shifted debates on true nebulosity to the larger public sphere of mass lectures and cheap print where vivid physiognomies were put to many persuasive uses. A social history of such vividness challenges standard models in which scientific accounts are allegedly produced in specialist settings, thence distributed for public consumption.[8] The distinction between specialists and the public on which this analysis relies was made, not taken for granted. Humboldt, for one, made sure that the techniques for picturing nature's face were connected with the right ways of picturing the sciences' standing. In his review of Humboldt's work, Herschel explained that "such a view of nature ought to be in the highest possible sense of the word, *picturesque*, nothing standing in relation to itself alone, but all to the general effect." The scientists' public must not demand access to detail but must be content with unified and distant pictures. "As in art, intense and elaborated beauty in any particular defeats picturesqueness by binding down the thought to a sensible object, annulling association, and saturating the whole being in its single perception."[9] This aesthetics was designed to reinforce the often fragile security of the gentlemen of science and the dependence of their audiences. Early Victorian astronomers' terrain was a network of heterogeneous sites each of quite unequal social legitimacy and hosting different forms of labor. The vividness of nebular imagery was made in the spaces around the great telescopes, where observers, draftsmen, engineers, and managers contested the proper ways of making physiognomies of the heavens; imperial zones of political and racial conflict, where liminal knowledges were forged of populations that lay just beyond the borders of metropolitan power; and the world of print culture, where ingenious juxtapositions of image and text were used to forge caricatures and physiognomies of the messages it was claimed the heavens taught.

The next section connects astronomical picturing in colonial South Africa, where John Herschel proved his astronomical worth in the 1830s, with the aesthetic empires of natural history. This discussion is followed by an account of the world of metropolitan engravers, whose struggle for status made the authorship of images a vital sociopolitical problem for early Victorians, astronomical image makers not least. The latter sections of the chapter then shift to the preeminent reflecting telescopes of the period, the giant three-foot and six-foot instruments commissioned by the Earl of Rosse in Ireland in the 1830s and 1840s. They were used to resolve the Orion nebula, an object *not* considered to be resolvable by modern astronomers, but which Rosse and his colleagues *did* eventually claim had been seen as stars. Resolution was supposed to remove evidence for truly nebular fluid in space. Careful techniques were used to tell whether resolved stars were really visible; whether the nebula's shape had changed; and, from the 1860s, whether the nebula showed the bright spectral lines of a truly luminous gas. So these different picturing techniques might spell doom for materialist evolutionism, or else provide evidence of its truth. Such evolutionist doctrines counted rather intensely in Anglo-Irish relations, another imperially critical setting, and helped define Rosse's team's antievolutionist graphics of nebular astronomy. Physiognomic conventions were of exceptional importance in debates on evolution, progress, and degeneration and form parts of this labor history of picturing. Distinctions between reliable pictures and deceptive artifacts were made in the varying public spheres of Victorian society.

"THE EUROPEAN FACE IS QUITE LOST"

Telescopes and engravings used techniques crafted to make their craftiness vanish. Hence arose a problem of calibration. If there was little prior agreement about whether nebulae were truly stellar, or whether their shapes had changed, then astronomers needed extrinsic judgments of the reliability of reports about their resolution and physiognomy. Astronomical competence could only be checked against the instrument and drawings of a known object, but drawings were used to check whether instruments were reliable and celestial objects had changed.[10] So preexisting lists and trained practice directed the experienced gaze to and around carefully selected objects. This is how a Victorian astronomer, surrounded by support staff and equipped with a range of instruments, worked nightly at Rosse's big reflector:

> I shall suppose that we are ready to commence a night's work. The assistants . . . are already at their posts. Up we climb to the lofty gallery, taking with us a chronometer, our observing book, various eye-pieces, and a lamp. The "working list" as it is called, contains a list of all the nebulae we want to observe. A glance at the book and the chronometer shows which of these is coming into the best position at the time. The necessary instructions are immediately given to the attendants. The observer, standing at the eye-piece, awaits the appointed moment, and the object comes before him. He carefully scrutinises it to see whether the great telescope can reveal anything which was not discovered by instruments of inferior capacity. A hasty sketch is made in order to record the distinctive features as accurately as possible. One beautiful object having been observed, the telescope is moved back to the meridian to be ready for the next vision of delight. . . . I would point out that the work of observing in the manner above described is extremely trying and fatiguing. It should be remembered, however, that the nights on which the nicer astronomical observations can be made are few and far between.[11]

More than twenty years of such "nice observations" were in fact needed to produce the single published image of the Orion nebula's structure made at Rosse's observatory. The astronomer at the reflector's eyepiece selected and sketched from a "working list" made by other observers, such as the preeminent stellar astronomer Wilhelm Struve at the Russian imperial observatory at Pulkovo, or by John Herschel. Decisive in making Herschel's status was his astronomical expedition to the Cape Colony, where from early 1834 he organized the construction of a private observatory for his father's twenty-foot reflector to survey the southern heavens (Figure 1). He'd made preliminary drawings of the Orion nebula a decade earlier, and from the Cape he saw "the necessity of executing a redelineation of it," because the nebula rose much higher above the horizon there and he found it hard to tell whether changes in its shape were due to real shifts. He later admitted that the "supposed changes" in Orion "have originated partly from the difficulty of correctly drawing, and, still more, engraving such objects, and partly from a want of sufficient care in the earlier delineators themselves

Figure 1.   John Herschel's twenty-foot telescope set up at Feldhausen
in South Africa in 1834. From Herschel, Results of Astronomical
Observations at the Cape of Good Hope (1847).

in faithfully copying that which they really did see." So nebular astronomers needed a reliable physiognomy with which to depict astronomical appearances.[12]

Herschel was among the backers of a wide range of new techniques of "faithful copying," photography among them. Nebular objects long remained beyond such automatic processing, principally because of the long exposure times required and the notorious instability of telescope clock drives. With the huge light-grasp of unwieldy reflectors, many celestial objects only remained in sight for very short times indeed.[13] In these efforts to produce images of unfamiliar and exotic objects, Herschel was in the position of other naturalists at the colonial periphery who constructed manageable versions of the worlds they encountered using imported methods of ordering to travel colonial landscapes. Racist physiognomy, conventions of imaging faces as signs of moral and natural development, came with them from Britain. Such visions sought, often ineffectively, to command landscapes and populations under the gaze of the moving naturalist's eye. In February 1834, soon after setting up the family's telescope, Herschel turned it on the moon: "as it rises it presents a round, dull, blotchy human face, with broad nose sulky mouth and standing perpendicularly has just the effect of some preternatural being—Demon—or god of some barbarous nation looking down on his African territory & sniffing with sullen pleasure the scent of some bloody rite or looking down on the whole region as a scene of carnage agreeable to his nature & will. The *European* face is quite lost, by the reversal of its position." The astronomical point was that in the southern hemisphere the moon was seen inverted. The physiognomic concern instead centered on the encounter with a recalcitrant and exotic world.[14]

The inspiration for such surveys of "demonic" phenomena was at least partly the pursuit of a Humboldtian "aesthetic empire." In 1834–36 the British colony dispatched an expedition, for which Herschel acted as scientific advisor, to survey lands and populations across its eastern border. Among this expedition's most striking products were the pictures made by its artist, Charles Davidson Bell, a frequent visitor to Herschel's observatory whose physiognomies of indigenous peoples and terrains were praised enthusiastically by the keen young astronomical assistant at the Cape Town observatory, Charles Piazzi Smyth. "Here was the great interior's physical geography . . . depicted again and again, either in brilliant colour, or chiaroscuro force of black and white, and almost perfect truth of outline, with the very atmosphere also before one to look into, as it shimmered and boiled in the vividness of solar light."[15] The imagery of imperial power and knowledge of the theretofore "unknown" were combined in new "black and white" images of the torrid southern lands and wondrous skies.

Smyth, like Herschel the son of an eminent astronomer, was something of an amateur draftsman, and worked with Herschel on geodetic and celestial surveys and on nebular sightings. He produced some fine caricatures of the astronomers at work and instructed his Cape Town friends that "the object of painting as well as of literature is to present something more perfect than that which is commonly seen, to give a local name and habitation to those abstract images of ideal & perfect beauty which though derived from nature herself are never to be seen entire in any one of her forms."[16] This was certainly true of astronomical imaging. In late 1841, before his departure from the Cape for the chief astronomer's job at Edinburgh, Smyth delivered a paper at the South African Literary and Scientific Institution discussing the technology of astronomical drawing. Subsequently published in London by the Royal Astronomical Society, the talk was prompted by Smyth's commission to give engravers adequate images of Halley's comet as seen by Herschel and his colleagues from the Cape. In South Africa, Smyth engraved copper plates he had brought from England for astronomical and cartographic publication, and apparently worked with mezzotints and aquatints. In what he conceded was a "very hasty production," Smyth summarized all the problems that faced nebular representation in the 1840s.[17] He argued that nebulae posed special representational problems, both cultural and descriptive. They could not be subject to exact measurement; there was as yet no public observatory for extra-meridional astronomy; photography could not be used on these telescopic objects; above all, there was no secure inference to be made about the evolutionary course of nebulae simply because the Herschels were able to arrange nebular species in a series of increasingly resolvable objects.[18]

So Smyth set out to rule on the aesthetics and technique of nebular pictures, because comparisons between reliably made images would be the only means of getting at changes in nebular form. His principal theme was the need for "faithful imitation, for want of which none of the so-called high finishing can ever atone, and which can only be accomplished by correctness of eye, facility of hand, and a due appreciation of the subject." These principles told against most engraving convention, such as engravers' "misplaced attempt to produce a splendid figure." Like the publicist Nichol, Smyth commended what he called "positive representations," in which lights were represented as white. The "negative method," he judged, in which stars were represented as black marks on a white ground "is extremely likely to puzzle and to create misconception." Yet William Herschel's positive images of nebulae printed in 1811, which relied on ruling crossed lines and then a combination of stippling with mezzotint (so that the artist worked from a very dark background to highlights) "gives the objects a much better definition than they really possess," while John Herschel's celebrated early aquatints of the Orion nebula, acid etchings where varnished parts would end up white, produced "the disagreeable effect of a net thrown over the whole."[19]

These were conventional concerns among colonial artists. In the same year as Smyth's lecture, the drawing master at the East India Company's military college issued an authoritative engraving manual in which he, too, discussed the artistic skill required from the mechanisms of engraving: "strokes should never be crossed too much," and the engraver must "make the strokes wider and fainter in the lights and looser and firmer in the shades."[20] Smyth stressed that astronomers must be their own artists, especially in the colonies ("the more necessary the farther he is removed from Europe"). They must attend closely to aesthetics. Mezzotint was the preferred technique, because it gave a good black ground by roughening a steel plate with a curved rocker, and thus approached such models of skill as Rembrandt, the cynosure of fine chiaroscuro. "Though the study of these masters may not bear very closely on the point, still not a stone should be left unturned when our object is, or ought to be not the mere imitation, but the rivalship of nature." Trying to make publicly accessible the means for translating "to future ages those signs and appearances in the heavens which admit of no direct application of measure or number," Smyth was driven to increased mechanization and to ever more explicit aestheticization of representation. "The laying of the ground is the most tedious and troublesome part of the process." He proposed a machine to accelerate and render reliable the preparation of the ground by ruling parallel lines on the plate: "it is needless to enlarge on the rapidity with which astronomical drawings would increase in value." As astronomers tried making reliable representations of the nebulae and send them from the remotest sites to the imperial capital, they found themselves enmeshed in the world of artists, engravers, and graphic print.[21]

THE PHYSIOGNOMIC TELESCOPE

Savants such as the Cape astronomers and naturalists returned to a turbulent Britain armed with invaluable images of the southern world, faced with the puzzles of rendering reliable public reproductions of what they'd seen. Not everyone commanded the resources of Herschel, who got ducal subsidy for the publication of his astronomical surveys. Their dilemma was linked with the division of labor inside print culture. Engravers earned most of their living by selling reproductions of paintings, their skill lying in the degree to which these counted as worthy versions of some splendid original. Herschel and Smyth, for example, were optimistic that their own new work in South Africa on calotypes and other photochemical processes might scotch many difficulties in reproduction. In Ireland, the astronomically ambitious Earl of Rosse also started experiments on daguerrotypes in the 1840s. But nebular images were peculiarly hard to manage this way, thus especially susceptible to the vagaries of engravers' interpretation.[22]

Astronomers long depended on engravers' skills to make good scales on the brass fittings of their survey instruments. But engravers' capacity accurately to render fine landscapes, natural historical specimens, or astronomical drawings relied on changing social status and attributions of skill. In summer 1848, for example, Herschel and Sabine discussed recruiting an artist to work for Rosse on his new nebular surveys in Ireland. Rosse was considering a move to London as president of the Royal Society, but told Sabine that the six-foot telescope "has been inactive whenever it was out of my power to attend to it myself and I have found great difficulty in procuring an assistant who was likely to observe effectively." Herschel held that "some previous acquaintance with astronomical observations, or at least some degree of habit of seeing stars and nebulae in other telescopes" would be desirable. Sabine ran a Humboldtian survey of Ireland in the 1840s and Rosse looked to Sabine's staff for a good draftsman. Sabine judged nebular objects, of indeterminate physiognomy and challenging detail, would require "the power of faithful representation—perhaps a sufficient instrumental practice might soon be organized." So as the British Association's secretary he consulted the gentlemen of science for recommendations among the metropolitan engravers. The Dublin geology professor John Phillips, who'd been an astonished member of a British Association visit to Rosse's observatory in 1843, suggested a young microscopist and lithographer, Samuel Leonard, "a person of small means and very moderate expectations," fit to draw and engrave reliable pictures of natural historical and nebular figures. Leonard invented an eyepiece for binocular microscopes and worked for the London physiology professor William Carpenter as part of the Association's survey of the microscopical structure of shells in 1844. Good artists were rare. In autumn 1846 Charles Darwin, wanting drawings of his *Beagle* specimens of molluscs and corallines, also asked Carpenter "whether you think your artist [Leonard] would do such things well." In comparison with the shell drawings "it is a very different style, hard and precise."[23] The shells, seen through high-powered lenses, did closely resemble the natural history of the nebulae Rosse proposed to picture through his huge telescopes. Leonard's morals also fitted the bill. "He has a good moral character and is stated to be full of zeal and desires above all things to be engaged in some work of enduring reputation." Leonard was not in the end employed by Rosse, but worked for Darwin on drawings of cirripedes. In 1848 Rosse instead hired the mathematician George Stoney from Trinity College Dublin, then his brother Bindon Stoney, "a highly educated civil engineer well accustomed to use his pencil," and later the Dubliner Samuel Hunter, to whose "estimates we may attach much importance, as [he] had the advantage of a considerable amount of training as an artist." The factors judged crucial by observatory managers well showed how matters of workers' subordination, morality, and skill counted in the organization of nebular picturing and engraving.[24]

Throughout the early Victorian period, such workers sought status as creative artists, worthy of recognition by the corporatist Royal Academy and its scientific cognates, thus able to command high incomes from dealers and customers. Good pictures were pricey. In the 1840s Leonard earned five shillings per hour for his shell lithographs; Rosse later offered to subsidize the cost to the Royal Society of seven guineas for each of the engraver John Basire's remarkable plates of nebular engravings in his most famous astronomical paper.[25] Engravers argued in the late 1830s before Parliament that "no attention or respect is paid to engraving in this country. The public consider engravers only as a set of ingenious mechanics." For one of their leaders engraving was "more a translation of a picture than a copying." Even "translation," however, could not necessarily grant the engravers the status they sought. The Royal Academy riposted bluntly that "Engraving is wholly devoid" of "those intellectual qualities of Invention and Composition, . . . its greatest praise consisting in translating with as little loss as possible the beauties of these original Arts of Design." As an 1844 guide explained to print connoisseurs, "the print is, in truth, not a work of individual art, but a manufacture." The activity of copying and the very existence of a truly original image were both puzzles for the nebular astronomers too.[26] Replication of reliable images, it seemed, hinged on the mechanization of a workforce traditionally keen on its autonomous rights to "invention and composition." Mechanization also governed the outputs and the consumers of the engravers' work. The development of the steam press in 1814 and, from 1827, of stereotyping, which allowed the mass reproduction of images and texts by casting off metal impresses of type forms, accelerated accessible graphic work and the concentration of capital in the print trade. Cheap graphics, normally based on wood-engraved blocks that were uniquely capable of being juxtaposed with type, had been the standard resource of the loose alliance of pornographers, radicals, and publicists who dominated Regency London's Grub Street world, whence flowed hosts of broadsides, caricatures, and cheap shockers.[27]

During the 1830s the condition of British graphic print was transformed. Sustained by the fixed capital vested in their new machines and increasingly self-confident ideologies of moral and intellectual reformism, journals such as the *Penny Magazine* (1833), *Punch* (1841), and the *Illustrated London News* (1842) all in different ways exploited engravings and text to create unprecedentedly large markets for their remarkable images of improving knowledge, moralized physiognomies, or spectacular natural and political events. Charles Knight's *Penny Magazine*, with an initial readership estimated at one million and patronized by the genteel Society for the Diffusion of Useful Knowledge, soon became a key medium of moral education. Knight wrote that "ready and cheap communication breaks down the obstacles of time and space."[28] Less-cheap lithographs, like those Leonard made of shells, were for the elite. Nichol disliked them: "the worst of lithography," he told Herschel in 1838, "is that the impressions are most unequal, at least of any lithography I have been able to command." Conservatives decried the new magazines' use of cheaper mass-produced engravings. The ambitious journalist William Thackeray complained of the subservience of lithography, which he lauded as manual creativity, to the "machinery" of engraving, spawned "by the aid of great original capital and spread of sale." Willy-nilly, by 1842 Thackeray himself soon joined *Punch* and the mechanical-graphical world. In 1846 the Tory laureate William Wordsworth even penned a sonnet against the *Illustrated London News*: "Now prose and verse, sunk into disrepute, / must lacquey a dumb Art that best can suit / The taste of this once-intellectual land."[29]

Wordsworth, like other critics, here assailed caricature and physiognomy in the public press. Physiognomy became part of the Victorian "period eye." Linked with the developmental hierarchies of race and class, physiognomy provided a kind of medium for images made in colonial observatories, field stations and London printshops. The mass distribution of images of the big reflecting telescopes, and the nebulae, can be compared with the equally widespread publication of *Punch* caricatures, or such noteworthy hits as William Frith's "picture of the age," *Derby Day* (1858), a vast emblem of mid-Victorian characters that challenged its viewers to *resolve* each individual into their peculiar moral or social quality. As Mary Cowling has demonstrated in her analysis of Frith's genre pieces, physiognomic codes drawn from late-eighteenth-century manuals such as Lavater and Camper allowed interpretation of character, hierarchy, and fate. Victorian novels were larded with physiognomic accounts designed to cue their readers to characters' types.[30] Images such as *Punch*'s cartoons were exactly *stereotypes*, a term that shifted in the mid-nineteenth century from its earlier sense as a means for accelerating the reproduction of print and graphics to its current sense of banalized caricature. The *Punch* journalist Albert Smith's *The Natural History of Stuck-up People* (1847) was a witty example of the genre. Newspaper articles on the "moral physiognomy" of the "wandering tribes" of London first published in 1849–50 by another *Punch* journalist, the social explorer Henry Mayhew, also used evolutionist accounts of physiognomic development. Mayhew began his survey of metropolitan vagrants with physiognomic evidence from the 1830s surveys of South Africa: "in each of the classes above mentioned, there is a greater development of the animal than of the intellectual or moral nature of man, and they are all more or less distinguished for the high cheek-bones and protruding jaws—for their lax ideas of property." Physiognomy helped place ambiguous classes in the right series of natural types.[31] As Gillian Beer has suggested, stereotypes helped Victorian audiences make sense of the evolutionarily crucial "missing link," a means through which continuities between higher and lower orders could be asserted or denied. *Vestiges of the Natural History of Creation* (1844), an anonymous work on the nebular hypothesis and evolutionism by the Edinburgh publisher Robert Chambers, was only the most notorious text that hitched physiognomy to evolutionist strategy.[32]

Moral development, physiognomy, and evolutionism were stereotypically linked, especially in the representation of Irish affairs. From the early 1840s journals such as *Punch* simianized the physiognomy of the indigenous Irish (Figure 2). At periods of catastrophic famine and widespread insurrection in Ireland, British ethnographers underwrote physiognomic judgments that placed the Celts at an inferior, degenerate level in social development. In his remarks on the relation between "Paddy and Mr Punch" in the 1840s and 1850s, Roy Foster shows how physiognomic caricature of the Irish stayed in place: "one cannot get away from the intellectual and scientific (or pseudo-scientific) context of such stereotypes."[33] Using Pieter Camper's doctrine of the role of facial angle in discriminations between animals and humans, mid-Victorian caricaturists portrayed insurrectionary Celts as apelike and prognathous. Contemporary controversies about the link between human origins and the higher primates fueled appetite for these images. As Adrian Desmond puts it, "the concept of the facial angle was easily moulded into ideological shape."[34] From the 1850s, when concern with nebular resolution was also intense, public shows of apes, especially gorillas, and the publishing offensives of modish evolutionism, helped the Irish connection work. In 1861, *Punch* published a cartoon of "Mr G O'Rilla" and his extinction. The same month, its rival magazine *Fun* juxtaposed an image of an apelike Irish MP with a drawing of the notorious evolutionist Thomas Henry Huxley walking arm in arm with a gorilla. A more ferocious account followed a year later, when *Punch* carried a story on "The Missing Link" desciribing the habits of the "Irish Yahoo," "a creature manifestly between the Gorilla and the Negro."[35] The simian link was also apparent to Charles Kingsley,

Figure 2. "*The Fenian Guy Fawkes,*" *John Tenniel's cartoon of a simianized Irishman,* Punch, *December 28, 1867.*

expert on physiognomy's use in science and fiction. Kingsley prepared for his inauguration as Cambridge history professor by visiting a private astronomical observatory at Markree in Ireland, then reported as "the most richly furnished private observatory known." After hunting stars and salmon, the English physiognomist turned his gaze on the Irish peasantry: "I am haunted by the human chimpanzees I saw along that hundred miles of horrible country. . . . To see white chimpanzees is dreadful. . . . It is a land of ruins and the dead."[36] Such ways of seeing Ireland were common, and helped map the status of its astronomy too. In 1846–47, at the height of the Great Famine in Ireland, Thackeray contributed to *Punch* a long series of satirical physiognomies decorated with his own drawings. A few weeks after telling his London readers that "the shams of Ireland are more outrageous than those of any country," a land "where no-one believes anybody," he explicitly envisaged a physiognomic rival of "Lord Rosse's telescope," which "enables you to see a few houndred [sic] thousand of miles farther." What was truly required (from the British Association, he guessed) was "some telescopic philosopher to find the laws of the great science" of physiognomic caricature.[37]

Astronomical images were certainly current in the "dumb" visual culture Wordsworth bemoaned and Thackeray satirized. They had long been the stock of public lectures. Smog-bound Londoners could glimpse the Herschels' discoveries at London opera houses. In Easter 1839 one lecturer there showed "the Moon as viewed through Herschell's telescope" and huge machines of heavenly motions. In 1846 Humboldt himself proposed erecting "large panoramic buildings . . . thrown freely open to the people" to show "the physiognomy of nature."[38] Graphic print exploited these physiognomies. While Herschel was in South Africa, newspapers skittishly reported that through his reflector he had seen, amid suitably Humboldtian moonscapes, the physiognomies of lunar inhabitants. Lithographs showed the lunarians' "face . . . was a slight improvement upon that of the large orang-outang, being more open and intelligent in its expression and having a much greater expansion of forehead."[39] Like the Royal Academicians on whom they modeled their own organizations, the astronomers tried to secure public status by a judicious balance of distinction and management in the world of print and showmanship. The issue was by no means one of making a simple contrast between coolly secluded astronomical vision and the vulgar physiognomies fit for public consumption. It was rather regrettable, according to Herschel's letters from South Africa, that the widespread stories of his sights of "landscapes of every colouring, extraordinary scenes of lunar vegetation and groups of the reasonable inhabitants of the Moon" were "not true." Notes in Herschel's *private* observing notebook described the moon as an "African savage." In *public,* in his very successful astronomy textbook, Herschel produced a stunning physiognomy of the Orion nebula (Figure 3) designed to capture public interest: "in form, the brightest portion offers a resemblance to the head and yawning jaws of some monstrous animal, with a sort of proboscis running out from the snout."[40]

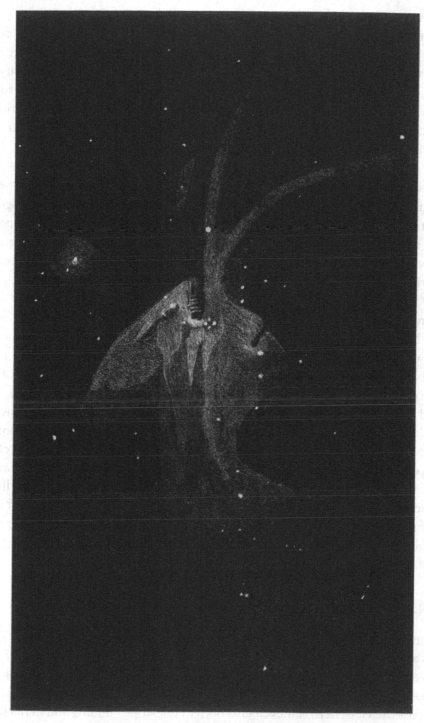

*Figure 3.   John Herschel's picture of the Orion nebula as reproduced
in John Pringle Nichol's* System of the World *(1846).*

The monstrous nebula, like other bizarre natural specimens, became public currency. In the 1840s Nichol hosted the scarred veteran of metropolitan print culture, Thomas de Quincey, at his new Glasgow observatory. One result of de Quincey's stay was a remarkable contribution to the journal for which Nichol acted as scientific consultant, *Tait's Edinburgh Magazine*. De Quincey told the magazine's readers to open Nichol's book, his *System of the World* (1846), where Herschel's picture of Orion was reproduced, then to turn the book upside down, as though reading the nebula were a Lavaterian exercise in resolving the morality of a hideous face.

> The following is the dreadful creature that will then reveal itself. . . . You see a head thrown back, and raising its face (or eyes if eyes it had) in the very anguish of hatred, to some unknown heavens. . . . The mouth, in that stage of the apocalypse which Sir John Herschel was able to arrest in his eighteen-inch mirror, is amply developed. Brutalities unspeakable sit upon the upper lip, which is confluent with a snout; for separate nostrils there are none.[41]

De Quincey linked the monstrous physiognomy of the nebula to that of the sublime head of Memnon brought from Egypt to the British Museum in 1818, and thence to reflections on the primal scene, on the Miltonic Fall, and on the politics of orientalism.[42] "Now when further examinations by Sir John Herschel at the Cape of Good Hope have filled up the scattered outline with a rich umbraceous growth, one is inclined to regard them as the plumes of a sultan." Nichol, stout materialist, was rather worried by de Quincey's "resolution of the Nebula into something different from Matter . . . notwithstanding its singular and undeniable power." De Quincey claimed that "as one belonging to the laity, and not to the *clerus* in the science of astronomy, I could scarcely have presumed to report minutely, or to sit in the character of dissector upon the separate details of Dr Nichol's works . . . had there not been room left disposable for such a task."[43] This question of *room* for diagnostic dissection and physiognomic play was crucial. The spaces of early Victorian print culture dominated the ways of nebular vision.

HIGH-BROW ASTRONOMY

Like so many of Victorian Britain's predicaments, the problem of the nebulae's meaning was at least temporarily focused on Ireland. There from the 1830s, on his Parsonstown estate, the wealthy Protestant nobleman William Parsons, Earl of Rosse, commissioned two unprecedentedly vast reflecting telescopes. His aim was to match and surpass the Herschels' achievements both in telescope building and in nebular resolution. Work performed at his observatory, especially attempts to resolve the Orion nebula into stars, explain away its shape changes, and thus undermine the nebular hypothesis, prompted the effusions of Nichol, Thackeray, and de Quincey, attracting audiences throughout the United Kingdom and beyond. Rosse got the same treatment as Herschel. In 1844 a London paper juxtaposed cheap woodcuts of ghost stories and engineering schemes with gossip that the earl's great reflector revealed "the skeleton of a gigantic animal" lying in a lunar crater. The lifelessness of the moon, the magazine ironized, was due to the fact that "the Monster, after ravaging all around, had at last perished by famine," an apt theme in 1840s Ireland, home of famine and Rosse's monster telescopes. "There are mysterious whisperings of some supposed ethereal or angelic beings discovered," according to the enthusiastic reporter, "nothing but the dread of attracting a concourse of peasantry prevented our joining in a general shout."[44]

The larger six-foot mirror installed by Rosse in 1842 in what was soon called the Leviathan of Parsonstown became a topic of journalism because of its technological virtuosity and its political significance (see Figure 4). Built at the astronomical cost of £12,000, immured between massive gothic walls that constricted the tube securely in the meridian and made it look like a medieval castle, the Leviathan and its owner were made into an emblem of benevolent rule in famine-struck Ireland. Parsonstown

Figure 4.   The "Leviathan of Parsonstown" in 1844 (the six-foot mirror installed
by William Parsons, Earl of Rosse, at his estate in Ireland in 1842).
From Illustrated London News 6 (1845): 253.

itself suffered at least 30 percent mortality rates in its fever sheds in the wake of the potato famine of the mid-1840s. The Countess of Rosse, a wealthy Yorkshire heiress, paid both for the telescope project and for tenants to build forts and earthworks on the estate as a job-creation scheme during the Great Hunger. The Edinburgh natural philosopher David Brewster, an active Protestant journalist in spreading the news of Rosse's work, wrote in 1844 that "it is a matter of no ordinary satisfaction that the intellectual energy of Ireland is concentrated in men of like faith with ourselves."[45] The new Earl of Rosse—he inherited the title in 1841 from his father, a distinguished critic of the Union—and his aide, the aggressive Ulster episcopalian Thomas Romney Robinson, son of a distinguished portrait painter and now director of the Armagh Observatory, were pillars of the Church of Ireland and the Ascendancy. Robinson's campaigns involved making the Irish telescopes better than their predecessors—he even suggested sending "a Reflector on Irish principles" to the Cape to "put an extinguisher on Herschel's labors." He also preached against the threats of evolution and materialism. Nichol's version of the nebular hypothesis was but one example of a general tide of unbelief, notoriously embodied in *Vestiges* and endorsed by the philosophic radical John Stuart Mill in his *System of Logic* (1843). Nichol and Mill also penned many journal articles on the evils of the Irish church and land establishment. The radical materialist tide was to be countered with the divine lessons of the Leviathan.[46]

The Leviathan's master made no secret of his views on the solution to the immediate Irish emergency of famine and potential revolt. In 1847 Rosse sent a series of letters to the *Times* "as an Irish landed proprietor" recommending "emigration, and on a great scale, and there must be very stringent regulations to prevent the subdivision of land." A committed Whig and Malthusian, Rosse held that the irrational Irish habit of parceling tenancies into ever smaller holdings simply encouraged excessive population growth. Rosse's family friend, the political economist and government advisor Nassau Senior, was reported as fearing the Irish Famine of 1848 "would not kill more than a million people, and that would be scarcely enough to do much good." Political economic principles, Rosse insisted, were as sure as those of physics. "Nothing is more difficult than to change the habits of a people—the only chance of effecting a change quickly would be by example," so plantation by "a few English or Scotch farmers" would help an otherwise essentially disastrous predicament. His collaborator Robinson was no less fierce in defence of the Protestant establishment, no less sceptical of English incomprehension of the Irish landlords' troubles and equally sure of their racial origin. Robinson shared the view that the original population of the island had degenerated from a high southern European stock into their current peasant occupants. Only Protestant Ulstermen, he reckoned, escaped "the mean vices which lower a large proportion of the people of this and the sister island."[47]

These views at Parsonstown during the tense 1840s help illuminate the reports of what they saw there, especially their ability to see in the nebulae a telling lesson *against* the law of natural and social progress. One foreign journalist judged it "remarkable that it is an Irish earl, placed at the centre of an island where, unfortunately, still reign so much poverty and ignorance, who has made so much progress in this difficult and important part of optics." The Tory Charles Piazzi Smyth, who in 1852 had visited Parsonstown with enthusiasm, wrote from Edinburgh to express his view that Rosse's achievements should become "a species of textbook."[48] Rosse's virtues, and the means he used to recruit local peasants to his workshops, underscored the political message of the Parsonstown mission and its telescope, immured as it was in a suitably Norman keep. "All these gigantic constructions . . . have been executed in Lord Rosse's workshops," Robinson commented, "by persons taken from the surrounding peasantry, who, under his teaching and training, have become accomplished workmen, combining with high skill and intelligence the yet more important requisites of steady habits and good conduct." As Rosse himself explained from the president's chair to the British Association in 1843, "the children of the fields" always displayed "a more determined tendency to religion and piety than amongst the dwellers in towns and cities." His estate was a haven of tranquil labor and divine science in a sea of economic, political, and racial catastrophe.[49]

This image was affirmed in one of the most striking travel reports about Parsonstown, produced in *Chambers' Edinburgh Journal* in late 1846. As Jim Secord has pointed out, the *Journal* was "the most public forum in early Victorian Britain." It was broadly secular, reformist, and meliorist in tone, and attracted fierce clerical hostility and envious glances from its competitors. Robert Chambers had been converted to nebular astronomy by his close ally Nichol and had visited Ireland with him in 1837. In some versions of his *Vestiges* and its sequel *Explanations*, especially in the first half of 1846, Chambers did his best to accommodate the attack on true nebulosity propagated from Parsonstown.[50] This gave his *Journal*'s report on Parsonstown in late 1846 considerable point. The article's look conveyed the moral of a contrast between stereotypical Irish peasant idiocy and British noble success. The first half of the story summarized a hackneyed series of "Irish" jokes: the craziness of coach timetables, the poverty of the houses and the towns, the absurd pretension of Catholic peasants, the ridiculous execution of justice: "the whole affair more resembled a scene in Tom and Jerry [a popular English comic strip of the period] than the proceedings of a well-appointed tribunal." Above all, the *Journal* made the Irish the exemplary time-wasters. Punctuality, it seemed, was the prerogative of sound British science. Parsonstown was "almost as neat and brisk a town as could be seen in England," entirely because of Rosse's "liberality." There was an important connection between the Irish "aptitude for instruction" and Rosse's "accomplishments in practical science." The power of the telescopes as representatives lay in their cultural setting, in the contrast between this order and success and the disasters, however humorous, of the Irish.[51]

In his final commentaries on Irish politics and economics Rosse reemphasised both the idyll of the well-managed estate and the threat posed by Irish peasant mores and English incomprehension. "The landlord," he explained, "is the centre of a little community who have all that is necessary for their decent maintenance and . . . there are self-righting principles which prevent things from going very wrong, which restrain the *perturbations* within certain fixed limits, and restore things again to a normal state." This was both a polemic against indefinite evolutionary change and overwrought plebeian fertility. The "tyranny of unbridled democracy," he argued, was the principal perturbation to class harmony. "Forced emigration" and land consolidation remained the cure for Ireland's ills. As one of his fiercest critics, the Home Ruler Isaac Butt, put it, Rosse betrayed his class' "fear of the presence of human beings," and was "rapidly completing the extermination of the Old Celtic Race." Rosse singled out another opponent, John Stuart Mill, not only because of his economic assaults on the landlords but because he had released a "people's edition," mass-marketing his radical subversive principles and deriving the law of social progress from the Nebular Hypothesis. False populism, the highbrow astronomer Lord Rosse held, was as bad in economics as in physics: "It is very much as if a treatise on applied mechanics were to introduce . . . an essay on the overthrow of the Newtonian philosophy." The public seemed unaccountably intrigued by the pseudoscientific images of social progress and reform. The Nebular Hypothesis of Mill, Nichol, and the *Vestiges* was a step to destruction. So too was their model of land, property, and the social order. It was important to produce a different picture of the face of the heavens and of humanity.[52]

This picture was profoundly divine. A note of 1852 in the Parsonstown astronomical diary recorded that "no-one who has had the privilege of viewing the . . . nebulous systems under these mighty aids to vision can well fail in realizing under deeply impressive convictions the force of the Sacred Aphorism, 'The Heavens declare the glory of God!'" Robinson was also much concerned with the industrial sublime, writing in the 1840s of his admiration for "Milton's splendid description of the infernal palace" as an image of contemporary metallurgical foundries. He was familiar with the stunning apocalyptic of the biblical illustrator John Martin and his fellow artists of the divine industrial vision. Several journalists compared the telescope with "that artillery described by Milton as pointed by the rebellious angels against the host of Heaven" and praised its "quiet victory over space." Between 1840 and the spring of 1842 Rosse commissioned a huge foundry with three furnaces and, after five separate trials, cast a perfect six-foot mirror for the new reflector. The drama of this moment was an occasion for Robinson's oratory.[53] Speaking at the Royal Irish Academy two weeks after the casting of the Leviathan's mirror, Robinson emphasized that in this Irish Pandemonium, noble mastery was confidently visible:

On this occasion, besides the engrossing importance of the operation, its singular and sublime beauty can never be forgotten by those who were so fortunate as to be present. Above, the sky, crowded with stars and illuminated by a most brilliant moon, seemed to look down auspiciously on their work. Below, the furnaces poured out huge columns of nearly monochromatic yellow flame, and the ignited crucibles, during their passage through the air were fountains of red light, producing on the towers of the castle and the foliage of the trees, such accidents of colour and shade as might almost transport fancy to the planets of a contrasted double star. Nor was the perfect order and arrangement of every thing less striking: each possible contingency had been foreseen, each detail carefully rehearsed; and the workmen executed their orders with a silent and unerring obedience worthy of the calm and provident self-possession in which they were given.[54]

Charles Weld, the Royal Society's secretary, plagiarized just this passage in his travelogue *Vacations in Ireland*, a work dedicated to Rosse with an engraving of the Leviathan in pride of place. The Leviathan attracted pilgrims. The local physician, Thomas Woods, published a guidebook, *The Monster Telescopes erected by the Earl of Rosse* (1844). The *Illustrated London News* dutifully carried a full-page spread of views of Parsonstown and its workshops: "a visit to the noble lord's demesne will amply repay any trouble attendant on it." John Timbs, subeditor at the *News*, also included the Leviathan as the very frontispiece of his best-selling *Curiosities of Science*. The preeminent publicist of Christian astronomy, Thomas Dick, used these reports for a special appendix on the Leviathan in his 1845 guide for amateur astronomers. Mary Somerville, eminent science writer, told the earl that "my only knowledge of this extraordinary instrument is from public report which so much exceeds all that any one has dared to hope for."[55]

The monster telescope and the vast nebulae belong to the history of the sublime and the gigantic in Victorian science, arts, and society. John Herschel was well aware that the astronomers' claims to see stars and nebulae "may be thought to savour of the gigantesque." At Parsonstown in 1843, according to the president of the British Association, "whatever met the eye was on a gigantic scale . . . structures of solid masonry . . . more lofty and massive than those of a Norman keep." During the nineteenth century the gigantesque shifted its application from the vast formations of nature to include the display of material production within the society of the spectacle. The work performed to resolve distant nebulae with great reflectors, then show these portraits to the public, followed such a path between natural history, machinery, and spectacular display.[56] The sublimity of the Leviathan of Parsonstown, and the power of its observations, can thus be situated within the public sphere of Victorian print and picturing. The physiognomy of the nebulae slowly produced from the Irish telescope was supposed to falsify a progressive science of change and evolution.

"THE EYE MAY IN SOME DEGREE
BE INFLUENCED BY THE MIND"

Initially Rosse's team sought to win over their public with little more than a quick glance at their stories of resolution. In 1845, when the Leviathan was ready, Robinson almost at once recorded in the observing book, and told the public, that "no REAL nebula seemed to exist among so many of these objects chosen without bias: all appeared to be clusters of stars."[57] Herschel was unprepared to make the inference that this destroyed the nebular hypothesis. In some cases he could not make sense of the drawings Rosse's team produced: "is this really the appearance in the telescope, or has the artist intended to express his conception of its solid form by this shading?" Herschel sent Rosse his own drawings of Orion, pointed out how dreadful was the Irish sky, and told the British Association that it was "a general law" that resolvability was limited to *spherical* nebulae. Rosse turned up at the Association's meeting in Cambridge in 1845, where Herschel verbally crucified John Stuart Mill's speculative versions of the Nebular Hypothesis but maintained that a nebular origin for stars remained plausible. The Irish earl handed round his startling drawing of a spiral nebula (M51), and his team's amazing pictures of such spirals became a prized result of his great telescopes: "he did not think the drawing would be found to need much future correction." Later, Rosse sent a "notebook of drawings of nebulae" to the London elite, to Michael Faraday, Sabine, and to Herschel himself.[58] A climax of the early campaign against true nebulosity was reached after Christmas 1845, when Orion itself was at last clearly observable from Parsonstown. Some foreign science journals prematurely reported that Rosse's team had successfully resolved the key object.[59] On March 19, 1846, Rosse told Nichol and others that "there can be little, if any, doubt as to the resolvability of the nebula."[60]

However, in 1846 there was simply no publicly available picture of the Orion nebula in its resolved state. It took *two decades* to make one for public release. To remove doubt of resolution it was important painstakingly to make a physiognomy of the Orion nebula. First, to place well-positioned stars on their maps, the Parsonstown team needed a reliable working list. They used the authoritative catalogues made by Wilhelm Struve and his son Otto at Pulkovo, the world center of stellar astronomy. Otto Struve came to Parsonstown in 1850, then proposed coordinating star surveys with the Irish observers. Like the contemporary campaigns to measure the personal equations of separate astronomical observers, this would need "direct experiments" to make sure Russians and Irish observers were comparable: "the relative sharpness of the eyes of both observers must be taken into account and this can only be deduced from direct comparisons made on the heavens by the same instrument." So, as Otto Struve explained, "a personal interview with your Lordship and your assistant and experiments made directly on the heavens in your company and with your instrument appears to be the only way to secure a perfect success of our combined labors."[61] This scheme for combined star gauges at Pulkovo and Parsonstown never happened. Instead, Rosse's group used Russian charts of Orion, but some seemed to show that "visible changes in the nebular parts have very probably happened." Such changes would imply that there was indeed true nebulosity in Orion, a conclusion of which the Irish group remained rather sceptical: "these are probably to be attributed in a great measure to the difference of power in the instruments used and the amount of labor expended on the drawings."[62]

Much was therefore made of the sheer amount of labor involved in producing the physiognomy of Orion. The Night Book kept by George Stoney in 1848–49 recorded sights of Orion "far beyond my greatest expectation." Such enthusiastic glances were not enough for the time-consuming work of resolution. On February 17, 1849, Stoney saw a "multitude of stars . . . but when they came to be drawn only got in 9 certain and 5 uncertain, the state of the air having become worse." Such comments were carefully selected for eventual publication. According to the published version, Bindon Stoney's drawing in 1851–52 of the Huygenian region of the nebula, the zone the Parsonstown team judged most likely to be resolved, "was made with great care, and he was engaged upon it the whole season." Several other observers were called in to check what Bindon Stoney had done. His draft showed "strong indications of change" when compared with the later drawing made by Samuel Hunter between February 1860 and 1864.[63] (See Figure 5.)

Changes in stellar positions and traces of milkiness would persistently affect observers' judgment that they were seeing stars instead of nebulous fluid. Rosse told Herschel in early 1849 that "in Orion we have nowhere seen a resolution into stars without intervening nebulosity, and in many parts the indications of resolvability are almost wholly wanting, still I think upon the evidence we are fully justified in concluding that it is a resolvable nebula." Bindon Stoney reminded his former boss why observers might wrongly fail to record successful resolutions. "Faint nebulosity might sometimes be erroneously inferred to exist" either "in the close neighbourhood of bright stars associated with nebulous matter" or "in an interval surrounded by bright nebulosity." So then "an observer might suppose milkiness to exist on unfavourable nights for definition or with an imperfect speculum, where in fact no nebulosity would appear with a freshly polished speculum free from all tarnish and on a first class night." Hunter recorded the hard work involved in making any judgment of "resolution." Thus on February 22, 1862, using the less-powerful three-foot reflector ("faint details cannot be made out with it"), Hunter recorded that the Huygenian region "looks just like fine flour scattered over a grey surface so that I have no hesitation in saying it is composed of stars, many small ones seen in it." Hunter also tellingly noted the rarity of "fine nights." "It may seem strange that it required so long a time" to make such pictures of the Orion nebula. But because of the high walls holding the Leviathan steady and limiting its field, observation could only last fifty minutes per night. In four years there were only five "really good" nights and twelve "fair nights" to gaze at Orion.

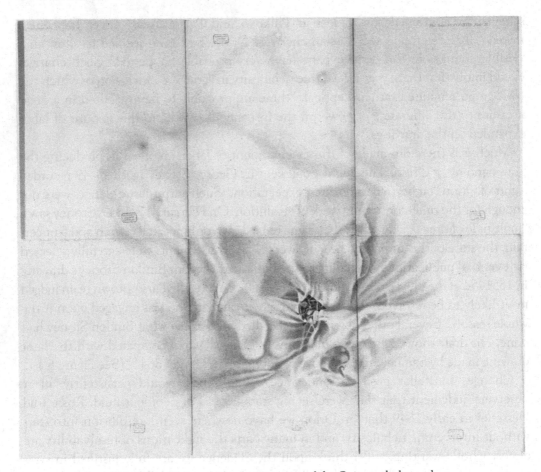

Figure 5.   Samuel Hunter's picture of the Orion nebula made
between 1860 and 1864 at Parsonstown. From Lord Oxmantown,
"Account of the Observations on the Great Nebula in Orion,"
Philosophical Transactions of the Royal Society (1868).

"Any details of which we were not confident were returned to night after night, until satisfied we had got the true form."[64]

The completed steel engravings, published by the Royal Society in 1868, which purported to show the "true form" of many resolved areas of the Orion nebula, were very widely distributed and debated. Some astronomers complained that Rosse's catalogues of his nebular surveys showed most of the images as negatives, with stars as black on white. The earl's son, Lord Oxmantown, now in charge of the Leviathan, reported that "the engraving is on the whole very accurate; a little more softening off in the faint outlying parts would have been desirable, but Mr. Basire [the engraver] did not think that it would be practicable consistent [sic] with the reasonable durability of the plate." It proved hard to keep all the draftsmen in line. When Oxmantown noted that he was "unable to find" some stars in Hunter's drawings of Orion, Hunter himself wrote from Dublin to complain that they would indeed be found "either in the Night Books or in the original drawing. I think they are in both of these. . . . Your lordship doubtless is aware that I frequently examined the object in its various positions in order to familiarize myself with its details and note its general character." The astronomer Robert Ball, who worked at Parsonstown in 1866–67, right at the end of the Orion campaign, confirmed that the published image of the nebula was "an exquisite piece of work . . . corrected or altered until accuracy was attained. Never before was so much pains bestowed on the drawing of a celestial object, and never again," in the new era of astrophotography, "will equal pains be devoted to the same purpose."[65]

The authority of the new picture of Orion relied on the Parsonstown astronomers' pains. Yet these labors could also call such images into doubt. Every time the Leviathan's mirror was repolished, its form changed and then tarnished in bad weather. "It would have been a hopeless task," the earl reported to the Royal Society, "to attempt to keep it in a state fit for the resolution of nebulae and the attempt was not made." His son eventually acknowledged that "the reflector of this year may be as to defining power practically a totally different instrument from what it may be in the next." Stoney privately agreed that even when "the speculum was quite fresh from the polisher" the "effect was lost in a very short time."[66] And then the sketching took place under "very feeble lamp-light." Hunter privately noted that after laying down Struve's grid of stars on his charts, Orion's nebulosity was "inserted with reference to the stars at the telescope by the light of a lamp, so held that the *direct* rays from it should not enter the eye, but even with this precaution the sensitiveness of the eye was impaired, so that for a minute or so faint details could not be seen." So, according to Rosse, "to see the sketch as we proceed, it is often necessary to mark it too strongly." Contrasts betwen faint and bright portions in the steel engraving could not be trusted, and "the well-marked confines of the nebula on paper" did not "really represent the boundaries of the object in space in all cases."[67] Variations in ink quality, the difficulty of introducing micrometric measures of star position, and the time needed for the eye to recover all vitiated the long-drawn-out task. Lord Oxmantown conceded in his analysis of Orion in 1867 that in the key Huygenian zone of the nebula, where resolution was most apparent, "it was found almost impossible to reproduce this difference of appearance in the engraving, since the whole of the surface consists of minute black dots." All the problems common to the engravers' shops in London and discussed at the start of the 1840s by Smyth and his colleagues were here canvassed in the learned journals of the astronomical elite.[68]

The collective labors of the Parsonstown astronomers looked rather fragile. To defend the Irish pictures against their critics, Robinson characteristically turned the argument about the artifactual quality of nebular drawings against rival telescopes. When he learned that the American astronomer William Bond had apparently already resolved the Orion nebula with his new (and expensive) German refractor at Harvard, a triumph scarcely achieved at Parsonstown, Robinson claimed "that this success must be in great measure due to that precise knowledge of the phenomenon and of the points where it might be looked for, which is afforded by Dr Nichol's work," thus damning his American colleague with the sins of Robinson's personal bête-noir.[69] But this appeal to the role that incautious expectation might play in claims to resolution was a double-edged weapon. The earl himself was quite frank: "the eye may in some degree be influenced by the mind," he conceded in 1850. On several occasions it was acknowledged that "sketches originally made in the gallery of the telescope" would "represent the objects placed as they appeared, not as they actually exist in space" and, as Rosse put it in 1861, "these descriptions, however accurately conveying the impressions made upon the eye at the time, cannot be taken as in all cases representing real facts."[70]

The "real facts" produced at Parsonstown about the Orion nebula were scarcely ever stable within the astronomical community. It is not now believed that this nebula is stellar. Its resolution by Rosse's team may have been due to the interposition of many small telescopic stars.[71] Where it served local interests, drawings of Orion as stellar were acceptable as a means of calibrating other instruments. In the 1840s, Bond used the capacity of his expensive new German refractor to resolve Orion as a means of telling Bostonians that the telescope was worth its high price. Inside his Harvard observatory, however, the "winning" of the nebula was never very clear. After fifteen years a "steel engraving of the Nebula of Orion" made at Harvard was eventually delivered to Parsonstown to help Rosse's team "should it not come to hand too late."[72] The grandeur of the Leviathan of Parsonstown was a negotiable asset. According to Humboldt in 1849, "even stronger telescopes, after having resolved what remains to us at present of nebulosity, will create nebulosity anew because they will penetrate further stellar layers which hitherto have escaped the observer."[73] Government astronomers such as the Astronomer Royal George Airy simply denied the inference that all nebulae could be resolved into stars, while at Pulkovo Otto Struve publicly stated that "the alleged miracles of resolution are nothing but illusions."[74] In the 1850s, evolutionist journalists on the *Westminster Review* challenged Rosse's results, and in 1864–65 the London amateur William Huggins, apparently moved by their stories, announced that astrospectroscopy showed that objects like Orion were really gaseous, with bright line emission spectra, and not stellar at all. Huggins reported that thanks to his new spectroscopes "the detection in a nebula of minute closely associated points of light which has hitherto been considered as a certain indication of a stellar constitution can no longer be accepted as a trustworthy proof that the object consists of true stars." Sabine, then the Royal Society's president, and others in the London elite, welcomed this new way of seeing true nebulosity.[75]

Rosse and his new assistant Robert Ball went to visit Huggins in London and see his remarkable spectroscopes. In early 1865 Humphry Lloyd, expert optical theorist at Trinity College Dublin, counseled the Parsonstown group about the difference between nebular emission lines and stellar absorption lines, and that they would have to change their apparatus with accurate cross-wires "to make it available for *rapid estimations*," a new style of seeing, reliant on accurate if speedy glances rather than a lengthy though careful gaze. Ball and Oxmantown privately noted during 1866 the troubles they had with this novel pattern of work. They "could not quite manage to illuminate the field to see the wires without rendering the spectral lines too faint." The Leviathan lacked a clock drive to make stable spectral measures. The battery used to generate comparison spectra kept freezing and was stored in a tin of hot water. But in public they were almost certain that there was in addition to three bright lines (the telltale mark of a truly gaseous object) "a faint, continuous spectrum" (so the object might really be stellar). They reckoned that stellar, continuous spectra with dark absorption lines were necessarily fainter, thus harder to see. On January 30, 1866 they privately "suspected . . . a much fainter bright band gradually fading away . . . perhaps this was a continuous spectrum from the nebula or from the diffused light of stars in neighbourhood—probably the former." In any case, Oxmantown announced in the *Philosophical Transactions* in 1868 that Hunter had, after all, really seen the main region of the nebula "clearly resolved." With the right equipment, personnel, and technique, the laborious "facts" of resolution might be compelling. Yet these very resources were rather too obviously local to command wide assent.[76]

It was becoming ever harder to keep the Leviathan's status secure. In the summer of 1869, Struve wrote privately to Oxmantown, now the new Earl of Rosse, admitting "how much more bright all features must appear in your instrument" compared with his own refractor, but then defining carefully what "resolution" should truly mean: "if a nebula is resolvable it will offer the same appearance on any occasion when the images are sufficiently favourable." So because of the many changes in their accounts of the nebula, the Parsonstown team had no right to claim that its pictures showed Orion resolved. Later that year the astronomical journalist Richard Proctor publicly alleged that the Leviathan was incapable of distinct imaging. Robinson promptly ransacked his old observing notes for records of its successes. Back in February 1845 he had recorded the "fact" that "the resolution of the flocky part of Orion's nebula cannot escape any eye." But Johnstone Stoney agreed that only about one-third of all mirrors had worked well and that it "is to be regretted that we seldom looked at the class of objects that are known as test objects," Struve's star lists used to calibrate telescope performance. A few years later it was even reported in the *Times* that Struve had fiercely denied the Leviathan's superiority. The Russian was forced to send Rosse a terse denial ("I am sorry my name is abused in such a manner by people who probably have a design of their own in depreciating the performance of the instrument.")[77] The period when the Orion nebula was judged to have been resolved was over. Its images gained what authority they possessed from the public culture in which they were distributed and used. De Quincey had already summed up the point. When a hostile critic in the *Westminster Review* told him that Parsonstown pictures of the Orion nebula were utterly different from the monstrous physiognomy he had seen in Herschel's drawing, de Quincey answered that "the reviewer says that this appearance had been dispersed by Lord Rosse's telescope. True, or at least so I hear. But for all this, it was originally created by that telescope." Though it was in fact John Herschel, not Lord Rosse, who was responsible for the fearsome physiognomy de Quincey had first interpreted, this was nevertheless a palpable hit. The power of the Leviathan to produce images could not guarantee its power to reproduce those pictures' interpretations.[78]

"Who is there who has not heard of Lord Rosse's telescope?" asked the Dublin natural philosophy professor, John Jellett, eulogizing Rosse at his funeral in 1867. This was the right question. Reputation mattered in nebular astronomy. The astronomers might be hostile to "people's editions" and graphic caricature. As de Quincey exclaimed, "how serene, how quiet, how lifted up above the confusion, and the roar and the strifes of earth, is the solemn observatory." Smyth's call for astronomical autonomy in making images was, in some sense, realized by the century's end. Draftsmen and engravers recruited from the public press had become rather dispensable. To "rival nature," in Smyth's evocative phrase, astronomers replaced artisans with photomechanical apparatuses and trained assistants who could presumably do better than "mere imitation" in image making. But in another sense the porosity of political, artisanal, and iconographic boundaries around celestial picturing never quite allowed astronomers' secure withdrawal. Vivid nebular pictures were made in Britain, South Africa, and Ireland with an unwieldy combination of widely distributed public resources and ways of seeing. Jellett argued that "none dare mock at Astronomy. And then, by an easy transition, we come to attribute to the workman something of the grandeur of the sphere in which he toils." Attribution of grandeur helps explain the authoritative, if temporary, credibility of astronomical images at which many publics glanced. Yet seeing all too clearly how grandeur was achieved could sometimes make the toil of picturing look like crafty ingenuity. Then gigantic visions could dissolve into showy caricatures.[79]

Notes

1. I am grateful for permission to quote from the Rosse papers (Birr), Herschel papers (Royal Society, London), and Whewell papers (Trinity College, Cambridge). Thanks for their generous help to Will

Ashworth, Jim Bennett, Mike Dettelbach, Adrian Lane, Elizabeth Green Musselman, Ann Secord, and Jim Secord. Richard Gregory, *The Intelligent Eye* (Weidenfeld and Nicolson: London, 1970), pp. 119–23. For picturing judgments in astronomy see Michael Lynch and Samuel Edgerton, "Aesthetics and Digital Image Processing," in *Picturing Power*, ed. Gordon Fyfe and John Law (Routledge: London, 1988), pp. 184–220; Alex Soojung-Kim Pang, "Victorian Observing Practices, Printing Technology and Representations of the Solar Corona," *Journal for the History of Astronomy* 25 (1994): 249–74 and 26 (1995): 63–75.

2. J. L. E. Dreyer, ed., *Collected Scientific Papers of William Herschel*, 2 vols. (Royal Society: London, 1912), vol. 2, pp. 654–56; M. A. Hoskin, *William Herschel and the Construction of the Heavens* (Oldbourne: London, 1963), p. 115. See J. H. Brooke, "Natural Theology and the Plurality of Worlds," *Annals of Science* 34 (1977): 221–86, 268–73; Stephen G. Brush, *Nebulous Earth* (Cambridge: Cambridge University Press, 1996), pp. 29–42, 67–72.

3. Herschel to Watson, 1782, in C. Lubbock, *The Herschel Chronicle* (Cambridge: Cambridge University Press, 1933), pp. 99–101; Bernard Smith, *European Vision and the South Pacific*, 2nd ed. (New Haven: Yale, 1985), pp. 203–12; Simon Schaffer, afterword in *Visions of Empire*, ed. David Philip Miller and Peter Hanns Reill (Cambridge: Cambridge University Press, 1996), pp. 335–52, 340–42.

4. Alexander von Humboldt, "Ideas for a Physiognomy of Plants," in *Views of Nature* (1808) (London: Bohn, 1850), pp. 210–352, 220–21 and *Essai sur la Géographie des Plantes* (1807) (Paris: Levrault, 1805), pp. 31–33.

5. Michael Dettelbach, "Global Physics and Aesthetic Empire: Humboldt's Physical Portrait of the Tropics," in *Visions of Empire: Voyages, Botany and Representations of Nature*, ed. David Philip Miller and Peter Hans Reill (Cambridge: Cambridge University Press, 1996), pp. 258–92, p. 271; Malcom Nicolson, "Alexander von Humboldt, Humboldtian Science and the Origins of the Study of Vegetation," *History of Science* 25 (1987): 167–94, pp. 181–82. Humboldt praises Herschel to Arago, February 19, 1840, in *Correspondance d'Alexandre de Humboldt avec François Arago*, ed. E. T. Hamy (Paris: Guilmoto, 1909) and is criticized for his commitment to the Nebular Hypothesis in [J. Forbes], "Humboldt's Cosmos," *Quarterly Review* 77 (1845): 154–91, p. 166. For Humboldt's cosmology see Jacques Merleau-Ponty, *La Science de l'Univers à l'Age du Positivisme* (Paris: Vrin, 1983), pp. 200–03.

6. John Herschel, *Essays from the Edinburgh and Quarterly Reviews* (London: Longman, 1857), pp. 268, 288; S. F. Cannon, *Science in Culture: The Early Victorian Period* (New York: Dawson, 1978), pp. 76–82.

7. John Pringle Nichol, *Architecture of the Heavens* (Tait: Edinburgh, 1837), p. 127; Simon Schaffer, "The Nebular Hypothesis and the Science of Progress," in *History, Humanity and Evolution*, ed. J. R. Moore (Cambridge: Cambridge University Press, 1989), pp. 131–64.

8. Ludwik Fleck, *Genesis and Development of a Scientific Fact* (Chicago: University of Chicago Press, 1979), pp. 115–17. For vividness see Carlo Ginzburg, "Montrer et citer," *Le Débat* 56 (1989): 43–54 and Michael Wintroub, "The Looking Glass of Facts," forthcoming.

9. John Herschel, *Essays*, pp. 260–61.

10. H. M. Collins, *Changing Order* (Berkeley: Sage, 1985), pp. 100–01.

11. W. Valentine Ball, ed., *Reminiscences and Letters of Sir Robert Ball* (London: Cassell, 1915), pp. 67–69.

12. John Herschel, *Results of the Astronomical Observations made at the Cape of Good Hope* (London: Smith, Elder, 1847), p. 25; David S. Evans et al., eds., *Herschel at the Cape* (Austin: University of Texas, 1969), p. 50; John Herschel, *Outlines of Astronomy*, 4th ed. (London: Longman, 1851), p. 609. For the politics of Herschel's Cape astronomy see Elizabeth Green Musselman, "Swords into Ploughshares: John Herschel's Progressive View of Astronomical and Imperial Governance," *British Journal for the History of Science*, 31 (1998), forthcoming.

13. Pang, "Victorian Observing Practices"; Holly Rothermel, "Images of the Sun: Warren de la Rue, George Biddell Airy and Celestial Photography," *British Journal for the History of Science* 26 (1993): 137–69.

14. Evans, *Herschel at the Cape*, p. 49. For these kinds of vision in South Africa in the 1830s, see Mary Louise Pratt, "Scratches on the Face of the Country," in *Race, Writing and Difference*, ed. Henry Louis Gates (Chicago: University of Chicago Press, 1986), pp. 138–62, pp. 141–43 and David Bunn, "Our Wattled Cot: Mercantile and Domestic Space in Thomas Pringle's African Landcapes," in *Landscape and Power*, ed. W. J. T. Mithell (Chicago: University of Chicago Press, 1994), pp. 127–73, p. 142.

15. Evans, *Herschel at the Cape*, pp. 41, 58, and 248; Brian Warner, *Charles Piazzi Smyth, Astronomer-Artist: His Cape Years 1835–1845* (Cape Town: Balkema, 1983), pp. 109–10.

16. Warner, *Smyth*, p. 111.

17. Warner, *Smyth*, pp. 112–20 ; H. A. Brück and M. T. Brück, *The Peripatetic Astronomer: The Life of Charles Piazzi Smyth* (Bristol: Adam Hilger, 1988), pp. 5–6.

18. Charles Piazzi Smyth, "On Astronomical Drawing," *Memoirs of the Royal Astronomical Society* 15 (1846): 71–82, pp. 71–73.

19. Ibid., pp. 73–76. See Paul Goldman, *Looking at Prints* (London: British Museum, 1981), pp. 1, 9.

20. Theodore Fielding, *The Art of Engraving* (London, 1841), pp. 34–37.

21. Smyth, "On Astronomical Drawing," pp. 79, 71; Fielding, *Art of Engraving*, pp. 32–33.

22. Pang, "Victorian Observing Practices," pp. 251, 255, and 258. For the Duke of Northumberland's subsidy of Herschel see Günther Buttmann, *The Shadow of the Telescope: A Biography of Sir John Herschel* (Guildford: Lutterworth, 1974), p. 157. For Rosse and photography see David H. Davidson, *Impressions of an Irish Countess* (Birr: Birr Scientific Heritage Foundation, 1989), p. 3.

23. Rosse to Sabine, March 23, 1848, Royal Society papers 259.1112. Sabine to Rosse, July 8, 1848, Birr papers, K6.1; Phillips to Whewell, Trinity College, Cambridge, Whewell papers, Add a.210/148. For Leonard's work see William Carpenter, "On the Microscopic Structure of Shells," *British Association Reports* (1844): 1–24 and (1847): 93–34, especially plate 1, fig. 1, and Gerard Turner, *The Great Age of the Microscope* (Bristol: Hilger, 1989), p. 170. For Leonard and Darwin see Darwin to Carpenter, 1846 and Darwin to Hooker, February 8, 1847, in *Correspondence of Charles Darwin*, 9 vols., ed. S. Smith and F. Burckhardt (Cambridge: Cambridge University Press, 1985-), vol. 3, p. 344 and vol. 4, p. 10.

24. Sabine to Rosse, July 8, 1848, Birr papers, K6.1. For Stoney and Hunter see Charles Parsons, ed., *Scientific Papers of William Parsons, Third Earl of Rosse* (London: Lund, Humphries, 1926), pp. 190–91, p. 199 ("Account of the Observations on the Great Nebula in Orion," 1868).

25. Leonard's wages are mentioned in Darwin to Carpenter 1846, *Correspondence of Darwin*, vol. 3, p. 344; the Royal Society costs are set out in Stokes to Rosse, June 25, 1862, Birr papers K13.10 and Royal Society Manuscripts MC.6.243 (August 1, 1862).

26. Charles Partington, *The Engravers' Complete Guide* (London, 1825), p. 124; Anthony Dyson, *Pictures to Print: The Nineteenth Century Engraving Trade* (London: Farrand, 1984), p. 57; Celina Fox, "The Engravers' Battle for Professional Recognition in Early Nineteenth Century London," *London Journal* 2 (1976): 3–31, p. 11; Gordon J. Fyfe, "Art and Reproduction: Some Aspects of the Relations between Painters and Engravers in London 1760–1850," *Media, Culture and Society* 7 (1985): 399–425, pp. 414–15.

27. Celina Fox, *Graphic Journalism in England during the 1830s and 1840s* (New York: Garland, 1988), pp. 29–33; Patricia Anderson, *The Printed Image and the Transformation of Popular Culture 1790–1860* (Oxford: Clarendon, 1991), pp. 2–3, 43–49; Iain McCalman, *Radical Underworld: Prophets, Revolutionaries and Pornographers in London 1790–1840* (Cambridge: Cambridge University Press, 1988), pp. 205–31.

28. [Charles Knight], "Preface," *Penny Magazine* 1 (1832): iii–iv. See Anderson, *Printed Image*, pp. 52–83; Fox, *Graphic Journalism*, pp. 50–55, 215–48, 267–80. For Knight and the SDUK see Steven Shapin and Barry Barnes, "Science, Nature and Control," *Social Studies of Science* 7 (1977): 31–74.

29. Nichol to Herschel, November 4, 1838, Royal Society, Herschel papers, 13.131; Fox, *Graphic Journalism*, pp. 16, 57.

30. Michael Baxandall, *Painting and Experience in Fifteenth Century Italy* (Oxford: Oxford University Press, 1972), p. 40; Mary Cowling, *The Artist as Anthropologist: The Representation of Type and Character in Victorian Art* (Cambridge: Cambridge University Press, 1989), pp. 157, 191, and 61–63.

31. Henry Mayhew, *London Labour and the London Poor*, 2 vols. (London: Griffin, Bohn, 1861), vol. 1, pp. 1–3.

32. Gillian Beer, *Open Fields: Science in Cultural Encounter* (Oxford: Oxford University Press, 1996), pp. 131–33; [Robert Chambers], *Vestiges of the Natural History of Creation* (London: John Churchill, 1844), pp. 306–10.

33. L. Perry Curtis, *Apes and Angels: the Irishman in Victorian Caricature* (Newton Abbott: David and Charles, 1971), pp. 11, 19–20; James Urry, "Englishmen, Celts and Iberians," in *Functionalism Historicized*, ed. George W. Stocking (Madison: University of Wisconsin Press, 1984), pp. 83–105, pp. 85–86; R. F. Foster, *Paddy and Mr Punch* (Harmondsworth: Penguin, 1993), p. 192.

34. Curtis, *Apes and Angels*, pp. 23–57; Adrian Desmond, *The Politics of Evolution* (Chicago: Chicago University Press, 1989), pp. 288–291.

35. Curtis, *Apes and Angels*, pp. 32 and 100; Alvar Ellegard, *Darwin and the General Reader* (Goteborg: Elander, 1958), p. 295; Adrian Desmond, *Archetypes and Ancestors: Palaeontology in Victorian London 1850–1875* (London: Blond and Briggs, 1982), pp. 74–81.

36. Charles Kingsley, *Letters and Memories of his Life*, 2 vols. (London: Macmillan, 1891), vol. 2, pp. 111–12; Susan McKenna-Lawlor and Michael Hoskin, "Correspondence of Markree Observatory," *Journal for the History of Astronomy* 15 (1984): 64–68, pp. 64–66. For Kingsley and physiognomy, see Cowling, *Artist as Anthropologist*, pp. 75, 277–79.

37. William Thackeray, *The Book of Snobs* (1848) (Gloucester: Alan Sutton, 1989), pp. 78–79, 104–05. For Thackeray and Irish satire see Foster, *Paddy and Mr Punch*, pp. 172–73.

38. Henry C. King and John Millburn, *Geared to the Stars* (Bristol: Hilger, 1978), pp. 317–18; Alexander von Humboldt, *Incitements to the Study of Nature*, ed. Edward Sabine (1846) (London: Bell and Daldy, 1866), pp. 91, 100.

39. Michael Crowe, *The Extraterrestrial Life Debate 1750–1900* (Cambridge: Cambridge University Press, 1986), pp. 210–15.

40. Herschel, *Outlines of Astronomy*, p. 609; Evans, *Herschel at the Cape*, p. 237.

41. Alexander Japp, *Thomas de Quincey: His Life and Writings* (new edition, London: John Hogg, 1890), pp. 236–37; Thomas de Quincey, "System of the heavens as Revealed by Lord Rosse's Telescopes," *Tait's Edinburgh Magazine* 8 (1846): 566–79, p. 571; Jonathan Smith, "De Quincey's Revisions to 'The System of the Heavens,'" *Victorian Periodicals Review* 26 (1993): 203–12.

42. John Barrell, *The Infection of Thomas de Quincey* (New Haven: Yale University Press, 1991), pp. 105–25 and Grevel Lindop, "English Reviewers and Scotch Professors: de Quincey's debts to Edinburgh Life and Letters," *Times Literary Supplement* 4839 (December 29, 1995): 9–10.

43. de Quincey, "Lord Rosse's telescopes," p. 575; Nichol to de Quincey, April 16, 1854, in *De Quincey Memorials*, ed. Alexander Japp, 2 vols. (London, 1891), vol. 1, pp. 276–77, recalls his remarks in 1846. See Smith, "De Quincey's Revisions," p. 207.

44. Charles Kilgour, "Discoveries in the Moon by Aid of the Monster Telescope Lately Erected by the Earl of Rosse," *Guide to Life* 17 (May 11, 1844): 129–31 (in Rosse papers, K5.79).

45. [David Brewster], "The Earl of Rosse's Reflecting Telescopes," *North British Review* 2 (November 1844): 175–212, p. 199; Brewster to Rosse, July 2, 1850, Rosse papers, K17.5. For deaths in Parsonstown see John O'Rourke, *The Great Irish Famine* (1874) (Dublin: Veritas, 1989), p. 245. For the Countess's funding, see Davison, *Impressions*, p. 2.

46. On Robinson see J. A. Bennett, *Church, State and Astronomy in Ireland* (Belfast: Armagh Observatory, 1990), pp. 59–138. Robinson directly attacks Nichol in Robinson to Rosse, April 7, 1841, Rosse papers, K5.4. For Nichol on Ireland, see his anonymous "Ireland in the Nineteenth and Scotland in the Sixteenth Century," *Tait's Edinburgh Magazine* 2 (1832): 84–92 and "Parliamentary Report on the

State of Ireland," *Tait's Edinburgh Magazine* 3 (1833): 1–22. For Robinson's campaign against Herschel see Forbes to Whewell, December 12, 1849, Trinity College, Cambridge, Whewell papers, Add.a.204/90.

47. Rosse, *Letters on the State of Ireland*, 2nd ed. (London: Hatchard, 1847), pp. 5, 24, and 27; Parsons, *Rosse Papers*, p. 42 (Robinson, "Contents of an Ancient Bronze Vessel," 1848); Bennett, *Church, State and Astronomy*, pp. 142–45. For Senior's opinion see Cecil Woodham-Smith, *The Great Hunger* (New York: Signet, 1964), p. 373.

48. Alfred Gautier, "Notice sur les Grands Téléscopes de Lord Rosse," *Bibliothèque Universelle de Genève* 57 (1845): 342–57, pp. 342–43; Smyth to Rosse, June 18, 1862, Rosse papers, K13.8.

49. Parsons, *Rosse Papers*, p. 48 ("Presidential Address," 1843); Thomas Romney Robinson, "On Lord Rosse's Telescopes," *Proceedings of the Royal Irish Academy* 3 (1845–7): 114–33 (read April 25, 1842, and April 14, 1845), p. 119. Compare Foster, *Paddy and Mr Punch*, p. 216: "From the inside of the demesne wall, a sense of threat was inevitable. As the nineteenth century wore on, Ascendancy marginalisation was reflected in their relation to architecture as well as to landowning."

50. J. A. Secord, "Behind the Veil: Robert Chambers and *Vestiges*," in Moore, *History, Humanity and Evolution*, pp. 165–94, pp. 168 and 175; Marilyn Ogilvie, "Robert Chambers and the Nebular Hypothesis," *British Journal for the History of Science* 8 (1975): 214–32. A detailed list of changes with respect to the Parsonstown news is in Robert Chambers, *Vestiges of the Natural History of Creation and other Evolutionary Writings*, ed. James Secord (Chicago: University of Chicago Press, 1994), pp. 218–19.

51. "Travels in search of Lord Rosse's telescopes," *Chambers' Edinburgh Journal* 154 and 156 (December 1846): 369–71, 401–04.

52. Rosse, *A Few Words on the Relation of Landlord and Tenant in Ireland* (London: John Murray, 1867), pp. 36–38, 50, and 31; Isaac Butt, *The Irish People and the Irish Land* (Dublin: Falconer, 1867), pp. 165, 172.

53. Astronomical Diary, September 28, 1852, Birr papers, L2.1; Parsons, *Rosse Papers*, p. 42 (Robinson, "Contents of an Ancient Bronze Vessel in the Collection of the Earl of Rosse," 1848); Kilgour, "Discoveries in the Moon," p. 129. See Francis Klingender, *Art and the Industrial Revolution* (Paladin: Frogmore, 1972), pp. 106–09.

54. Robinson, "On Lord Rosse's Telescopes," pp. 116–17.

55. Charles Weld, *Vacations in Ireland* (London: Longmans, 1857), 263–64; Thomas Dick, *The Practical Astronomer* (London: Seeley, 1845), pp. 548–62; Thomas Woods, *The Monster Telescopes Erected by the Earl of Rosse* (Parsonstown: Sheilds, 1844); John Timbs, *Curiosities of Science, Past and Present* (London: Kent, 1859), pp. 1, 96–99; "The Earl of Rosse's Great Telescope at Parsonstown," *Illustrated London News* (September 9, 1843): 165; Somerville to Rosse, November 11, 1843, Rosse papers, K17.16.

56. Herschel, *Outlines of Astronomy*, p. 593; George Peacock, "Address," *British Association Reports* (1844), pp. xxxi–xlvi, p. xxxi. For the gigantic in natural history and commodity production see Susan Stewart, *On Longing* (Durham: Duke University Press, 1993), pp. 70–76.

57. Robinson, "On Lord Rosse's telescopes," p. 130; Michael Hoskin, "Rosse, Robinson and the Resolution of the Nebulae," *Journal for the History of Astronomy* 21 (1990): 331–44, p. 339.

58. Hoskin, "Rosse, Robinson," p. 340; Herschel to Rosse, March 9, 1845, Rosse papers, K2.2; John Herschel, "Address," *British Association Report* (1845), pp. xxvii–xliv, p. xxxvii. For Rosse's drawing in Cambridge see M. A. Hoskin, "The First Drawing of a Spiral Nebula," *Journal for the History of Astronomy* 13 (1982). For Herschel versus Mill at Cambridge, see Silvan S. Schweber, "Auguste Comte and the Nebular Hypothesis," in *In the Presence of the Past: Essays in Honour of Frank Manuel*, ed. R. T. Bienvenu and M. Feingold (Dordrecht: Kluwer, 1990), pp. 131–91, pp. 164–65. For the dispatch of Rosse's notebook to London, see Sabine to Rosse, July 8, 1848, Rosse papers, K6.1.

59. John Pringle Nichol, *Thoughts on Some Important Points Relating to the System of the World*, 2nd ed. (Edinburgh: Johnstone, 1848), p. 109. For premature reports of resolution, see Gautier, "Notice sur les Grands Téléscopes de Lord Rosse," p. 354n.

60. Nichol, *Thoughts*, pp. 110–13.

61. Struve to Rosse, February 4, 1852, and September 20, 1853, Rosse papers, K36A.S3 and S7. For Pulkovo's status see Mari Williams, "Astronomical Observatories as Practical Space: the case of Pulkowa," in *The Development of the Laboratory*, ed. Frank James (London: Macmillan, 1989), pp. 118–36. The Parsonstown team also used Herschel's Cape observations to make their grids: see Rosse to Herschel, January 29, 1849, Royal Society papers HS.13. 228.

62. Struve to Rosse, September 20, 1853, and July 13, 1880, Rosse papers K36A.S7 and K36.2; Parsons, *Rosse Papers*, p. 199 ("Observations on the Great Nebula of Orion," 1868).

63. Parsons, *Rosse Papers*, pp. 190–91 ("Observations on the Great Nebula of Orion," 1868). George Stoney's observations of 1849 are in Rosse papers, L2.2 and selectively published in *Rosse Papers*, p. 202.

64. Rosse to Herschel, January 29, 1849, Royal Society papers HS.13. 228. Stoney to Rosse, July 27, 1867, Birr papers L6.1; Hunter's notes on Orion nebula 1860–64, Birr papers L2.2.

65. T. W. B., "Lord Rosse on the Nebulae," *Astronomical Register* 1 (1863): 33–35, 49–51, p. 51; Parsons, *Rosse Papers*, p. 206 ("Observations on the Great Nebula of Orion," 1868); Hunter to Rosse, July 27, 1868, Birr papers, L6.1; Ball, *Reminiscences and Letters*, p. 69.

66. Parsons, *Rosse Papers*, p. 147 ("On the Construction of Specula of 6-feet Aperture," 1861); Fourth Earl of Rosse, "Observations of Nebulae and Clusters of Stars made with the Six-foot and Three-foot Reflectors," *Scientific Transactions of the Royal Dublin Society* 2 (1881): 1–178, p. 4; Stoney to Rosse, July 27, 1868, Birr papers, L6.1. See J. A. Bennett, "A Viol of Water or a Wedge of Glass," in *The Uses of Experiment*, ed. David Gooding, et al. (Cambridge: Cambridge University Press, 1989), pp. 105–14, pp. 111–13.

67. Parsons, *Rosse Papers*, p. 119 ("Observations on the Nebulae," 1850); Hunter's notes, Birr papers, L2.2.

68. Parsons, *Rosse Papers*, p. 189 ("On the Construction of Specula of 6-feet Aperture," 1861); p. 203 ("Observations on the Great Nebula of Orion," 1868).

69. Parsons, *Rosse Papers*, p. 35n. (Robinson, "On Lord Rosse's Telescope," 1848); Robinson was responding to William Bond, "Description of the Nebula about the Star Θ Orionis," *Memoirs of the American Academy of Sciences* 3 (1848): 87–96.

70. Parsons, *Rosse Papers*, p. 114 ("Observations on the Nebulae," 1850); p. 147 ("On the Construction of Specula of 6-feet Aperture," 1861).

71. Hoskin, "Rosse, Robinson," p. 342.

72. The Harvard resolution is accepted in Herschel, *Outlines*, p. 609 and denied in Agnes Clerke, *A Popular History of Astronomy in the Nineteenth Century* (London: Black, 1908), p. 119; G. P. Bond's gift of the engraving to Rosse, May 14, 1863, Rosse papers, L6.1. For the Harvard telescope, see Adrian Lane, "Resolving to Show One's Worth: The Harvard College Observatory and its Audience 1840–1850," (M. Phil. essay, History and Philosophy of Science, University of Cambridge, 1995).

73. Humboldt to Arago, November 9, 1849, in Hamy, *Correspondance d'Alexandre de Humboldt avec Arago*, pp. 304–05.

74. George Airy, "Address," *British Association Report* (1851): xxxix–liii, p. xli; ; Struve (1853), in M. A. Hoskin, *Stellar Astronomy: Historical Studies* (Chalfont St. Giles: Science History, 1982), p. 150.

75. Herbert Spencer, *Essays Scientific, Political and Speculative*, 3 vols. (London: Williams and Norgate, 1891), vol. 1, pp. 111 and 114 ("The Nebular Hypothesis," 1858); William Huggins, *Scientific Papers* (London: Wesley, 1909), p. 118 ("On the Spectrum of the Great Nebula in the Sword Handle of Orion," 1865); Edward Sabine, 'President's Address," *Proceedings of the Royal Society* 13 (1864): 499–502, p. 502. See Barbara Becker, *Eclecticism, Opportunism and the Evolution of a New Research Agenda: William and Margaret Huggins and the Origins of Astrophysics* (Ph.D. thesis, Johns Hopkins University Press, 1993), pp. 126–39.

76. Ball, *Reminiscences and Letters*, pp. 73–74; Lloyd to Oxmantown, March 6, 1865, Rosse papers, K8.1 (1); Orion notebook (1866) Rosse papers, L2.9; Parsons, *Rosse Papers*, pp. 203–06 ("Account of the Observations on the Great nebula in Orion," 1868).

77. Struve to Rosse, July 28, 1869, Rosse papers, L6.1; Richard Proctor, "The Rosse Telescope set to New Work," *Fraser's Magazine* 80 (1869): 754–60; Rosse, "Observations of Nebulae and Clusters of Stars made with the Six-foot and Three-foot Reflectors" (1881), appendix; Struve to Rosse, April 14, and April 18, 1880, Rosse papers, K36.1 and K36A.S1.

78. Thomas de Quincey, *Posthumous Works*, ed. Alexander Japp, 2 vols. (London: Heinemann, 1891), vol. 1, pp. 276–77 and Smith, "De Quincey's Revisions," p. 207.

79. John Jellett, *The Immortality of the Intellect* (Dublin: Hodges, Smith, 1867), pp. 14–15; de Quincey, "System of the Heavens," p. 575. For mechanized picturing see Rothermel, "Images of the Sun," pp. 157–58; Pang, "Victorian Observing Practices," pp. 63–72.

Jonathan Crary

Attention and Modernity in the Nineteenth Century

O ne of the most important developments in the history of visuality in the nineteenth century was the relatively sudden emergence of models of subjective vision in a wide range of disciplines during the period 1810–1840. Dominant discourses and practices of vision, within the space of a few decades, effectively broke with a classical regime of visuality, and grounded the truth of vision in the density and materiality of the body.[1] One of the consequences of this shift was that the functioning of vision became dependent on the complex and contingent physiological makeup of the observer, rendering vision faulty, unreliable, and, it was sometimes argued, arbitrary. Even before the middle of the century, an extensive amount of work in science, philosophy, psychology, and art was a coming to terms in various ways with the understanding that vision, or any of the senses, can no longer claim an essential objectivity or certainty. By the 1860s, the work of Helmholtz, Fechner, and many others defined the contours of a general epistemological uncertainty in which perceptual experience lost the primal guarantees that had once upheld its privileged relation to the foundation of knowledge. A widespread crisis in perception, in which new truths and doubts about perception were being contested, was one of the continuing effects of this epochal break with classical epistemology in the early 1800s. This essay examines some of the responses to that crisis in the later nineteenth century, through a consideration of the increasing importance of *attention* as an apparent guarantee of psychic and perceptual unity.

The idea of subjective vision—the notion that our perceptual and sensory experience depends less on the nature of an external stimulus and more on the makeup and functioning of our sensory apparatus—was one of the conditions for the historical emergence of notions of autonomous vision, that is, for a severing (or liberation) of perceptual experience from a necessary relation to an exterior world. Equally important, the rapid accumulation of knowledge about the workings of a fully embodied observer disclosed possible ways that vision was open to procedures of normalization, of quantification, of discipline. Once the empirical truth of vision was determined to lie in the body, the senses and vision in particular were able to be annexed and controlled by external techniques of manipulation and stimulation. This was the decisive achievement of the science of psychophysics in the mid-nineteenth-century, which, by apparently rendering sensation measurable, embedded human perception in the domain of the quantifiable and the abstract. Conceived in this way, vision became compatible with so many other processes of modernization, even as it also opened up the possibility of visual experience that was intrinsically non-rationalizable and that exceeded any procedures of normalization. This is part of a critical historical threshold in the second half of the nineteenth century when any significant qualitative difference between a *biosphere* and a *mechanosphere* began to evaporate. This disintegration of an indisputable distinction between interior and exterior became a condition for the emergence of spectacular modernizing culture, and for a dramatic expansion of the terrain of aesthetic experience. The relocation of perception (as well as process and functions previously assumed to be "mental") into the physiological thickness of the body was a precondition for the instrumentalizing of human vision into a component of new machinic arrangements; but it was also a precondition for the astonishing efflorescence of visual invention and experimentation in Europe during the second half of the nineteenth century.

More specifically, since the late nineteenth century, and increasingly during the last two decades, one crucial dimension of capitalist modernity is a constant remaking of the conditions of sensory experience, in what could be called a revolutionizing of the means of perception. For the last one hundred years, perceptual modalities have been and continue to be in a state of perpetual transformation, or some might claim, a state of crisis. That is, if vision can be said to have any enduring characteristic within twentieth-century modernity, it is that is has no enduring features. Rather it is embedded in an accelerating rhythm of adaptability to new technological relations, social configurations, and economic imperatives. Paradoxically, in the late nineteenth century the dynamic logic of capital began to undermine dramatically any stable or enduring structure of perception just as this same logic simultaneously attempted to impose a disciplinary regime of attentiveness. It was at this moment, within the human sciences and particularly the nascent field of scientific psychology, that the question of attention became a fundamental issue. It was a problem whose centrality was directly

related to the emergence of a social, urban, psychic, industrial field increasingly saturated with sensory input. Inattention, especially within the context of new large-scale
forms of industrialized production, began to be treated as a danger and a serious problem, even though it was often the very modernized arrangements of labor that produced inattention.[2] It is possible to see one crucial aspect of modernity as an ongoing
crisis of attentiveness in which the changing configurations of capitalism continually
push attention and distraction to new limits and thresholds—with unending introduction of new products, new sources of stimulation, and streams of information—and
then respond with new methods of managing and regulating perception. But at the
same time, attention itself is hardly reducible to simply a component of disciplinary
strategies and practices. As I shall argue, the empirical articulation of a docile subject
in terms of attentive capacities simultaneously disclosed a subject incapable of con-
forming to such disciplinary imperatives.

Of course since Kant, part of the epistemological predicament of modernity has
hinged on the human capacity for synthesis amid the fragmentation and atomization
of a cognitive field. That dilemma became especially acute in the second half of the
nineteenth century alongside the development of various techniques for imposing
specific kinds of perceptual synthesis, from the mass diffusion of the stereoscope in the
1850s to early forms of cinema in the 1890s. The nineteenth century saw the steady
demolition of Kant's transcendental standpoint and its synthetic a priori categories,
detailed in his first *Critique*. Kant insisted that all possible perception could occur only
in terms of an original synthetic unification principle that stood over and above any
empirical sense experiences such as vision. "Unity of synthesis according to empirical
concepts would be altogether accidental, if these latter were not based on a transcendental ground of unity. Otherwise it would be possible for appearances to crowd in
upon the soul. . . . Since connection in accordance with universal and necessary laws
would be lacking, all relation of knowledge to objects would fall away."[3] For as soon as
the philosophical guarantees of any a priori cognitive unity collapsed, the problem of
"reality maintenance" became a function of a contingent and merely psychological
capacity for synthesis or association.[4] Schopenhauer's substitution of the will for
Kant's transcendental unity of apperception was an event with many aftershocks, for it
implied that the perceived wholeness of the world no longer had an apodictic or law-
like character, but depended on a potentially variable *relation of forces*.[5] It became
imperative for thinkers of all kinds to discover what faculties, operations, or organs
produced or allowed the complex coherence of conscious thought.[6] For it was the failure or malfunction of a capacity for synthesis, often described as dissociation, that
became linked in the late nineteenth century with psychosis and other mental
pathologies. But what was often labeled as a regressive or pathological disintegration
of perception was also evidence of a fundamental shift in the relation of the subject to
a visual field. In Bergson, for example, new models of synthesis involved the binding of

immediate sensory perceptions with the creative forces of memory. Wilhelm Dilthey discussed at length the creative forms of synthesis and fusion that are specific to the human imagination. For Nietzsche, the will to power was linked to a dynamic mastering and synthesizing of forces.

G. Stanley Hall, writing in 1883, pessimistically indicated what some of the stakes were, once this contingency was accepted as a condition of knowledge: "Does life cultivate the mind only in spots or nodes, and are these so imperfectly bound together by associative and apperceptive processes that special stress upon one of them causes it to isolate itself still more till the power of self-direction is lost, and devolution and disintegration slowly supervene?"[7] For institutional psychology in the 1880s and 1890s, part of psychic normality was the ability to bind synthetically perceptions into a functional whole, thereby warding off the threat of dissociation, or the dangers of what Kant saw as perceptions "crowding in upon the soul." And attention became an imprecise way of evaluating the relative capacity of a subject to isolate selectively certain contents of a sensory field at the expense of others in the interests of engaging an orderly and productive world. The German psychologist Oswald Külpe insisted that without a capacity for attention "consciousness would be at the mercy of external impressions . . . thinking would be made impossible by the noisiness of our surroundings."[8]

Max Nordau was perhaps the most notorious writer to link a failure of attentiveness with sociopathic behavior but his diatribes are not far removed from the social determinations which underpinned the work of more sober, "scientific" authorities:

> Untended and unrestrained by attention, the brain activity of the degenerate and hysterical is capricious and without aim or purpose. Through the unrestricted play of association representations are called into consciousness and are free to run riot there. They are aroused and extinguished automatically; and the will does interfere to strengthen or to suppress them. . . . Weakness or want of attention, produces, then, in the first place false judgements respecting the objective universe, respecting the qualities of things and their relations to each other. Consciousness acquires a distorted and blurred view of the external world. . . . Culture and command over the powers of nature are solely the result of attention; all errors, all superstition, the consequence of defective attention.[9]

Attention for Nordau, and in a less extreme way for many others, was a repressive and disciplinary defense against all potentially disruptive forms of free association. The words of British psychologist James Cappie in the 1880s are perhaps more typical:

> It is unnecessary to enlarge on the psychological importance of this function. It may be said to underlie every other mental faculty. It is the bringing of the

consciousness to a focus in some special direction. . . . [W]ithout it meaningless reverie will take the place of coherent thought.[10]

Armed with the quantitative and instrumental arsenal of psychophysics, the study of attention purported to rationalize what it ultimately revealed to be unrationalizable. Clearly specific questions were asked—how does attention screen out some sensations and not others, how many events or objects can one attend to simultaneously and for how long (i.e., what are its quantitative and physiological limits), to what extent is attention an automatic or voluntary act, to what extent does it involve motor effort or psychic energy? For researchers at the end of the century its significance as an "interior" activity diminished and it became a quantity or set of effects that could be observed or measured externally. In most cases though, attention designated some process of perceptual or mental organization in which a limited number of objects or stimuli were isolated from a larger background of possible attractions. But however it was described—organization, selection, isolation—attention implied an inevitable fragmentation of a visual field in which the unified and homogenous coherence of classical models of vision were impossible.

Once an observer is understood in terms of the essential subjectivity of vision, attention is an inescapable component of that organization of visuality. But what in fact determined how attention operated as a narrowing and focusing of conscious awareness? What forces or conditions caused an individual to attend to some limited aspects of an external world and not others? The camera obscura model of vision in the eighteenth century described an ideal relation of self-presentness between observer and world. Attention as a process of selection necessarily set up perception as an activity of *exclusion*, of rendering parts of a visual field unseen. Within the wider frame of nineteenth-century modernity, the articulation of this attentive subject is obviously crucial: Was attention an expression of the conscious will of an autonomous subject for whom the very activity of attention, as choice, was part of that subjects' self-constituting freedom? To what extent was attention a function of biologically determined instincts, unconscious drives, a remnant, as Freud believed, of our archaic evolutionary heritage, which shaped the practical texture of our lived relation to an environment?[11] Or can an observer configured around attentive capacities more appropriately be said to coincide with the emergence of knowledge and practices that imagine the human individual as a site of procedures of management, control, and subjectification including a wide ranging technology of "attraction."[12]

Clearly this problem was elaborated within an emergent economic system that demanded attentiveness of a subject in terms of a wide range of new productive and spectacular tasks. But it was at the same time a system whose internal movement was continually eroding the basis of any disciplinary attentiveness. Part of the cultural logic of capitalism demands that we accept as *natural* switching our attention rapidly from one thing to another.[13] Capital, as accelerated exchange and circulation, necessarily produces this kind of human perceptual adaptability and becomes a regime of reciprocal attentiveness and distraction. Helmholtz's account of subjective vision in his *Physiological Optics* established the truth of an observer in terms of an innate compatibility with this organization of experience:

> It is natural for the attention to be distracted from one thing to another. As soon as the interest in one object has been exhausted, and there is no longer anything new in it to be perceived, it is transferred to something else, even against our will. When we wish to rivet it on an object, we must constantly seek to find something novel about it, and this is especially true when other powerful impressions of the senses are tugging at it and trying to distract it.[14]

Unlike any previous order of visuality, mobility, novelty, and distraction are singled out as constituent elements of perceptual experience. Even some of the most avid defenders of technological progress acknowledged that subjective adaptation to new perceptual speeds and sensory overload would not be without difficulties. Nordau predicted that

> the end of the twentieth century, therefore, will probably see a generation to whom it will not be injurious to read a dozen square yards of newspapers daily, to be constantly called to the telephone, to be thinking simultaneously of the five continents of the world, to live half their time in a railway carriage or in a flying machine and . . . know how to find its ease in the midst of city inhabited by millions.[15]

What he and others could not grasp then was that modernization was not a onetime set of changes but an ongoing and perpetually modulating process that would never allow time for individual subjectivity to accommodate and "catch up" with it.

The problem of attention is interwoven, although not coincident, with the history of visuality in the late nineteenth century. In a wide range of institutional discourses and practices, within the arts and human sciences, attention becomes part of a dense network of texts and techniques around which the truth of perception is organized and structured. I use the word perception to indicate both visual and auditory perception or an amalgam of several senses.[16] In the context of laboratory experimentation subjects were described in terms of tactile and even olfactory attentiveness as well. It is through the frame of attentiveness, a kind of inversion of Foucault's Panoptic model, that the seeing body was deployed and made productive, whether as students, workers, consumers, or patients. Beginning in the 1870s but developing fully in the 1880s, there was an explosion of research and debate on this issue—it was a major issue in the influential work of Fechner, Wilhelm Wundt, Edward Titchener, Theodor Lipps, Carl Stumpf, Oswald Külpe, Ernst Mach, William James, and many others, with questions about the empirical and epistemological status of attentiveness.[17] Also, the pathology of a supposedly normative attentiveness was an important part of the inaugural work in France of researchers like Charcot, Alfred Binet, Pierre Janet, and Théodule Ribot. In the 1890s attention became a major issue for Freud and was a pivotal problem in the *Project for a Scientific Psychology* before his move to new psychical models.

Obviously, notions of attention and attentiveness exist in many different places long before the nineteenth century and even a summary outline of their history would be enormous. My aim here is simply to indicate how in the second half of the nineteenth century attention becomes a fundamentally new kind of object within the modernization of subjectivity. In most cases before the nineteenth century, even when attention was an object of philosophical reflection it was a marginal and at best secondary problem within explanations of mind and consciousness that did not constitutively depend on it. Or it had a local importance in matters of education, self-fashioning, etiquette, pedagogical and mnemonic practices, or scientific inquiry.[18] Eighteenth-century British philosophy with its models of a mind that was a *passive* receiver of sensation had no need of such an idea, and the word is almost entirely absent from the work of Locke, Hume, and Berkeley. Attention as it was conceived in the later nineteenth century was radically alien to an eighteenth-century notion of mental activity as a stamp or a mold that will somehow preserve the constancy of objects, in an act that cannot but be an immobilization. In historical discussions of the problem of attention one often encounters the claim that the modern psychological category of attention is continuous with (though more rigorously developed than) notions of apperception that were important in very different ways for Leibniz and Kant.[19] But in fact what is crucial is the unmistakable historical discontinuity between the problem of attention in the second half of the nineteenth century and its place in European thought in previous centuries. Only by the 1870s does it become, in Europe and North America, a problem that traverses an entire social and cultural field. Before then, in no sense did it subsist as an interrelated social, cultural, economic, psychological, and philosophical issue that was central to the most powerful determinations of the nature of human subjectivity. Edward Bradford Titchener, the premier importer of German experimental psychology into America (who moved from Leipzig to Ithaca, New York), asserted categorically in the 1890s that "the problem of attention is essentially a modern problem," although he had no sense of how the particular perceiving subject he was helping to delineate was to become a crucial component of institutional modernity.[20]

For attention is not just one of the many topics examined experimentally by late-nineteenth-century psychology.[21] It can be argued that a certain notion of attention is in fact the fundamental condition of its knowledge. That is, most of the crucial areas of research—whether of reaction times, sensory and perceptual sensitivity, mental chronometry, reflex action, or conditioned responses—all presupposed a subject whose attentiveness was the site of observation, classification, and measurement, and thus the point around which knowledge of many kinds was accumulated. Fechner's attempts in the 1850s to quantify subjective experience by measuring external stimulation is one of the early instances of this emerging model of attention. Fechner's famous unit of measurement, "a just noticeable difference," (or JND) was possible only through an experimental practice in which a test subject was required to be attentive to various magnitudes of sensory stimulation, and judged at what level differences between stimuli were perceptible. Fechner explicitly acknowledged the intrinsic unreliability of subjective testimony and the variability of attentiveness itself, but through what he called "the method of average error" he made the undependableness of human subjects fully compatible with statistical computations based on very large amounts of data.

The dominant model of an attentive human observer in the empirical sciences from the 1880s on was also inseparable from a radically transformed notion of what constituted sensation for a human subject. Within the increasingly sophisticated laboratory environment, sensation became an effect or set of effects that are technologically produced and are used to describe a subject who is compatible with those technical conditions. In particular, attention was studied in terms of response to machinically produced stimuli, often electrical in nature and abstract in content, that allowed a quantitative determination of the sensory capacities of a perceiving subject.[22] But within this vast project, which began perhaps with Fechner's *Elements of Psychophysics* (1860), the irrelevance of an older model of sensation as something belonging to a subject became clearer. Sensation had empirical significance only in terms of magnitudes that corresponded to specific quantities of energy (e.g., light) on one hand and to measurable reaction times and other forms of performative behavior on the other. Equally significant, along with the development of X-rays, photometry, and many other forms of artificial vision, was that the idea of sensation ceased to be a significant component in the cognitive picture of nature.[23]

But just as the rise of psychometry (i.e., any attempt at quantification or measurement of mental processes) in the human sciences was diminishing or altering the significance of subjective sensation, a very different challenge to the classical notion of sensation can be seen in the work of a wide range of thinkers, such as William James, Nietzsche, Bergson, and Peirce but also Seurat, Cézanne, and other artists. James and Bergson, in particular, are crucial for their questioning the notion of a pure or simple sensation, on which associationism had depended. Both insisted that any sensation, no matter how seemingly elemental, is always a compounding of memory, desire, will, anticipation, and immediate experience.[24] Peirce also argued against the idea of "immediate" sensations, asserting that they were irreducible complexes of association and interpretation.[25] Ernst Mach continued to employ the word "sensations" but in fact refashioned it to indicate psychic "elements" that could not provide knowledge of a "true" external world.[26] Important within this reorganization of perceptual experience, the contours of which I have only hinted at, was a struggle over how sensation and stimuli were interpreted, attended to, and made productive.

The problem of attention, then, was not a question of a neutral, timeless activity like breathing or sleeping but rather about the emergence of a specific model of behavior with a historical structure, which was articulated in terms of socially determined norms and was part of the formation of a specifically modern technological milieu. Anyone familiar with the history of modern psychology knows the symbolic importance of the date 1879—the year when Wilhelm Wundt established his laboratory at the University of Leipzig.[27] Irrespective of the specific nature of Wundt's intellectual project, this laboratory space, with its procedures and apparatuses, became the model for the whole modern social organization of psychological experimentation around the study of an observer attentive to a wide range of artificially produced stimuli. To paraphrase Foucault, this has been one of the practical and discursive spaces within modernity in which human beings "problematize what they are."[28]

While it is easy and appropriate to situate the wide-ranging research on attention within the imperatives of larger disciplinary and administrative apparatuses for the management and control of human subjects, it is also important to insist on another interrelated dimension of the knowledge accumulated within the newly configured human sciences in the nineteenth century. Foucault has taken us through what he calls the great eschatological dream of the nineteenth century, which was

> to make this knowledge of man exist so that man could be liberated by it from his alienations, liberated from all the determinations of which he was not the master, so that he could, thanks to this knowledge of himself, become again or for the first time master of himself, self-possessed. In other words, one made of man an object of knowledge so that man could become subject of his own liberty and of his own existence.[29]

Thus the attempt to determine empirically the specific physiological and practical conditions under which a perceiving subject could be most acutely attentive to the world, or could, through an exercise of a sovereign and attentive will, stabilize and objectify the contents and relations within that world—the determining of these conditions would also be a claiming of that subject's self-possession as potential master and fully conscious organizer of that perceptible world.[30]

But scientific psychology never was to assemble knowledge that would guarantee a full copresence of the world to an attentive subject. Instead, the more one investigated, attention was shown to contain within itself the conditions for its own undoing—

that attentiveness was fully continuous with states of distraction, reverie, dissociation, and trance.

I must insist that my project is not concerned with whether or not there is some empirically identifiable mental or neurological capacity for attention. It is an object for me only in terms of a massive accumulation of *statements* and concrete social *practices* during a specific historical period. It cannot be hypostatized as a substantive object. So if I repeatedly use the term attention, it refers to the field of those statements and practices and to a network of effects that they produced. Also, given the centrality of attentiveness as a scientific object, it must be emphasized that the 1880s and 1890s produced sprawling diversity of often contradictory attempts to explain it. For many of the thinkers for whom attention was an issue represent opposed or completely irreconcilable intellectual and philosophical positions, such as Wundt and Mach, Dilthey and Ebbinghaus, Freud and Janet, Delboeuf and Binet, Helmholtz and Hering, and so on. Since then the problem of attention has remained more or less within the center of institutional empirical research, though throughout the twentieth century various positions in philosophy and psychology have rejected it as a relevant or even meaningful problem.[31] One might with some justification insist rather strictly that during the hegemony of behaviorism beginning in the early twentieth century that attention, along with the idea of a "mental process," disappeared as an explicit object of research for a few decades. But in fact, regardless of terminological polemics, the entire regime of stimulus-response research is founded on the attentive capacities of a human (or animal) subject. More recently, within the context of a dramatically transformed space of knowledge and neurological research, it is not uncommon to encounter claims such as the following by Antonio Damasio: "Without basic attention and working memory there is no prospect of coherent mental activity."[32] Thus I would insist on the remarkable persistence of attention as a problem within the generalized disciplinary setup of the social and behavioral sciences.[33] At the same time, the preceding remarks should make clear that I am not suggesting or proposing that there was any single or dominant model of an attentive observer. Nor am I describing attention as constituting a particular regime of power but as part of a space in which new conditions of subjectivity were articulated and thus a space in which effects of power operated and were deployed. And in fundamental ways, the newly mapped out attentive body, like the sexualized body of the nineteenth century, was a composite of forces that inherently resisted territorialization and organization.

From a certain vantage point the use of the problem of attention to frame an investigation of modernity in the late nineteenth century may seem out of synch with a whole legacy of critical practice. That is, attention might seem superficially to be a return to traditional problems of an epistemological nature, problems that were radically transformed or made irrelevant by the whole modern shift to semantic and/or semiotic frameworks of analysis, what Richard Rorty has described as a move "from epistemology to hermeneutics."[34] It is that shift demonstrated most vividly in the work of, for example, Mallarmé, Nietzsche, and Peirce (and later, Wittgenstein and Heidegger), thinkers operating in a terrain where it is no longer a question of how an already constituted subject knows or perceives the objectivity of an external world but how a subject is constructed provisionally through language and other systems of social meaning and value. Within this syntactic-semantic remaking of epistemology the study of the function of various psychic *faculties* became increasingly irrelevant. My project is based on the proposal that the emergence of attention as a way of describing or explaining a perceiving subject is in fact a crucial sign of the same general epistemological crisis, the termination of various analyses that took the problem of consciousness as a starting point, and the increasing irrelevance of the dualistic models within which classical epistemology had operated. The very uncertainty and vagueness about what precisely attention was is an indication of the precariousness of older theories of perception. Attention implied that cognition could no longer be posed in terms of the unmediated givenness of sense-data. To use Peircian terms, it made a previously dyadic system of subject-object into a triadic one, with the third element consisting of a shifting and intervening space of socially articulated physiological functions, institutional imperatives, and a wide range of techniques, practices, and discourses relating to the perceptual experience of a subject in time. Attention here is not reducible to attention *of* something. Thus attention within modernity is inseparable from these forms of *exteriority*, not the intentionality of an autonomous subject. Rather than a faculty of some already constituted subject, it is a sign not so much of the subject's disappearance, but of its precariousness, contingency, and insubstantiality.

As I suggested earlier, there were two important conditions for the emergence of attention as a problem and as a crucial part of an account of subjectivity. The first was the collapse of classical models of vision and of the stable, punctual subject those models presupposed. The second was the untenability of a priori solutions to epistemological problems. What this entailed, of course, was the loss of any permanent or unconditional guarantees of mental unity and synthesis. Jan Goldstein has detailed the significance of the problem of the unity of the self for Victor Cousin and others in the 1820s, who held to the general principle: "Character is unity." Cousin's eclecticism "combined a limited reliance on sensationalism with a priori belief in the self, or *moi*, a repository of self-initiated mental activity and free will known through introspection."[35] Especially during the period from 1840 to the mid-1860s there are a variety of often convoluted "systemic" attempts to propose new principles from which to deduce an effective unity of mind or thought, including the work of J. S. Mill, Herbert Spencer, Hermann Lotze, and the early work of Alexander Bain, in which attention does not play any significant part.[36] It is only by the 1870s that one finds attention consistently being accorded a *central* and *formative* role in accounts of how a practical or knowable world of objects came into being for a perceiver. It would be difficult to find before 1850 a statement making similar unconditional claims to what Henry Maudsley wrote in the early 1880s: "Whatever its nature, [attention] is plainly the essential condition of the formation and development of mind."[37] I do not want to belabor this point or insist on some precise historical dividing line, but one telling piece of evidence is in the work of the enormously important physiologist William B. Carpenter, whose texts were widely read and cited as authoritative in Europe and North America from the 1840s well into the 1880s. In the 1853 edition of his standard textbook, attention is covered in a single paragraph, and discussed as merely one of many mental faculties such as observation, reflection, and introspection; by the 1874 edition, he devotes more than fifty pages to the topic of attention, and references to it are scattered throughout many other sections of the book. Attention in 1853 was noted almost in passing as "That state in which the consciousness is actively directed to a sensorial change"; by 1874 attention has an effect "on each principal form of Mental activity," it is indispensable "for the systematic acquirement of Knowledge, for the control of the Passions and Emotions, and for the regulation of the Conduct."[38]

By the 1880s the similarity between will and attention became a central issue in work of many kinds, and highlighted how far removed psychological thought was now from Mill's associationism and his "psychic chemistry" of laws regarding regularities of sensations or from Spencer's work in the 1850s, which had defined experience as the *passive* response to external order. In a very general way the shift that took place in the 1870s was from the *structural* psychology of associationism to various kinds of *functional* psychological accounts.[39] The change was, in part, the product of the increasing importance and richness of a physiological understanding of the human subject; the poverty and inadequacy of associationist theories of knowledge became evident in the face of a widespread coming to terms with the subject as an active, motor center of behavior, of will, as a composite of processes unfolding in time.

At the same time it should be stressed that attention flourished and persisted as a problem independent of the obsolescence or fashionability of various scientific approaches or systems of thought. For example, in the 1870s and 1880s, many writers, both social thinkers and psychologists, either closely associated or identified attention with will. But as Lorraine Daston has convincingly shown, the movement toward a more rigorously "scientific psychology," which gathered momentum and institutional potency in the 1890s, was a joining of forces "in the campaign against consciousness, volition, introspection and other distinctive aspects of mind." By the turn of the century, "the theory of the will became the common target of an attack launched by several different schools of American and British psychology."[40] But if the will, the mind, and introspection were superfluous elements, attention remained a fluctuating but nonetheless durable component of an institutional construction of subjectivity. Hugo Munsterberg and James McKeen Cattell can stand as examples of this jettisoning of any notion of an active will, while still retaining attention as an important problem in various attempts to align psychology with strategies of social control. In a related way today, attention remains an indispensable category for institutional discourses and techniques of the subject, not only in its obvious social manifestations like the current debate around a so-called Attention Deficit Disorder but also within much of the sprawling terrain of the cognitive sciences, even as the relevance or existence of "mind" and "consciousness" are questioned in those same domains. Both "attention" and "consciousness" are historically constructed notions but the last century has shown very vividly how they have a variable and independent relation to each other: that is, attention as part of an account of subjectivity is not inherently synonymous with consciousness.

The work of Wilhelm Wundt can serve as a key instance of the many ways in which Kant's transcendental unity of apperception was replaced with merely psychological processes of synthesis and integration. Attention, for Wundt, because of its essential (but not a priori) role in producing an effective unity of consciousness and perception, was the single most important psychic category. Wundt explained attention in terms of a distinction (derived from Leibniz) between perception and apperception: perception described habitual, ordinary, automatic responses to external stimuli while apperception involved focusing one's attention on a stimulus, assessing, evaluating, and interpreting it. In most of Wundt's work apperception is effectively a synonym for voluntary attention. His postulation of an attention center located in the frontal cerebral lobes was particularly influential.[41] His account thus posed attention as one of the highest integrating functions (distinct from the automatic functions of the lower brain and spinal column) within an organism whose makeup was emphatically hierarchical. Through the notion that "ontogeny repeats phylogeny," work on attention was suffused with many of the social assumptions of evolutionary thought in 1870s and 1880s. The groundbreaking neurological work of John Hughlings Jackson was a related articulation of this hierarchical model, in which different functions were associated with specific areas of the nervous system: Jackson distinguished so-called "higher" functions like voluntary attentiveness from more automatic and "lower" forms of motor behavior. Perhaps more significantly, Wundt's model of attention, which he effectively equated with will, was founded on the idea that various sensory, motor, and mental processes were necessarily *inhibited* in order to achieve the restricted clarity and focus that characterized attention.[42] Others posed models with related explanations. Charles Féré and Alfred Binet described "the simple fact of attention" as "a concentration of the whole mind on a single point, resulting in the intensification of the perception of this point and producing all around it *a zone of anaesthesia*; attention increases the force of certain sensations while it weakens others." They insisted on the *"negative effects of attention."* [43]

The idea that inhibition or anesthesia is a constitutive part of perception is an indication of a dramatic reordering of visuality, implying the new importance of models based on an economy of forces rather than an optics of representation. Freud's formulations on the relation between perception and repression (from the "Project" in 1895 to the essay on psychogenic visual disturbances in 1910), are only the more widely known products coming out of speculation and research by others in the 1870s and 1880s. [44] It is another sign of the irrelevance of the camera obscura model of vision, in which an ideal observer had the capacity to apprehend the unedited contents of a visual field. Thus, a normative observer in the late nineteenth century begins to be conceptualized, not only in terms of the isolated objects of attention, but equally in terms of what is not perceived, or only dimly perceived, of the distractions, the fringes and peripheries that are excluded or shut out of a perceptual field.

However, it should be emphasized that the themes of inhibition, exclusion, and fringe do not necessarily support a Freudian model of an unconscious actively denying certain contents to attentive awareness. Jonathan Miller has argued recently that an alternate European tradition in the nineteenth century posed the unconscious as part of a system in which *automatic* behavior was reciprocally intertwined with the changing needs of conscious activity, including attention. In contrast to the "custodial" Freudian interpretation, many nineteenth-century psychologists saw the unconscious as

> actively generating the processes which are integral to memory, perception, and behavior. Its contents are inaccessible not, as in psychoanalytic theory, because they are held in strenuously preventive detention but, more interestingly, because the effective implementation of cognition and conduct does not actually *require* comprehensive awareness. On the contrary, if consciousness is to implement the psychological tasks for which it is best fitted, it is expedient to assign a large proportion of psychic activity to automatic control; if the situation calls for a high level management decision, the unconscious will freely deliver the necessary information to awareness. [45]

Helmholtz, for example, proposed a quasi-utilitarian functioning of the mind in which sensory information that is unlikely to be useful or necessary is involuntarily unattended to. To become aware of such information (like the blind spot in our visual field) requires a special effort at reorienting one's attention.

A wide range of studies on attention, then, defined it in various, sometimes related ways as an activity of *selection*. [46] Its importance within the course of human evolution was emphasized by many, following Darwin's insistence on it as a survival mechanism:

Hardly any faculty is more important for the intellectual progress of man than the power of attention. Animals clearly manifest this power, as when a cat watches a hole and prepares to spring on its prey. Wild animals sometimes become so absorbed when thus engaged, that they may be easily approached.[47]

A certain kind of reactive attention was believed to be an essential part of human biology. It triggered a systemic response to novel stimuli, whether visual, olfactory, or auditory, in which the organism was instantly able to shut down (or inhibit) ongoing motor activity while focusing mental effort exclusively on the relevant stimuli, usually either potential predator or prey. Parallel to Wundt's work in the 1870s were the neurological researches of the Scottish physician Sir David Ferrier, who was one of the champions of the idea of localization of brain function. Ferrier developed the hypothesis of inhibitory centers in specific parts of the brain, which were effectively the physiological basis of will and attention. He demonstrated how attention and volition depended on the physiological *suppression* of movement, that is, how paradoxically certain forms of sensory-motor activity inhibited other motor activity.[48] Thus an attentive observer might appear motionless but was in fact the site of a ferment of physiological (and motor) occurrences, upon which that relative "stasis" depended. But this state of heightened alertness and of intense focus on a restricted area of a visual field could be understood in many ways. For example, it could be transposed from the animal realm of sheer survival into a biological adaptation of the organism to disciplined and productive labor within a social realm. But attention, as a shutting out, a powerful filter, also could be seen as a model of a Nietzschean forgetting, a forgetting that is an essential precondition not merely for subsistence but for affirmation of the self through *action*.[49] Attention here has less to do with a model of consciousness than with an ideo-motor network of *forces*. It is paradoxically that which immobilizes, but if seen as a part of a biological heritage, it is inseparable from mobility.

Thus, as part of the larger physiological reconfiguration of subjectivity that occurs during the nineteenth century, attention, in most of the varied ways it is theorized, is inseparable from physical effort, movement, or action. During the period I am examining, attentiveness is generally synonymous with an observer who is fully embodied and for whom perception coincides with physiological and/or motor activity. To specify further, there were three particularly important models through which attention was understood. Occasionally elements of these models overlapped, but for the most part they stood for relatively incompatible positions.

(1) Attention as a *reflex* processes, part of a mechanical adaptation of an organism to stimuli in an environment. Important here is the evolutionary legacy of attention, and its origins in *involuntary* and instinctive perceptual responses. (2) Attention as determined by the operations various *automatic* or unconscious

processes or forces, a position articulated in many ways, beginning with Schopenhauer, Janet, Freud, and numerous others. (3) Finally, attention as a decisive, voluntary activity of the subject that is an expression of the autonomous power of the subject to actively organize and impose itself on a perceived world.

It was these physiological conceptions of attention that so much late-nineteenth- and early-twentieth-century aesthetic theory attempted to escape from, by posing various modalities of "contemplation" and vision that were radically cut from the processes and activities of the body.[50] The whole neo-Kantian legacy of a disinterested aesthetic perception, from Konrad Fiedler, T. E. Hulme, and Roger Fry to more recent "formalisms," has been founded on the desire to escape from physiological time and its vagaries. Hulme, for example, insisted that the artist was someone in whom "nature had forgotten to attach their faculty for perception to their faculty for action," and outlined an aesthetic attentiveness that is "emancipated" from the physiological.[51] Much modernist art and music theory has been about the invention of dualistic systems of perception in which a rapt, timeless presence of perception is contrasted with lower, mundane, or quotidian forms of seeing or listening.[52] Within the visual arts, Rosalind Krauss argues that modernism imagines two orders, the first of which is "empirical vision, the object as it is 'seen,' the object bounded by its contours, the object modernism spurns. The second is that of the formal conditions of the possibility of vision itself, the level at which 'pure' form operates as a principle of coordination, unity, structure: visible but unseen," and Krauss outlines how temporality is necessarily excluded from it.[53] Modernist vision with its "all-at-oneness," she insists, is founded on the cancellation of the empirical conditions of perception, including the experience of successiveness implicit in any motor activity.

What became clear, though often evaded, in work of many different kinds on attention was what a volatile concept it was. Attention always contained within itself the conditions for its own disintegration; it was haunted by the possibility of its own excess—which we all know so well whenever we try to look at or listen to any one thing for too long.[54] In any number of ways, attention inevitably reaches a threshold at which it breaks down. Usually it is the point at which the perceptual identity of its object begins to deteriorate and in some cases (like certain sounds) disappears altogether. Or it can be a limit at which attention imperceptibly mutates into a state of trance or even autohypnosis. In one sense attentiveness was a critical feature of a productive and socially adaptive subject but the border that separated a socially useful attentiveness and a dangerously absorbed or diverted attention was profoundly nebulous and could be described only in terms of performative norms. Attention and distraction were not two essentially different states but existed on a single continuum, and thus attention was, as most increasingly agreed, a dynamic process, intensifying and diminishing, rising and falling, ebbing and flowing according to an indeterminate set of variables.[55] Attention thus had certain thermodynamic qualities by which a given force could assume more than one form. Ernst Mach was one of many who, in the 1880s, grasped its apparently paradoxical nature:

> Where the development of intelligence has reached a high point, such as is presented now in the complex conditions of human life, representations may frequently absorb the whole of attention, so that events in the neighborhood of the reflecting person are not noticed, and questions addressed to him are not heard;—a state which persons unused to are wont to call absent-mindedness, although it might with more appropriateness be called present-mindedness.[56]

In this sense my work moves away from some assumptions that have been part of a long-established characterization of modernity in terms of experiences of distraction. In particular the work (over several decades) of Georg Simmel, Walter Benjamin, Siegfried Kracauer, Theodor Adorno, and others insisted that various kinds of distracted perceptual or cognitive states are central to any account of subjectivity within modernity.[57] The German word *Zerstreuung* became central to numerous critical analyses; it suggests its generalized intellectual legacy in Kantian theories of knowledge, where *Zerstreuung* referred to a dispersion, a scattering of perceptions outside of any necessary synthesis, that were "merely a blind play of representations, less even than a dream."[58] One of the enduring legacies of this work has been accounts of modernity as a process of fragmentation and destruction in which premodern forms of wholeness and integrity were irretrievably broken up or degraded through technological, urban, and economic reorganizations. More specifically, it is a generalized historical narrative in which premodern modalities of looking and listening are either implicitly or explicitly set up as richer, deeper, or more valuable forms. Exemplary would be Simmel's account of how modern urban life as "the swift and continuous shift of external and internal stimuli" contrasts with "the slower, more habitual, more smoothly flowing rhythm of the sensory-mental phase" of premodern social life. Or a related position sees the fragmentation implicit in modernity as destructive to a whole set of traditional artistic and cultural values, but in this view distraction is seen as part of a process of overcoming the bankruptcy of bourgeois aesthetics. Nonetheless, there is the overriding sense of distraction as the product of "decay" or "atrophy" of perception within a larger deterioration of experience.[59] Adorno, for example, writes about distraction as "regression," as perception that has "arrested at the infantile stage" and for which deep "concentration" is no longer possible. My position is, on the contrary, that if distraction emerges as a problem in the late nineteenth century, it is fully inseparable from the parallel construction of an attentive observer in various domains. That is, modern distraction was *not* a disruption of some stable or "natural" kinds of sustained, value-laden perception that had existed for centuries but was an *effect* and in many cases a constituent element of the vast range of attempts to produce attentiveness in human subjects. Even though Benjamin, in some of his work, makes affirmative claims for distraction (suggesting that the disruption inherent in shock and distraction held forth the possibility of new modes of perception) he does so within the frame of a fundamental duality in which an absorbed attentiveness was the other term. "Distraction and concentration form polar opposites," declares Benjamin in his well-known discussion of architecture and film as two paradigms of modern "reception in a state of distraction."[60] Instead, attention and distraction cannot be thought of outside of a continuum in which the two are ceaselessly flowing into one another, as part of a social field in which the *same* imperatives and forces incite one and the other.

Much of the discourse on attention attempted to salvage some relatively stable notion of consciousness and some form of a distinct subject/object relation, but it tended rather to describe only a fleeting immobilization of a "subject effect" and an ephemeral congealing of a changing sensory manifold into a cohesive real world. Ribot acutely observed that attention "is an exceptional, abnormal state, which cannot last long, for the reason that it is in contradiction to the basic condition of psychic life, namely change."[61] Earlier, Helmholtz had similarly insisted "An equilibrium of the attention, persistent for any length of time, is under no circumstances attainable. The natural tendency of attention when left to itself is to wander to ever new things."[62] Attention was described as that which prevents our perception from being a incoherent flood of sensations, yet research showed it to be an undependable defense against such disorder. It was an indispensable component of the "normal" and "rational" subject of late-nineteenth-century industrial society, yet was clearly also an opening onto "pathological" and "irrational" effects. In spite of the importance of attention in the organization and modernization of production and consumption, most studies implied that attention rendered perceptual experience into something labile, continually undergoing change, and finally dissipative.[63] Attention seemed as if it should be about perceptual fixity and the certainty of presence, but was instead about duration and flux within which objects and sensation had a mutating provisional existence, and it was ultimately that which obliterated its objects through an entropic decay of its own energy. From the classical model of a mental stabilization of perceptions into a fixed mold, attention in the nineteenth century effectively became a continuum of variation, a temporal modulation.[64] That which seemed to hold the possibility of building up stable and orderly (though not necessarily truthful) cognitions also seemed to contain within itself uncontrollable forces that would put that organized world in jeopardy. Attention would always be inseparable from absence and the impossibility of presence.

Notes

1. See my *Techniques of the Observer: On Vision and Modernity in the Nineteenth Century* (Cambridge, Mass.: MIT Press, 1990). The present essay is drawn from my forthcoming book on attention and modern culture in the late nineteenth century.

2. Marx discusses how even by the 1840s, factory management understood that "the extent of vigilance and attention on the part of the workmen was hardly capable of being increased" and that the shortening of the working day, by being less taxing on the worker's attentiveness, resulted in greater productivity and profits. See Karl Marx, *Capital* (New York: International Publishers, 1967), vol. 1, pp. 410–12. For a groundbreaking study of the parallel modernization of labor and accumulation of knowledge about the productive body, see Anson Rabinbach, *The Human Motor: Energy, Fatigue and the Origins of Modernity* (New York: Basic Books, 1990).

3. Immanuel Kant, *Critique of Pure Reason*, trans. Norman Kemp Smith (New York: St. Martins, 1965), p. 138.

4. Victor Cousin exemplifies a wider sense of dismay at the rise of "psychological" explanation within epistemology: "Now as soon as the laws of reason are degraded to being nothing but laws relative to the human condition, their whole compass is circumscribed by the sphere of our personal nature, and their widest consequences, always marked with an indelible character of subjectivity, engender only irresistible persuasions, if you please, but no independent truths." *Elements of Psychology*, trans. Caleb Henry (New York: Ivison & Phinney, 1856), pp. 419–20. See Jan Goldstein's essay on Cousin, "Eclectic Subjectivity and the Impossibility of Female Beauty," in this volume.

5. Schopenhauer, *The World As Will and Representation*, vol. 2, p. 137.

6. By the 1850s, a range of interpretations of Kant "turned the a priori forms into 'innate laws of the mind,'" often with a neurological substrate, according to Klaus Köhnke, *The Rise of Neo-Kantianism: German Academic Philosophy Between Idealism and Positivism*, trans. R. J. Hollingdale (Cambridge: Cambridge University Press, 1991), p. 98. Köhnke provides a valuable discussion of the persistent question of "apriority," particularly in the work of the neo-Kantians Alois Riehl and Hermann Cohen in the 1870s.

7. G. Stanley Hall, "Reaction Time and Attention in the Hypnotic State," *Mind* 8 (1883): 171–82.

8. Oswald Külpe, *Outlines of Psychology* (1893), trans. E. B. Titchener (New York: Macmillan, 1901), p. 215.

9. Max Nordau, *Degeneration* (1892) (New York: Appleton, 1895), p. 56. Nordau's work had been preceded by numerous more "scientific" studies of his subject. For example, mental degeneration, manifested as defective attentiveness, is discussed in the context of larger cosmic and devolutionary processes of decline in Henry Maudsley, *Body and Will* (New York: Appleton, 1884). Both these texts are evaluated historically in Daniel Pick, *Faces of Degeneration: A European Disorder, c.1848–c.1918* (Cambridge: Cambridge University Press, 1989).

10. James Cappie, "Some Points in the Physiology of Attention, Belief and Will," *Brain* 9 (July 1886), p. 201.

11. Freud, *The Origins of Psycho-Analysis*, trans. Eric Mosbacher and James Strachey (New York: Basic Books, 1954), p. 417.

12. The work of Tom Gunning has been crucial for insisting that one of the formative components of a modernized mass visual culture the West, as it took shape in the late 1880s and 1890s, was a technology of "attraction." Discussing early cinema, Gunning demonstrates that what was at stake was not primarily representation, imitation, narration, or the updating of theatrical forms. Rather it was a strategy of engaging an attentive spectator: "From comedians smirking at the camera, to the constant bowing and gesturing of conjurors in magic films, this is a cinema that displays its visibility, willing to rupture a self-enclosed fictional world for a chance to solicit the attention of the spectator." "The Cinema of Attractions: Early Film, Its Spectator, and the Avant-Garde," in *Early Cinema: Space, Frame Narrative*, ed. Thomas Elsaesser (London: BFI, 1990), pp. 56–62, p. 57.

13. See the related discussion in Fredric Jameson, "Regarding Postmodernism: A Conversation with Fredric Jameson," in *Postmodernism, Jameson, Critique,* ed. Douglas Kellner (Washington, D.C.: Maisonneuve Press, 1989), pp. 43–74, p. 46.

14. Hermann von Helmholtz, *Physiological Optics,* vol. 3, ed. James P. C. Southall (New York: Dover, 1962), p. 498.

15. Max Nordau, *Degeneration,* p. 541.

16. Two recent studies that insist on the importance of the auditory within problematizations of modernity are Douglas Kahn, "Introduction: Histories of Sound Once Removed," in *Wireless Imagination: Sound, Radio and the Avant-Garde,* ed. Douglas Kahn and Gregory Whitehead (Cambridge, Mass.: MIT Press, 1992), pp. 1–29, and Steven Connor, "The Modern Auditory I," in *Rewriting the Self: Histories from the Renaissance to the Present,* ed. Roy Porter (London: Routledge, 1997), pp. 203–23.

17. A few of the very large number of works that treat this subject during this period are William James, *The Principles of Psychology,* vol. 1 (1890) (New York: Dover, 1950), pp. 402–58; Théodule Ribot, *La psychologie de l'attention* (Paris: F. Alcan, 1889); Edward Bradford Titchener, *Experimental Psychology: A Manual of Laboratory Practice* (New York: Macmillan, 1901), pp. 186–328; Henry Maudsley, *The Physiology of Mind* (New York: Appleton, 1883), pp. 310–24; Oswald Külpe, *Outlines of Psychology* (1893), trans. E. B. Titchener (London: Sonnenschein, 1895), pp. 423–54; Carl Stumpf, *Tonspsychologie,* vol. 2 (Leipzig: S. Hirzel, 1890), pp. 276–317; F. H. Bradley, "Is There Any Special Activity of Attention," *Mind* 11 (1886): 305–23; Angelo Mosso, *Fatigue* (1891), trans. Margaret Drummond (New York: G. P. Putman), pp. 177–208; Lemon Uhl, *Attention* (Baltimore: Johns Hopkins University Press, 1890); George Trumbull Ladd, *Elements of Physiological Psychology* (New York: Scribners, 1887), pp. 480–97, 537–47; Eduard von Hartmann, *Philosophy of the Unconscious* (1868), trans. William C. Coupland (New York: Harcourt Brace, 1931), pp. 105–08; G. Stanley Hall, "Reaction Time and Attention in the Hypnotic State," *Mind* 8 (April 1883): 170–82; Georg Elias Müller, *Zur Theorie der sinnlichen Aufmerksamkeit* (1873) (Leipzig: A. Adelmann, n.d.); James Sully, "The Psycho-Physical Processes in Attention," *Brain* 13 (1890): 145–64; John Dewey, *Psychology* (New York: Harper, 1886), pp. 132–55; Henri Bergson, *Matter and Memory* (1896), trans. W. S. Palmer and N. M. Paul (New York: Zone Books, 1988), pp. 98–107; Theodor Lipps, *Grundtatsachen des Seelenlebens* (Bonn: M. Cohen, 1883), pp. 128–39; Léon Marillier, "Remarques sur le mécanisme de l'attention," *Revue philosophique* 27 (1889): 566–87; Charlton Bastian, "Les processus nerveux dans l'attention et la volition," *Revue philosophique* 33 (1892): 353–84; James McKeen Cattell, "Mental Tests and their Measurement," *Mind* 15 (1890): 373–80; Josef Clemens Kreibig, *Die Aufmerksamkeit als Willenserscheinung* (Wien: Alfred Hölder, 1897); Walter B. Pillsbury, *Attention* (1906) (London: Sonnnenschein, 1908); Heinrich Obersteiner, "Experimental Researches on Attention," *Brain* 1 (1879): 439–53; Pierre Janet, "Etude sur un cas d'aboulie et d'idées fixes," *Revue philosophique* 31 (1891): 258–87, 382–407; Theodor B. Hyslop, *Mental Psychology especially in its Relations to Mental Disorders* (London: Churchill, 1895), pp. 291–304; William B. Carpenter, *Principles of Mental Psychology* (New York: D. Appleton, 1886), pp. 130–47; Giuseppe Sergi, *La psychologie physiologique* (1885, Italian) (Paris: F. Alcan, 1888), pp. 237–48; Theodor Ziehen, *Introduction to Physiological Psychology,* trans. C. C. van Liew (London: Sonnenschein, 1892), pp. 206–14; James Cappie, "Some Points in the Physiology of Attention, Belief and Will," *Brain* 9 (July 1886): 196–206; Alfons Pilzecker, *Die Lehre von sinnliche Aufmerksamkeit* (Munich: Akademische Buchdruckerei von F. Straub, 1889); André Lalande, "Sur un effet particulier de l'attention appliquée aux images," *Revue philosophique* 35 (March 1893): 284–87; John Grier Hibben, "Sensory Stimulation by Attention," *Psychological Review* 2, 4 (July 1895): 369–75; Jean-Paul Nayrac, *Physiologie et psychologie de l'attention* (Paris: F. Alcan, 1906); Charles Sanders Peirce, "Some Consequences of Four Incapacities [1868]," *Writings of Charles S. Peirce,* vol. 2 (Bloomington, Ind.: University of Indiana Press, 1982), pp. 211–42; Sigmund Freud, "Project for Scientific Psychology," in *The Origins of Psycho-analysis,* trans. Eric Mosbacher and James Strachey (New York: Basic Books, 1954), pp. 415–45; Edmund Husserl, *Logical Investigations,* vol. 1 (1899–1900), trans. J. N. Findlay (New York: Humanities Press, 1970), pp. 374–86.

18. Descartes' discussion of *admiration* or wonderment in *The Passions of the Soul* defines some of the terms of a fundamentally different historical regime of attention. See *The Philosophical Writings of Descartes*, vol. 1, trans. John Cottingham et al. (Cambridge: Cambridge University Press, 1985), pp. 354–56.

> Of wonder, in particular, we may say that it is useful in that it makes us learn and retain in our memory things of which we were previously ignorant. For we wonder only at what appears to us unusual and extraordinary. . . . [W]hen something previously unknown to us comes before our intellect or our senses for the first time, this does not make us retain it in our memory unless our idea of it is strengthened in our brain by some passion, or perhaps also by an application of our intellect as fixed by our will in a special state of attention and reflection.

For a superb account of this tradition of admiration/wonderment, see Lorraine Daston, "Curiosity in Early Modern Science," *Word & Image* 11, 4 (October-December 1995): 391–404:

> Seventeenth-century natural philosophers regularly paired "inquisitive" with "industrious;" "attention" with "diligence." By the mid-eighteenth century, it had become the moral criterion by which to distinguish the serious savant from the frivolous amateur, for only the former was capable of converting "noble curiosity" into "work and continued application" by "use of attention." . . . The unswerving, penetrating attention scientific investigation was thought to require slackened without curiosity, and curiosity was triggered by wonder. Attention screwed to this virtuoso pitch amounted to intellectual possession. (p. 401)

19. See, for example, Gardner Murphy and Joseph K. Kovach, *Historical Introduction to Modern Psychology*, 3rd ed. (San Diego: Harcourt, Brace, Jovanovich, 1972), pp. 23–24.

20. Edward Bradford Titchener, *Experimental Psychology: A Manual of Laboratory Practice*, vol. 1 (New York: Macmillan, 1901), p. 186. Elsewhere Titchener insists that late-nineteenth-century "experimental psychology discovered attention" and recognized "its separate status and fundamental importance; the realization that the doctrine of attention is the nerve of the whole psychological system." In Titchener, *Lectures on the Elementary Psychology of Feeling and Attention* (New York: Macmillan, 1908), p. 171.

21. On the particular status of psychology in the nineteenth century and its special relation to philosophy see Katherine Arens, *Structures of Knowing: Psychologies of the Nineteenth Century* (Dordrecht: Kluwer, 1989); Elmar Holenstein, "Die Psychologie als eine Tochter von Philosophie und Physiologie," in *Das Gehirn, Organ der Seele? Zur Ideengeschichte der Neurobiologie*, ed. Ernst Florey and Olaf Breidbach (Berlin: Akademie Verlag, 1993), pp. 289–308; David E. Leary, "The Philosophical Development of the Conception of Psychology in Germany," *Journal of the History of the Behavioral Sciences* 14 (1978): 113–21.

22. On the technological transformation of physiology and psychology in the nineteenth century see, Timothy Lenoir, "Models and instruments in the development of electrophysiology, 1845–1912," *Historical Studies in the Physical and Biological Sciences* 17 (1986): pt. 1, 1–54. See the suggestive footnote on the possibility of a cultural history of electricity "that would address the specific ways in which it has shaped subjectivity," in Felicia McCarren, "The 'Symptomatic Act' Circa 1900: Hysteria, Hypnosis, Electricity, Dance," *Critical Inquiry* 21 (Summer 1995): 748–74, p. 763.

23. See the important historical problematization of "mechanical objectivity" in the nineteenth century and the related orientation of the observer "beyond the limits of the human senses" in Lorraine Daston and Peter Galison, "The Image of Objectivity," *Representations* 40 (Fall 1992): 81–128.

24. James was, however, convinced that "pure sensations" could be realized in the first days of life by an infant. *Principles of Psychology*, vol. 2 (New York, 1890), p. 7.

25. Charles S. Peirce, "Some Consequences of Four Incapacities," in *Peirce on Signs: Writings on Semiotic by Charles Sanders Peirce*, ed. James Hoopes (Chapel Hill: University of North Carolina Press, 1991), pp. 71–75.

26. See the discussion of Mach's remaking of scientific objectivity and parallel disintegration of the subject, in Theodore Porter, "The Death of the Object: Fin-de-Siècle Philosophy of Physics," in *Modernist Impulses in the Human Sciences 1870–1930*, ed. Dorothy Ross (Baltimore: Johns Hopkins University Press, 1994), pp. 128–51.

27. On Wundt and the beginnings of the psychology laboratory see Kurt Danziger, *Constructing the Subject: Historical Origins of Psychological Research* (Cambridge: Cambridge University Press, 1990) pp. 17–33. See also Didier Deleule, "The Living Machine: Psychology as Organology," in *Incorporations*, ed. Jonathan Crary and Sanford Kwinter (New York: Zone, 1992), pp. 203–33.

    Occasionally, the priority of Wundt's laboratory is challenged by those who point to William James's "laboratory" in Laurence Hall at Harvard in 1875, where he performed demonstrations for his students although did not conduct or initiate any systematic research program.

28. Michel Foucault, *The Use of Pleasure*, trans. Robert Hurley (New York: Random House, 1985), p. 10.

29. Michel Foucault, "Foucault Responds to Sartre," in *Foucault Live*, trans. John Johnston (New York: Semiotexte, 1989), p. 36. Interview originally published in *La Quinzaine littéraire* (March 1–15, 1968).

30. Nietzsche made this link between attention and the will to mastery:

    That which is termed "freedom of the will" is essentially the affect of superiority in relation to him who must obey: "I am free, 'he' must obey"—this consciousness is inherent in every will; and equally so the straining of the attention, the straight look that fixes itself exclusively on one aim, the unconditional evaluation that "this and nothing else is necessary now," the inward certainty that obedience will be rendered—and whatever else belongs to the position of the commander. *Beyond Good and Evil*, pp. 25–26 (sec. 19).

31. See, for example, the devaluation of attention as a problem in Maurice Merleau-Ponty, *The Phenomenology of Perception*, trans. Colin Smith (New York: Routledge, 1962), pp. 26–31. Many studies since the mid-twentieth century have worked with notions of cognitive processing and channel capacity borrowed from information theory. One influential mid-century account of attention was Donald Broadbent's "filter theory" in his *Perception and Communication* (New York: Pergamon, 1958). Some of the various kinds of recent research on attention are represented in *Varieties of Attention*, ed. Raja Parasuraman and D. R. Davies (Orlando: Academic Press, 1984). See also Julian Hochberg, "Attention, Organization and Consciousness," in *Attention: Contemporary Theory and Analysis*, ed. D. I. Mostofsky (New York: Appleton-Century-Crofts, 1970); Alan Allport, "Visual Attention," in *Foundations of Cognitive Science*, ed. Michael Posner (Cambridge, Mass.: MIT Press, 1989), pp. 631–82; Gerald Edelman, *Bright Air, Brilliant Fire: On the Matter of Mind* (New York: Basic Books, 1992), pp. 137–144; Patricia Smith Churchland, *Neurophilosophy: Toward a Unified Science of the Mind-Brain* (Cambridge, Mass.: MIT Press, 1986), pp. 474–478.

32. Antonio R. Damasio, *Descartes' Error: Emotion, Reason and the Human Brain* (New York: Putnam, 1994), p. 197.

33. A very different understanding of attention, remote from the concerns of this book, can be encountered within some areas of twentieth-century analytic philosophy, where distinctions are made between various concepts such as "noticing," "interest," "awareness," and "heed." See, for example, the discussion in Gilbert Ryle, *The Concept of Mind*, pp. 135–44. For Ryle "heed" refers to "the concepts of noticing, taking care, attending, applying one's mind, concentrating, putting one's heart into something, thinking what one is doing, alertness, interest, intentness, studying and trying." See also the general overview in A. R. White, *Attention* (Oxford: Blackwell, 1964).

34. Richard Rorty, *Philosophy and the Mirror of Nature* (Princeton: Princeton University Press, 1979), pp. 315–56.

35. Jan Goldstein, "Foucault and the Post-Revolutionary Self," in *Foucault and the Writing of History*, ed. Jan Goldstein (Oxford: Blackwell, 1994), p. 102. See also Goldstein's important related argument in her "The Advent of Psychological Modernism in France: An Alternative Narrative," in *Modernist Impulses in the Human Sciences*, ed. Dorothy Ross (Baltimore: Johns Hopkins University Press, 1994), pp. 190–209.

36. The place of attention in the thought of Thomas Reid, Dugald Stewart, and James Mill is differentiated from modern speculation and research in Charlton Bastian, "Les processus nerveux dans l'attention et volition," *Revue philosophique* 32 (April 1892): 353–84, pp. 354–55.

37. Henry Maudsley, *The Physiology of Mind* (New York: D. Appleton, 1883), p. 310.

38. William B. Carpenter, *Principles of Human Physiology*, (Philadelphia: Blanchard and Lea, 1853), p. 780, and *Principles of Mental Physiology* (1876), 4th ed. (London: Kegan Paul, 1896), pp. 130–31. The later volume is a retitled expansion of the earlier.

39. See George Herbert Mead, *Movements of Thought in the Nineteenth Century*, vol. 2 (Chicago: University of Chicago Press, 1936), pp. 386–87. Mead writes: "The structure of the act is the important character of conduct. This psychology is also called motor psychology, as over against the older psychology of sensation; voluntary psychology, as over against the mere association of ideas with each other."

40. Lorraine J. Daston, "The Theory of Will versus the Science of Mind," in *The Problematic Science: Psychology in Nineteenth Century Thought*, ed. William R. Woodward and Timothy G. Ash (New York: Praeger, 1982), pp. 88–115.

41. Wilhelm Wundt, *Grundzüge der physiologischen Psychologie*, vol. 3 (1874), 6th ed. (Leipzig: Engelmann, 1908), pp. 306–64; in English as *Principles of Physiological Psychology*, trans. Edward Bradford Titchener (New York: Macmillan, 1904). See the general intellectual characterization of Wundt in Kurt Danziger, "Wundt and the Two Traditions of Psychology," in *Wilhelm Wundt and the Making of a Scientific Psychology*, ed. R. W. Rieber (New York: Plenum, 1980), pp. 73–87.

42. For a detailed overview of this problem in the nineteenth century, see Roger Smith, *Inhibition: History and Meaning in the Sciences of Mind and Brain* (Berkeley and Los Angeles: University of California Press, 1992). But the relation between attention and inhibition is also articulated in many places uninformed by neurological ideas. See, for example, F. H. Bradley, "On Active Attention," *Mind* 11 (1902): 1–30: "Attention will thus consist in the suppression of any psychical fact which would interfere with the object, and its essence therefore is not positive at all, but merely negative" p. 5.

43. Alfred Binet and Charles Féré, *Le magnétisme animal* (Paris: Félix Alcan, 1888), p. 239, emphasis added.

44. On the likely influence of John Hughlings Jackson on Freud in the 1890s, see Anne Harrington, *Medicine, Mind, and the Double Brain: A Study in Nineteenth Century Thought* (Princeton: Princeton University Press, 1987), pp. 235–47.

45. Jonathan Miller, "Going Unconscious," *New York Review of Books* (April 20, 1995): 59–65, p. 64. Miller discusses the work of Sir William Hamilton, William B. Carpenter, and Thomas Laycock.

46. Studies on attention, like almost all important work within experimental psychology in the late nineteenth century, obviously involved human test subjects with specific demographic and sociological features such as age, gender, and social class. It is well known, for example, that in the first ten years of the operation of Wundt's Leipzig laboratory his subjects were almost exclusively his own male students. See the valuable discussion in Kurt Danziger, "A Question of Identity: Who Participated in Psychological Experiments," in *The Rise of Experimentation in American Psychology*, ed. Jill G. Morawski (New Haven: Yale University Press, 1988), pp. 35–52.

47. Charles Darwin, *The Descent of Man, and Selection in Relation to Sex* (1871) (Princeton: Princeton University Press, 1981), p. 44. Angelo Mosso, for example, begins his chapter on attention by citing Darwin, in his *Fatigue*, p. 177. On the epistemological impact of Darwin's work, see Robert J. Richards, *Darwin and the Emergence of Evolutionary Theories of Mind and Behavior* (Chicago: University of Chicago Press, 1987), pp. 275–94.

48. See David Ferrier, *The Functions of the Brain* (London: Smith Elder, 1876). See the valuable discussion of Ferrier in Smith, *Inhibition*, pp. 116–21.

49. This sense of attention as a forgetting that is a condition for the affirmation of the organism persisted well into the twentieth century in Bergson, but also, for example, in Abraham H. Maslow's notion of the "peak-experience," which was widely popularized in the 1960s. He describes a mode of "total attention" in which it is "as if the world were forgotten, as if the percept had for the moment become the whole of Being." In *Toward a Psychology of Being* (New York: Van Nostrand Reinhold, 1968), p. 74. The enduring (or recyclable) nature of such formulations is evident now in the 1990s in such

bestselling self-improvement handbooks as Mihaly Csikszentmihalyi, *Flow: The Psychology of Optimal Experience* (New York, Harper, 1990): "Attention is our most important tool in the task of improving the quality of our experience" (p. 33).

50. In his 1909 "An Essay in Aesthetics," Roger Fry specifically set up the aesthetic faculty as a form of perception that was fully cut off from "the complex nervous machinery" of the body and the instincts. "The whole of animal life, and a great part of human life, is made up of these instinctive reactions to sensible objects, and their accompanying emotions," while for Fry "imaginative life" is about contemplation disconnected from the possibility of action. *Vision and Design* (Cleveland: Meridian, 1956), pp. 16–38, pp. 17–18. Fry elsewhere insisted on "the a priori case for the existence in all aesthetic experiences of a special orientation of the consciousness, and, above all, a special focussing of the attention, since the act of aesthetic apprehension implies an attentive passivity to the effects of sensations apprehended in their relations." *Transformations: Critical and Speculative Essays on Art* (London: Chatto and Windus, 1927), p. 5.

51. T. E. Hulme, *Speculations* (New York: Harcourt, Brace and Co., 1924), pp. 154–57.

52. See, for example, the opposition between "free" artistic perception and "unfree" nonartistic perception in Konrad Fiedler, *On Judging Visual Works of Art* (1876), trans. Henry Shaeffer-Simmern (Berkeley and Los Angeles: University of California Press, 1949), pp. 40–43.

53. Rosalind Krauss, *The Optical Unconscious* (Cambridge, Mass.: MIT Press, 1993), p. 217.

54. See Théodule Ribot, *Psychology of Attention* (1889) (Chicago: Open Court, 1896), p. 3: "Attention is a state that is fixed. If it is prolonged beyond a reasonable time . . . everybody knows from individual experience, that there results a constantly increasing cloudiness of the mind, finally a kind of intellectual vacuity, frequently accompanied by vertigo."

55. Gustav Fechner was one of the first to articulate this continuum with some specificity. He outlines a reciprocal relation between attention and "partial sleep" in his *Elemente der Psychophysik*, vol. 2 (1860), p. 452–57. Kurt Goldstein has written that unless attention has "a differential emphasis" it will shift into "a pathological boundness to stimuli," and he insists "that distractibility and abnormal fixation are expressions of the same functional change under different conditions." "The Significance of Psychological Research in Schizophrenia," *Journal of Nervous and Mental Disease* 97, 3 (March 1943): 261–79, p. 272.

56. Ernst Mach, *Contributions to the Analysis of the Sensations* (1885), trans. C. M. Williams (La Salle, Ill.: Open Court, 1890), p. 85.

57. See, for example, now canonical texts such as Georg Simmel, "The Metropolis and Mental Life," in *On Individuality and Social Forms* (Chicago: University of Chicago Press, 1971); Walter Benjamin, "On Some Motifs in Baudelaire," in *Illuminations*, trans. Harry Zohn (New York: Schocken, 1969); Siegfried Kracauer, "The Cult of Distraction," in *The Mass Ornament*, trans. Thomas Y. Levin (Cambridge: Harvard University Press, 1995), pp. 75–88; Theodor Adorno, "On the Fetish Character in Music and the Regression of Listening" in *The Essential Frankfurt School Reader*, ed. Andrew Arato and Eike Gebhardt (New York: Urizen, 1978), pp. 270–99.

58. Kant, *Critique of Pure Reason*, p. 139.

59. See the excellent discussion of Benjamin's ambivalent historicization of perception in Miriam Hansen, "Benjamin, Cinema and Experience," *New German Critique* 40 (Winter 1987): 179–224.

60. Walter Benjamin, *Illuminations*, pp. 239–40.

61. Théodule Ribot, *The Psychology of Attention*, p. 2.

62. Hermann Helmholtz, quoted in William James, *Principles of Psychology*, vol 1, p. 422.

63. From Wilhelm Wundt, *Outlines of Psychology* [1893], 4th rev. ed., trans. Charles H. Judd (Leipzig: Englemann, 1902), p. 233.

> The successive movement of attention over a number of objects appears accordingly to be a *periodic* process, made up a number of separate acts of apperception following one another. Such *a periodic rise and fall of attention* can under favorable conditions be directly demonstrated. . . . Thus, if we allow a weak continuous impression to act on a sense organ and remove as far as

ATTENTION AND MODERNITY IN THE NINETEENTH CENTURY   499

possible all other stimuli, it will be observed when the attention is concentrated upon this impression that at certain generally irregular intervals, the impression becomes for a short time indistinct, or even appears to fade out entirely, only to reappear the next moment (emphasis in original). (From Wundt, *Outlines of Psychology*, p. 233.)

Angelo Mosso noted that "attention involves modifications of a complex nature. . ." involving periodic oscillations. "Experiments have shown that attention is not a continuous but an intermittent process proceeding almost by bounds." *Fatigue* (1891) trans. Margaret Drummond (New York: G. P. Putnam, 1906), pp. 183–84. Attention is described as a periodic, wave-like form in Thaddeus Bolton, "Rhythm," *American Journal of Psychology* 6, 2 (January 1894): 145–238. See the proposal of an anti-modernist model of vision as a rhythmic "pulse" in Rosalind Krauss, *The Optical Unconscious*, pp. 213–25. An account of attention also could also be developed through Hegel's description of "sense-certainty" as a self-canceling form of apprehension, as a rhythm of "appearing" and "melting away." See G. F. W. Hegel, *The Phenomenology of Mind*, trans. J. B. Baillie (New York: Harper and Row, 1967), pp. 149–61.

64. On the distinction between mold and modulation see Gilbert Simondon, *L'individu et sa genèse psycho-biologique* (Paris: P.U.F., 1964), pp. 39–44. See the related discussion in Gilles Deleuze, *The Fold: Leibniz and the Baroque*, trans. Tom Conley (Minneapolis: University of Minnesota Press, 1993), pp. 19–21.

Contributors

SVETLANA ALPERS is Professor Emerita of the History of Art, University of California, Berkeley. She is the author of *The Art of Describing* (Chicago: 1983), *Rembrandt's Enterprise* (Chicago, 1988), *The Making of Rubens* (Yale, 1995), and, with Michael Baxandall, *Tiepolo and the Pictorial Intelligence* (Yale, 1994).

JONATHAN CRARY is Associate Professor in the Art History Department at Columbia University. A founding editor of Zone Books, he is the author of *Techniques of the Observer* (MIT, 1990) and coeditor of *Incorporations* (Zone, 1992). He is currently completing a book on late nineteenth-century visual culture, *Suspensions of Perception*.

LORRAINE DASTON is Director at the Max Planck Institute of the History of Science, Berlin. She is the author of *Classical Probability in the Enlightenment* (Princeton, 1988) and (with Katharine Park) *Wonders and the Order of Nature* (Zone, 1998). Her current research concerns the history of scientific objectivity; she is completing a book, *Images of Objectivity*, with Peter Galison.

ARNOLD I. DAVIDSON is Professor of Philosophy, Divinity, and the Conceptual Foundations of Science at the University of Chicago, where he also serves as Executive Editor of *Critical Inquiry*. He is coeditor of *Questions of Evidence* (Chicago, 1991), and *Foucault and His Interlocutors* (Harvard, 1997). His forthcoming book is entitled *The Emergence of Sexuality: Historical Epistemology and the Emergence of Concepts* (Harvard).

After completing his term as Andrew W. Mellon Professor at the National Gallery of Art's Center for Advanced Study in the Visual Arts, DAVID FREEDBERG will return to his position as Professor of Art History at Columbia University in New York. His publications include *The Power of Images* (Chicago, 1986) and *The Paper Museum of Cassiano dal Pozzo* (London, 1996); his forthcoming book focuses on the Academy of the Lynx in the Italian Renaissance.

PETER GALISON is the Mallinckrodt Professor of the History of Science and of Physics at Harvard University. He is coeditor of *Big Science: The Growth of Large-Scale Research* (Stanford, 1992) and *The Disunity of Science* (Stanford, 1996). Author of *How Experiments End* (Chicago, 1987) and *Image and Logic: A Material Culture of Microphysics* (Chicago, 1997), he is currently completing a book with Lorraine Daston, *Images of Objectivity.*

Among other books, CARLO GINZBURG is author of *The Cheese and the Worms* (Johns Hopkins, 1980), *The Enigma of Piero* (Verso, 1985), *Clues, Myths, and the Historical Method* (Johns Hopkins, 1989), *Il Giudice e lo storico* (Einaudi, 1991), and *Occhiacci di legno: Nove riflessioni sulla distanza* (Feltrinelli, 1998). He is Franklin D. Murphy Professor of Italian Renaissance Studies at the University of California at Los Angeles.

JAN GOLDSTEIN is a professor of history and a member of the Committee on the Conceptual Foundations of Science at the University of Chicago. She is the author of *Console and Classify: The French Psychiatric Profession in the Nineteenth Century* (Cambridge, 1987), editor of *Foucault and the Writing of History* (Blackwell, 1994), and is completing *The Post-Revolutionary Self: Competing Psychologies in Nineteenth-Century France.*

DONNA HARAWAY is a professor in the History of Consciousness Department at the University of California at Santa Cruz, where she teaches feminist theory, science studies, and women's studies. She is the author of *Crystals, Fabrics and Fields* (Yale, 1976), *Primate Visions* (Routledge, 1989), *Simians, Cyborgs, and Women* (Routledge, 1991), and *Modest_Witness@Second_Millennium.FemaleMan© Meets OncoMouse™* (Routledge, 1997).

PERRY HOBERMAN's site-specific, interactive installations have been commissioned by museums, galleries, and festivals of computer graphics and media art throughout the world. Recently given a retrospective at the Otso Gallery in Espoo, Finland, Hoberman is represented in New York by Postmasters Gallery; his work can also be seen at the Zentrum für Kunst und Media in Karlsruhe, Germany. He currently lives and works in New York, where he teaches at the School of Visual Arts.

CAROLINE A. JONES teaches contemporary art and theory in the Art History Department at Boston University, where she is also director of Museum Studies. Author of *Modern Art at Harvard* (Abbeville, 1985), *Bay Area Figurative Art* (California, 1990), and *Machine in the Studio* (Chicago, 1996), she is also the curator/essayist of *Painting Machines* (Boston University Art Gallery, 1997). Her forthcoming book will focus on the art critic Clement Greenberg.

JOSEPH LEO KOERNER is Professor of Fine Arts at Harvard University. His books include *Die Suche nach dem Labyrinth* (Suhrkamp, 1983), *Caspar David Friedrich* (Yale, 1990), and *The Moment of Self-Portraiture* (Chicago, 1993). He is currently completing a study of Lutheran images and the routines of modern belief.

Although specializing mainly in the sociology of science, BRUNO LATOUR was first trained in philosophy and theology. Author of *Microbes: The Pasteurization of France* (Harvard, 1988) and *We Have Never Been Modern* (Harvard, 1993), he has worked extensively on scientific and religious images. He is professor at the Ecole des Mines in Paris and visiting professor at the London School of Economics.

KATHARINE PARK is Zemurray Stone Radcliffe Professor of the History of Science and Women's Studies at Harvard University. Her recent publications include *Wonders and the Order of Nature*, with Lorraine Daston (Zone, 1998) and essays on the history of medicine and the life sciences in medieval and Renaissance Europe.

KRZYSZTOF POMIAN, born in Warsaw in 1934, studied philosophy at Warsaw University, where he held a teaching position until 1968, when he was dismissed because of his political opinions. Since 1973 he has been a professor at the Centre National de la Recherche Scientifique in Paris. Author of *Collectors and Curiosities* (Blackwell, 1990), his current research concerns European cultural history and historical epistemology.

SIMON SCHAFFER is Reader in History and Philosophy of Science at the University of Cambridge. Coauthor of *Leviathan and the Air-Pump* (Princeton, 1985), he has also published several studies of Victorian astronomy and its organization, most recently in *Victorian Science in Context*, edited by Bernard Lightman (Chicago 1997) and *Scientific Practice*, edited by Jed Buchwald (Chicago 1996).

LONDA SCHIEBINGER is Professor of History of Science and Women's Studies at Pennsylvania State University. She is author of *The Mind Has No Sex?* (Harvard, 1989) and *Nature's Body* (Beacon Press, 1993). Her new book, *Has Feminism Changed Science? If So How, If Not Why Not?* will appear with Harvard University Press in 1998. She is currently at work on the gendered production of knowledge and ignorance in European voyages of discovery.

AMY SLATON is Assistant Professor in the Department of History and Politics at Drexel University. She has written on the role of materials science in the history of American construction, and is completing a book on the social origins and impacts of industrial quality control measures.

JOEL SNYDER is professor and chair of the Department of Art History at the University of Chicago. He has published extensively on issues in the history and theory of photography, the theory of representation, and on aspects of the history, theory, and use of linear perspective. Snyder is a coeditor of the journal *Critical Inquiry*, and coauthor of *On the Art of Fixing a Shadow* (Bullfinch Press, 1991).

IRENE J. WINTER is William Dorr Boardman Professor of Fine Arts at Harvard. Her principal work is in the art and archaeology of the ancient Near East. Two books, on Mesopotamian 'aesthetics' and on the great Victory Stele of Naram-Sin of Agade, are in preparation; she has also published on "The Royal Image and the Visual Dimensions of Assyrian Ideology."

Index

Note: Italicized citations indicate references to illustrations.

Printed in the United States
by Baker & Taylor Publisher Services